Non-Halogenated Flame Retardant Handbook
2nd Edition

Scrivener Publishing
100 Cummings Center, Suite 541J
Beverly, MA 01915-6106

Publishers at Scrivener
Martin Scrivener (martin@scrivenerpublishing.com)
Phillip Carmical (pcarmical@scrivenerpublishing.com)

Non-Halogenated Flame Retardant Handbook
2nd Edition

Edited by
Alexander B. Morgan
University of Dayton Research Institute, Ohio, USA

Scrivener
Publishing

WILEY

This edition first published 2022 by John Wiley & Sons, Inc., 111 River Street, Hoboken, NJ 07030, USA and Scrivener Publishing LLC, 100 Cummings Center, Suite 541J, Beverly, MA 01915, USA
© 2022 Scrivener Publishing LLC
For more information about Scrivener publications please visit www.scrivenerpublishing.com.

1st edition (2014), 2nd edition (2022)
Wiley Global Headquarters
111 River Street, Hoboken, NJ 07030, USA

For details of our global editorial offices, customer services, and more information about Wiley products visit us at www.wiley.com.

Limit of Liability/Disclaimer of Warranty
While the publisher and authors have used their best efforts in preparing this work, they make no representations or warranties with respect to the accuracy or completeness of the contents of this work and specifically disclaim all warranties, including without limitation any implied warranties of merchantability or fitness for a particular purpose. No warranty may be created or extended by sales representatives, written sales materials, or promotional statements for this work. The fact that an organization, website, or product is referred to in this work as a citation and/or potential source of further information does not mean that the publisher and authors endorse the information or services the organization, website, or product may provide or recommendations it may make. This work is sold with the understanding that the publisher is not engaged in rendering professional services. The advice and strategies contained herein may not be suitable for your situation. You should consult with a specialist where appropriate. Neither the publisher nor authors shall be liable for any loss of profit or any other commercial damages, including but not limited to special, incidental, consequential, or other damages. Further, readers should be aware that websites listed in this work may have changed or disappeared between when this work was written and when it is read.

Library of Congress Cataloging-in-Publication Data

ISBN 978-1-119-75056-7

Cover image: Wikimedia
Cover design by Russell Richardson

Set in size of 11pt and Minion Pro by Manila Typesetting Company, Makati, Philippines

Contents

Preface   xiii

1   **Regulations and Other Developments/Trends/Initiatives Driving Non-Halogenated Flame Retardant Use**   **1**
Alexander B. Morgan
1.1   Regulatory History of Halogenated vs. Non-Halogenated Flame Retardants   1
1.2   Regulations of Fire Safety and Flame Retardant Chemicals   6
1.3   Current Regulations   8
    1.3.1   International – United Nations   8
    1.3.2   United States (Federal vs. State)   9
    1.3.3   Canada   10
    1.3.4   European Union   10
    1.3.5   Asia   11
    1.3.6   China   11
    1.3.7   Japan   12
    1.3.8   Korea   12
    1.3.9   Australia   13
1.4   Fire Safety and Non-Fire Safety Issues Requiring Non-Halogenated Flame Retardants   13
1.5   Regulatory Outlook and Future Market Drivers   16
    References   17

2   **Phosphorus-Based Flame Retardants**   **23**
Sergei Levchik
2.1   Introduction   24
2.2   Main Classes of Phosphorus-Based Flame Retardants   25
2.3   Red Phosphorus   26
2.4   Ammonium and Amine Phosphates   29
2.5   Metal Hypophosphites, Phosphites and Dialkyl Phosphinates   36
2.6   Aliphatic Phosphates and Phosphonates   40

2.7 Aromatic Phosphates and Phosphonates 48
2.8 Aromatic Phosphinates 58
2.9 Phosphine Oxides 62
2.10 Phosphazenes 64
2.11 Environmental Fate and Exposure to Organophosphorus FRs 65
2.12 Conclusions and Further Trends 67
   References 68

3 Mineral Filler Flame Retardants 101
   Reiner Sauerwein
3.1 Introduction 101
3.2 Industrial Importance of Mineral Flame Retardants 102
   3.2.1 Market Share of Mineral FRs 103
   3.2.2 Synthetic Mineral FRs within the Industrial Chemical Process Chain 104
   3.2.3 Natural Mineral FRs 106
3.3 Overview of Mineral Filler FRs 107
   3.3.1 Mineral Filler Flame Retardants by Chemistry 107
   3.3.2 Classification by Production Process 109
      3.3.2.1 Crushing and Grinding 109
      3.3.2.2 Air Classification 110
      3.3.2.3 Precipitation and Their Synthetic Processes 110
      3.3.2.4 Surface Treatment 113
   3.3.3 Physical Characterisation of Mineral FRs 114
      3.3.3.1 Particle Shape/Morphology/Aspect Ratio 115
      3.3.3.2 Particle Size Distribution 116
      3.3.3.3 Sieve Residue 117
      3.3.3.4 BET Surface Area 118
      3.3.3.5 Oil Absorption 118
      3.3.3.6 pH-Value/Specific Conductivity 119
      3.3.3.7 Bulk Density and Powder Flowability 120
      3.3.3.8 Thermal Stability/Loss on Ignition/ Endothermic Heat 122
   3.3.4 General Impact of Mineral FRs on Polymer Material Properties 123
      3.3.4.1 Optical Properties 123
      3.3.4.2 Mechanical Properties 123
      3.3.4.3 Water Uptake and Chemical Resistance 124
      3.3.4.4 Thermal Properties 126
      3.3.4.5 Electrical Properties 127

          3.3.4.6   Rheological Properties   128
    3.4 Working Principle of Hydrated Mineral Flame Retardants   128
          3.4.1   Filler Loading, Flammability and Flame Propagation 130
          3.4.2   Smoke Suppression   132
          3.4.3   Heat Release   134
    3.5 Thermoplastic and Elastomeric Applications   136
          3.5.1   Compounding Technology   136
          3.5.2   Compound Formulation Principals   138
          3.5.3   Wire & Cable   140
          3.5.4   Other Construction Products   147
          3.5.5   Special Applications   150
          3.5.6   Engineering Plastics for E&E Applications   152
    3.6 Reactive Resins/Thermoset Applications   154
          3.6.1   Production Processes for Glass Fiber-Reinforced
                  Polymer Composite   156
                  3.6.1.1   Paste Production   156
                  3.6.1.2   Hand Lamination/Hand-Lay-Up   157
                  3.6.1.3   SMC and BMC   157
                  3.6.1.4   Pultrusion   158
                  3.6.1.5   RTM/RIM   158
          3.6.2   Formulation Principles   159
          3.6.3   Public Transport Applications of GFRP   160
          3.6.4   E&E Applications   161
          3.6.5   Construction and Industrial Applications   163
    3.7 Conclusion, Trends and Challenges   164
          References   165

4   **Intumescence-Based Flame Retardant**   **169**
    Serge Bourbigot
    4.1 Introduction   169
    4.2 Fundamentals of Intumescence   172
    4.3 Intumescence on the Market   179
    4.4 Reaction to Fire of Intumescent Materials   180
    4.5 Resistance to Fire of Intumescent Materials   204
    4.6 Conclusion and Future Trends   224
          References   225

5   **Nitrogen-Based Flame Retardants**   **239**
    Alexander B. Morgan and Martin Klatt
    5.1 Introduction   239
    5.2 Main Types of Nitrogen-Based Flame Retardants   240

5.3 Ammonia-Based Flame Retardants 241
    5.3.1 Ammonium Polyphosphate 242
    5.3.2 Other Ammonia Salts 246
5.4 Melamine-Based Flame Retardants 247
    5.4.1 Melamine as Flame Retardant 248
    5.4.2 Melamine Salts 250
    5.4.3 Melamine Cyanurate 251
    5.4.4 Melamine Polyphosphate 254
    5.4.5 Melamine Condensates and Its Salts 256
5.5 Nitrogen-Based Radical Generators 257
5.6 Phosphazenes, Phospham and Phosphoroxynitride 261
5.7 Cyanuric-Acid Based Flame Retardants 263
5.8 Summary and Conclusion 264
    References 265

6 Silicon-Based Flame Retardants 271
Alexander B. Morgan and Mert Kilinc
6.1 Introduction 271
6.2 Basics of Silicon Chemistry 272
6.3 Industrial Applications of Silicones 274
6.4 Silicon-Based Materials as Flame Retardant Materials 277
    6.4.1 Inorganic Silicon-Based Flame Retardants 278
        6.4.1.1 Silicon Dioxide (SiO_2) (Silica) 278
        6.4.1.2 Wollastonite 280
        6.4.1.3 Magadiite 281
        6.4.1.4 Sepiolite 281
        6.4.1.5 Kaolin 282
        6.4.1.6 Mica 283
        6.4.1.7 Talc 283
        6.4.1.8 Halloysite 284
        6.4.1.9 Layered Silicate Nanocomposites 285
        6.4.1.10 Sodium Silicate 289
        6.4.1.11 Silsesquioxane 289
    6.4.2 Organic Silicone-Based Flame Retardants 290
        6.4.2.1 Polyorganosiloxanes 290
        6.4.2.2 Silanes 292
    6.4.3 Other Silicone-Based Flame Retardants 293
    6.4.4 Silicone/Silica Protective Coatings 294
6.5 Mode of Actions of Silicone-Based Flame Retardants and Practical Use Considerations 294
    6.5.1 Silicon Dioxide 294

6.5.2   Silicate-Based Minerals   295
6.5.3   Silicones   296
6.6   Future Trends in Silicon-Based Flame Retardants   296
6.7   Summary and Conclusions   298
      References   299

7   **Boron-Based Flame Retardants in Non-Halogen Based**
    Polymers   **309**
    Kelvin K. Shen
7.1   Introduction   309
7.2   Major Functions of Borates in Flame Retardancy   310
7.3   Major Commercial Boron-Based Flame Retardants
      and Their Applications   311
7.4   Properties and Applications of Boron-Base Flame Retardants   311
      7.4.1   Boric Acid $[B_2O_3 \cdot 3H_2O/B(OH)_3]$, Boric Oxide (B_2O_3) 311
      7.4.2   Alkaline Metal Borate   314
              7.4.2.1   Borax Pentahydrate $(Na_2O \cdot 2B_2O_3 \cdot 5H_2O)$,
                        Borax Decahydrate $(Na_2O \cdot 2B_2O_3 \cdot 10H_2O)$ 314
              7.4.2.2   Disodium Octaborate Tetrahydrate
                        $(Na_2O \cdot 4B_2O_3 \cdot 4H_2O)$ 315
      7.4.3   Alkaline-Earth Metal Borate   317
              7.4.3.1   Calcium Borates $(xCaO \cdot yB_2O_3 \cdot zH_2O)$ 317
              7.4.3.2   Magnesium Borate $(xMgO \cdot yB_2O_3 \cdot zH_2O)$ 317
      7.4.4   Transition Metal Borates   317
              7.4.4.1   Zinc Borates $(xZnO \cdot yB_2O_3 \cdot zH_2O)$ 317
      7.4.5   Nitrogen-Containing Borates   324
              7.4.5.1   Melamine Diborate $[(C_3H_8N_6)O \cdot B_2O_3 \cdot 2H_2O)]/$
                        $(C_3H_6N_6 \cdot 2H_3BO_3)$ 324
              7.4.5.2   Ammonium Pentaborate
                        $[(NH_4)_2O \cdot 5B_2O_3 \cdot 8H_2O)]$ 325
              7.4.5.3   Boron Nitride (h-BN)   325
              7.4.5.4   Ammonium Borophosphate   327
      7.4.6   Phosphorus-Containing Borates   327
              7.4.6.1   Boron Phosphate (BPO_4) 327
              7.4.6.2   Metal Borophosphate   328
      7.4.7   Silicon-Containing Borates   328
              7.4.7.1   Borosilicate Glass and Frits   328
      7.4.8   Carbon-Containing Boron or Borates   329
              7.4.8.1   Graphene (Boron-Doped)   329
              7.4.8.2   Boric Acid Esters $[B(OR)_3]$ 329
              7.4.8.3   Boronic Acid $[ArB(OH)_2]$ 330

7.4.8.4   Boron Carbide (B_4C) 331
7.5   Mode of Actions of Boron-Based Flame Retardants 331
7.6   Conclusions 332
References 333

8   Non-Halogenated Conformal Flame Retardant Coatings 337
Federico Carosio
List of Acronyms 337
8.1   Introduction to Conformal Coatings: The Role of Surface
During Combustion 339
8.2   Fabrics 346
8.2.1   Natural Fabrics 347
8.2.2   Synthetic Fabrics and Blends 358
8.2.3   Process Equipment and Related Patents 371
8.3   Porous Materials 373
8.3.1   Open Cell PU Foams 374
8.3.2   Other Porous Substrates 386
8.3.3   Process Equipment and Related Patents 391
8.4   Other Substrates 393
8.5   Future Trends and Needs 396
References 397

9   Multicomponent Flame Retardants 413
Bernhard Schartel
9.1   The Need for Multicomponent Flame Retardants 413
9.2   Concepts 419
9.3   Combination with Fillers 424
9.4   Adjuvants 428
9.5   Synergists 431
9.6   Combinations of Different Flame Retardants 435
9.7   Combinations of Different Flame-Retardant Groups
in One Flame Retardant 437
9.8   Conclusion 439
References 439

**10 Other Non-Halogenated Flame Retardants and Future Fire
Protection Concepts & Needs** 475
Alexander B. Morgan, Paul A. Cusack and Charles A. Wilkie
10.1   The Periodic Table of Flame Retardants 475
10.2   Transition Metal Flame Retardants 478
10.2.1   Vapor Phase Transition Metal Flame Retardants 478

10.2.2   Condensed Phase Transition Metal Flame
          Retardants   479
          10.2.2.1   Metal Oxides   480
          10.2.2.2   Metal Complexes   481
10.3   Sulfur-Based Flame Retardants   484
10.4   Carbon-Based Flame Retardants   485
       10.4.1   Cross-Linking Compounds – Alkynes,
                 Deoxybenzoin, Friedel-Crafts, Nitriles,
                 Anhydrides   486
                 10.4.1.1   Alkynes   486
                 10.4.1.2   Deoxybenzoin   488
                 10.4.1.3   Friedel-Crafts   488
                 10.4.1.4   Nitriles   490
                 10.4.1.5   Anhydrides   490
       10.4.2   Organic Carbonates   491
       10.4.3   Graft Copolymerization   492
       10.4.4   Expandable Graphite   493
10.5   Bio-Based Materials   494
10.6   Tin-Based Flame Retardants   496
       10.6.1   Introduction   496
       10.6.2   Zinc Stannates   497
       10.6.3   Halogen-Free Applications   498
                 10.6.3.1   Polyolefins   499
                 10.6.3.2   Styrenics   500
                 10.6.3.3   Engineering Plastics   500
                 10.6.3.4   Thermosetting Resins   503
                 10.6.3.5   Elastomers   503
                 10.6.3.6   Paints and Coatings   505
                 10.6.3.7   Textiles   506
       10.6.4   Novel Tin Additives   506
                 10.6.4.1   Coated Fillers   507
                 10.6.4.2   Tin-Modified Nanoclays   509
                 10.6.4.3   Mechanism of Action   510
                 10.6.4.4   Summary   512
10.7   Polymer Nanocomposites   513
10.8   Engineering Non-Hal FR Solutions   513
       10.8.1   Barrier Fabrics   514
       10.8.2   Coatings   515
                 10.8.2.1   Inorganic Coatings   515
                 10.8.2.2   IR Reflective Coatings   517

10.8.2.3   Nanoparticle Coatings 517

10.8.2.4   Conformal/Integrated Coatings 518

10.9   Future Directions 519

10.9.1   Polymeric Flame Retardants and Reactive Flame Retardants 521

10.9.2   End of Life Considerations For Flame Retardants 524

10.9.3   New and Growing Fire Risk Scenarios 528

10.9.4   Experimental Methodology for Flame Retardant Screening 531

References 532

Index 555

Preface to the 2nd Edition of the Non-Halogenated Flame Retardant Handbook

Since the writing of the first edition in 2014, and the writing/publishing of this book in the 2020-21 timeframe, there have been some notable changes to the non-halogenated flame retardant field. Mostly, there have been an increase in regulations of flame retardant (FR) chemicals, further de-selection of chlorinated/brominated flame retardants, and some diversification of the flame retardants that exist in the "non-halogenated" category. Therefore, a need for a 2nd edition existed. This is actually quite rapid for the FR field and for generation of a 2nd edition of a book only 6 years after the publishing of the first, vs. the typical 10-15 years for FR book updates. FR technology tends to move at a slow pace due to flame retardant material science being a reactive field in response to regulations. Regulations take a notable amount of time to develop, be debated, and put into law, and so developments in FR chemistry in response to those regulations takes equally long. It can take even longer to take those developments out of the lab and scientific literature and then commercialize them.

The need of a book dedicated to non-halogenated flame retardant chemistry remains as strong as it did with the 1st edition, and this book includes chapters on FRs including Phosphorus-based (Chapter 2), Mineral-based (Chapter 3), Intumescent-based (Chapter 4), Nitrogen-based (Chapter 5), Silicon-based (Chapter 6), Boron-based (Chapter 7), and all the other non-halogenated FR systems which are not based upon the above elements (Chapter 10). All of these chapters existed in the 1st edition of the book, but each of these chapters has been updated for this book, and in some cases, significantly so. Chapter 10 is significantly revised, and includes bio-based FR chemicals, as well as some newer concepts. There are three new chapters in this book, including a chapter on regulations (Chapter 1), a chapter on conformal non-halogenated flame retardant coatings (Chapter 8), and a chapter on multi-component flame retardant systems (Chapter 9). Overall, the revised content in the chapters from the 1st edition, and the new chapters to this edition create an excellent contribution to flame retardant materials science. Or I hope you'll come to that conclusion after you read the book.

As stated in the first edition, the fire threat for materials has not changed just because the regulatory environment around FR chemicals has. There is still a pressing need for flame retardant solutions for materials throughout modern society. One could argue that in some cases, with an increase in electrification of vehicles and buildings, that there may be more need, rather than less, for fire protection, and so non-halogenated solutions are part of that solution space. So with that, I believe you'll find that this book is a single source of practical non-halogenated FR technology and information that will guide whatever research, development, testing, and evaluation (RDT&E) is needed for new fire safe solutions. I further believe that even with the many recent changes in regulations occurring as we speak, and in the years to come, that the book will still be very useful for many years beyond the 2021 publishing of this book.

As with all prefaces, I would like to thank those who helped make this book possible, especially the authors of the individual chapters who have taken time out of their busy lives to write the chapters. I also want to thank Scrivener Publishing again for their willingness to publish the 2nd edition. Special thanks goes out to Dr. Anteneh Worku of FR Adviser LLC and Dr. Adrian Beard of Clariant GmbH for helping review chapters in this book. I certainly must thank my (now retired) colleague, Prof. (emeritus) Charles Wilkie of Marquette University, for getting me started into flame retardant book publishing many years ago. This is now the sixth book I've edited, and my first one on my own. Chuck taught me how to navigate this area, and for this I'm quite grateful. Finally, I want to thank my wife, Julie Ann G. Morgan, for her continual support during my career, and for teaching me enough grammar that I can finally write on my own, although she may still take issue that I actually learned those lessons.

Alexander B. Morgan
Dayton, Ohio, USA
October, 2021

Regulations and Other Developments/ Trends/Initiatives Driving Non-Halogenated Flame Retardant Use

Alexander B. Morgan

University of Dayton Research Institute,
Center for Flame Retardant Materials Science, Dayton, OH, USA

Abstract

Fire safety of materials is regulated via laws like building codes or product safety laws, which in turn refer to standards of performance and testing needs to meet various fire risk scenarios. As such, fire safety of materials, and the individual components and chemicals involved in these materials are highly regulated. Indeed, as a field of materials science, it is performance needs together with regulatory requirements or voluntary schemes like ecolabels and market trends which drive chemical selection for fire safety needs – in addition to economical constraints. This chapter will discuss the regulatory requirements which affect the choice flame retardant chemicals. Specific regulations of the chemicals themselves, and new regulatory issues that are driving selection and de-selection of specific flame retardant chemicals for fire safety needs will be presented as well.

Keywords: Regulations, codes, standards, fire risk scenarios, fire safety, environment, chemical regulation, politics, ecolabels

1.1 Regulatory History of Halogenated vs. Non-Halogenated Flame Retardants

Before beginning any book on non-halogenated flame retardants, it is important to understand the history of why there is such a book focused

---

Email: alexander.morgan@udri.udayton.edu

Alexander B. Morgan (ed.) Non-Halogenated Flame Retardant Handbook 2nd edition, (1–22)
© 2022 Scrivener Publishing LLC

on non-halogenated flame retardants. Prior to the late 20th century, flame retardants were not necessarily singled out by any particular chemistry. Indeed, prior to the 1930s, halogen was not used at all as a flame retardant chemical, and even after its discovery and use, it was just another class of chemicals used to impart fire safety to other materials.

To begin with, a flame retardant chemical is a chemical that shows the ability to retard flame growth and spread in a particular material in a particular fire risk scenario. This flame retarding function can be achieved with very diverse chemistries based on the elements bromine, chlorine, phosphorus, nitrogen, and aluminum to name the most prominent. Many of these elements, other than halogen, are discussed throughout this book. On top of these elements, both organic and inorganic substances are being used. The only common feature is that the flame retardant interferes with some of the chemical reactions which are necessary for sustained burning of a material and generally raises the energy that is necessary to ignite a material – flame retardants do not make materials non-combustible.

Not all flame retardants are universally able to flame retard all polymers in all fire risk scenarios. A particular chemical may be very effective in one polymer, but not in another. This is really no different than most chemicals in use throughout the world today: each has its specific chemistry it is capable of, and its own chemical structure-property relationships that yield certain end effects when a chemical reaction occurs. There can be simplicity in grouping chemicals by general structural class and similarity due to how they chemically react. For example, halogenated flame retardants tend to have very similar flame retardant mechanisms of vapor phase combustion inhibition, regardless of chemical structure. There are exceptions where aliphatic and aromatic halogenated compounds can have different reactivity in fire events, as well as additional fuel/chemical interactions that one class will show and the other will not, but some general mechanisms of flame retardancy can be assigned to a group of similar chemicals. As will be discussed, some general classes of flame retardant chemicals include halogenated, phosphorus-based, mineral fillers, nitrogen-based, silicon-based, boron-based, and a wide range of other niche chemicals ranging from transition metal materials to metalloids to carbon-based structures. So while it is possible to group chemicals by flame retardant activity and mechanism, it becomes more complicated to group those same chemicals for reactivity in non-flame retardant scenarios. For example, one mineral filler used as a flame retardant, magnesium hydroxide, works as a flame retardant chemical for wire and cable applications. It also is the active ingredient in "Milk of Magnesia", which is an oral antacid for heartburn and digestive issues. Other mineral fillers with flame retardant effect may not have this same dual effect and further, may not be safe for ingestion at all. Environmental chemical effects, as well as chemical persistence, bioaccumulation, and toxicity (PBT) profiles are very chemical structure dependent when interacting with humans and the natural environment. All mineral fillers may be persistent (and it is debatable that persistence for minerals is really a problem or not), but they will have very different bioaccumulation and toxicity profiles dependent upon their chemical structure. Likewise, all chemicals and chemical flame retardants will have different PBT profiles, even if they are in the same general chemical class. With this in mind, we can discuss some regulatory history of halogenated vs. non-halogenated flame retardant chemicals.

Halogenated flame retardants began use in earnest in the 1930s and onwards, as they were found to be potent flame retardant additives for flammable materials, as well as strong extinguishing agents such that liquid halogenated solvents were used in fire extinguishers. Indeed, there are reports of fire extinguishing "hand grenades" that were glass globes filled with carbon tetrachloride (now known to be a potent carcinogen) that firemen would lob into fires to help put them out, and, this same halogenated chemical was used in hand-held fire extinguishers [1]. As hazards of these liquid chemicals were found, these liquid halogenated flame retardants were pulled from service and other active extinguishing agents were instead put into fire extinguishers. Halon gas extinguishers were used for severe fire situations, but even these have been pulled from service due to ozone depletion issues. Their relative chemical stability made them non-toxic and therefore a preferred choice, however, for the same reason the chemicals were able to reach the stratosphere where they finally reacted with ozone. Halogenated flame retardant additives put into plastics began to be under regulatory scrutiny in the late 1990s to early 2000s as part of a move to prevent dioxin formation when end-of-life plastics (and other household waste) would be sent to incinerators. Incineration of waste is commonly carried out in Europe due to the lack of landfill space there, and for waste-to-energy efforts that are present in some European countries, especially in Scandinavia. Waste is difficult to presort, and so large amounts of polyvinyl chloride (PVC), as well as other halogenated compounds, ended up in the waste and large amounts of dioxin were formed as part of the emissions from these incineration facilities. As this was discovered, regulations were put in place to mitigate and cease dioxin formation via two methods. The first was with improved emissions capture and cleanup systems (baghouses, scrubbing systems, afterburners), and the second was to remove halogen from the waste stream. The second approach was where regulations against halogenated flame retardants began in earnest, with two well-known directives, the Reduction of Hazardous Substances (RoHS) [2, 3] and Waste Electrical and Electronic Equipment (WEEE) [4, 5]. These initiatives sought to reduce and eliminate the use of halogenated additives in consumer products, namely electronics, which would in turn reduce the amount of halogenated additives going to incinerators, or, accidentally released to the environment. The directives also aim at eliminating legacy brominated flame retardants from recycle streams, so that they do not end up in new E&E equipment via recycling.

Another reason for banning or limiting use of halogenated flame retardant additives in flammable materials (such as polymers) is the corrosive gases that form from these flame retardants as they activate in a fire. The vapor phase flame inhibition mechanism of halogenated flame retardants is well known to produce acid gases (HF, HCl, HBr) [6–9] which can present some secondary health effects (irritation of eyes and lungs) which can exacerbate the toxicity situation caused by the primary toxicant in fires, carbon monoxide [10–14]. Additionally, the acid gases can cause significant economic damage to materials that are sensitive to corrosive gases. Modern electronics are particularly sensitive to corrosive gas damage, and so there have been new regulations banning halogenated flame retardants from computer server facilities computer chip fabrication sites for this very reason. There are also some acidic gas regulations for aerospace, maritime, and mass transportation which also limit or effectively ban halogenated flame retardants from use.

Other European Union (EU) regulations have come into effect banning specific brominated flame retardant molecules found to have negative persistence, bioaccumulation, and toxicity (PBT) profiles, especially as new information comes to light indicating that a particular chemical structure is hazardous. This is how things evolve from a chemical use perspective, and is how it should occur. With new information about hazards, hazardous materials should be removed from use and commerce. However, as new information comes along, sometimes the regulatory picture becomes clouded. Going back to the main issue with dioxin formation, it is now well known that with halogen being naturally present everywhere in our environment, any time you have a fire or combustion event where halogen is present, you will form dioxins. Halogenated dioxins can be found in forest fires [15] as well as from electrical/electronic fires [16]. Unless you have capture systems and afterburners, dioxins will be emitted. The amount of dioxins formed depends on the materials involved in the fire event, as well as combustion conditions. It's impossible to remove halogen from the environment, and indeed, fires themselves, especially accidental ones involving modern materials, produce all sorts of toxins and pollutants including sub-lethal gasses, lethal gases, and carcinogens such as polyaromatic hydrocarbons (PAHs) [17–22]. These toxins can be found in fires where flame retardants are present, as well as those without flame retardants, although the total volume of pollutants produced is less if the fire growth is lowered by the presence of effective flame retardants [23–29]. Therefore, the original regulatory reason behind halogenated flame retardant regulation and use (to prevent dioxin formation) is still correct, but with new information, the benefits and drawbacks of said regulations are now not as clear as they once were.

Stepping aside from the emission issue of hazards from halogenated flame retardant in fire events, there is the non-fire "emission" of the halogenated flame retardant when it gets into the environment. Going back to the above mentioned PBT issue, any chemical will be of concern in the environment if it should be emitted, spilled or introduced outside of controlled situations and the chemical is persistent (lasts for a long time), bioaccumulates (enters and concentrates in living organisms), and is toxic. Halogenated flame retardants of old are by design persistent due to their chemical structure, and the fact that one wants the flame retardant to last for years inside the product. One does not want to buy something with a 20 year lifetime only to have the fire protection wear out in the first year. This persistence has found halogenated flame retardants in many different places in the environment [30–39], and it is rightfully troubling. Many of the older halogenated flame retardants are small lipophilic molecules, meaning they can also be bioaccumalative (in the fatty tissue of many organisms), and some have also been found to be toxic. These negative PBT issues are why polybrominated diphenyl ethers (PBDEs), which are small molecule halogenated flame retardants, have been banned from use in the EU and US, as well as many other countries [3, 5, 33–38] By extension, several countries and US states have started to extend the bans on PBDEs to all halogenated flame retardants, regardless of chemical structure. It is important here to note that small molecule flame retardants are of concern when they migrate out of the plastic, but polymeric brominated flame retardants are of high molecular weight and while they are persistent, current data indicates they are not bioaccumulative or toxic. Likewise, reactive flame retardants which covalently bond into a polymer structure cannot get into the environment and cannot become bioaccumulative or toxic, even if they may be persistent. So wholesale bans on entire classes of chemicals may not be merited, but regardless of the lack of scientific merit, these wholesale bans are being implemented. Further, the volume of data against small molecule halogenated flame retardants having negative PBT profiles is such that even when halogenated flame retardants are polymeric or reactive, market conditions shy away from their use. Still, technology moves forward, as do opinions and personal/market tastes, and so there is still a need for fire safety protection/flame retardant chemistry, and therefore the market moves to non-halogenated flame retardants. Hence the reason for this book to guide materials scientists toward how to use non-halogenated flame retardant chemicals to provide fire safety, and to guide them on the newest information available.

1.2   Regulations of Fire Safety and Flame Retardant Chemicals

With some basic history about halogenated and non-halogenated flame retardants in place, we can now discuss more detailed regulation of flame retardants. In general, regulations are mostly reactive to information and events, rather than proactive to potential or perceived hazards. There are exceptions, but this reactive mode of regulation is applied in the majority of regulatory cases.

Modern fire safety regulations are often found within various legal codes, especially building codes, aviation regulations, and federal registers that describe particular requirements and test methods to ensure fire safety in a structure, vehicle, component, sub-component, or material. These regulations do not require any particular flame retardant chemistry to be used, but instead prescribe a particular level of performance. In fact, regulations really do not mandate flame retardants to be used at all. Flame retardants get used because it is one of many ways to provide fire protection, and may be selected depending upon all the other "non-fire" requirements for a functional item, including cost, thermal/mechanical/electrical performance, manufacturing requirements, intellectual property, and so on. It is important to emphasize this point as there is some perception that fire safety regulations mandate or push the use of flame retardants. This is not correct. The only time a particular chemical will be mandated for use is when it is prescribed in a manufacturer requirement document after certifications for use have been achieved. For example, a composite part inside the cabin of an aircraft that meets flame spread and heat release requirements and has been deemed "airworthy" may have manufacturer requirements to hold to a particular polymer formulation to ensure the part meets the requirements and does not have to be recertified for use. This requirement may then specify specific flame retardant chemicals, and loading levels, to meet the performance. But again, if one reads the original fire safety requirements, the original laws will not mandate any particular approach or chemical to be used.

Fire safety regulations will seek to mimic a particular fire risk scenario where there has been a notable hazard identified, and some probabilities of that hazard occurring with notable loss of life or property. Within the regulation is a test method that seeks to mimic the fire risk scenario, and validate, in a reproducible way, that the item does meet the fire safety goals of the regulation. This typically means pass/fail test methods, but sometimes it can be a quantitative test that assigns levels of fire safety to the item tested depending upon that measured quantity. For example, different fire safety classes may be assigned to some building materials depending upon their ability to resist various heat sources, as well as levels of flame spread and smoke release. Therefore, for anyone to be able to sell a product into an application that has a fire safety requirement, one must test their materials via the regulatory test method. If the material should not pass the test, then flame retardant or fire protection methodology may be required. This is where flame retardants often get introduced into products, when the product tested does not meet the fire safety test. Flame retardants will not be added to a material if the material already passes a fire test, as it just adds cost and complexity to a material. Flame retardants will be added to the material if it enables that material to pass the particular regulatory test *and* it meets all the other product requirements. Sometimes, flame retardants are not needed if simple engineering controls can be used to provide fire protection for the item. Examples of engineering controls can be isolating the flammable material from ignition sources or using sprinkler systems. However, when flame retardant additives are used, they are tailored for each fire risk scenario and for each material – they are not universal and cannot be swapped from material to material without careful consideration. Therefore, one must study each specific material in each specific fire risk scenario to know what flame retardant chemical to use. This chapter will not see to cover the wide range of fire risk scenarios and test methods, as there are other excellent resources for this [9, 14, 40–42, 57]. Instead, keeping in mind that specific flame retardants get used for specific materials in specific fire risk scenarios, we can discuss flame retardant chemical regulations.

Returning to the historical perspective of flame retardancy for a moment, many of the older flame retardants now banned were used for decades because they worked very well in a particular material to provide fire protection against a particular fire risk scenario. Just as new information can come to light on the PBT profile of a chemical which will affect its use, fire risk scenarios can change over time. However, in other cases, the fire risk scenario may remain the same, but particular chemicals or classes of chemicals may be regulated differently. As discussed previously, halo-genated flame retardants have been heavily regulated in recent years due to concerns about their dioxin formation, as well as specific PBT issues. So in more recent times, there are regulatory changes to which chemicals may be used, while not changing the regulatory fire test, and in other cases, the regulatory change is made to the fire test *and* to the chemicals allowed to be used. As will be discussed below, there have been approaches taken to dis-incentivize the use of flame retardant chemicals through other product regulation, while maintaining the need for particular fire safety, or, to change the fire safety regulations themselves. When the latter is chosen, the current approach has been to lessen the fire safety require-ments. While there can be changes in fire risk scenario that can support this approach, as will be discussed below, sometimes the change in fire risk scenario is driven by perceptions and political considerations, and not actual fire safety requirements. Fundamentally, the assessment of fire risks for certain products like upholstered furniture should be done separately from the chemical safety assessment of flame retardants which might be used. Reducing fire safety requirements to get rid of "unwanted" FRs is the wrong approach, as one should rather restrict the use of any problematic chemicals directly and promote the use of safer alternatives (see detailed discussion below).

1.3   Current Regulations

As previously discussed, regulations are often reactive based upon past his-torical events in a particular location where local or national fire events drive new requirements to prevent a particular fire event from happening again. Likewise, local cultural uses of building products, building styles, and operating of technology may drive particular fire safety requirements, especially if there are local population density issues, or environmental effects (earthquakes, wildfires) that may drive fire safety requirements in one direction or another. Therefore, regulations are be best discussed at the national and regional level.

1.3.1   International – United Nations

Legacy halogenated flame retardants have meanwhile been restricted under the United Nations Persistent Organic Pollutants (POP) convention: HBCD, PBDEs including DecaBDE, and short-chain chlorinated paraffins (SCCP) [43].

1.3.2   United States (Federal vs. State)

In the United States (US), federal government regulations overrule state regulations. However, if there is no specific federal regulation on a particular topic or chemical, then state regulations apply. This can mean that a product sold in the US could have to meet 50 different state regulations if they are different. Currently, most chemicals are regulated by the Toxic Substances Control Act (TSCA, 1976) which was "updated" by the Frank R. Lautenberg Chemical Safety Act for the 21st Century in 2016. Under TSCA, only very few chemicals were banned and it generally took many years. Regarding flame retardant chemicals, there have been voluntary phase outs of brominated diphenyl ethers in the US due to rulemaking and agreements with the US Environmental Protection Agency (EPA), and some scrutiny of hexabromocyclododecane (HBCD), [44–49]. The US EPA set up a workplan on flame retardants already in 2012 but with slow progress. In March 2019 they concluded TCEP, TBBPA and TPP as "high priority substance" candidates for risk assessments.

In addition to these regulatory workstreams, from 2005 to 2015, the US EPA did run a serious of extensive Design for Environment (DfE) projects which evaluated alternatives to the legacy brominated flame retardants pentabromo- and decabromo diphenylether, hexabromocyclo dodecane and tetrabromo bisphenol-A [50]. The conclusion was that often halogen free alternatives exist with a better environmental and health profile. Furthermore, in 2017 the US Consumer Product Safety Commission (CPSC) voted to initiate rulemaking based on a petition to protect consumers from "toxic" flame retardant chemicals commonly referred to as organohalogens (OFRs), under the Federal Hazardous Substances Act [51]. The initiative refers to children's products, furniture, mattresses, and electronic device casings. CPSC further advised setting up a Chronic Hazard Advisory Panel to further study the effects of OFRs as a class of chemicals on consumers' health. The petition lists 24 organohalogens including decabromodiphenyl ether and several chlorinated phosphate esters, believed to be toxic, that tend to migrate out of products, and can bioaccumulate.

At the state level however, there has been a lot of regulatory movement to ban flame retardant chemicals by broad chemical class, rather than by specific molecule. Most of the bans are focused around keeping flame retardant chemicals out of mattresses and furniture, but some bans on manufacture and use of flame retardant chemicals are broader in scope than just furniture and mattresses. The wide range of state regulations is far too much to cover in this chapter, and a reasonable summary of each rule with links to each state law is available online [52]. That being said, the emphasis of most state laws is to ban flame retardants by class (halogen, phosphorus, nitrogen, etc.) in specific consumer products (mostly furniture and mattresses) rather than by specific chemistry. If TSCA change does occur which lists particular flame retardants as safe/not safe to use, that TSCA change would overrule all of the individual state laws. Otherwise, it is highly recommended that material scientists work with their respective regulatory experts if they are planning on using any flame retardant chemical for products in the US, whether halogenated or not. As of the writing of this chapter, the situation is still very uncertain how these state laws will move forward, or if they will get challenged and found to be unworkable by generating broad bans of chemical classes vs. specific negative PBT profile chemicals.

1.3.3   Canada

Chemical regulation in Canada is governed by the Canadian Environmental Protection Act (CEPA) [53] as well as new substances/existing substance under its Chemical Management Plan. Flame retardant chemicals which are regulated under this law include brominated diphenyl ethers (BDPEs), hexabromocyclododecane (HBCD), and tetrabromobisphenol A. As per the law, new chemicals are investigated and added to the regulatory list as PBT data becomes available. Similar to laws in the US, known brominated flame retardants with known negative PBT profiles are banned from use and import into Canada. In 2019 Environment Canada stated that decebromo diphenylethane (DBDPE) may contribute to the formation of persistent, bioaccumulative, and inherently toxic transformation products, such as lower brominated BDPEs, in the environment. A ban on the manufacture, sale or import of the brominated FR DBDPE has been proposed (pending as of 2021-03). This is remarkable in so far as DBDPE has often been cited as an example of regrettable substitution, where a regulated substance (decabromodiphenylether, DBDE) is replaced by industry with a molecule that is just slightly modified, so evading the regulatory restriction whilst still having similar environmental properties.

1.3.4   European Union

The European Union (EU) has been at the forefront of chemical regulation for chemicals used in commerce. Relevant to flame retardant use, the Restriction of Hazardous Substances in Electrical and Electronic Equipment (RoHS) [4, 5] and Waste Electrical and Electronic Equipment (WEEE) [6, 7] laws have forced out the legacy brominated flame retardants

PBDEs and PBBs from use. The newest chemical regulation which governs all chemicals, including non-halogenated, is the Registration, Evaluation, Authorisation and Restriction of Chemicals (REACH) [54]. REACH introduced the concept of Substances of Very High Concern (SVHC), based on PBT and CMR (carcinogenic, mutagenic and reprotoxic) properties. SVHC are supposed to be phased out and substituted unless there is authorization for specific uses. The following flame retardants are identified as SVHC or on the candidate list (as of 2021-03): Penta-, Octa- and DecaBDE, HBCD, Short chain chlorinated paraffins, Tris(2-chloroethyl) phosphate, Boric acid (toxic for reproduction) and Trixylyl phosphate. Since 2021, manufacturers of finished articles have to provide information on SVHCs in their products in a public database (SCIP, substances of concern in products) maintained by the European Chemicals Agency (ECHA).

Because commerce is global, REACH will likely affect flame retardant use in multiple countries, especially those which import to the EU, and export or manufacture within the EU. It is highly likely that as flame retardant chemicals with negative PBT profiles are found they will be banned or regulated under REACH, and this guidance will likely lead to other countries following suit for their own regulations. It is important to note here that the EU, as of the writing of this chapter, does have harmonized regulations across EU member states, but, there is some disagreement and discord between member states where a particular member state would want stricter or lesser regulation on chemicals. There is a long and deliberate mechanism in place in the EU to resolve these disputes, but the disputes can take years to address. Of final note, the United Kingdom has left the EU, but is still sorting out its regulations and commercial connections and collaborations with the EU. How UK independence will affect regulation of flame retardant chemicals in that country is not clear at this time.

1.3.5   Asia

There are many sovereign countries in Asia such that the potential regulations from country to country can be quite different. The three main markets with chemical regulations related to flame retardants are China, Japan, and Korea, but it is likely that other Asian countries have or will develop chemical regulations that also cover flame retardants.

1.3.6   China

China released its own version of RoHS in 2007 which is based upon the EU RoHS [55], but it only applies to imported materials, not exported electronics. It's important to note that items exported from China to other parts of the world may have chemicals of concern that are banned from use in those countries. In the US for example, there have been several cases of imported goods from China containing chemicals (flame retardants and otherwise) that were banned from use in the US. There are even companies in China which produce flame retardants that are no longer produced in the US and EU because of their negative PBT profiles. It is unclear at this time how the China flame retardant regulations are enforced for domestic vs. export items, but local translation and guidance on Chinese environmental regulations is strongly recommended prior to selling into the Chinese market, or, getting exports of items potentially containing flame retardant chemicals from the Chinese market. An updated version of the China RoHS was issued in 2016, restricting the same six substances as the original EU RoHS. Products and parts that contain restricted substances exceeding limits can still be sold in China but need to be marked as such. A peculiar concept of China RoHS is the "Environment Friendly Use Period" (EFUP) designating the time before any of the RoHS substances might to leak out, causing possible harm to health and the environment. Every product that contains RoHS substances above the maximum permitted concentration is carries an orange circle label composed of two arrows containing a number that indicates the EFUP in years.

1.3.7   Japan

In Japan, the Ministry of Economy, Trade, and Industry manages the Chemical Substances Control Law (CSCL) [56] that would govern any use of flame retardant chemicals in that country, both in regards to manufacturing for domestic use and for export. The list of controlled chemicals on the CSCL is extensive, and does include some of the older flame retardants banned in the US and EU, such as brominated diphenyl ethers (BDPEs) and hexabromocyclododecane (HBCD) [57]. The CSCL and list of chemicals is updated from time to time and should be monitored for changes.

1.3.8   Korea

In South Korea (Republic of Korea), chemicals (including flame retardants) are governed by the Toxic Chemicals Control Act [58]. This act controls the manufacture and use of chemicals in Korea, and new chemicals introduced into commerce in this country as well as any new chemicals made domestically in South Korea. At the time of writing this chapter, gaining access to this list of chemicals in English was not possible for the author of this chapter, so it is unclear if flame retardants are on the list of controlled chemicals. However, given that this law conducts analyses of chemical safety similar to other countries which have banned HBCD, BDPEs, and various metals like Cd and Cr, it seems highly likely that flame retardants with known/established negative PBT profiles are regulated and banned from use in South Korea. Those selling into the South Korean market are encouraged to check the local regulations and see which chemistries are banned from use. Non-halogenated chemistries like red phosphorus may be limited in use in South Korea due to the potential for PH_3 formation when red phosphorus is exposed to humidity (which can be avoided by encapsulation of the red phosphorus).

1.3.9   Australia

In July of 2020, Australia introduced a sweeping new regulation for chemicals called the Australian Industrial Chemicals Introduction Scheme (AICIS) [59]. This law looks at all chemicals imported into Australia, as well as those created/used domestically, and covers applications that use those chemicals as well. There is an extensive chemical inventory to see what is known about a particular chemical (and where there are gaps) in a searchable database, and a list of those chemicals which are banned/restricted from use. The Australian law looks to international law as a basis for chemical bans, including chemicals found on the Stockholm Convention of Persistent Organic Pollutants, which includes HBCD, BDPEs, and some perbrominated/perchlorinated aromatic compounds. Tris(2,3-dibromopropyl)phosphate is also subject to regulation due to its negative PBT profile. As of 2020, no other non-halogenated flame retardants are on the regulated list with Australia.

1.4   Fire Safety and Non-Fire Safety Issues Requiring Non-Halogenated Flame Retardants

Depending upon the nature of a fire risk scenario and the codes/standards involved in regulating and testing materials with that particular scenario, there are cases where halogenated flame retardants cannot meet the fire safety requirements. These cases may result in non-halogenated FRs being favored for fire safety use. Typically, fire requirements around smoke and corrosive gas release limit the use of halogenated flame retardants. Smoke release may be limited in some compartment fires as the smoke may block visibility of exit signs or just visibility in general. Many halogenated flame retardants produce high levels of smoke upon burning since the halogenated flame retardants inhibit vapor phase combustion, thus leading to higher levels of smoke release [60–62]. Additionally, halogenated flame retardants often release corrosive gases during combustion (namely HBr and HCl), which can cause significant damage to electronics and electrical systems. Locations holding large numbers of computer servers, electronic data storage, and power switching systems typically have requirements against corrosive gases, and small compartments with sensitive electronics (aviation, maritime, aerospace) are likely to have similar requirements that limit or prevent the release of corrosive gases in a fire event. This can be compounded by heating, ventilation, and air conditioning (HVAC) systems that feed air into rooms holding these items, such that components in the ductwork or feeding air to the HVAC systems must also be non-halogenated to meet the corrosive gas requirement. Computer chip fabrication facilities also have corrosive gas emission requirements which often require the furniture, cabinetry, and items in the room to also be non-halogenated, or at least compliant by releasing little to not corrosive gases during a fire event [63]. Finally, there are some very strict combustion emission requirements for vehicles in extreme environments that have sealed atmospheres (submarines, spacecraft), and for that reason, corrosive gases are excluded with an emphasis on very low flammability first, and emissions second [64, 65].

Along with fire safety issues affecting the selection/de-selection of particular flame retardants, there are several non-fire safety issues that must be considered as well. Some of this was discussed in the previous section where specific chemicals are regulated, but it is worthwhile to review this again, as well as share some of the other requirements not specifically covered in chemical bans and chemical regulation. As mentioned previously, there are regulations which de-select halogenated flame retardants for environmental reasons, namely the WEEE and RoHS requirements covering electrical and electronic equipment. There are also voluntary labeling requirements which if met, allow additional branding of items to suggest that it has a better environmental profile than items that do not have this label. Examples of this include "ecolabels" like Blue Angel [66], European Union Ecolabel [67], Nordic Swan [68], Green Seal [69], Global GreenTag [70], and other relevant labels. These other labels may have a variety of non-fire related requirements which may de-select certain flame retardant chemicals. Recycling requirements and how a flame retardant chemical interacts with the recycling process (regrind & remelt, chemical depolymerization, waste-to-energy, etc.) may also prefer a particular flame retardant chemistry. Finally, there are issues with customer perception that will need to be addressed. For all the reasons mentioned above that are driving halogenated chemicals out of use where selected halogenated chemicals have negative PBT profiles, customers may demand non-halogenated even if the halogenated flame retardant can be made in such a way to have a positive PBT profile. The phrase "the customer is always right" is particularly true here if the customer is asking for non-halogenated flame retardant technology, even if it may be more expensive or the balance of properties in the final product is not as good as that possible with halogenated FRs. The same negative PBT data about halogenated FRs has led to consumer advocates/consumer influences asking for non-halogenated FRs in certain applications. To date, no one is clamoring for halogenated flame retardants, but there are many customer requests for non-halogenated FR use in consumer electronics and items that need fire protection, and have a short useful lifetime. The coming regulations seeming to ban all flame retardant chemicals by broad class may eventually lead to customer requests for no-flame retardant chemicals to be added to a product, and to instead use inherently low flammable materials. This latter request only holds if the cost is right, and for now, cost is still "king" in regards to what most consumers will pay. A consumer may specifically request non-halogenated chemistry and/or lower environmental impact items, but rarely will pay more for it.

Political considerations based upon perception of scientific information, rather than actual interpretation of data and studying the entire complex picture of fire safety and materials science, are indeed something that must be considered as a non-fire issue affecting flame retardant use. Perversely, these political motivations are moving to change the fire safety regulations to weaken them such that no flame retardant chemicals are needed. It is fair to state that sometimes regulations outlive their historical use as technology and consumer use trends change over time. But the materials do not change or lose their flammability with time. The main example of this political move is associated with polyurethane foam used in furniture. Polyurethane foam, due to its chemical structure, is a highly flammable material [71–75], and is just as flammable today as it was in the 1970s when fire safety regulations were first put in place around furniture. In the US, California Technical Bulletin #117 (TB-117) and Technical Bulletin #133 (TB-133) were put in place with small and large flame exposure tests to ensure the foams could not propagate a fire beyond that initial ignition source. As such, flame retardant chemicals were added to the polyurethane foam to allow them to pass this test and be sold into commerce. Some of those flame retardants were found to have negative PBT profiles, but instead of banning the critical flame retardants in question (which unfortunately takes a long time and is complicated), the fire safety standard was changed and morphed into a smolder-only test in 2013 [76]. This new test method may end up protecting against cigarette ignition sources causing furniture to ignite (although there is still a lot of debate about the effectiveness of this test) [77], but it provides no fire protection against open flame sources. While no flame retardants are required to pass the new TB-117 2013 test, the heat release of the polyurethane foam is still very high, and if it ignites, will still lead to major fire losses [24, 78]. To date, this is the only case of fire safety regulations becoming weakened to provide an incentive to not use flame retardant chemicals, but it is a dangerous precedent. Hopefully fire losses do not have to increase in California (and other places) before arguments are made to require higher levels of fire safety while also demanding no hazardous chemicals can be used.

1.5 Regulatory Outlook and Future Market Drivers

Based upon current information as of the writing of this chapter, the regulatory outlook appears to have more regulations of flame retardants, not less, in the coming decade. There will be regional pushes for regulations of chemicals, with the EU continuing to push the most regulation via existing protocols (RoHS, WEEE) and newer broader chemical registration programs (REACH). The US will remain fragmented at the state level since federal regulations will continue to be delayed until such time as the US legislative branch writes and passes new regulations. Some US states will continue to promote bans on flame retardant chemicals by class, rather than by specific structure. Other countries are expected to maintain their regulations, and perhaps selectively strengthen them where there is national will to do so. Because of the emphasis on PBT for chemical use, polymeric and reactive flame retardant use is expected to grow, while small-molecule additives/non-reactive flame retardant use is expected to shrink in the coming decade. Issues with plastic waste and end-of-lifetime/sustainability issues may further push some flame retardants out of use if they can migrate out of the plastic, or, prevent recycling or clean disposal of said plastic waste. Vigilance by fire safety scientists and fire safety engineers is required to push back against regulations which weaken fire safety under the guise of improving environmental and personal health and safety, as the pollution damage from accidental fires far outweighs the pollution from using flame retardant chemicals. That being said, we can have improved fire safety and eliminate chemicals of concern – it does not need to be one or the other. The correct choice of material and fire protection solution can deliver both, and fire safety regulations do not need to be weakened, nor do all chemicals with potential flame retardant benefit need to be regulated out of existence.

References

1. https://en.wikipedia.org/wiki/Carbon_tetrachloride (accessed 09/07/20).
2. http://en.wikipedia.org/wiki/Restriction_of_Hazardous_Substances_ Directive (accessed 09/07/20).
3. http://eur-lex.europa.eu/legal-content/EN/TXT/?qid=1399998664957&uri= CELEX:02011L0065-20140129 (accessed 09/07/20).
4. http://en.wikipedia.org/wiki/Waste_Electrical_and_Electronic_Equipment_ Directive (accessed 09/07/20).
5. http://eur-lex.europa.eu/legal-content/EN/TXT/?uri=celex%3A32012L0019 (accessed 09/047/20).
6. Westbrook, C.K., Inhibition of Laminar Methane-Air and Methanol-Air Flames by Hydrogen Bromide. *Combust. Sci. Technol.*, *23*, 191–202, 1980.
7. Westbrook, C.K., Numerical Modeling of Flame Inhibition by CF3Br. *Combust. Sci. Technol.*, *34*, 201–225, 1983.
8. Pitts, W.M., Nyden, M.R., Gann, R.G., Mallard, W.G., Tsang, W., Construction of an Exploratory List of Chemicals to Initiate the Search for Halon Alternatives, in: *NIST Technical Note 1279*, US National Institute of Standards and Technology, Gaitherburg, MD, USA, August 1990.
9. Weil, E.D., Levchik, S.V., *Flame Retardants for Plastics and Textiles: Practical Applications*, Hanser Publishers, Cincinnati, OH, 2009.
10. Neviaser, J.L., Gann, R.G., Evaluation of Toxic Potency Values for Smoke from Products and Materials. *Fire Tech.*, *40*, 177–199, 2004.
11. Butler, K.M., Mullholland, G.W., Generation and Transport of Smoke Components. *Fire Tech.*, *40*, 149–176, 2004
12. Gann, R.G., Estimating Data for Incapacitation of People by Fire Smoke. *Fire Tech.*, *40*, 201–207, 2004.
13. Gann, R.G., Babrauskas, V., Peacock, R.D., Hall, J.R., Fire Conditions for Smoke Toxicity Measurement. *Fire Mater.*, *18*, 193–199, 1994.
14. DiNenno, P.J., Drysdale, D., Beyler, C.L., Walton, W.D., Custer, R.L.P., Hall, J.R., Watts, J.M. (Eds.), *SFPE Handbook of Fire Protection Engineering*, 4th Edition, National Fire Protection Association, Quincy, MA, 2008.
15. Aurell, J., Gullet, B.K., Emission Factors from Aerial and Ground Measurements of Field and Laboratory Forest Burns in the Southeastern U.S.: $PM_{2.5}$, Black and Brown Carbon, VOC, and PCDD/PCDF. *Environ. Sci. Technol.*, 2013, https://pubs.acs.org/doi/abs/10.1021/es402101k.
16. https://www.epa.gov/saferchoice/partnership-evaluate-flame-retardants-printed-circuit-boards-publications (accessed 09/08/20).
17. Blais, M. and Carpenter, K., Flexible Polyurethane Foams: A Comparative Measurement of Toxic Vapors and Other Toxic Emissions in Controlled Combustion Environments of Foams With and Without Fire Retardants. *Fire Technol.*, *51*, 3–18, 2015.
18. Blomqvist, P., McNamee, M.S., Stec, A.A., Glyestam, D., Karlsson, D., Detailed study of distribution patterns of polycyclic aromatic hydrocarbons

and isocyanates under different fire conditions. *Fire Mater.*, *38*, 125–144, 2014.

19. Reisen, F., Bhujel, M., Leonard, J., Particle and volatile organic emissions from the combustion of a range of building and furnishing materials using a cone calorimeter. *Fire Saf. J.*, *69*, 76–88, 2014.

20. Hewitt, F., Christou, A., Dickens, K., Walker, R., Stec, A.A., Release of volatile and semi-volatile toxicants during house fires. *Chemosphere*, *173*, 580–593, 2017.

21. Keir, J.L.A., Akhtar, U.S., Matschke, D.M.J., Kirkham, T.L., Chan, H.M., Ayotte, P., White, P.A., Blais, J.M., Elevated Exposures to Polycyclic Aromatic Hydrocarbons and Other Organic Mutagens in Ottawa Firefighters Participating in Emergency, On-Shift Fire Suppression. *Environ. Sci. Technol.*, *51*, 12745–12755, 2017.

22. Fent, K.W., Evans, D.E., Babik, K., Striley, C., Bertke, S., Kerber, S., Smith, D., Horn, G.P., Airborne contaminants during controlled residential fires. *J. Occup. Environ. Hyg.*, *15*, 399–412, 2018.

23. Blais, M. and Carpenter, K., Combustion Characteristics of Flat Panel Televisions With and Without Flame Retardants in the Casing. *Fire Technol.*, *51*, 19–40, 2015.

24. Blais, M.S., Carpenter, K., Fernandez, K., Comparative Room Burn Study of Furnished Rooms from the United Kingdom, France and the United States. *Fire Technol.*, *56*, 489–514, 2020.

25. Simonson, M., Blomqvist, P., Boldizar, A., Möller, K., Rosell, L., Tullin, C., Stripple, H., Sundqvist, J.O., *Fire-LCA Model: TV Case Study*, SP Report 2000, SP Sveriges Tekniska Forskningsinstitut, Printed in 2000.

26. Simonson, M., Andersson, P., Bliss, D., Fire Performance of Selected IT-Equipment. *Fire Tech.2004*, *40*, 27–37, 2004.

27. Blomqvist, P., Rosell, L., Simonson, M., Emissions from Fires Part I: Fire Retarded and Non-Fire Retarded TV-Sets. *Fire Tech.*, *40*, 39–58, 2004.

28. Blomqvist, P., Rosell, L., Simonson, M., Emissions from Fires Part II: Fire Retarded and Non-Fire Retarded TV-Sets. *Fire Tech.*, *40*, 59–73, 2004.

29. Simonson-McNamee, M. and Andersson, P., Application of a Cost-benefit Analysis Model to the Use of Flame Retardants. *Fire Technol.*, *51*, 67–83, 2015.

30. Gomes, G., Ward, P., Lorenzo, A., Hoffman, K., Stapleton, H.M., Characterizing Flame Retardant Appications and Potential Human Exposure in Backpacking Tents. *Environ. Sci. Technol.*, *50*, 5338–5345, 2016.

31. Stapleton, H.M., Dodder, N.G., Offenberg, J.H., Schantz, M.M., Wise, S.A., Polybrominated Diphenyl Ethers in House Dust and Clothes Dryer Lint. *Environ. Sci. Technol.*, *39*, 925–931, 2005.

32. Tange, L. and Drohmann, D., Waste electrical and electronic equipment plastics with brominated flame retardants – from legislation to separate treatment – thermal processes. *Polym. Degrad. Stab.*, *88*, 35–40, 2005.

33. Unwelcome Guest: PBDEs in Indoor Dust. *Environ. Health Perspect.*, *116*, A202–209, 2008.

34. Vorkamp, K., Bester, K., Rigét, F.F., Species-Specific Time Trends and Enantiomer Fractions of Hexabromocyclododecane (HBCD) in Biota from East Greenland. *Environ. Sci. Technol.*, *46*, 10549–10555, 2012.

35. Crosse, J.D., Shore, R.F., Wadsworth, R.A., Jones, K.C., Pereira, M.G., Long-Term Trends in PBDEs in Sparrowhawk (Accipiter nisus) Eggs Indicate Sustained Contamination of UK Terrestrial Ecosystems. *Environ. Sci. Technol.*, *46*, 13504–13511, 2012.

36. Salamova, A. and Hites, R.A., Brominated and Chlorinated Flame Retardants in Tree Bark from Around the Globe. *Environ. Sci. Technol.*, *47*, 349–354, 2013.

37. Saini, A., Thyasen, C., Jantunen, L., McQueen, R.H., Diamond, M.L., From clothing to Laundry Water: Investigating the Fate of Phthalates, Brominated Flame Retardants, and Organophosphate Esters. *Environ. Sci. Technol.*, *50*, 9289–9297, 2016.

38. Liu, D., Lin, T., Shen, K., Li, J., Yu, Z., Zhang, G., Occurrence and Concentrations of Halogenated Flame Retardants in the Atmospheric Fine Particles in Chinese Cities. *Environ. Sci. Technol.*, *50*, 9846–9854, 2016.

39. Li, T.-Y., Ge, J.-L., Pei, J., Bao, L.-J., Wu, C.-C., Zeng, E.Y., Emissions and Occupational Exposure Risk of Halogenated Flame Retardants from Primitive Recycling of E-Waste. *Environ. Sci. Technol.*, *53*, 12495–12505, 2019.

40. Digges, K.H., Gann, R.G., Grayson, S.J., Hirschler, M.M., Lyon, R.E., Purser, D.A., Quintiere, J.G., Stephenson, R.R., Tewarson, A., Human survivability in motor vehicle fires. *Fire Mater.*, *32*, 249–258, 2008.

41. Hirschler, M.M., Polyurethane foam and fire safety. *Polym. Adv. Technol.*, *19*, 521–529, 2008.

42. Hirschler, M.M., Procedures for development and revision of codes and standards associated with fire safety in the USA. *Fire Mater.*, *41*, 1058–1071, 2017.

43. United Nations, *Stockholm Convention on persistent organic pollutions (POPs)*, 2017 latest revision, www.pops.int (accessed 03/29/21).

44. https://www.epa.gov/saferchoice/partnership-evaluate-flame-retardant-alternatives-decabde-publications (accessed 09/08/20).

45. https://www.epa.gov/saferchoice/partnership-evaluate-flame-retardant-alternatives-hbcd-publications (accessed 09/08/20).

46. https://www.epa.gov/assessing-and-managing-chemicals-under-tsca/poly brominated-diphenyl-ethers-pbdes (accessed 09/08/20).

47. https://www.epa.gov/assessing-and-managing-chemicals-under-tsca/poly brominated-diphenylethers-pbdes-significant-new-use (accessed 09/08/20).

48. https://www.epa.gov/assessing-and-managing-chemicals-under-tsca/risk-management-hexabromocyclododecane-hbcd (accessed 09/08/20).

49. https://www.epa.gov/sites/production/files/2018-06/documents/hbcd_problem_formulation_05-31-18.pdf (accessed 09/08/20).

50. https://www.epa.gov/saferchoice/design-environment-alternatives-assessments (accessed 03/28/21).

51. USA Federal Hazardous Substances Act (FHSA), *Consumer Product Safety Commission (CPSC)* 82 FR 45268.

52. https://www.saferstates.com/toxic-chemicals/toxic-flame-retardants/ (accessed 09/07/20).

53. https://www.canada.ca/en/environment-climate-change/services/canadian-environmental-protection-act-registry/related-documents.html (accessed 09/07/20).

54. https://en.wikipedia.org/wiki/Registration,_Evaluation,_Authorisation_and_Restriction_of_Chemicals (accessed 09/07/20).

55. https://en.wikipedia.org/wiki/China_RoHS (accessed 09/07/20).

56. https://www.meti.go.jp/policy/chemical_management/english/cscl/ (accessed 09/07/20).

57. https://www.nite.go.jp/en/chem/kasinn/lists.html (accessed 09/07/20).

58. http://www.cirs-reach.com/KoreaTCCA/Korea_Toxic_Chemicals_Control_Act_TCCA.html (accessed 09/07/20).

59. https://www.industrialchemicals.gov.au/ (accessed 09/07/20).

60. Wilkie, C.A. and Morgan, A.B. (Eds.), *Fire Retardancy of Polymeric Materials*, 2nd Edition, Taylor and Francis, Boca Raton, FL, 2010.

61. Morgan, A.B. and Gilman, J.W., An overview of flame retardancy of polymeric materials: application, technology, and future directions. *Fire Mater.*, *37*, 259–279, 2013.

62. Weil, E.D., Levchik, S., Moy, P., Flame and Smoke Retardants in Vinyl Chloride Polymers – Commercial Usage and Current Developments. *J. Fire Sci.*, *24*, 211–236, 2006.

63. Cleanroom Materials Flammability Test Protocol, in: *FM Global Standard 4910*, Factory Mutual Research Corporation, Norwood, MA USA, 2013.

64. *Fire and Toxicity Test Methods and Qualification Procedure for Composite Material Systems Used in Hull, Machinery, and Structural Applications Inside Submarines*, MIL-STD 2031, US Department of Defense, Washington, DC, USA, 26 February 1991.

65. *Flammability, Offgassing, and Compatibility Requirements and Test Procedures*, NASA-STD-6001, US National Aeronautics and Space Administration, US National Aeronautics and Space Administration, Washington, DC, USA, 2011-08-26.

66. https://en.wikipedia.org/wiki/Blue_Angel_(certification) (accessed 09/07/20).

67. https://en.wikipedia.org/wiki/EU_Ecolabel and www.ecolabel.eu (accessed 09/07/20).

68. https://en.wikipedia.org/wiki/Nordic_swan (accessed 09/07/20).

69. https://en.wikipedia.org/wiki/Green_Seal (accessed 09/07/20).

70. https://www.globalgreentag.com/ (accessed 09/07/20).

71. Sundström, B. (Ed.), *Fire Safety of Upholstered Furniture – The Full Report of the European Commission Research Programme CBUF*, Interscience Communications Ltd, London, 1995.

72. Cleary, T.G., Ohlemiller, T.J., Villa, K., The influence of ignition source on the flaming fire hazard of upholstered furniture. *Fire Saf. J.*, *23*, 79–102, 1994.

73. Kramer, R.H., Zammarano, M., Linteris, G.T., Gedde, U.W., Gilman, J.W., Heat release and structural collapse of flexible polyurethane foam. *Polym. Degrad. Stab*, *95*, 1115–1122, 2010.

74. Lefebvre, J., Le Bras, M., Bastin, B., Paleja, R., Delobel, R., Flexible Polyurethane Foams: Flammability. *J. Fire Sci.*, *21*, 343–367, 2003.

75. Pau, D.S.W., Fleischmann, C.M., Spearpoint, M.J., Li, K.Y., Thermophysical properties of polyurethane foams and their melts. *Fire Mater.*, *38*, 433–450, 2014.

76. *Technical Bulletin 117-2013, Requirements, Test Procedure and Apparatus for Testing the Smolder Resistance of Materials Used in Upholstered Furniture*, Department of Consumer Affairs, State of California, January 2019, https://bhgs.dca.ca.gov/laws/tb117_2013.pdf (accessed 04/14/20).

77. Butry, D.T. and Thomas, D.S., Cigarette Fires Involving Upholstered Furniture in Residences: The Role That Smokers, Smoker Behavior, and Fire Standard Compliant Cigarettes Play. *Fire Technol.*, *53*, 1123–1146, 2017.

78. Storesund, K., Steen-Hansen, A., Bergstrand, A., Fire safe upholstered furniture: Alternative strategies to the use of chemical flame retardants, in: SP Fire Research Report SPFR Report A15 20124:2, December 12, 2015.

2

Phosphorus-Based Flame Retardants

Sergei Levchik

ICL-IP America, 769 Old Saw Mill River Rd., Tarrytown, NY, USA

Abstract

Because phosphorus chemistry is very diverse there are many classes of phosphorus-based flame retardants with specific applications. Red phosphorus is a unique flame retardant which is used in its elemental form. Despite being very flammable in air, red phosphorus is a very efficient flame retardant mostly for thermoplastic polyesters and polyamides. Most inorganic phosphates are water soluble and because of this they are used as non-durable treatment for textiles and wood. Water insoluble ammonium polyphosphate, piperazine polyphosphate and melamine phosphates are very efficient flame retardants especially for polyolefins. Aluminum and calcium hypophosphites and aluminum diethyl phosphinate were introduced to the market about two decades ago but are still actively researched for new applications by industrial labs and for mechanisms of action by academic institutions. Aliphatic phosphates and phosphonates and aromatic phosphates are the oldest classes of organophosphorus flame retardants and plasticizers with well-established applications. Aromatic bisphosphates are broadly used in polycarbonate-based and polyphenylene ether-based blends but their market share grows mostly because these types of thermoplastics grow fast. Fueled by a fast growing sector of high speed and high frequency printed wiring boards, aromatic phosphinates, phosphine oxides and phosphazenes are the most active areas of research both in industry and in academia.

Keywords: Phosphorus flame retardant, intumescent, char, plastic, textile, epoxy resin, polyurethane foam

Email: Sergei.levchik@icl-group.com

Alexander B. Morgan (ed.) Non-Halogenated Flame Retardant Handbook 2nd edition, (23–100)

2.1 Introduction

It is generally accepted that the most efficient flame retardants provide their action both in the condensed and gas phases. Although halogen- and phosphorus-based flame retardants exhibit these two mechanisms of action, the difference is that halogen flame retardants can promote charring of most organic polymers by bromine radicals abstracting hydrogen atoms from polymer chains resulting in formation of double bonds or cross-links [1]. Phosphorus flame retardants are more specific to the polymer chemistry than halogen ones and they are mostly effective in the oxygen- or nitrogen-containing polymers due to the fact they need to react with the polymer e.g., phosphorylate it and thus involve it in the charring. The char impedes the heat flux to the polymer surface and retards diffusion of the volatile pyrolysis products to the flame.

If conditions are right, the phosphorus-based molecules or fragments can volatilize and be oxidized producing active moieties in the flame. Volatile phosphorus compounds can be as effective as halogen radicals in the flame, but even here phosphorus surprisingly works well only in heteroatomic polymers. It has always been challenging to design phosphorus-based flame retardants, which will volatilize into the flame at relatively low temperatures and at the same time will not be lost during polymer processing. Therefore, there are not many commercial phosphorus-based flame retardants that provide mostly gas phase action.

The author of this chapter has published a similar chapter on phosphorus-based flame retardants in the first edition of this handbook [2] and he also co-authored two earlier reviews on phosphorus-based flame retardants [3, 4]. This current chapter is an update and extension of the previous publications. This chapter does not cover the large class of chloroalkyl phosphates since they are not halogen-free, but these products were reviewed previously. Although there is a large body of academic publications and patent literature on new phosphorus flame retardants, this chapter focuses only on flame retardants which, to the best of the author's knowledge, are in commercial use or in advanced commercial development. A recent review on some commercial phosphorus-based and intumescent flame retardants was published elsewhere [5]. There are also broader non-selective reviews on phosphorus flame retardants [6, 7]. Mechanisms of action of phosphorus flame retardants were recently reviewed by Shartel [8].

2.2   Main Classes of Phosphorus-Based Flame Retardants

The ammonium phosphate treatment of cellulosic materials (canvas, wood, textiles, etc.) has been known for almost three centuries [9]. However, only with commercialization of synthetic polymeric materials in the twentieth century, organophosphorus compounds have become an important class of flame retardants.

All phosphorus-based flame retardants can be separated into three large classes:

- Inorganic represented by red phosphorus, ammonium phosphates and metal phosphites and hypophosphites.
- Semi-organic represented by amine and melamine salts of phosphoric acids, metal salts of organophosphinic acids and phosphonium salts.
- Organic represented by phosphates, phosphonates, phosphinates, phosphine oxides and phosphazenes.

Water-soluble phosphorus flame retardants are mostly used for topical treatment of wood, textile and other cellulosic products. Some water soluble FRs can be further reacted with cross-linkers (cured) which provides durable water resistant treatment. Water-insoluble phosphorus FRs find a very broad range of applications in thermoplastics, thermosetting resins, synthetic foams, coatings, etc.

Phosphorus flame retardants have certain advantages over other flame retardants (mostly halogen based) but also have some disadvantages which are both listed below:

Advantages:

- Low specific gravity which results in light plastic parts
- Achieve flame retardant efficiency at lower phosphorus content compared to the halogen content needed for the same rating
- High comparative tracking index (CTI) test performance
- Better UV stability than most halogen-based FRs
- Less tendency to intensify smoke obscuration
- Less acidic smoke compared to halogen FRs
- Most phosphorus FRs are biodegradable and therefore not persistent or less persistent than halogen FRs

Disadvantages:

- Low efficiency in polyolefins, styrenics and elastomers unless charring agent is added.
- Absence of a good general synergist that works in most polymers.
- Many phosphorus FRs are hydrophilic and possibly cause moisture uptake, limiting use in some applications.
- May hydrolyze to give acids which decrease the molecular weight of acid-sensitive plastics (polycarbonates, polyesters, polyamides, etc.)
- Recycling of acid sensitive polymers is problematic due to the hydrolytic instability of organophosphates.
- Some phosphates are toxic to aquatic organisms. Some phosphates exhibit a certain degree of neurotoxicity.
- Apart from a few selected cases, the cost/efficiency of phosphorus FRs is higher than halogen based FRs.

2.3   Red Phosphorus

Red phosphorus is a polymeric form of elemental phosphorus consisting of non-periodic five- and six-membered rings with some crosslinks. Red phosphorus is made by prolonged heating of white phosphorus at about 250-300°C in anaerobic conditions. At the end of the production red phosphorus comes out in the form of cake like solid chunk which needs to be broken and milled. Being exposed to moist air, non-stabilized red phosphorus slowly reacts with water and oxygen producing phosphine and various phosphorous acids. Oxidation starts on the sharp edges which are considered to be active sites. Smaller dusty particles are the most susceptible to oxidation and because of this red phosphorus is usually sieved out to different fractions with the removal of dusty particles. In the late 80s a process of producing mostly spherical particles of red phosphorus [10] which are considered less prone to oxidation was developed in Japan. Red phosphorus is very thermally stable and environmentally benign [11] because being exposed to weathering conditions it eventually converts to phosphoric acids.

Red phosphorus is a controversial flame retardant because on one hand it is very efficient, allowing achievement of a V-0 rating in glass-filled polyamides or polyesters at only 6-12% wt. loading and on the other hand it is very combustible, and the powder can easily ignite if heated in the open air.

This doesn't create problems in molded plastics, but storage and processing of red phosphorus should be monitored very carefully. Because of possible accumulation of phosphine, the containers with red phosphorus should be opened under a fume hood or ventilation canopy. Direct compounding of red phosphorus requires an inert gas blanket to the extruder and all feeding equipment, which can be a very costly investment. Another disadvantage of red phosphorus is its red color which is difficult to overcome even with a high concentration of other colorants like TiO_2 [12, 13]. Because of this, red phosphorus containing molded parts are typically pigmented black [14].

Historically, red phosphorus was mostly used in Europe and Asia and almost not used in North America [15]. For safe handling, red phosphorus is usually coated, for example with urea-formaldehyde resin [16], and stabilized with metal oxides [17] or metal hydroxides[18] or metal salts [19] or hydrotalcite [20] which can react and scavenge the phosphine as it forms. In order to ensure safe handling red phosphorus is sold as a wet filter cake, or as a dispersion in an epoxy resin or as a dispersion in castor oil. Red phosphorus is also available in the form of masterbatches with a variety of polymers [21].

Red phosphorus is particularly useful in glass-filled polyamide 6.6 where a high processing temperature (> 280°C) excludes the use of less stable phosphorus compounds [11]. Many industrial studies have been done to find synergists for red phosphorus. It has been found helpful to combine red phosphorus with phenolic resins [22]. It is believed that under burning conditions, red phosphorus-phenolic combinations form a cross-linked network which eliminates flaming drips by reducing melt flow. A typical polyamide 6.6 formulation which contains 25 wt.% glass fibers, 7% wt. red phosphorus and 5% wt. phenolic resin gives a V-0 rating. It is interesting that polyethylene terephthalate (PET) is synergistic with red phosphorus when used in glass-filled polyamide 6.6 [23]. Recently synergistic combinations with other phosphorus flame retardants like hexaphenoxytricyclophosphazene [24] and aluminum phosphite were patented [25]. Combinations of red phosphorus, aluminum diethyl phosphinate and melamine salts also allow boosting the glow wire ignition temperature [26].

Early work on the flame-retardant mechanism of red phosphorus in polyamide 6 suggests a mostly gas phase mechanism of action due to depolymerization into white phosphorus and volatilization in the oxygen depleted atmosphere of the pre-flame zone [27]. Later Levchik *et al.* [28] studied thermal decomposition of polyamide 6 flame-retarded with red phosphorus in nitrogen and found that phosphate esters are formed even in an inert atmosphere. Similarly, it was found that interaction of red phosphorus with PET leads to the formation of phosphate esters [29]. A more recent study also confirms an increased solid residue of polyamide

6.6 in the presence of red phosphorus [30]. Although an interaction of red phosphorus with traces of O_2 and absorbed moisture [31] as well as H_2O formed during the thermal decomposition of the polyamide [32] is a possible explanation of the formation of phosphate esters, it is also possible that red phosphorus reacts directly with polyamide via a free radical mechanism as an electron spin resonance study suggests [28].

Despite the fact that red phosphorus is effective in both condensed and gas phase [33] it achieves V-0 rating only in nitrogen and oxygen containing heteropolymers, mostly engineering thermoplastics. In polyolefins, red phosphorus was found useful for a V-2 rating and a high limiting oxygen index (LOI) especially in polyethylene [34]. It is believed there is better match between the temperature of the thermal decomposition of polyethylene and the volatilization of red phosphorus compared to polypropylene [22]. It was shown that a V-2 rating at 1.6 mm could be obtained at a level as low as 2.5% finely divided red phosphorus (5 μm) [35]. Melamine-formaldehyde coating improves the efficiency of red phosphorus in polyolefins because it provides charring to enhance the flame-retardant effect of the phosphorus [36]. Although red phosphorus is not efficient alone in poly(acrylonitrile-butadiene-styrene) (ABS) it allows achievement of a V-0 rating at 8 wt. % loading when combined with 15 wt. % PET [37] or 8 wt. % magnesium hydroxide and 6 wt. % polyamide 6 [38].

Apart from thermoplastics red phosphorus finds applications in polyurethane foams, polyurethane elastomers [39], epoxy resins and textiles [40]. Typically, polyurethane foams are flame retarded with chloroalkyl phosphates, but in some applications where stringent fire tests are required red phosphorus can be used. For example, only 5 wt. % red phosphorus with 5 wt. % of melamine is required to pass the British BS 5852 Crib 5, furniture test [41] versus almost 20 wt. % of chloroalkyl phosphate combined with melamine. Red phosphorus combined with expandable graphite is used in automotive sound insulation foam under the hood [42].

In the mid 90s production of epoxy-molding compounds flame retarded by stabilized and coated red phosphorus for encapsulation of electronic devices started in Japan [43]. Because encapsulating epoxy is very heavily filled with silica only few percent of red phosphorus is enough to achieve a V-0 rating [44]. In the late 90s and early 2000s electronics manufacturers using red phosphorus containing molding epoxy started to receive massive recalls due to electronics failure. Further investigation showed that despite stabilization and coating, red phosphorus was still reacting with oxygen producing acids and phosphine which damaged semiconductor circuits [43]. In recent years more failures were reported in cables [45] and connectors [46] due to the presence of red phosphorus. Recently some electronics manufacturers announced a ban on red phosphorus in their products and resin manufacturers start reducing their red phosphorus-based lines of products. The only potential applications for red phosphorus in epoxy which are still being researched are light weight composites [47] and coatings [48].

2.4 Ammonium and Amine Phosphates

Because this Handbook has a separate chapter on Intumescent Flame Retardants (see Chapter 4 of this book), in this chapter only a short overview of commercial intumescent FRs is given. Numerous academic publications on intumescent flame-retardant systems for polyolefins, elastomers and rubbers are out of the scope of this chapter. The broad subject of intumescent flame retardants was earlier discussed in a book [49], a number of reviews [50, 51] and a book chapter [52]. The mechanism of char formation in the pentaerythritol-ammonium polyphosphate (APP) systems was very extensively studied and described in great detail [53].

Water soluble monoammonium dihydrogen phosphate (MAP) and diammonium hydrogen phosphate (DAP) or short chain ammonium polyphosphate is the oldest class of flame retardants which are used on cellulosic materials such as wood, paper and cotton. A large volume of water-soluble ammonium polyphosphate is used in forest fire control, usually by aerial application, often in combination with ammonium sulfate. MAP and DAP can be applied to cotton or cotton-based blends by soaking, padding or spraying and then drying which will result in a FR finish non-durable for laundering. Non-durable finishes are most often used for disposable goods, for example medical gowns, party costumes, and sometimes wall covering. Ammonium phosphate finishes are resistant to dry-cleaning solvents but not to laundering or to leaching by water. However, some degree of durability of MAP can be achieved by combining with hydrophobic poly(methylhydrogen siloxane) or poly(dimethyl siloxane) [54]. Interestingly, cotton treated with MAP or DAP or water-soluble APP, urea and tauramine oxide and cured for 2 minutes at 170°C achieves durability of up to 15 washes [55].

Depending on fabric weight and density, 1-2% of phosphorus provides self-extinguishing performance and effectively prevents afterglow. Some organic co-additives can be added to the solution [56] to improve textile wetting and inhibit crystallization upon drying in order to avoid the formation of visible crystals of ammonium phosphates. Urea is used as a synergistic co-additive which helps to significantly increase the oxygen index and decrease char length as measured in ASTM D6413 flame test [57] and decrease the heat release rate [58]. Chitosan alone [59] or combined with sodium stannate [60] was also explored as a synergistic nitrogen source in combination with DAP. It was reported [61] that ammonium phosphates phosphorylate cellulose, which changes its mechanism of thermal decomposition suppressing the evolution of levoglucosan, a major fuel source, and increasing charring. It is also believed [61] that stannic oxide being a Lewis acid catalyzes the dehydration reaction of cellulose and increases char.

MAP and DAP combined with boric acid or other inorganic borates or sulfates are commonly used for flame retardant treatment of wood. Numerous formulations with water soluble phosphates for wood treatment can be found in the patent literature [62, 63]. Grexa and Utike [64] studied various inorganic flame retardants and their combination on particleboard using cone calorimeter and concluded that MAP and boric acid is the most efficient combination. In order to provide decay resistance to the wood treated with ammonium phosphates, apart from boric acid other antifungal additives can be applied [65]. Guanyl urea phosphate [66] or urea phosphate [67] in combination with boric acid are other systems for wood treatment which are somewhat less leachable than ammonium phosphate-based systems. Even more durable flame retardants applicable for outdoor use are achieved by the use of dicyandiamine and urea formaldehyde pre-condensates together with phosphoric acid and further polymerized after treatment [68]. DAP in combination with boric acid can be applied to paper which needs further treatment of the phenol-formaldehyde resole resin to preserve paper integrity [69].

Diguanidine hydrogen phosphate or monoguanidine dihydrogen phosphates are also used for non-durable cotton treatment [70]. These salts are particularly synergistic with 3-aminopropylethioxysilane [70]. Hexaammonium (nitrilotris(methylene))trisphosphonate (Formula 2.1) is another water-soluble ammonium salt which is used in nondurable automotive and aircraft upholstery to minimize the effect on "hand" (texture).

$$
\begin{array}{ccc}
& O & O \\
& \| & \| \\
NH_4O - P - CH_2 - N - CH_2 - P - ONH_4 & \\
| & | & | \\
ONH_4 & CH_2 & ONH_4 \\
& | & \\
NH_4O - P - ONH_4 & \\
\| & \\
O &
\end{array}
\tag{2.1}
$$

When MAP and DAP are heated under ammonia pressure and preferably in the presence of urea, relatively water-insoluble ammonium polyphosphate is produced [71]. At least five crystalline forms of APP have been reported in the literature [72] but only forms I, and II are commercially sold as flame retardants. It is believed that form I has a linear chain structure, relatively low molecular weight (from 30 to about 150 repeating units), relatively low thermal stability (onset of weight loss at about 240°C) and relatively high water solubility. It is mostly used in coatings. Form II is higher molecular weight (700-1000 repeating units), probably cross-linked [73], with onset of weight loss at ~270°C, and much more water-resistant. Form II is used in coatings, thermoset resins, PU foams, textile backcoatings and thermoplastics. APP is the principal component of many intumescent FR systems.

Finely divided ammonium polyphosphate is the major flame retardant for intumescent paints and mastics [74]. When the intumescent coating is exposed to a high temperature, APP yields polyphosphoric acid that then interacts with an organic component such as a pentaerythritol to form a carbonaceous char. This chemistry has been described in detail by Camino and Costa [75] and is covered in detail in this book in Chapter 4. A blowing (gas-generating) agent, typically melamine, is also added to foam the char, thus forming a fire-resistant insulating barrier to protect the substrate. In addition, the intumescent formulations typically contain resinous binders, pigments, plasticizers, and other additives. Mastics are related but more viscous formulations, intended to be applied in thick layers to girders, trusses, and decking for structural fire protection; these generally contain mineral fibers to increase coherence.

The intumescent systems concept originally developed for flame retardant coatings [76] was later adapted for thermoset low temperature processed polyolefins, elastomers and rubbers. The decomposing polymer can produce enough gaseous product for foaming and therefore use of melamine is unnecessary, but a charring agent is still needed. Although intumescent systems based on APP were intensively studied the main factors limiting their broad application are thermal stability and water solubility. Both thermal stability and water solubility can be improved by increasing the chain length (molecular weight) of the polyphosphate. Many varieties of APP form II with various coatings/encapsulations like a silane surface-reacted, melamine surface-reacted, melamine formaldehyde resin coated which further decrease the water solubility are commercially available [77]. These surface treatments allow decreasing the water solubility of form II from 0.5 to 0.01-0.1 g/liter. These coatings can also provide a synergistic effect to APP because they can work as charring agents to further enhance the activity of APP. Although cable manufacturers are trying to adopt APP or APP formulated systems in the cable jacketing [78] it seems to still have a very limited application due to water absorption issues. Silicone surface-treated ammonium polyphosphate in combination with pentaerythritol and methylmethoxysiloxane made by reacting trichloromethylsilane with methanol and water provide high LOI = 34 and UL-94 V-0 ratings in thermoplastic polyurethane (TPU) [79].

Over many years, APP producers and compounders tried to develop flame retardant compositions (formulated packages) which included along with APP the charring and foaming agents. Nitrogen-containing low molecular weight or polymeric products behave the best because they combined both charring and foaming functions [80]. For example, patents suggest that some commercial systems may contain tris(hydroxyethyl)isocyanurate [81], or poly(triazinyl piperazine) [82]. In more recent developments the condensation product of melamine, morpholine and piperazine was suggested as a charring agent and synergist with APP [83]. Reportedly, APP combined with such a condensate provides a V-0 rating in polypropylene at only 20 wt. % loading [84]. Interestingly, aliphatic polyamides which are considered not charrable polymers can also be used along with APP as charring agents. For example, a group of French researchers [85] developed a formulated system of APP/polyamide 6/poly(ethylene-co-vinyl acetate) (EVA), where EVA is used as a compatibilizer for polyethylenic polymers.

Alumina trihydrate (ATH) suppresses the intumescent performance of APP in polyolefins, elastomers and rubbers and therefore these two FRs are almost never used together in these polymers, but it seems not to be the case with unsaturated polyesters (UPE) [86]. In order to decrease the loading of ATH it can be partially or completely replaced with more efficient APP. For example, 15-25 parts APP and 50 parts ATH (per 100 parts resin) will provide a UL-94 V-0 rating in UPE [87]. One academic publication [88] shows that silane treated APP at 35 wt. % loading results in a decrease of the heat release rate by 70%, but more importantly total smoke released decreases by 50% as measured by a cone calorimeter test. Another publication suggests [89] use of diatomite/APP encapsulated in triphenyl phosphate as an effective flame retardant for UPE providing a V-0 rating at 20 wt. % loading. It was also shown that a combination of APP with expandable graphite is beneficial and probably shows a synergistic effect [90].

Although the polyurethane foam (PUF) industry prefers dealing with liquid FRs similar to other foam components, sometimes the use of solid FRs is a more economical way of achieving high flammability standards even if it requires the installation of special equipment. The fire-retardant effect of APP is more pronounced in dense foams, rather than in light foams [91]. For example, use of finely divided APP combined with ATH and a cyclic phosphonate allows achievement of class I in E-84 tunnel test [92] and a combination of APP, ATH and zinc borate allows passing the UL 790 roof assembly [93] test of spray foam roofing. Very efficient combinations of APP with expandable graphite were reported in the literature [94–96]. Guanidinium phosphate can be used instead of APP again in combination with expandable graphite or red phosphorus [97]. Use of lignin-modified polyol along with phenol-formaldehyde coated APP allows a significant improvement in charring of rigid PUF [98]. Usually, triethyl phosphate or some other low viscosity liquid FR is used in combination with solids to improve processability. Interestingly, one recent publication reports on the use of soluble APP in semi-rigid water blown foam [99].

It has been long recognized [100] that addition of a small amount (typically 2-3 wt. %) of multivalent metal salts or oxides provides a synergistic effect in APP based intumescent systems in polyolefins. Zinc borate [101] or some natural products like talc, zeolites [102] and clays show similar behavior. The synergistic effect is observed in a very narrow concentration range and it is believed to be due to the formation of cross-links in polyphosphoric acid involving multivalent metals [103]. Increasing the concentration of the synergist results in the formation of stochiometric crystalline phosphates which negatively affect intumescence and the effect switches from synergistic to antagonistic. In the academic literature there are numerous publications on the benefit of the addition of organically modified clays to the intumescent systems. Synergistic effects are often, perhaps erroneously, attributed to the physical effect of the clay reinforcing char, whereas it could be the same effect of chemical interaction with polyphosphoric acid and cross-linking.

Backcoating is a very common and cost efficient method of flame retarding cotton or synthetic textiles or their blends. The phosphorus-based backcoatings are more limited to cellulosics because their efficiency relies mostly on charring. The durability of backcoating in laundering depends on the binder and the hydrolytic stability of the flame retardant. Horrocks et al. [104] studied a wide range of phosphate salts and some phosphate esters and concluded that ammonium polyphosphate is the most efficient FR for cotton and cotton polyester blends because APP decomposes to polyphosphoric acid and involves cotton in charring [105–107]. In textile backcoatings coated ammonium polyphosphate are more preferred over untreated APP because of better water resistance. There are numerous patents [108] on the use of APP in cellulose based barrier fabric for mattresses in the USA which need to pass the severe Consumer Product Safety Commission (CPSC) 1633 open flame test.

Melamine phosphate also has been originally developed for intumescent coatings but found some use in polyolefins. Melamine phosphate is converted to the pyrophosphate and further to the polyphosphate with the loss of water on heating. The pyrophosphate is reported to be only soluble in water to the extent of 0. 09 g/liter water, whereas melamine orthophosphate is soluble to 0.35 g/liter. More thermally stable melamine pyrophosphate and melamine polyphosphate ensured safe processing even in polyamides and polyesters. Different applications of melamine phosphate and pyrophosphate were reviewed by Weil and McSwigan [109, 110]. A detailed study of the thermal decomposition of melamine phosphates has been published [111].

In intumescent thermoplastic formulations, melamine phosphates have been shown to have an advantage over ammonium polyphosphate by causing less mold deposition and having better water resistance [112]. Further encapsulation of a melamine polyphosphate/pentaerythritol system with thermoplastic polyurethane improves compatibility, water resistance and flame-retardant performance in polyethylene [113]. Melamine phosphates are typically less efficient than APP, because they are more thermally stable and have a lower phosphorus content. However interestingly, melamine pyrophosphate combined with a triazine based charring agent made from cyanuric chloride, ethanolamine, and diethylenetriamine provides a V-0 rating in polypropylene at only 25 wt. % FR loading [114, 115]. Encapsulation of melamine polyphosphate with 4,4'-oxydianiline-formaldehyde also helps to boost the oxygen index of polyurethane composites [116].

A further improvement of the thermal stability of melamine polyphosphate was done by the partial replacement of some melamine groups with Al, Zn or Mg [117, 118]. These show enhanced performance because of increased fire residues, notably in polyamides and epoxies [119]. The same group of inventors [120] also synthesized melamine mixed trimethylene amine phosphonate salts (Formula 2.2), but their commercial status is unknown.

$$(2.2)$$

The ethylenediamine salt of phosphoric acid (1:1) (EDAP) having a phosphorus content of 63 wt. % was introduced to the market in the early 90s [121, 122]. In contrast to APP and melamine salts, EDAP shows a self-intumescent behavior because it melts at about 250°C, right around where its thermal decomposition starts and because it contains aliphatic carbons which undergo charring. EDAP is very efficient because it quickly activates as an intumescent FR once it reaches this temperature. EDAP is more soluble in water compared to the form II of APP and is less thermally stable which limits its applications to polyolefins. A flame retardant made from diethylene triamine and polyphosphoric acid by heating at 200°C, has a higher thermal stability with the beginning of decomposition at about 300°C [123].

In order to improve thermal stability and decrease water solubility, EDAP has often been sold as a mixture with melamine or melamine phosphates. Some of these mixtures are also synergistic because the temperature of thermal decomposition of EDAP and melamine phosphates are different, and the extended temperature interval better matches the thermal decomposition of the host polymer. Some synergists, such as phase transfer catalysts (quaternary ammonium salts) or spirobisamines may further enhance the action of EDAP and melamine pyrophosphate or APP combinations [124, 125]. Coating of EDAP with amine cured epoxy allows for a decrease in water sensitivity [126].

Intumescent systems based on the mixed salts of melamine and piperazine phosphates were first developed in Italy [127] and marketed for wire and cable applications [128]. Later, an improved method of synthesis of polymeric piperazine pyrophosphate (Formula 2.3), which results in a product with superior thermal stability [129] was developed in Japan. It is more resistant to water than coated ammonium polyphosphate. Another patent [130] shows the milling of piperazine pyrophosphate together with melamine pyrophosphate and the addition of some polymethylsiloxane oil for decreasing dusting and improving processability. This intumescent flame retardant is said to be effective in polypropylene at about 25 wt. %, and in low density polyethylene (LDPE), high density polyethylene (HDPE) or EVA at about 30 wt. % [131]. It is stable enough to permit extrusion and molding at 220-240°C and it is effective in cable jackets [132]. Recently many patents were filed on TPU formulations for cable application based on piperazine pyrophosphate and melamine pyrophosphates combined with bisphosphates [133] and stabilized with epoxidized novolac [134]. Addition of a small amount of silica improves dispersion and boosts flame retardant performance [135]. With a stabilizing amount of hydrotalcite or zinc oxide [136] or calcium glyceorate [137] or zinc cyanurate or calcium cyanurate [138] or boehmite [139] it is a particularly effective intumescent flame retardant for unreinforced and glass-reinforced polypropylene [140]. A recent publication showed synergism of piperazine pyrophosphate and aluminum hypophosphite in glass-filled polyamide 6 [141]. A mixed salt of piperazine and aluminum diphosphate was also found to be efficient in polypropylene [142].

$$^+H_2N \overset{}{\underset{}{\bigcirc}} NH_2{}^+O^- \overset{\overset{O}{\underset{}{\parallel}}}{\underset{\underset{O^-}{|}}{P}} - O - \overset{\overset{O}{\underset{}{\parallel}}}{\underset{\underset{O^-}{|}}{P}} - O^- \tag{2.3}$$

2.5   Metal Hypophosphites, Phosphites and Dialkyl Phosphinates

The requirements for flame retardants in polyesters and polyamides are stringent because of high processing temperatures and sensitivity to hydrolytic degradation catalyzed by possible acids or catalytic decomposition assisted by some metals. Since the most common use of flame retardant polyesters and polyamides is in connectors, there is a requirement for long-term dimensional stability, which means minimal water absorption which is especially difficult to maintain with polyamides. Because polyesters and polyamides are semicrystalline and a flame retardant can be accommodated only in the amorphous regions, there is an issue of exudation ("blooming") of low molecular weight flame retardants. For many years brominated flame retardants dominated in this market sector and many phosphorus-based flame retardants were not considered for polyesters and polyamides.

However, some time ago it was discovered that calcium hypophosphite (Ca-Hypo, $Ca(HPO_2)_2$) combined with melamine cyanurate provides V-0 in glass-filled poly(butylene terephthalate) (PBT) at 25 wt. % of total loading [143]. Ca-Hypo was also found efficient in the polycarbonate (PC) blends PC/ABS [144] and PC/poly(butylene terephthalate) (PBT) [145]. Later it was found that aluminum hypophosphite (Al-Hypo, $Al(HPO_2)_3$) alone or in combination with melamine cyanurate [146, 147] or melamine polyphosphate [148] is a more effective flame retardant because it requires only a 15-20 wt. % total loading for achieving V-0 rating in glass filled PBT. Al-Hypo starts to decompose (disproportionate) at about 300°C with the evolution of phosphine which provides gas phase action [149, 150], whereas the remaining aluminum pyrophosphate is believed to provide a condensed phase action [151]. Calcium hypophosphate also releases phosphine, but at a higher temperature. Because of the risk of the evolution of phosphine during compounding, Al-Hypo and Ca-Hypos are available as coated or double coated grades. For example, Ca-Hypo can be surface reacted with succinic or phthalic or oxalic acid [152] or melamine cyanurate [153] which allows the use of this FR in high melting polyamides. Significant effort was put into the development of a process for producing a purer and more stable version of Ca-Hypo [154]. It was found that Ca-Hypo can be also stabilized with a small addition of zinc borate [155].

A recent study compares the flame retardant efficiency of Ca-Hypo in polyamide 6, polylactic acid, thermoplastic polyurethanes and poly(methyl methacrylate) [156]. Interestingly, Al-Hypo was found to be efficient in polyolefins and styrenics. In these polymers phosphorus non intumescent flame retardants are typically inefficient. For example, Al- or Ca-Hypos combined with hindered N-alkoxyamines (NOR) stabilizers shows a V-2 rating in polyethylene, polypropylene and EVA [157]. More efficient are combinations of hypophosphite salts with brominated FRs, the most efficient of which seems to be melamine hydrobromide [158]. These synergistic blends allow achievement of a V-2 rating in PP copolymers at a level below 3 wt. % especially when combined with a free-radical initiator or NOR. At such a low loading, the content of bromine in the polymer is below 900 ppm, which qualifies it as halogen-free according to IEC 61249-2-21. Al-Hypo was found to be synergistic in combination with APP and a triazine based intumescent system in polypropylene [159] and with piperazine pyrophosphate in thermoplastic elastomers [160] and polyamide 6 [161]. Al-Hypo, combined with melamine cyanurate, liquid bisphosphate and about 2.5 wt. % phenolic novolac as a charring agent was found efficient in polyester thermoplastic elastomers [162]. 17.5 wt. % Al-Hypo, 7.5 wt. % resorcinol bis(2,6-xylenol phosphate) (Formula 2.26) and 0.3 wt. % polytetrafluoroethylene (PTFE) as an antidrip agent provides a V-0 rating at 3.2 mm thickness in ABS [163]. APP, melamine cyanurate, cyclic phosphonate, magnesium hydroxide and even antimony trioxide were found synergistic with stabilized Ca-Hypo in achieving a V-0 rating in ABS[164]. Because Ca- and Al-Hypo provide gas phase flame retardant action, they can replace antimony trioxide in bromine-based systems in ABS [165] and glass-filled PBT [166].

The development of alkylphosphinate salts as flame retardants goes back to the late 70s and early 80s when various metal salts of dialkylphosphinates were tested in poly(ethylene terephthalate)[167] and in polyamide 6[168]. Later zinc, aluminum and calcium dialkylphosphinate salts were tested in glass filled polyamides and PBT. Aluminum and calcium ethylmethylphosphinates were found to give V-0 at 15 wt. % in plain PBT, at 20 wt. % in glass-filled PBT [169], and at 30 wt. % in glass-filled polyamides [170]. Because key raw material methyldichlorophosphine is strictly regulated, methylethylphosphinates were never commercialized but instead less expensive and safer to manufacture aluminum diethylphosphinate (DEPAL, Formula 2.4(a)) and zinc diethylphosphinate (DEPZN, Formula 2.4(b)) were developed [171].

$$Al \left[O - \overset{\overset{\displaystyle O}{\|}}{\underset{\underset{\displaystyle C_2H_5}{|}}{P}} - C_2H_5 \right]_3 \qquad Zn \left[O - \overset{\overset{\displaystyle O}{\|}}{\underset{\underset{\displaystyle C_2H_5}{|}}{P}} - C_2H_5 \right]_2 \qquad\qquad (2.4)$$

(a)   (b)

Although DEPAL is only moderately efficient in polyamides it was found to be synergistic with nitrogen-containing flame retardants such as melamine cyanurate [172], melamine phosphate or melamine polyphosphate[173]. Further addition of a few percent of zinc borate as a stabilizer is required for high temperature processing of polyamide 6.6 [174]. These products provide UL-94 V-0 ratings in glass-filled polyamides at 15-20 wt. % loading down to 0.4 mm thickness [175]. Melamine polyphosphate with some melamine replaced by Mg, Zn or Al were also tested as more thermally stable synergists to DEPAL [176]. Another synergistic combination of DEPAL with aluminum phosphite ($AlPO_3$, PHOPAL) was found to be highly efficient in polyamides [177, 178] and it is believed to be used commercially. More thermally stable boehmite (AlOOH) in combination with zinc borate was recommended for high melting semiaromatic polyamides [179] and also shows synergism in PBT [180]. Surprisingly, high aspect ratio talc was also found to be synergistic with DEPAL in polyamide 6 [181]. Since DEPAL is efficient in polyamides it was also found to be efficient in polyphenylene ether (PPE) polyamide blends [182].

It seems there is no actual synergism between DEPAL and melamine salts in glass-filled PBT, but about 1/3 of DEPAL can be replaced with melamine cyanurate or melamine polyphosphate without loss of the V-0 rating [175], which is probably beneficial for cost saving. Compared to brominated flame retardants used in the same application, DEPAL allows a high Comparative Tracking Index (CTI) > 500 volt [183, 184]. On the other hand, DEPAL and DEPAL-based synergistic combinations show significant wear (corrosion) of processing equipment [185], which can be decreased by using acid scavengers.

Based on academic studies there is a strong indication that DEPAL mostly operates in the gas phase by a flame inhibiting mechanism [186, 187]. It was believed that DEPAL decomposes with the evolution of phosphinic acid which evaporates to the flame, however there is other evidence showing that aluminum alkylphosphinates can volatilize without decomposition. Interestingly, under thermooxidative conditions of thermogravimetric analysis in air DEPAL doesn't volatilize, but mostly oxidizes to aluminum phosphates [188]. The higher volatility of the salt results in a higher flame retardant efficiency as demonstrated by aluminum diisobutylphosphinate [189, 190].

Melamine polyphosphate provides a condensed phase action by increasing the charring of the polymer [191] and thus provides a synergistic action with DEPAL [186]. In contrast, melamine cyanurate mostly volatilizes and provides a cooling effect to the flame [187], therefore its action is mostly adjunctive but not synergistic. A comparative study of DEPAL and Al-Hypo plus resorcinol bis(2,6-xylenol phosphate) (Formula 2.6) showed that DEPAL is more efficient at 20 wt. % loading in glass filled PBT [192].

Apart from traditional uses of DEPAL in polyamides and polyesters it was also found alone [193] or in combination with melamine polyphosphate [194] or with aromatic bisphosphates [195] to be efficient in PPE/styrene-ethylene-butadiene-styrene (SEBS) blends typically used in cable jackets. About 30 wt. % DEPAL is required to achieve a VW-1 rating in thermoplastic elastomers (TPE) wire insulation [196] and the same rating can be achieved if about 1/3 of the DEPAL is replaced with melamine polyphosphate [197] or melamine cyanurate [198]. Similarly, in thermoplastic polyurethane (TPU) about 30 wt. % DEPAL and some melamine salt allow achievement of a VW-1 rating [199].

Apparently, DEPAL alone [200] or in combination with ATH, APP or melamine [201] or melamine polyphosphate[202] shows high efficiency in unsaturated polyesters (UPE). For example, a combination of 10 wt. % aluminum diethylphosphinate and 10 wt. % melamine polyphosphate provides a V-0 rating in a 30% glass-filled composite and shows an LOI of 42. DEPAL can also be pre-dispersed in a polyester/styrene prepolymer [203] which results in higher LOI values compared to freshly added DEPAL. Because DEPAL doesn't dissolve in epoxy resin and behaves as a flame-retardant filler, its finely milled grade is useful in low loss factor compositions [204]. Pre-dispersion of DEPAL in epoxy [205] or use of multifunctional highly charrable epoxy resin [206] or combinations with melamine polyphosphate [207] help to boost the efficiency of DEPAL. Because non epoxy based printed wiring boards have an even lower loss factor, DEPAL became a popular flame retardant in compositions based on polyphenylene ether[208] or bismaleimide [209].

In contrast to DEPAL which melts with decomposition above 400°C, zinc diethylphosphinate (DEPZN) melts at about 220°C and therefore it was claimed [210] to be particularly suitable for fiber and film additive applications, for example to decrease the heat release rate of PET textile [211]. A recent patent [212] suggests that DEPZN is useful in poly(trimethylene terephthalate) fibers for carpeting. It can also be melt processed with semi-aromatic polyamide which is blended with rayon fibers to pass the ASTM D6413 test [213] or with TPU to produce wrapping tape passing the automotive FMVSS 302 test [214]. A recent academic publication [215] shows that 8 wt. % DEPZN can decrease the ATH loading from 60 wt. % to 37 wt. % in UPE in order to achieve a V-0 rating.

2.6   Aliphatic Phosphates and Phosphonates

Polyurethane (PU) foams encompass a wide range of foamed materials with very different properties starting from low density open cell flexible and rigid foams all the way to high density isocyanurate closed cell foams. From the point of view of response to flame, PU foams are considered to be thermally thick materials. This means that the heat applied to the foam doesn't dissipate deeply but stays in the surface layer. The surface reaches a high temperature quickly and therefore PU foams are easy to ignite. The ignition of the rigid foams is an interesting phenomenon because the flame flashes over the surface and then quickly retreats. If the heat flux to the surface is not high enough the flame can extinguish. The foam may reignite again if the heating is continued. Since the rigid foam is more densely cross-linked compared to the flexible foam, it doesn't melt away but undergoes charring. In such a scenario the best strategy to flame retard foam is to leverage both the gas phase and the condensed phase modes of action. This is achieved by combining phosphate ester flame retardants and reactive bromine-based flame retardants. Polyisocyanurate foams (PIR) are made with a significant 2.0-3.5 times excess of isocyanate over polyol. An excess of isocyanate forms an isocyanurate cross-linked network rich in nitrogen which is thermally more stable than the urethane groups. PIR foam is intrinsically more flame retardant than rigid spray PU foams and typically does not require help with brominated flame retardants.

Tris(isopropyl-2-chloro)phosphate (TCPP) is the largest commercially produced phosphorus flame retardant, but it is out of the scope of this chapter because it contains chlorine. Dimethyl methylphosphonate (DMMP, Formula 2.5(a)) for many years was used in rigid PU foams [216], but it was effectively removed from the market in the USA and Europe because it was categorized as a suspected mutagen. It is still used in China for passing stringent fire test requirements for high rise buildings. Diethyl ethylphosphonate (DEEP, Formula 2.5(b)) and dimethyl propylphosphonate (DMPP, Formula 2.5(c)) [217] were sold as replacements of DMMP but didn't gain a large market share because of a higher cost. Triethyl phosphate (TEP, Formula 2.5(d)), now produced only in Asia, is used in rigid PU foam as a co-additive with TCPP or brominated FRs as a viscosity cutter. TEP also helps with decreasing smoke, however, in fact it doesn't reduce smoke, but just doesn't increase it as much as halogen-containing FRs tend to do. For example, 9 parts TEP provides a B-2 rating in DIN 4102 in high density rigid PU foam and shows lower smoke [218] compared to TCPP. Interestingly, TEP allows production of translucent rigid PU foam [219].

$$
\begin{array}{cccc}
& \mathrm{O} & \mathrm{O} & \mathrm{O} & \mathrm{O} \\
& \| & \| & \| & \| \\
\mathrm{CH_3O-P-OCH_3} & \mathrm{C_2H_5O-P-OC_2H_5} & \mathrm{CH_3O-P-OCH_3} & \mathrm{C_2H_5O-P-OC_2H_5} \\
| & | & | & | \\
\mathrm{CH_3} & \mathrm{C_2H_5} & \mathrm{C_3H_7} & \mathrm{OC_2H_5} \\
\textbf{(a)} & \textbf{(b)} & \textbf{(c)} & \textbf{(d)}
\end{array}
$$

(2.5)

One study [220] compared DMMP, DEEP, DMPP and TEP with TCPP and tris(chloroethyl phosphate) (TCEP, removed from the market a decade ago). It was surprisingly found that the halogen-free phosphates and phosphonates show a higher LOI, 25-26.5 compared to chloroalkyl phosphates. It seems that the high volatility of halogen-free FRs compensated for a lack of chlorine. TEP, DEEP and DMPP showed good compatibility with blowing agents n-pentane and water, which resulted in an overall better shelf life of the mixed composition. On the negative side, halogen-free FRs showed lower compression strength and elastic modulus, probably due to stronger plasticization of the PU polymer. Another study [221] found similar FR efficiency of TEP (phosphate) and TCPP (phosphonate) confirming that the volatility of the FR plays an important role, but not the oxidative state of the phosphorus atom. Interesting research involving reactive FRs for rigid PU foams was reported from Korea [222]. A large amount of TEP or trimethyl phosphate or TCPP was added to waste PU foam and the mixture was heated to 190°C for 6 hours. At this temperature PU decomposes and the polyol fragments transesterify phosphate ester thus producing phosphorylated polyol. Rigid foam produced with the addition of this recycled polyol showed a decrease in peak heat release rate as measured by cone calorimeter.

For years low molecular weight phosphates TCEP and DMMP were used in highly filled ATH unsaturated polyester (UPE) systems or in glass-fiber composites with the main purpose of viscosity reduction [223, 224]. For example, 55-60 wt. % ATH and 1-2 wt. % DMMP allows passing the UL 723 test with class I for ventilation stacks [225]. Researchers at the Industrial Technology Research Institute (Taiwan) showed transesterification of simple phosphorus compounds such as DMMP to form phosphorylated unsaturated polyester resins [226]. Similar work was performed in China [227], where it was found that addition of about 15 wt. % DMMP to the reactive mixture in the synthesis results in UPE composites with a V-0 rating. Because the use of TCEP and DMMP was significantly restricted in North American and European markets the use of TCPP or TEP or DMPP was promoted for viscosity reduction in UPE. For example, it was suggested to use 5-10 wt. % DMPP as a viscosity reducer and synergist with APP and ATH [228]. Surprisingly only 10 wt. % ATH, 4 wt. % EDAP and 1 wt. % DMPP provide a V-0 rating in a glass-filled UPE composite [229].

Flexible PU foams have mostly open cell structures. Because of this, flexible foams are very combustible with an LOI in the range of 16-18 [230], and they show fast flame spread and a high heat release rate [231, 232]. The flammability of PU foams strongly depends on the foam density and the openness of the cells (air flow). Light foam with open cells burns very fast. Flexible PU foam is the main and most combustible component of upholstered furniture, mattresses [233] and car seats. Fires involving PU foams are the deadliest. "No ignition – no fire" is the best strategy to mitigate the fire hazard of flexible PU foams. Paradoxically, although PU foams are easy to ignite it is also easy to extinguish the fire when the flame is still small. This relates to the same inherent property of the PU foam being a thermally thick material. Because the heat cannot penetrate to the depth of the foam the heated layer where the foam decomposes and produces combustible gases is shallow. Such small flames can be extinguished by small changes in the fuel supply or by decreasing the heat by means of incomplete combustion. Flame retardants added to the flexible PU foams are specifically designed to extinguish small accidental fires [234]. However, if small flame doesn't extinguish the foam begins to liquefy and collapses in the liquid pool [235] which creates dangerous conditions for fire spread.

The most common flame retardants used in flexible PU foams are chlorinated phosphate esters. However, in recent years oligomeric or reactive flame retardants which don't contribute to VOC and do not migrate from the foam started taking market share. An oligomeric ethyl ethylene glycol phosphate (Formula 2.6) has been on the market for two decades. Because of the high 19% phosphorus content, it is quite efficient and as little as 4-8 php is effective in passing FMVSS 302 in a 1.5-1.8 lb/cu.ft. foam [236]. This oligomeric FR has been especially of interest in Europe and Japan, particularly with respect to the low-fogging low-volatiles-emission requirements of the automotive industry. It has been recommended for use in combination with alkylphenyl phosphates, which improve the flame retardant performance and also decrease the additive viscosity [237]. A number of recent patents [238, 239] indicate that a similar oligomeric phosphate but with a diethylene glycol bridging group (Formula 2.7) is in significant commercial development in Europe.

$$C_2H_5O \underset{\underset{OC_2H_5}{|}}{\overset{\overset{O}{||}}{P}} \left[O - (CH_2)_2 - O - \underset{\underset{OC_2H_5}{|}}{\overset{\overset{O}{||}}{P}} \right]_n OC_2H_5 \tag{2.6}$$

$$C_2H_5O \underset{\underset{OC_2H_5}{|}}{\overset{\overset{O}{||}}{P}} \left[O - (CH_2)_2 - O - (CH_2)_2 - O - \underset{\underset{OC_2H_5}{|}}{\overset{\overset{O}{||}}{P}} \right]_n OC_2H_5 \tag{2.7}$$

In manufacture of flexible polyurethane foams, if the foam reaches an excessively high temperature, "scorch" can occur. Scorch is, at the least, a discoloration of the interior of the slab or bun, and more seriously the loss of mechanical properties because of polymer degradation. Some of the commonly used flame retardants can aggravate scorch. Mechanistic studies showed [240, 241] that scorch is largely the result of the oxidation of aromatic amino groups arising from the hydrolysis of isocyanate groups which became isolated in the PU network. The formation of chromophoric groups is aggravated by the presence of flame retardants with alkylating capabilities such as chloroalkyl or alkyl phosphates because alkylated aminophenyl structures are more easily oxidized to quinoneimines. Ethyl ethylene glycol polyphosphate causes some scorch, especially in low density water blown foam, therefore the foam needs to be stabilized [242].

There is some market interest in reactive flame retardants for rigid and flexible PU foams. The advantage of a reactive FR is its permanence in the foam which is especially important in roofing applications in hot desert and tropical climates where the temperature of the roof can be very high and non-reactive FRs can be lost. Diethyl N,N bis(2-hydroxyethyl) aminomethylphosphonate (Formula 2.8) has been on the market for long time. The main application of this FR is in roofing spray foam and in the insulation foam of large refrigerators. A mechanistic study on this product showed [243] that even though most of the phosphorus splits off and volatilizes from the foam during combustion, it still helps with significant char increase which indicates that this reactive FR provides both condensed phase and gas phase modes of action.

$$C_2H_5O - \underset{\underset{\underset{HO - (CH_2)_2 - N - (CH_2)_2 - OH}{|}}{\overset{\overset{CH_2}{|}}{}}}{\overset{\overset{O}{||}}{P}} - OC_2H_5 \tag{2.8}$$

Another long time in the market reactive product is a diol mixture obtained by the reaction of propylene oxide and dibutyl acid pyrophosphate (Formula 2.9) [244]. The product contains 11% phosphorus and it is a mixture of isomers. Its recommended use is in flexible and rigid PU foams and polyurethane based coatings and adhesives.

$$
\begin{array}{ccccc}
 & O & & CH_3 & & O \\
 & || & & | & & || \\
HO-C_3H_6O-P-O-CH_2-CH-O-P-OC_3H_6-OH \\
 & | & & & & | \\
 & OC_4H_9 & & & & OC_4H_9
\end{array}
\tag{2.9}
$$

Recently realization came that a reactive flame retardant doesn't need to be difunctional, but a monofunctional can still be anchored to the rigid PU network to prevent migration and it can be easily released to the gas phase during the thermal decomposition of the foam due to reverse scission of one urethane bond. A recent patent [245] shows an improved process of manufacturing diethyl hydroxymethyl phosphonate (DEHMP, Formula 2.10) and a number of patents claim advantages of use of this product in rigid PU [246] and low smoke release PIR foams [247].

$$
\begin{array}{ccc}
 & O & \\
 & || & \\
C_2H_5O & -P-CH_2OH \\
 & | & \\
 & OC_2H_5 &
\end{array}
\tag{2.10}
$$

Similar to rigid foams there is a market desire to have a reactive phosphorus based flame retardant for flexible foams. However, technical development of such a product is more difficult because the cell structure of flexible foams is more sensitive to the variations in the composition compared to rigid foams. For example, diethyl N,N bis(2-hydroxyethyl) aminomethylphosphonate (Formula 2.8) broadly used in rigid PU foams, can be used in flexible foams only as a co-additive at the levels of 1-2 phr because the hydroxyl (OH) number is very high compared to typical flexible foam polyols.

In recent decades significant attempts were made to commercialize halogen-free phosphorus-containing diols for flexible foams. One of these diols is a hydroxyethyl terminated ethyl ethylene glycol phosphate oligomer (Formula 2.11) with about 17% phosphorus content [248, 249]. It is noticeable that this product is like the ethyl ethylene glycol oligomer of Formula 2.6, but it has terminal OH groups. It is primarily recommended for use in molded and high density slabstock flexible foams, where it passes the FMVSS302 test at 7.5 parts. The main advantage of this product is permanency in the flexible foam which allows achievement of low volatile organic compounds (VOC). Another phosphorus ester with about 12% P is a reactive phosphonate [250] made by reacting methyl phosphonic acid with ethylene oxide (Formula 2.12). It is mostly used in automotive flexible PU foams where it reacts in and becomes part of the PU network. It is highly efficient especially in high density foam where passing of the FMVSS302 test is achieved at < 4 parts.

$$HO - C_2H_4O - \overset{\overset{\displaystyle O}{\|}}{\underset{\underset{\displaystyle OC_2H_5}{|}}{P}} \left[O - (CH_2)_2 - O - \overset{\overset{\displaystyle O}{\|}}{\underset{\underset{\displaystyle OC_2H_5}{|}}{P}} \right]_n OC_2H_4OH \qquad (2.11)$$

$$HO - (C_2H_4O)_n - \overset{\overset{\displaystyle O}{\|}}{\underset{\underset{\displaystyle CH_3}{|}}{P}} - (OC_2H_4)m - OH \qquad (2.12)$$

One of the limitations of phosphorus containing diols is their tendency to create closed cell foams, which is not desirable in flexible PU. That is why these diols are used only in high density foams at low concentration. Some time ago it was discovered that monofunctional reactive flame retardants are easier to formulate in flexible PU foams [251]. Because monofunctional flame retardants are anchored on the PU foam chains they do not contribute to VOC and therefore are mostly targeted for automotive foam. An example of such phosphorus containing monohydric alcohol is the product made by reacting cyclic neopentyl acid phosphate with propylene oxide [252] (Formula 2.13) developed in Japan. In spite of a low phosphorus content of 11 % it allows passing FMVSS302 test at 8 parts which is similar to the chloroalkyl phosphates widely used in automotive foam. The market penetration of this cyclic flame retardant was limited because it is a solid, but the PU industry likes to operate with liquid flame retardants.

$$\underset{H_3C}{\overset{H_3C}{>}}\!\!\!\underset{O}{\overset{O}{\diagdown}}\!\!\underset{O}{\overset{O}{\diagup}}\!\!\underset{O}{\overset{P}{\diagdown}}\!\!O - (CH - CH_2O)_{1.2} - H \atop CH_3 \qquad (2.13)$$

For many years mixed methyl phosphate methylphosphonate ethylene glycol oligomer was sold as a flame retardant for paper automotive filters.

However later it was taken from the market and replaced with a chain end hydroxyl terminated version (Formula 2.14) but for textile finishing. It has been shown that this oligomeric product can be curable on cotton or blends using dimethyloldihydroxyethyleneurea (DMDHEU) and trimethylolmelamine [253] or melamine-formaldehyde [254] to obtain a durable finish with low formaldehyde odor. It is also efficient on cotton-nylon [255] and cotton-polyester [256] blends. It can also be used in non-formaldehyde finishes where the bonding to cellulose is achieved by using a polycarboxylic acid such as butanetetracarboxylic acid or citric acid [257]. Now this product is mostly sold in China for military uniforms.

$$H\left[O-(CH_2)_2-O-\underset{\underset{OCH_3}{|}}{\overset{\overset{O}{\|}}{P}}\right]_{2n}\left[O-(CH_2)_2-O-\underset{\underset{CH_3}{|}}{\overset{\overset{O}{\|}}{P}}\right]_{n}O-(CH_2)_2-OH$$

(2.14)

Similarly to PU foams, phosphorus based FR for textile finishing don't need to be di- or multifunctional. For many years the product of the addition of dimethyl phosphite to acrylamide followed by methylolation (Formula 2.15) was marketed for cotton and cotton-based blends [258]. This product is fixed on the cellulose using an amino resin and an acid curing catalyst. A recent academic study [259] shows that use of titanium dioxide as a co-catalyst for cotton textile treatment improves the flame retardant efficiency especially after laundering. It has a mild formaldehyde odor because it contains some components with less well bound formaldehyde [260]. This product is not used in the USA and has limited use in Europe because of potential formaldehyde exposure. There are methods of decreasing formaldehyde release [261] and it is believed that they are used commercially. A recent patent application [262] shows use of this product on lyocell fibers where it is introduced in the spinning solution.

$$CH_3O-\underset{\underset{OCH_3}{|}}{\overset{\overset{O}{\|}}{P}}-(CH_2)_2-\overset{\overset{O}{\|}}{C}-NH-CH_2OH$$

(2.15)

Thermosol finishes with phosphorus-based flame retardants have been used for many years in PET textiles [263] and probably in polyamides. The major product used in the thermosol treatment of polyesters is a liquid cyclic phosphonate (Formula 2.16). It is a mixture of diphosphonate and triphosphonate with the ratio mostly shifted towards diphosphonate x=1. Usually, a small concentration of phosphorus 0.3-0.5 wt. %, in PET is needed to pass the textile flammability test NFPA 701 and 0.7 wt.% is needed to pass the vertical FAR 25.853 test for use in aviation airbags [264]. Even being highly soluble in water, after the phosphonate is trapped under the fibers surface it is resistant to laundering and doesn't leak out. Because this phosphonate has a high phosphorus content it is also attractive for use in rigid or flexible PU foams. However, it has high viscosity and needs to be diluted with an aromatic phosphate [265] or bisphosphate [266]. A version of the spirophosphonate where the ratio is mostly shifted to triphosphonate (Formula 2.16, x is mostly 0) has shown promise as an additive in polyamide fibers via a melt process [267].

$$(2.16)$$

About a decade ago pentaerythritol spirobis(methylphosphonate) (Formula 2.17) was introduced for use in polyamides, including melt-spun fibers [268] and for use in combinations with intumescent flame retardants in polyethylene [269]. When combined with a free-radical generator, it is effective for flame-retarding polyethylene foam [270] and thin polyethylene films [271]. Another application of spirobis(methylphosphonate) seems to be in polyurethane based textile backcoatings [272]. In a recent academic publication [273] it was found that spirobis(methylphosphonate) has a higher flame retardant efficiency in molded PET compared to a bisphosphate, bisphosphinate and bis-n,n-naphthtylimide of similar structure. This high efficiency was attributed to the phosphorus gas phase action.

$$(2.17)$$

Similarly pentaerythritol spirobis(benzylphosphonate) (Formula 2.18) has been developed in Japan and introduced as a flame retardant for thermoplastic polyesters and styrenics [274]. However, based on later patents the main applications of this product seem to be in polylactic acid [275] and its blends [276] and in bio-based polycarbonate [277]. It doesn't affect the transparency and the clarity of the polycarbonate. Other potential applications are poly(methyl methacrylate) and its blends [278], backcoating for polyester textiles [279] and polyurethane based artificial leather [280].

$$(2.18)$$

1,3,2-dioxaphosphorian-2,2-oxy-bis-(5,5-dimethyl-2-sulphide) (Formula 2.19) is a solid flame retardant additive developed and commercialized in Europe for use in viscose rayon [281]. Despite the anhydride structure, it is remarkably stable, surviving addition to the highly alkaline viscose, the acidic coagulating bath, and also resisting multiple laundering of the rayon fabric. The unusual stability may be attributed to the sulfur atoms, which enhance hydrophobicity, and to the sterically hindering neopentyl groups that retard hydrolysis. The process of producing flame-retardant viscose fibers has been improved in recent years [282] because of increasing use of rayon as a fire barrier in mattresses.

$$(2.19)$$

Other commercial aliphatic phosphates e.g., tributyl phosphate, triethoxybutyl phosphate and tri-2-ethylhexyl phosphate can be used as a flame retardants or part of a flame-retardant mixture, but their major use is in other areas. Therefore, they are out of the scope of this chapter.

2.7   Aromatic Phosphates and Phosphonates

Triphenyl phosphate (TPP, Formula 2.20(a)), a white low melting (48°C) solid, and tricresyl phosphate (TCP, Formula 2.20(b)), a liquid, were introduced into commercial use early in the twentieth century, initially for cellulose nitrate and later for cellulose acetate. TPP is usually produced in the form of flakes or shipped in heated vessels as a liquid. TPP has also been used as a flame-retardant additive for engineering thermoplastics such as polyphenylene ether–high impact polystyrene (PPE/HIPS) [283] and polycarbonate-ABS (PC/ABS)[284] blends. The largest use of TCP has been as a plasticizer for PVC [285]. In recent years some adverse aquatic toxicity data and later discovery of endocrine disruption effects of TPP led to a significant decrease in its use. Manufacturers of aromatic phosphates are now in the process of eliminating it from other products where it is present as a blend component or as an impurity. Originally tricresyl phosphate was made from petroleum-derived or coal-tar-derived cresols. Discovery of the toxicity of ortho-cresyl phosphates led manufacturers to switch to synthetic cresols having very little o-cresol, but this significantly increased the cost of TCP and almost eliminated it from the market. Another product from the same family, cresyl diphenyl phosphate (CDP, Formula 2.20(c)), took the market share of TCP, but mostly in Europe.

(a)   (b)   (c)

$$(2.20)$$

Typical applications of TCP and CDP are in PVC tarpaulins, mine conveyer belts, air ducts, cable insulation, and vinyl films. These phosphates are usually used in blends with phthalates. The proportion of the more expensive phosphate is usually chosen such as to permit the product to reliably pass the flammability specifications. Other uses of CDP are in various rubbers, polyisocyanurate foams [286], semi-rigid PU foams in combination with expandable graphite [287], in phenolic and epoxy laminates [288] and as a plasticizer in epoxy-based coatings [289].

In the late 60s, the use of more economical synthetic isopropyl- and tert-butylphenols as alternatives to cresols was developed [290, 291]. Commercial triaryl phosphates are based on partially isopropylated or tert-butylated phenols. Made from the product of isopropylation of phenol by propylene, isopropylphenyl phenyl phosphate (Formula 2.21(a)) is a mixture of mainly ortho- and para-isomers and contains a distribution of different levels of alkylation [292, 293]. The plasticizer performance of isopropylphenyl phenyl phosphate is close to that of TCP. Mixed tert-butylphenyl phenyl phosphate (Formula 2.21(b)), is a slightly less efficient plasticizer for PVC by itself but it is quite effective in blends with phthalate plasticizers. Both commercial isopropylated and tert-butylated phosphates contain a significant amount of TPP. tert-Butylphenyl phenyl phosphate is the least volatile and the most oxidatively stable in the family of alkylphenyl phosphates [294].

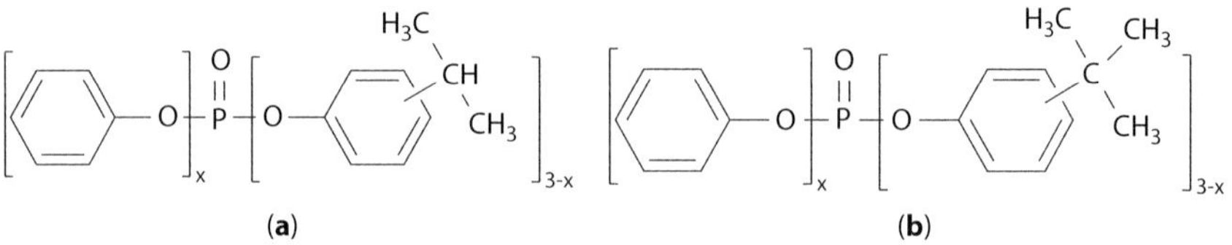

(2.21)

Apart from PVC both of these mixed phosphates found some use in flexible foam formulations [295] sometimes in combination with bromine-containing additives [296]. Even though these phosphates are not very efficient they have good hydrolytic stability and low volatility which is important for automotive foams. *tert*-Butylphenyl phenyl phosphate was also shown to be a viable flame retardant in high index PIR foams [297]. It has also been used as a flame retardant in PPE/HIPS pallets to pass the large scale UL 2335 test [298], in PPE/elastomer blends [299] and in PC/ABS where it shows good stress cracking resistance [300]. *tert*-Butylphenyl phenyl phosphate is often added to PPE-based blends as a processing aid even if flame retardancy is not required. Another large application of alkylphenyl phosphates is a plasticizer for thick intumescent coating for offshore oil rigs [301].

Alkyl diphenyl phosphates are products developed to provide improved low temperature flexibility, a fault of triaryl phosphate plasticizers in PVC [302]. There are three commercial phosphates in this family (Formula 2.22), e.g., 2-ethylhexyl diphenyl phosphate, isodecyl diphenyl phosphate and diphenyl phosphate with a mixture of longer (C_{12}-C_{14}) chains. These phosphates generally provide slightly less flame-retardant efficacy but generally produce less smoke compared to triaryl phosphates when the PVC formulation burns [303]. 2-Ethylhexyl and isodecyl diphenyl phosphate find their use in PVC sheet applications, PVC and TPU based artificial leathers and PVC/nitrile rubber tapes for insulative wrap of conduits. C_{12}-C_{14}-alkyl diphenyl phosphate has lower volatility compared to the other two phosphates and therefore it is used in PVC cable jacketing. Other applications of alkyl diphenyl phosphates are in casted polyurethane-polyurea goods [304] and in combination with intumescent flame retardants in thermoplastic elastomers cable jackets [305].

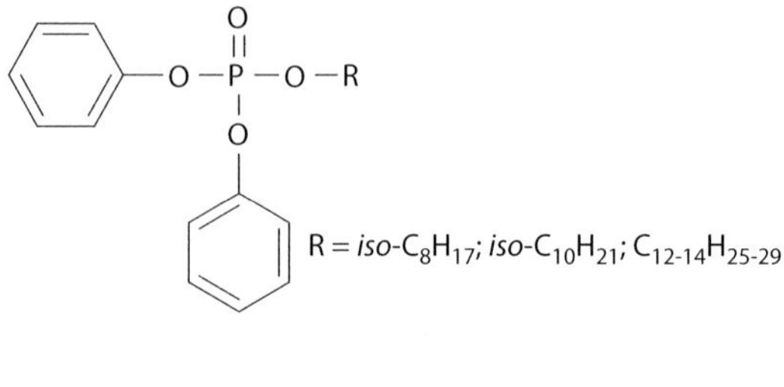

$R = \textit{iso-}C_8H_{17};\ \textit{iso-}C_{10}H_{21};\ C_{12\text{-}14}H_{25\text{-}29}$

(2.22)

Aromatic phosphates or aromatic phosphate oligomers (mostly diphosphates) are very widely used in PC/ABS and PPE/HIPS blends. Historically triphenyl phosphate (TPP) was the first phosphorus-based flame retardant used in these blends. Although TPP is soluble in these resins and it doesn't bloom out at room temperature, it deposits on the mold surfaces during molding. Because of the low melting point of TPP (48°C), it leads to bridging at extrusion feeding ports. The next generation of aromatic phosphate FR in PC/ABS and PPE/HIPS was *tert*-butylphenyl phenyl phosphate, which is still used nowadays in old formulations. However, now oligomeric aromatic phenyl phosphates (mainly diphosphates) are finding broader application than monophosphates because of better thermal stability and lower volatility.

The first product which became commercial was resorcinol bis(diphenyl phosphate) (RDP, Formula 2.23) which is a mixture of oligomers with two to five phosphorus atoms, but with the distribution heavily shifted towards the diphosphate [306]. In commercial PC/ABS blends where ABS content normally does not exceed 25%, RDP gives a V-0 rating at 8-12 wt. % loading [307]. Poly(tetrafluoroethylene) (PTFE) is a necessary ingredient in the formulation, which is usually added at <0.5 wt.% to retard dripping. Since the glass transition temperature of PTFE is below room temperature, it is soft and difficult to handle. To improve PTFE feeding it can be added during the production of ABS so that it is embedded in the polymer [308], or it can be specially treated to become free flowing [309], or pre-processed as a masterbatch [310]. RDP is somewhat less hydrolytically stable compared to other bisphosphates, which limits its application in humid environments and may cause a problem in recycling. This shortcoming of RDP can be alleviated by adding acid scavengers such as epoxies, oxazolines, or ortho esters [311].

(2.23)

Bisphenol A bis(diphenyl phosphate) (BDP) (Formula 2.24) was introduced to the market in the late 90s as an alternative to RDP [312]. Since bisphenol A is less expensive that resorcinol, BDP is more cost efficient despite a lower phosphorus content (8.9 % P for BDP vs. 10.7 % P for

RDP). BDP is significantly more viscous than RDP (12500 cP for BDP vs. 600 cP for RDP at 25°C) and therefore it requires a heated storage tank and heated transfer lines, whereas RDP needs only heated transfer lines. Because of the high viscosity, the oligomers content (n>1, Formula 2.24) in the BDP mixture is usually limited to only 10-15% which creates a problem of potential crystallization of BDP during transportation. Despite the many disadvantages over RDP, BDP became the major product used in PC/ABS and the second largest phosphorus-based flame retardant produced. On the positive side BDP has better hydrolytic stability than RDP [313] and can be used in high humidity applications especially if it is further stabilized by adding epoxy [314] as an acid scavenger. PC/ABS with an ABS content less than 25 wt. % usually needs more than 12 wt. % BDP plus a small co-addition of PTFE in order to assure a V-0 rating [315].

(2.24)

BDP and RDP are also used in PC/PBT and PC/PET but further addition of an impact modifier, for example polyethylene copolymer [316] or core-shell copolymer [317] is needed. Recently, new flame-retardant blends of PC/PMMA [318] (copolymer of methyl methacrylate and phenyl methacrylate) which produce very high gloss and have excellent scratch resistance were introduced to the market. New FR blends using as one component a bio-based polymer PC/PLA [319] (polylactic acid) are also being explored for use in electronic equipment. Further addition of talc improves the heat stability of PC/PLA [320]. The content of bisphosphate in these blends depends mostly on PC content, the higher the PC content, the less bisphosphate required to achieve a V-0 rating.

Although major compounders of PC based blends are likely to be well equipped with liquid feeding systems, small and medium size compounders prefer to use solid bisphosphates even if they cost more than BDP. Very close to RDP, hydroquinone bis(diphenyl phosphate) (HDP, Formula 2.25) when made relatively pure with low TPP content and low oligomers content is a solid with a melting point of 105-108°C. It can be fed into the extruder without extensive cooling of the feeding zone and therefore some large compounders also use this product where flexibility of changing extrusion lines is desirable. HDP has a phosphorus content of 10.8%,

similar to BDP, hydrolytic stability and requires 8-12 wt.% loading in PC/
ABS [321] and other PC based blends to achieve a V-0 rating. In talc filled
PC only 7 wt. % HDP is needed for a V-0 rating [322].

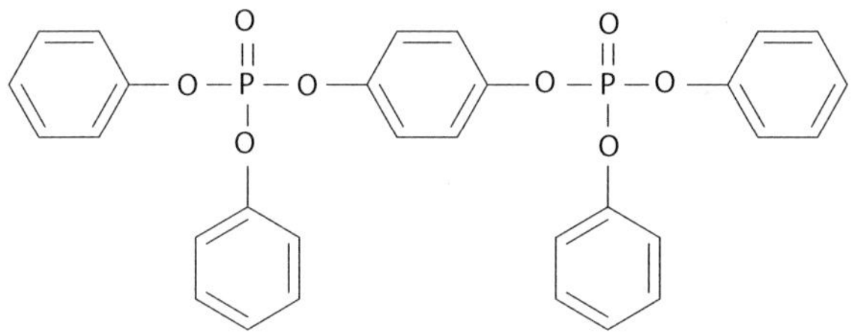

(2.25)

Resorcinol bis(di-2,6-xylyl phosphate) (RXP, Formula 2.26) [323] has
been on the market for over 20 years, but mostly in Asia. Similar to HDP,
RXP is mostly pure bisphosphate with very little oligomers present.
Because of the specific chemical structure and high purity RXP is a solid
with a melting point of 95°C. The steric hindrance provided by the 2,6-
xylyl groups makes this product more hydrolytically stable than BDP. RXP
has a phosphorus content of 9.0 % and its fire-retardant efficiency is similar
to that of BDP; it provides a V-0 rating in PC/ABS at 12–16 wt. % loading
[324] and about 7-10% wt. % loading in mineral filled PC/ABS [325].

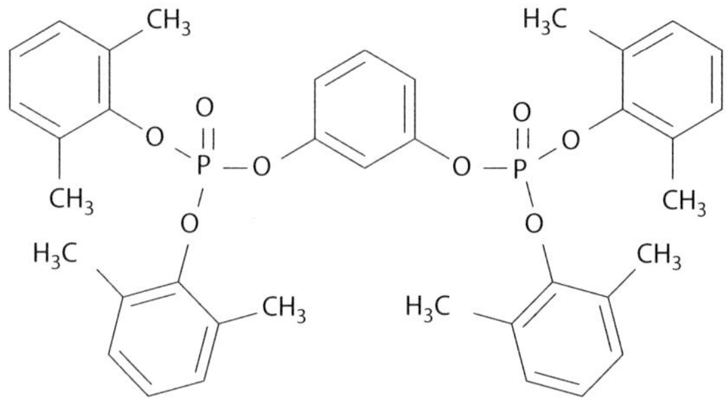

(2.26)

4,4'-Biphenyl bis(diphenyl phosphate) (Formula 2.27) is a specialty bisphosphate for high temperature molding of glass-filled PC and PC/ABS [326].
It has a melting range of 65-85°C [327] and a phosphorus content of about
9.5 %. It gives a V-0 rating in PC at 3.5wt. % loading and 0.3 wt. % PTFE
and at 10 wt. % loading it gives a V-0 rating in PC/ABS and a comparative
tracking index (CTI) of 600 V [328]. When used without PTFE it preserves
the transparency of polycarbonate [329]. As measured by thermogravimetry, 4,4'-biphenyl bis(diphenyl phosphate) shows a 5 wt. % loss at about
405°C which is significantly higher than other bisphosphates. Because of its
low melting point this bisphosphate requires a significant cooling system in

order to avoid bridging in the extruder feeding ports. A recently developed variation of the same bisphosphate, but with a significantly higher content of oligomeric fraction (n>1, 30-40%) is a viscous liquid [330].

$$(2.27)$$

Academic studies [331, 332] on the mechanism of the flame-retardant action of aromatic phosphates in a PC based blend revealed that BDP shows mostly condensed phase action, RDP shows a mixed condensed phase and gas phase, whereas TPP is mostly gas-phase-active. This was attributed to the temperature of decomposition of PC and phosphates, e.g., TPP evaporates at a relatively low temperature and doesn't have a chance to react with PC, whereas bisphosphates react with PC [333, 334]. RDP or BDP tend to cause PC to produce more char, decreasing the fuel supply to the flame and decreasing the flame temperature. TPP, which has gas phase activity, becomes more effective in the gas phase with a decrease in the flame temperature. HDP shows significant gas phase efficiency and when mixed with a mostly condensed phase active BDP exhibits a synergistic effect [335]. Another study showed [336] that the hindered structure of RXP slows down the reaction with PC and therefore it shows less condensed phase action compared to RDP. Interestingly talc improves the flame retardant efficiency of bisphosphates because of the better protective properties of the char [337] and on the other hand glass fiber reinforcement deteriorates the flame retardant efficiency because it increases the combustion surface (candlewick effect) [338].

Another large application of aromatic bisphosphates and oligomers is in polyphenylene ether (PPE) based blends. Polyphenylene ether cannot be processed alone because of its very high melting temperature, but it is readily compatible with many polymers and can be processed as a blend. Depending on the molecular weight and chain ends PPE can be blended with HIPS, polyamides, styrenic elastomers and even epoxy. Apart from improving the physical properties of the host polymer, PPE is an excellent charring polymer due to its specific thermal decomposition mechanism (Formula 2.28). PPE undergoes Fries isomerization [339] and forms a phenolic type of resin with numerous OH groups which are reactive with phosphorus FRs.

$$(2.28)$$

Commercial PPE/HIPS blends, also known as modified PPE, contain from 35 to 65 wt. % PPE. Similarly to PC/ABS, the first FR used in PPE/HIPS was TPP, which was later replaced with RDP and BDP [340]. Typically, between 9 and 15 wt. % of a phosphate ester is needed to achieve V-0 rating; the lower the PPE content in the blend, the higher the phosphate loading required. PTFE is required to prevent dripping. A copolymer of polydimethyl- and polydiphenyl siloxane can prevent dripping and is at the same time synergistic with RDP [341]. Addition of polysiloxane also helps to decrease smoke formation allowing the achievement of HL3 rating in the European mass transit test EN 45545 in PPE/HIPS based glass fiber composites [342]. Apart from flame retardancy, phosphate esters also play an important role in plasticization and resin flow improvement. Therefore, some phosphate esters can be added to PPE/HIPS even if flame retardancy is not needed. Because PPE is not sensitive to hydrolysis, any bisphosphate can be used in high humidity applications, as for example water pipes [343].

Apart from HIPS, PPE is also compatible with styrene based thermoplastic elastomers, such as styrene-ethylene-butylene-styrene (SEBS) block copolymer. A copolymer of SEBS and maleic anhydride is used as a compatibilizer [344]. These blends are mostly used in electric wire jackets and often polyolefins and HIPS are also included in the blends. As the patent literature indicates, aromatic bisphosphates RDP [345] and BDP [346] were originally used as flame retardants in PPE/SEBS blends. Bisphosphates are very compatible and soluble in PPE, but not in TPEs and polyolefins and therefore the total loading of bisphosphates is limited because of potential exudation. To overcome this problem in blends containing less than 50% PPE, solid flame retardants are added along with bisphosphates. The patent literature shows combinations of aromatic phosphates with magnesium hydroxide [347], melamine phosphates [348], ammonium polyphosphate [349] or DEPAL [350].

Mechanistic studies of the flame-retardant action of bisphosphates and TPP in PPE based blends showed that phosphates catalyze the Fries rearrangement [351] (Formula 2.28) and promote charring and improve the morphology of the char by making it intumescent-like [352]. The PPE charring capability is higher than PC, therefore PPE improves the fire retardant performance of RDP in PC/PBT blends [353]. One comprehensive study [354] looked at a large number of substituted aromatic phosphates and bisphosphates and compared them with red phosphorus and aliphatic phosphates. This study concluded that aliphatic phosphates are the least efficient because they decompose at temperatures much lower than the decomposition temperature of PPE. The efficiency of all aromatic phosphates and bisphosphates were similar in the range of experimental error and directly proportional to the phosphorus content. The efficiency of red phosphorus was similar to that of aromatic phosphates at the same phosphorus concentration. However, the strongest factor that controls the flammability of PPE/HIPS blends was the PPE content [355].

Another use of aromatic bisphosphates is in TPU. One of the common commercial halogen-free TPU formulations is based on about 25 wt. % melamine cyanurate and 5 wt. % RDP [356]. This TPU still drips, but the droplets do not ignite cotton and therefore it is rated V-0. Addition of some free isocyanate during processing creates additional cross-links and prevents dripping [357]. Many formulations based on RDP and ATH passing the VW-1 rating in the UL-1581 test for wire and cables applications were developed [358] and some were probably commercialized. Interestingly, the addition of only 2.5 wt. % novolac type epoxy resin provides robustness in passing the test [359] probably by cross-linking and decreasing the resin flow and dripping.

Various aromatic bisphosphates, more specifically RDP [360] can be incorporated by the exhaust method in PET textiles in the presence of polycaprolactone as a dispersing agent and polyethylene diamine as an auxiliary FR helping to retain RDP in the fiber. An add-on level > 10 wt. % was achieved and the textile passed the stringent DIN 54336 test with immediate extinguishment. A similar result was achieved by dispersing RDP, BDP or RXP in water using a non-ionic surfactant with a small addition of a cationic surfactant and then immersing the PET fibers at 130°C in an autoclave [361]. Emulsified RDP can also be applied as a backcoating to a nylon/cotton fabric blend [362]. In terms of combustion performance films are often close to textiles. About 8 wt. % RDP was used to pass the FMVSS 302 test in polyester films based on ethylene and 1,4-cyclohexanedimethane terephthalate [363] or 3 wt. % in a 45 degree flame spread test in poly(trimethylene terephthalate) film [364].

Independently of the physical form (liquid or solid), aromatic bisphosphates have very limited compatibility with polyolefins. Interestingly it was found [365] that aromatic bisphosphates can be loaded in PP plus EVA at 5 wt. % without visible exudation after heating for 72 hours at 70°C. A solid bisphosphate HDP showed a slightly better performance than liquid RDP. The films with 5% bisphosphate showed an HB rating in the UL-94 test. Interestingly, the maximum loading of triphenyl phosphate achievable in PP and EVA was only 3 wt. %. It is believed that bisphosphates can be used in PP fibers, films and foams to provide some level of flame retardancy. For example, 2.5 wt. % RDP or BDP combined with 1 wt. % aminophenyl disulfide provides a UL-94 HBF rating in PP foam [366]. About 3-8 wt. % of aromatic bisphosphate allows passing the 45 degree angle ISO 11925-2 test in HDPE/EVA flash spun sheets [367].

Some time ago it was discovered that the P-O-C bond in aromatic alkylphosphonates is reactive towards epoxy. Based on this discovery a new curing agent poly(1,3-phenylene methylphosphonate), (PMP, Formula 2.29(a)) for epoxy resins was developed [368]. It is semi-solid at room temperature, but it melts at about 45-55°C. The product is very rich in phosphorus (17.5%) and is thermally stable with a weight loss starting only above 300°C. PMP is qualified as an active ester and it cures epoxy by opening the epoxy group and insertion into the phosphonate ester linkage [369]. Because PMP doesn't produce secondary aliphatic alcohol groups as typical amine or phenolic curing agents, epoxy resin cured with PMP shows an improved thermal stability and a high glass transition temperature [370]. From 20 to 30 wt. % PMP provides a V-0 flammability rating in epoxy laminates.

(a)   (b)   (c)

(2.29)

Poly(bisphenol A methylphosphonate) (PAMP, Formula 2.29 (b)) was first developed in the '80s [371] but commercialized only a quarter of a century later [372]. The homopolymer can be used as an additive in PC or PC/ABS or co-polymerized with PC [373]. Being combined with potassium sulfonates, PAMP or its copolymers give a V-0 rating and good transparency in PC up to a 0.4 mm thickness [374]. Oligomeric and end chain functionalized PAMP are also suitable for special applications such as epoxy resins [375], cyanate resins [376], flexible PU foams [377] and TPU [378]. Despite its many potential applications, the main use of PAMP at the time of writing this chapter seems to be in PET fibers [379] for carpets and in PET films [380].

Interestingly, one of the first phosphonates used in PET fibers was poly (sulfonyldiphenylene phenylphosphonate) (Formula 2.29 (c)) produced in Japan. This oligomer is easily miscible with PET [381] up to 15 wt. % but for fiber applications typically less than 5 wt. % loading is needed. This product was discontinued in Japan in favor of reactive type phosphinates (see next subchapter), but it is reportedly produced now in China.

2.8   Aromatic Phosphinates

In general, the flame retardancy of phosphorus-containing polyester and polyamide fibers is mostly achieved by enhanced melt flow and melt drip, presumably catalyzed by phosphoric acid species produced in the process of oxidative degradation during combustion. Although it was a significant effort to try to introduce phosphate or phosphonate types of flame-retardant monomers into polyesters and polyamides, none of them led to a commercial produc [382]. The problem is that undesirable transesterification and hydrolysis reactions occur during the copolymerization. However, these side reactions do not seem to be a problem with phosphinates which have two non-reactive and not hydrolysable P-C bonds. For many years cyclic 2-methyl-2,5-dioxa-1,2-phospholane was copolymerized with ethylene glycol and dimethyl terephthalate to produce flame retardant PET fibers. About a decade ago this product was discontinued because one of the raw materials in its production was strictly regulated. This cyclic phosphinate was replaced with an adduct of benzenephosphinic acid and acrylic acid also known as CEPPA (Formula 2.30). CEPPA can be co-polymerized in the PET chain at 0.3-0.9 wt.% which leads to a significant increase in the LOI of PET fibers [383]. Interestingly, CEPPA also helps to improve the color stability of PET fibers [384]. Reportedly it can also be copolymerized in polyamide 6.6 fibers [385] to produce flame retardant carpets.

$$(2.30)$$

Some time ago it was discovered that the product of the reaction of o--phenyl phenol and phosphorus trichloride [386] followed by hydrolysis [387] resulted in a unique cyclic 9,10-dihydro-9-oxa-10-phosphaphenan-threne 10-oxide structure, also known as DOPO (Formula 2.31). Its P-H bond is more reactive than a similar bond in many other phosphinates or phosphonates and therefore DOPO can be reacted with alkenes, epoxies and aldehydes to form different flame retardants.

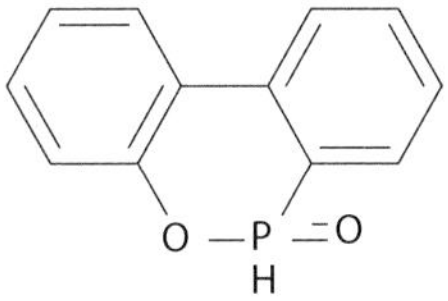

(2.31)

For example, the adduct of DOPO and dimethyl itaconate [388] (Formula 2.32 (a)) is a reactive flame retardant commercially used as a co-monomer in polyester fibers [389]. Like CEPPA, this product is efficient in PET fibers at a low concentration of 0.3-0.65 wt. % phosphorus and maintains good fiber properties [390]. It was found that placing the phosphorus ester link-age in the side chain, instead of the main chain, afforded superior hydrolysis resistance 391 and thermal stability [392, 393]. The adduct of DOPO and itaconic acid (Formula 2.32(b)) can be further copolymerized with diols and maleic anhydride to form an unsaturated ester prepolymer [394, 395] which can be cured with styrene to produce a thermoset resin.

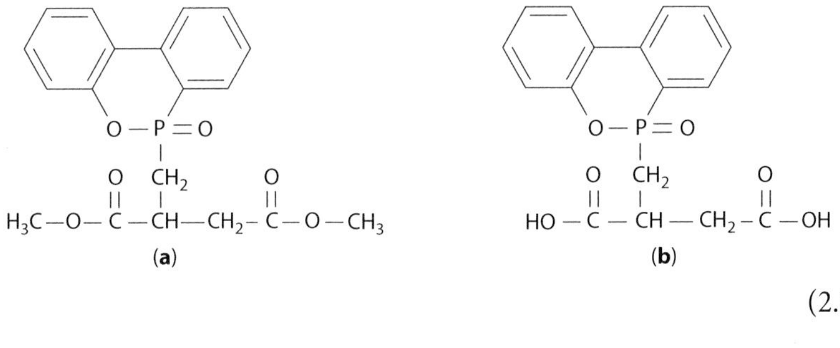

(2.32)

The printed wiring boards (PWB) which are produced with epoxy resins must pass the UL-94 test with a rating of V-1 or V-0. Phosphorus-based FRs can be added to epoxy as an additive or can be incorporated in the epoxy network by phosphorylation of the epoxy resin or in the form of phosphorus-based cross-linking agents [396]. Reactive FRs are more pre-ferred in epoxy because they show fewer negative effects on the physical properties, mostly glass-transition temperature and hydrolytic stability. Although DOPO is monofunctional, it was adopted by the industry for use in PWB laminates [397] and now it is the largest phosphorus FR used in epoxy. The common practice is to react DOPO with a multifunctional novolac type epoxy [398] in order to achieve a phosphorus content of about 3 wt. % (Formula 2.33). This still leaves on average 2-4 epoxy functionalities unconsumed which allows further curing of the phosphorylated epoxy resin.

(2.33)

Because DOPO is monofunctional it cannot be used with most common difunctional bisphenol A epoxy resins. In novolac type epoxies DOPO provides V-0 at a relatively low phosphorus content of 2.0-2.5 wt. %. [399]. The high efficiency of DOPO compared to other phosphorus FRs is partially attributed to its gas phase action [400]. DOPO can be combined with ATH [401] which is normally not the case with many phosphorus FRs showing mostly condensed phase action. When combined with ATH or fine silica, DOPO-based laminates require only 1 wt. % P or less to achieve a V-0 rating. The main disadvantage of DOPO modified epoxy is a challenge to achieve a high glass transition temperature $T_g > 150°C$ even when combined with multifunctional epoxy [402].

By reacting DOPO with quinone, a phenolic difunctional product can be made (DOPO-HQ, Formula 2.34(a)). It can be incorporated in an epoxy resin through a chain-extension process like tetrabromobisphenol A with difunctional epoxies [403]. Although it provides good physical properties and the required level of flame retardancy, it is not finding broad application because it is low in phosphorus and more expensive than DOPO. Because DOPO-HQ has poor solubility in the common solvents for epoxies, it is not used as a co-curing agent. DOPO-HQ can be co-polymerized into the polyester chain [404, 405], but this polyester seems not to have been commercialized. If naphthoquinone is used instead of quinone DOPO-NQ (Formula 2.34(b)) can be made. Because DOPO-NQ shows gas phase efficiency as well as good charring tendency it is efficient even at relatively low levels of addition [406]. DOPO-NQ is compatible with ATH and magnesium hydroxide (MDH) and when incorporated in multifunctional epoxy shows very good thermal and hydrolytic stability [407]. Because DOPO-NQ is a high melting temperature (295°C) solid [408] it can also be used as an additive in high frequency laminates based on polyphenylene ether where it allows maintaining a low dissipation factor [409]. Interestingly, DOPO-HQ and DOPO-NQ can be further functionalized with cyanate groups [410] instead of OH groups in order to be used in high end cyanate ester laminates [411]. DOPO-HQ can also be reacted with acetic anhydride and then transesterified with isophthalic acid to produce polymeric product which is an active ester that effectively cures epoxy [412].

(2.34)

By the reaction of DOPO with butoxymethylated bisphenol A [413] a mixture of phosphorylated bisphenols (DOPO-BPA) can be made with the major component presented in Formula 2.35. Because DOPO-BPA is a difunctional reactive FR and has a high phosphorus content of about 9% compared to phosphorylated epoxy of 3% (Formula 2.33) it allows production of laminates with a $T_g > 175°C$ and with good thermal stability as measured by a delamination test. It also has good solubility in solvents that are compatible with epoxy lamination processes. Another positive attribute of DOPO-BPA is good electrical properties in epoxy [414] and benzoxazine [415] laminates.

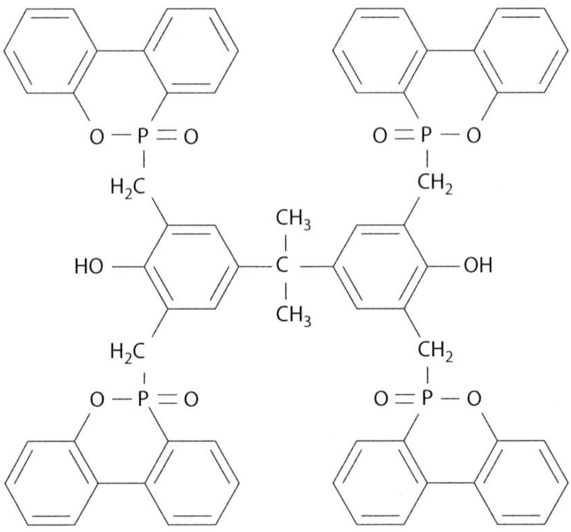

(2.35)

In order to achieve lower thermal expansion, improve heat dissipation and decrease the dissipation factor, a significant amount of silica is added to high end laminates. This new technology also opens the door for use of high melting, non-reactive and non-soluble flame retardants which further improve electrical properties. An example of such an FR is ethylene bis-DOPO phosphinate (Formula 2.36) made by reacting dichloroethane with DOPO [416] or reacting ethylene glycol with DOPO in the presence of sodium iodide [417]. This phosphinate provides a V-0 rating in novolac epoxy-based laminates at 20 wt. % loading. However, the main use of this flame retardant seems to be in non-epoxy polyphenylene ether (PPE) based laminates [418] or in hydrocarbon laminates based mostly on butadiene rubber and some PPE [419].

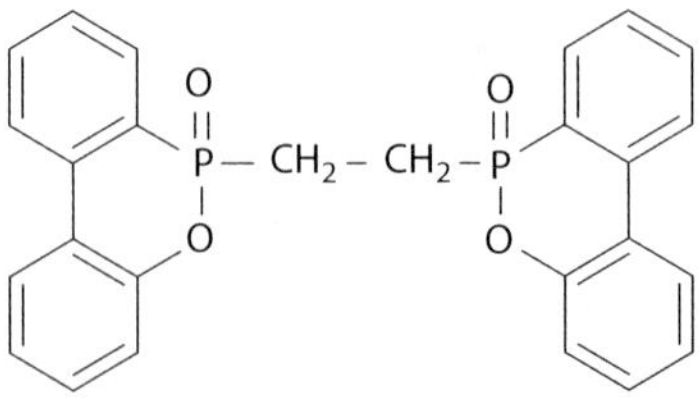

(2.36)

2.9   Phosphine Oxides

Phosphine oxides have three P-C bonds which are hydrolytically stable, and they seem to be ideal flame-retardant candidates for critical applications where there is exposure to moisture. On the other hand, phosphine oxides, especially aromatic ones are difficult to produce, and they tend to be more expensive than other organophosphates. This somehow limits the broad the use of phosphine oxides as flame retardants. One of the oldest applications of phosphine oxides is in textile finishing where the leading commercial products are tetrakis(hydroxymethyl) phosphonium chloride (THPC) or sulfate (THPS) [420]. THPC and THPS are water-soluble, but non-hydrolysable phosphonium salts that ensure exceptional durability. In the finishing process THPC or THPS is reacted with urea first and the product obtained is used to impregnate textile which is then dried and cross-linked with gaseous ammonia. At this stage some methylol groups react with cotton OH groups to permanently fix this finish on the textile [421]. Finally, the textile is treated with aqueous hydrogen peroxide which oxidizes phosphine into a more thermally stable phosphine oxide. The idealized structure [422] which doesn't have hydrolyzable bonds is shown in Formula 2.37. As a result, this finish is durable for 100 industrial launderings with alkaline detergent, more durable than any other flame-retardant cotton finish [423]. The need for using gaseous ammonia is the major disadvantage of this process and it requires special equipment.

$$\sim NH - CH_2 - \overset{\overset{O}{\|}}{P} - CH_2 - NH - \overset{\overset{O}{\|}}{C} - NH - CH_2 - \overset{\overset{O}{\|}}{P} - CH_2 \sim$$

(2.37)

Recently, another commercial phosphine oxide type flame retardant p-xylenebis(diphenyl phosphine oxide) (Formula 2.38) was introduced to the market. Although DOPO based flame retardants are most common in printed wiring boards the hydrolytic and thermal stability of DOPO sometimes is not sufficient for multiple pressing and reflow operations. Since phosphine oxide is not soluble in the common solvents used by PWB laminators it is applied as a filler in high frequency formulations [424]. p-Xylenebis(diphenyl phosphine oxide) provides a mostly gas phase flame retardant mode of action. In order to boost its efficiency it can be combined with resorcinol bis(di-2,6-xylyl phosphate) (Formula 2.26) which allows a decrease in the total FR loading [425].

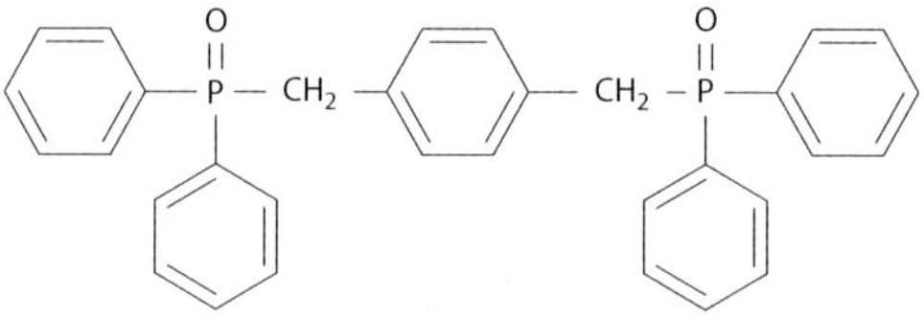

(2.38)

2.10 Phosphazenes

Cyclic phenoxyphosphazenes are thermally and hydrolytically stable phosphorus-nitrogen products. Typically, these are mixtures of hexaphenoxytricyclophosphazene as a major component (Formula 2.39) with some octaphenoxytetraphosphazene and a smaller fraction of linear oligomers. Because cyclic phenoxyphosphazenes have low polarity and favorable electrical properties they found use in high end epoxy [426], cyanate ester [427], benzoxazine [428] and PPE [429] based printed wiring boards. On the negative side the low melting point of cyclic phenoxyphosphazene of 110°C can cause excessive resin flow in the lamination process and as a result its loading level in the formulation is limited. Therefore, it is often applied in combination with other FRs. Another commercial use of cyclic phenoxyphosphazene is in PC/ABS where it is effective at 12-15 wt. % loading [430]. Reportedly, it shows a higher heat distortion temperature compared to aromatic bisphosphates [431]. It also provides excellent hydrolytic stability and low temperature ductility. However, the relatively high cost of cyclic phenoxyphosphazene compared to traditional bisphosphates limits its use in PC/ABS. Cyclic phenoxyphosphazenes can be further functionalized with cyanate groups [432] for specific applications with cyanate esters. Cyclic phenoxyphosphazenes functionalized with allylic groups [433] are used in acrylic terminated PPE laminates. However, at the time of writing this chapter the commercial status of these specialty phenoxyphosphazenes is unclear.

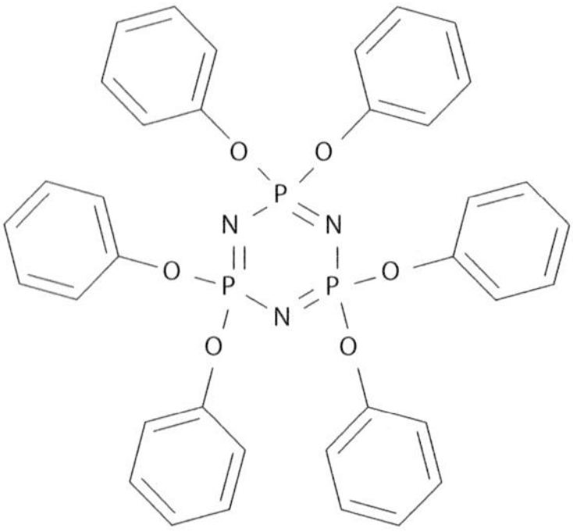

(2.39)

2.11   Environmental Fate and Exposure to Organophosphorus FRs

Tests in pure water, river water, and activated sludge showed that commercial triaryl phosphates, alkyl diphenyl phosphates and aromatic bis-phosphates undergo reasonably facile conversion to inorganic phosphates by hydrolysis and biodegradation [434–437]. Due to their low water solubility triaryl phosphates and in particularly TPP are rapidly absorbed into aquatic sediments [438]. Phosphate-cleaving enzymes are widespread in nature. Trialkyl phosphates are more resistant to hydrolysis [439], but they undergo easier photooxidation compared to aromatic phosphates [440]. However, the oxidation can be significantly slowed down if the aliphatic phosphates are absorbed on inert particles [441]. Phosphonates can undergo biodegradation of the P-C bond by certain microorganisms [442, 444]. Proper incineration at 600 - 800°C of flame-retardant plastics or thermosets containing organophosphorus flame retardants leads to quantitative conversion into phosphorus oxides or inorganic phosphates [445, 446], but flame retardants and byproducts of partial decomposition can be emitted in informal open flame recycling [447, 448]. The resultant inorganic phosphates from flame retardants would be orders of magnitude lower than phosphates from agricultural and municipal sources, and thus an inconsequential contributor to algae proliferation. Therefore, it is reasonable to believe that organophosphorus flame retardants and plasticizers cannot be persistent organic pollutants (POP).

However, recent development of more sensitive analytical techniques [449] allows the identification of very low concentrations of organophosphorus compounds in the environment [450, 451]. Some studies analyzed organophosphorus flame retardants in snow [452], rainwater [453] and surface waters [454] and found the highest concentration of thousands of ng/L in urban areas especially close to roads and airports [455] implying that FRs are emitted from the interior of cars and from lubricating oils. Organophosphorus flame retardants were found in Great Lakes atmosphere at levels of hundreds of pg/m^3 with the highest concentration close to large cities [456]. A recent study [457] measures the presence of organophosphorus flame retardants in seawater and sediments and despite the relatively low lipophilicity finds bioaccumulation in some marine species. Although organophosphorus FRs are less persistent than organohalogen FRs they still biomagnify in some biological and food nets [458]. Indoor measurements of organophosphorus flame retardants reveal that building insulation and electronic equipment are the main emission sources

[459] and in general the concentration is higher in offices than in residential houses. Like brominated FRs, organophosphorus FRs are mostly accumulated in dust [460, 461]. In some work-related environments such as electronic waste recycling facilities [462] or in construction during the application of foam insulation [463] the concentration of specific organophosphorus FRs is elevated. One unexpected route of transfer of organophosphorus FRs from indoors to the environment was suggested to be from clothing to laundry water [464].

A few recent reviews summarized the observed [465, 466] and computed [467] effects of phosphorus flame retardants on human health and the environment. The most concentrated compounds in every study were tris(1-chloro-2-propyl) phosphate (TCPP) and tris(2,3-dichloropryl) phosphate (TDCP) which are chlorinated phosphates and therefore are out of the scope of this chapter. The most common halogen free contaminants are tris(2-butoxyethyl) phosphate and tributyl phosphate which are mostly used as non-flame-retardant plasticizers or leveling agents in floor wax. These phosphates were also found in human urine along with triphenyl phosphate. Triphenyl phosphate is a common impurity in bisphosphates used in electronic and other consumer products [468], but it is also commonly used in nail lacquer. Finding all of these compounds in the environment especially indoors is not surprising because they are relatively volatile [469] and more mobile than most organohalogen FRs [470], whereas heavier and less mobile bisphosphates are very rarely detected in the environment. In general, aliphatic phosphates hydrolyze more slowly than aromatic phosphates and in addition to this TCPP and TDCP are sterically hindered which also slows down their hydrolysis and eventually mineralization [471].

There are many academic studies which erroneously attribute the detection of organophosphates in the environment as being due to the replacement of brominated flame retardants with organophosphorus FRs [472] and call this "regrettable substitution" [473]. This is not correct because chloroaliphatic phosphates, aliphatic phosphates and TPP which are most commonly found in the environment have been in commercial use for a longer time than common halogenated flame retardants. In fact, production of TPP significantly decreased during the last decade. Commercial use of tris(chloroethyl)phosphate (TCEP) was ceased in the USA and Europe about 20 years ago and its occurrence in the environment is sharply decreasing [474]. It is also important to stress that many very low detected concentrations of organophosphorus FRs in the environment are way below the safe threshold for many national regulations. For example, an average 880 ng/kg of body weight daily uptake of organophosphorus FRs found in food [475] is significantly lower than the reference safe dose given by the US EPA (> 5,000 ng/kg). In another study [459] it was found that the daily intake of TCPP from indoor air is only 11% of the tolerably daily intake (TDI) established by the Swiss Federal Health Ministry. Calculated emissions from operating computers is about 50 times lower than the safe suggested concentration by the German Federal Environmental Agency for TCEP, the most toxic organophosphorus FR.

2.12   Conclusions and Further Trends

The growth of phosphorus-based flame retardants as a class is often attributed to the replacement of halogenated flame retardants which is disputable. Although there is a general market trend to halogen-free flame retardants mostly dictated by original equipment manufacturers (OEMs) for their "green image", there are many areas where halogenated flame retardants cannot be replaced due to technical reasons. Since phosphorus flame retardants possess gas phase and condensed phase modes of flame-retardant action with the gas phase being mostly underutilized there are good prospects for the development of new FRs with mostly gas phase activity or the discovery of new synergistic combinations. Because phosphorus FRs are selectively active in only a handful of highly charrable and heteroatomic polymers there is a need for the development of more universal flame retardants. This research can progress either by development of new highly efficient and hydrolytically stable intumescent systems or of highly efficient gas phase active FRs or of a combination of both. Plastics containing phosphorus FRs are poorly recyclable and so there is interest in more hydrolytically and thermally stable phosphorus flame retardants that are favorable to recycling.

Phosphorus flame retardants have not escaped environmentalist concerns and there are a growing number of academic publications that detect organophosphorus flame retardants in the environment. However, it would be an error to say that phosphorus flame retardants are found more often because they replace halogens. Mostly relatively mobile organophosphorus flame retardants that were detected have been in use for at least half a century. Specific organophosphorus compounds raised sufficient concerns to warrant their discontinuance, but this was not generic to the entire class. Following the general trend of the flame-retardant industry towards polymeric and reactive products which show less negative effects to the final products and minimal exposure to humans and the environment, phosphorus flame retardants will follow this general strategy. Trends in the research and development of phosphorus flame retardants have been in the direction of less volatile, less toxic, and more stable compounds, and where feasible, in the direction of built-in phosphorus structures. At the same time, the existing phosphorus flame retardants are finding increased exploitation in the form of mixtures with synergists and adjuvants.

References

1. Georlette, P., Simons, J., Costa, L., Halogen-containing Fire-Retardant Compounds, in: *Fire Retardancy of Polymeric Materials*, A.F. Grand and C.A. Wilkie (Eds.), pp. 245–284, Marcel Dekker, 2000.
2. Levchik, S.V., Phosphorus-based FRs, in: *Non-halogenated Flame Retardant Handbook*, A.B. Morgan and C.A. Wilkie (Eds.), pp. 17–74, Wiley, New York, 2014.
3. Levchik, S.V. and Weil, E.D., A review of recent progress in phosphorus-based flame retardants, *J. Fire Sci.*, 24, 345, 2006.
4. Levchik, S.V. and Weil, E.D., Developments in phosphorus flame retardants, in: *Advances in Fire Retardant Materials*, A.R. Horrocks and D. Price (Eds.), pp. 41–66, Woodhead, Boca Raton, FL, 2008.
5. Hoerold, S., Phosphorus-based and intumescent flame retardants, in: *Polymer Green Flame Retardants*, C.D. Papaspyrides and P. Kiliaris (Eds.), pp. 221–254, Elsevier, Amsterdam, 2014.
6. Joseph, P. and Ebdon, J.R., Phosphorus-based flame retardants, in: *Fire Retardancy of Polymeric Materials*, C.A. Wilkie and A.B. Morgan (Eds.), CRC Press, Boca Raton, FL, pp. 107–127, 2009.
7. Jain, P., Choudhary, V., Varma, I.K., Flame Retarding Epoxies with Phosphorus, *J. Macromol. Sci., Polym. Rev.*, C42, 139, 2002.
8. Schartel, B., Phosphorus-based flame retardancy mechanisms - old hat or a starting point for future development? *Materials*, 3, 4710, 2010.
9. Flammability: A New Look at Age-old Problem. *Dimensions, Tech. News Bull. NBS*, 58, 6, 130, 1974.
10. I. Sakon, Red phosphorus flame retardant and nonflammable resinous composition containing the same, US Patent 4879067, assigned to Rinkagaku Kogyo, November 7, 1989.
11. Uske, K. and Ebenau, A., A plea for red phosphorus. flame-retardant polyamides, *Kunststoffe Int.*, 9, 130, 2013.
12. Y. Kinose, A. Inoue, T. Nagayama, Modified red phosphorus, method of producing the same, decolorized red phosphorus composition and flame-retardant polymer composition, US Patent 7045561, assigned to Nippon Chemical, May 16, 2006.

13. A. Koenig, T. Erdmann, M. Roth, K. Uske, J. Engelmann, A. Ebenau, M. Klatt, Pale-colored flame-retardant polyamides, US Patent 9388341, assigned to BASF, July 12, 2016.

14. P. Baierweck, D. Zeltner, G. Heiner, K. Ulmerich, K. Muehlbach, M. Gall, Flameproofed black polyamide molding materials, US Patent 5405890, assigned to BASF, April 11, 1995.

15. Weil, E.D., Formulation and modes of action of red phosphorus, in: *Proc. BCC Conf. Recent Advances in Flame Retardancy of Polymeric Materials*, Stamford, CT, May 2000.

16. M. Wagner, W. Podszun, H. Peerling, Microencapsulation of red phosphorus, US Patent 6846854, assigned to Bayer, January 25, 2005.

17. M. Suzuki, K. Hironaka, J. Haruhara, Fire-retardant resin compositions, European Patent 0978540, assigned to Teijin, December 7, 2005.

18. S. Hoerold and J. Laubner, Stabilized red phosphorus material and a process for its preparation, US Patent 6645625, November 11, 2003.

19. W. Nielinger, K.H. Herrmann, D. Michael, Stabilized red phosphorus and its use for flameproofing thermoplastic polyamide molding compositions, US Patent 4550133, assigned to Bayer, October 29, 1985.

20. Y. Bonin and J. LeBlanc, Fire-resistant polyamide compositions, US Patent 4985485, assigned to Rhone-Poulenc, January 15, 1991.

21. Gatti, N., New red phosphorus masterbatches find new application areas in thermoplastics, *Plast. Add. Compound.*, 34, April 2002.

22. Huggard, M.T., Phosphorus flame retardants – activation 'synergists, in: *Proc. BCC Conf. Recent Advances in Flame Retardancy of Polymeric Materials*, Stamford, CT, May 1992.

23. M.-K. Lim, Fire-retardant composition of an alloy of polyamide and polyester resins, US Patent 9765217, assigned to Rhodia, September 19, 2017.

24. A. Koenig, S. Xue, K. Uske, M. Roth, Flame-retardant thermoplastic molding composition, US Patent 8629206, assigned to BASF, January 14, 2014.

25. M. Roth, M. Heussler, K. Uske, C. Minges, Polyamide with phosphorus and Al phosphonates, PCT Patent Application WO 2018/234429, assigned to BASF, December 27, 2018.

26. M. Roth, K. Uske, C. Minges, Glow wire resistant polyamides, US Patent 9828503, assigned to BASF, November 28, 2017.

27. Alfonso, G.C., Costa, G., Pasolini, M., Russo, S., Ballistreri, A., Montaudo, G., Puglisi, C., Flame-resistant polycaproamide by anionic polymerization of e-caprolactam in the presence of suitable flame-retardant agents, *J. Appl. Polym. Sci.*, 31, 1373, 1986.

28. Levchik, S.V., Levchik, G.F., Balabanovich, A.I., Camino, G., Costa, L., Mechanistic study of combustion performance and thermal decomposition behaviour of nylon 6 with added halogen-free fire retardants, *Polym. Degrad. Stab.*, 54, 217, 1996.

29. Suebsaeng, T., Wilkie, C.A., Burger, V.T., Carter, J., Brown, C.E., Solid products from thermal decomposition of polyethylene terephthalate: investigation

by CP/MAS 13C-NMR and Fourier Transform-IR spectroscopy, *J. Appl. Polym. Sci.*, 22, 945, 1984.

30. Schartel, B., Kunze, R., Neubert, D., Red phosphorus-controlled decomposition for fire retardant PA 66, *J. Appl. Polym. Sci.*, 83, 2060, 2002.

31. Kuper, G., Hormes, J., Sommer, K., *Macromol. Chem. Phys.*, In situ x-ray absorption spectroscopy at the k-edge of red phosphorus in polyamide 6,6 during a thermo-oxidative degradation, 195, 1741, 1994.

32. Levchik, S.V., Weil, E.D., Lewin, M., Thermal Decomposition of Aliphatic Nylons, *Polym. Int.*, 48, 532, 1999.

33. Granzow, A., *Account. Chem. Res.*, Flame retardation by phosphorus compounds, 11, 177, 1978.

34. Peters, E.N., *J. Appl. Polym. Sci.*, Flame-retardant thermoplastics. I. polyethylene - red phosphorus, 24, 1457, 1979.

35. Gatti, N. and Costanzi, S., Is red phosphorus an effective solution for flame proofing polyolefin articles?, in: *Flame Retardants 2004*, pp. 133–138, Interscience Communication, London, 2004.

36. Wu, Q., Lu, J., Qu, B., *Polym. Int.*, Preparation and characterization of microcapsulated red phosphorus and its flame-retardant mechanism in halogen-free flame retardant polyolefins, 52, 1326, 2003.

37. K. Yamauchi, H. Matsuoka, H. Matsumoto, T. Inoue, Flame retardant resin composition, US Patent 6136892, assigned to Toray, October 24, 2000.

38. S. Oishi, M. Yoshi, W. Hiraishi, Magnesium hydroxide particles, process for producing the same, and resin composition containing the particles, US Patent 6676920, assigned to Kyowa, January 13, 2004.

39. Savas, L.A., Deniz, T.K., Tayfun, U., Dogan, M., Effect of microcapsulated red phosphorus on flame retardant, thermal and mechanical properties of thermoplastic polyurethane composites filled with huntite and hydromagnesite mineral, *Polym. Degrad. Stab.*, 135, 121, 2017.

40. Mostashari, S.M., The superiority of red phosphorus over polymetaphosphate as flame-retardants on cellulosic substrates, *Cellulose Chem. Technol.*, 43, 199, 2009.

41. B. Klesczewski, M. Otten, S. Meyer-Ahrens, Method for producing flame-retardant polyurethane foam materials having good long-term use properties, European Patent 2451856, assigned to Bayer, December 11, 2013.

42. H. Schmidt, G. Ng, L. Tang, P. Pepic, D. Ulman, Isocyanate-based polymer foam having improved flame retardant properties, European Patent 2922921, assigned to Proprietect, January 30, 2019.

43. Pecht, M. and Deng, Y., Electronic device encapsulation using red phosphorus flame retardants, *Microelectr. Rel.*, 46, 53, 2006.

44. Y. Kinose, R. Imamura, A. Inoue, T. Hata, E. Okuno, Red phosphorus-base flame retardant for epoxy resins, red phosphorus-base flame retardant compositions therefor, processes for the production of both, epoxy resin compositions for sealing for semiconductor devices, sealants and semiconductor devices, US Patent 6858300, assigned to Nippon Chemical, February 22, 2005.

45. Brown, D., The return of the red retardant, in: *Proc. 2016 Pan Pacific Microelectronics Symp*, Big Island, Hawaii, January, 2016.

46. Chen, Z., Du, J., Li, X., Xie, Z., Wang, Y., Wang, H., Zheng, J., Yang, R., Failure behavior of nylon products for red phosphorus flame retardant electrical connectors, *RSC Adv.*, 9, 24935, 2019.

47. L. Sequeira, Fire retardant epoxy resin formulations and their use, European Patent 2956497, assigned to Hexcel, January 2017.

48. J.J. Sang, F. Lenzi, J.E. Meegan, L. Macadams, Y. Zhao, D.K. Kohli, Multifunctional surfacing material with burn-through resistance, PCT Patent Application WO2017/117383, assigned to Cytec, July 6, 2017.

49. Le Bras, M., Camino, G., Bourbigot, S., Delobel, R. (Eds.), *Fire retardancy of polymers: the use of intumescence*, Royal Society of Chemistry Special Publication 224, Springer Verlag, Berlin, 1998.

50. Bourbigot, S., Le Bras, M., Duquesne, S., Rochery, M., Recent advances for intumescent polymers, *Macromol. Mater. Eng.*, 289, 490, 2004.

51. Alongi, J., Han, Z., Bourbigot, S., Intumescence: tradition versus vovelty. A comprehensive review, *Progr. Polym. Sci.*, 51, 28, 2015.

52. Bourbigot, S. and Duquesne, S., Intumescence-based fire retardants, in: *Fire Retardancy of Polymeric Materials*, C.A. Wilkie and A.B. Morgan (Eds.), pp. 129–162, CRC Press, Boca Raton, FL, 2009.

53. Camino, G. and Delobel, R., Intumescence, in: *Fire Retardancy of Polymeric Materials*, A.F. Grand and C.A. Wilkie (Eds.), pp. 217–243, Marcel Dekker, New York, 2000.

54. Jindasuwan, S., Sukmanee, N., Supanpong, C., Suwan, M., Nimittrakoolchai, O., Supothina, S., Influence of hydrophobic substance on enhancing washing durability of water soluble flame-retardant coating, *Appl. Surf. Sci.*, 275, 239, 2013.

55. J.R. Johnson, K.Y.M., Chan, C., Wang, Formaldehyde-free flame retardant compositions and their use for manufacturing durable formaldehyde-free flame retardant cellulosic materials, US Patent 9675998, assigned to Winnitex, June 13, 2017.

56. E. van der Meulen, Fireproof and fire-retardant composition, PCT Patent Application WO 2008/150157, assigned to Finifire, December 11, 2008.

57. Nama, S., Condon, B.D., White, R.H., Zhao, Q., Yao, F., Cintrón, M.S., Effect of urea additive on the thermal decomposition kinetics of flame retardant greige cotton nonwoven fabric, *Polym. Degrad. Stab.*, 97, 738, 2012.

58. Parikh, D.V., Nam, S., He, Q., Evaluation of three flame retardant (FR) grey cotton blend nonwoven fabrics using Micro-Scale Combustion Calorimeter, *J. Fire Sci.*, 30, 187, 2012.

59. El-Tahlawy, K., Chitosan phosphate: A new way for production of eco-friendly flame-retardant cotton textiles, *J. Textile Inst.*, 99, 185, 2008.

60. El-Tahlawy, K., Eid, R., Sherif, F., Hudson, S., Chitosan: a new route for increasing the efficiency of stannate/phosphate flame retardants on cotton, *J. Textile Inst.*, 99, 157, 2008.

61. Nair, G.P., Mechanisms of flame retardancy in cotton fabrics and studies on durability of the finish. A novel method to retain the efficacy of the finish wash after wash, *Colourage*, p. 29, June 2000.

62. J. Zhang, J. Horton, X.H. Gao, Method of conferring fire retardancy to wood and fire-retardant wood products, US Patent 9669564, assigned to Koppers Performance Chemicals, July 6, 2017.

63. D.-W. Son, Flame retardant composition for wood, flame retardant wood, and method for manufacturing the same, US Patent 10093813, assigned to Korean National Institute of Forest Science, October 9, 2018.

64. Grexa, O. and Utike, H., Flammability parameters of wood tested on a cone calorimeter, *Polym. Degrad. Stab.*, 74, 427, 2001.

65. Terzi, E., Kartal, S.N., White, R.H., Shinoda, K., Imamura, Y., Fire performance and decay resistance of solid wood and plywood treated with quaternary ammonia compounds and common fire retardants, *Eur. J. Wood Prod.*, 69, 41, 2011.

66. Gao, M., Yang, S., Yang, R., Flame retardant synergism of GUP and boric acid by cone calorimetry, *J. Appl. Polym. Sci.*, 102, 5522, 2006.

67. G. Sun, Flame-retardant and corrosion-resistant fiber bamboo substrate and preparation method thereof, US Patent 9751810, assigned to Wuxi Boda Bamboo and Wood Industrial, September 5, 2017.

68. Holmes, C.A. and Knispel, R.O., *Exterior weathering Durability of some leach-resistant fire-retardant treatments for wood shingles: a five-year report*, 1981, USDA Research Report FPL403.

69. A. Deholm, Fire retardant paper, US Patent 7510628, assigned to FF Seely Nominees Pty, March 31, 2009.

70. Vroman, I., Lecoeur, E., Bourbigot, S., Delobel, R., Guanidine hydrogen phosphate based flame-retardant formulations for cotton, *J. Ind. Tex.*, 34, 27, 2004.

71. Shen, C.Y., Stahlheber, N.E., Dyroff, D.R., Preparation and characterization of crystalline long-chain ammonium polyphosphate, *J. Am. Chem. Soc.*, 91, 62, 1969.

72. Liu, G., Liu, X., Yu, J., A novel process to prepare ammonium polyphosphate with crystalline form II and its comparison with melamine polyphosphate, *Ind. Eng. Chem. Res.*, 49, 5523, 2010.

73. Liu, G., Chen, W., Liu, X., Yu, J., Controllable synthesis and characterization of ammonium polyphosphate with crystalline form V by phosphoric acid process, *Polym. Degrad. Stab.*, 95, 1834, 2010.

74. Weil, E.D., Fire-protective and flame-retardant coatings - a state-of-the-art review, *J. Fire Sci.*, 29, 259, 2011.

75. Camino, G. and Costa, L., Mechanism of intumescence in fire retardant polymers. *Rev. Inorg. Chem.*, 8, 69, 1986.

76. Vandersall, H.L., Intumescent coating systems, their development and chemistry, *J. Fire Flammability*, 2, 97, 1971.

77. Futterer, T., A Review about the use of novel intumescent flame retardants in thermoplastic materials, in: *Proc. Fall FRCA Conference*, pp. 141–142, Cleveland, OH, October 2002.

78. Cogen, J.M., Jow, J., Lin, T.S., Whaley, P.D., New approaches to halogen free polyolefin flame retardant wire and cable compounds, in: *Proc. 52nd IWCS/ Focus International Wire & Cable Symposium*, Philadelphia, PA, November 2004.

79. T. Ihara, Y. Koshiro, S. Suzuki, Flame retardant polyurethane resin composition and making method, US Patent 8093318, assigned to Shin-Etsu, January 10, 2012.

80. Bertelli, G., Goberti, P., Marchetti, E., Camino, G., Luda di Cortemiglia, M.P., Costa, L., Structure-Char Forming Relationship in Intumescent Fire Retardant Systems, in: *Fire Safety Science, Proc. 3rd Int. Symp*, G. Cox and B. Langford (Eds.), Edinburgh, July, 1991, Elsevier, London, p. 537, 1991.

81. R. Nalepa and D. Scharf, Two-component intumescent flame retardant, US Patent 5204392, assigned to Clariant, April 20, 1993.

82. R. Pernice, M. Checchin, A. Moro, R. Pippa, Olefin or styrene (co)polymer, triazine polymer, ammonium phosphate and phosphorite, US Patent 5514743, assigned to Enichem, May 1996.

83. B.L. Kaul, Polytriazinyl compounds as flame retardants and light stabilizers, US Patent 8202924, assigned to MCA Technologies, June 19, 2012.

84. Enescu, D., Frache, A., Lavaselli, M., Monticelli, O., Marino, F., Novel phosphorous-nitrogen intumescent flame retardant system. Its effects on flame retardancy and thermal properties of polypropylene, *Polym. Degrad. Stab.*, 98, 297, 2013.

85. Le Bras, M., Bourbigot, S., Felix, E., Pouille, F., Siat, C., Traisnel, M., Characterization of a polyamide 6-based intumescent additive for thermoplastic formulations, *Polymer*, 41, 5283, 2000.

86. Hoerold, S., Walz, R., Zopes, H.-P., Halogen-free additives meet rail standards, *Reinforced Plast.*, 40, January 2000.

87. Reilly, T., Flame retarded thermoset materials based on phosphorus chemistry, Paper presented at *annual meeting of Thermoset Resin Formulators Association (TRFA)*, Pittsburgh, PA, September 2009.

88. Ricciardi, M.R., Antonucci, V., Zarrelli, M., Giordano, M., Fire behavior and smoke emission of phosphate-based inorganic fire retarded polyester resin, *Fire Mater.*, 36, 203, 2012.

89. Chen, Z., Jiang, M., Chen, Z., Chen, T., Yu, Y., Jiang, J., Preparation and characterization of a microencapsulated flame retardant and its flame-retardant mechanism in unsaturated polyester resins, *Powder Technol.*, 354, 71, 2019.

90. Shih, Y.-F., Wang, Y.-T., Jeng, R.-J., Wei, K.-M., Expandable graphite systems for phosphorus-containing unsaturated polyesters. I. Enhanced thermal properties and flame retardancy, *Polym. Degrad. Stab.*, 86, 339, 2004.

91. Yang, H., Liu, H., Jiang, Y., Chen, M., Wan, C., Density effect on flame retardancy, thermal degradation, and combustibility of rigid polyurethane foam

modified by expandable graphite or ammonium polyphosphate, *Polymers*, 11, 668, 2019.

92. J.L. Clatty and D.L. McCalmon, Water-blown, flame retardant rigid polyurethane foam, PCT Patent Application WO 2007/075251, assigned to Bayer, July 5, 2007.

93. J.W. Rosthauser, Process for the production of medium density decorative molded foams having good fire retardant properties with reduced mold times, fire retardant compositions and foams produced by this process, US Patent 8097658, assigned to Bayer, January 17, 2012.

94. Meng, X.-Y., Ye, L., Zhang, X.-G., Tang, P.-M., Tang, J.-H., Ji, X., Li, Z.-M., Effects of expandable graphite and ammonium polyphosphate on the flame-retardant and mechanical properties of rigid polyurethane foams, *J. Appl. Polym. Sci.*, 114, 853, 2009.

95. Hu, X.M. and Wang, D.M., Enhanced fire behavior of rigid polyurethane foam by intumescent flame retardants, *J. Appl. Polym. Sci.*, 129, 238, 2013.

96. Li, J., Mo, X., Li, Y., Zou, H., Liang, M., Chen, Y., Influence of expandable graphite particle size on the synergy flame retardant property between expandable graphite and ammonium polyphosphate in semi-rigid polyurethane foam, *Polym. Bull.*, 75, 5287, 2018.

97. Chen, X.-Y., Huang, Z.-H., Xi, X.-Q., Li, J., Fan, X.-Y., Wang, Z., Synergistic effect of carbon and phosphorus flame retardants in rigid polyurethane foams, *Fire Mater.*, 42, 447, 2018.

98. Xing, W., Yuan, H., Zhang, P., Yang, H., Song, L., Hu, Y., Functionalized lignin for halogen-free flame retardant rigid polyurethane foam: preparation, thermal stability, fire performance and mechanical properties, *J. Polym. Res.*, 20, 234, 2013.

99. Yao, W., Wang, H., Guan, D., Fu, T., Zhang, T., Dou, Y., The effect of soluble ammonium polyphosphate on the properties of water blown semirigid polyurethane foams, *Adv. Mater. Sci. Eng.*, 2017, ID 5282869.

100. Lewin, M. and Endo, M., Catalysis of intumescent flame retardancy of polypropylene by metallic compounds, *Polym. Adv. Technol.*, 14, 3, 2003.

101. Amigouet, P. and Shen, K., Talc/zinc borate: potential synergisms in flame retardant systems, in: *Proc Flame Retardants 2006*, pp. 155–162, Interscience Communication, London, 2006.

102. Bourbigot, S., Le Bras, M., Breant, P., Tremillon, J.-M., Delobel, R., Zeolites: new synergistic agents for intumescent fire retardant thermoplastic formulations - criteria for the choice of the zeolite, *Fire Mater.*, 20, 145, 1996.

103. Scharf, D., Nalepa, R., Heflin, R., Wusu, T., Studies on flame retardant intumescent char: part I, *Fire Saf. J.*, 19, 103, 1992.

104. Horrocks, A.R., Wang, M.Y., Hall, M.E., Sunmonu, F., Pearson, J.S., *Polym. Int.*, 49, 1079, 2000.

105. Kandola, B.K. and Horrocks, A.R., Complex char formation in flame-retarded fiber/intumescent combinations: physical and chemical nature of char, *Textile Res.*, *J.* 69, 374, 1999.

106. Kandola, B.K. and Horrocks, A.R., Complex char formation in flame-retarded fibre-intumescent combinations – IV. Mass loss and thermal barrier properties, *Fire Mater.*, 24, 265, 2000.

107. Horrocks, R., Developments in flame retardants for heat and fire resistant textiles - the role of char formation and intumescence, *Polym. Degrad. Stab.*, 54, 143, 1996.

108. M.P. Jones, J.D. Small, Jr., J.H. Walton, A.F. Baldwin, Jr., Z. Mikaelian, W.S. Kinlaw, Flame resistant filler cloth and mattresses incorporating same, US Patent 8513145, assigned to Precision fabrics, August 20, 2013.

109. Weil, E.D. and McSwigan, T., Melamine phosphates and pyrophosphates in flame retardant coatings: old products with new potential, *J. Coating Technol.*, 66, 75, 1994.

110. Weil, E.D. and McSwigan, B., Melamine phosphate flame retardants, *Plast. comp.*, p. 31, May–June 1994.

111. Costa, L., Camino, G., Luda di Cortemiglia, M.P., Mechanism of thermal degradation of fire-retardant melamine salts, in: *Fire & Polymers, Am. Chem. Soc. Symposium Ser. 425*, G. Nelson (Ed.), pp. 211–238, Washington, DC, 1990.

112. S. Imanishi, Flame retardant polyolefin resin composition, US Patent 6921783, assigned to Daicel, July 18, 2005.

113. Lai, X., Zeng, X., Li, H., Liao, F., Zhang, H., Yin, C., Preparation and properties of flame retardant polypropylene with an intumescent system encapsulated by thermoplastic polyurethane, *J. Macromol. Sci., Phys.*, 51, 35, 2012.

114. Chen, X. and Jiao, C., Flame retardancy and thermal degradation of intumescent flame retardant polypropylene material, *Polym. Adv. Technol.*, 22, 817, 2011.

115. Lai, X., Zeng, X., Li, H., Liao, F., Yin, C., Zhang, H., Synergistic effect between a triazine-based macromolecule and melamine pyrophosphate in flame retardant polypropylene, *Polym. Compos.*, 33, 35, 2012.

116. Liu, S.-H., Kuan, C.-F., Kuan, H.-C., Shen, M.-Y., Yang, J.-M., Chiang, C.-L., Preparation and flame retardance of polyurethane composites containing microencapsulated melamine polyphosphate, *Polymers*, 9, 407, 2017.

117. T. Dave and W. Wehner, Triazine compounds containing phosphorous as flame retardants, US Patent 8754154, assigned to JM Huber, June 17, 2014.

118. H.-G. Koestler, T. Dave, W. Wehner, Flame protection agent compositions containing triazine intercalated metal phosphates, US Patent 9708538, assigned to JM Huber, July 18, 2017.

119. Mueller, P. and Schartel, B., Melamine poly(metal phosphates) as flame retardant in epoxy resin: performance, modes of action, and synergy *J. Appl. Polym. Sci.*, 133, 43549, 2016.

120. T. Dave and W. Wehner, Triazine compounds containing phosphorous as flame retardants, US Patent 9200122, assigned to JM Huber, December 1, 2015.

121. M.T. Huggard and P.S. White, Flame retardant thermoplastic resin composition with intumescent flame retardant, US Patent 5137937, assigned to Rhodia, August 11, 1992.

122. Goin, C.L. and Huggard, M.T., AMGARD EDAP a new direction in fire retardants, in: *Proc. BCC Conf. Recent Advances in Flame Retardancy Polymeric Materials*, Stamfort, CT, May 1991.

123. R.V. Kasowski, Flame retardant and flame retardant uses, US Patent 10501602, assigned to R.V. Kasowski, December 10, 2019.

124. M.S. Rhodes, L. Israilev, J. Tuerack, P.S. Rhodes, Activated flame retardants and their applications, US Patent 6733697, assigned to Broadview Technologies, May 11, 2004.

125. J.D. Reyes, Flame Retardant Wire and Cable, US Patent 7709740, assigned to JJI Technologies, May 4, 2010.

126. R.V. Kasowski, Flame retardant and composition containing it, US Patent 8703853, assigned to R.V. Kasowski, June 20, 2013.

127. O. Cicchetti, A. Bevilacqua, A. Pagliari, Process for the production of flame retarding additives for polymer compositions, and products obtained from said process, US Patent 5948837, assigned to Montel, September 7, 1999.

128. S. Belli, D. Tirelli, P. Veggetti, A. Bareggi, Impact-resistant self-extinguishing cable, US Patent 7049524, assigned to Prysmian, May 23, 2006.

129. R. Kimura, H. Murase, M. Nagahama, T. Kamimoto, S. Nakano, High purity piperazine pyrophosphate and process of producing same, US Patent 7449577, assigned to Asahi Denka, November 11, 2008.

130. H. Murase, M. Nagahama, K. Yoshikawa, Y. Tanaka, T. Kaneka, A. Yamaki, Flame retardant composition with improved fluidity, flame retardant resin composition and molded products, US Patent 7465761, assigned to Asahi Denka, December 16, 2008.

131. Kamimoto, T., Murase, H., Yamaki, A., Nagahama, M., Kimura, R., Funamizu, T., Zingde, G., A highly efficient intumescent flame retardant for polyolefins, in: *Proc. International Conference on Polyolefins*, Houston, TX, February 2004.

132. T. Hatanaka, M. Nakamura, H. Sakurai, Flame-retardant resin composition and electric wire using same, US Patent 9240260, assigned to Adeka, January 19, 2016.

133. C.M. Makadia, Non halogen flame retardant thermoplastic polyurethane, US Patent 8957141, assigned to Lubrizol, February 17, 2015.

134. J.G. Chen, X.W. Yan, B. Li, L. Lu, W. Ma, H.L. Huang, H.D. Gao, Halogen free, flame retardant compositions, US Patent 8920929, assigned to Dow, December 30, 2014.

135. Y. Ni and T. Shimizu, Fire-resistant thermoplastic polyurethane elastomer composition, US Patent 9926434, assigned to Adeka, March 27, 2018.

136. T. Kamimoto, Y. Yonezawa, N. Nakamura, Y. Okamoto, K. Omori, Flame-retardant composition and flame-retardant synthetic resin composition, European Patent 3037502, assigned to Adeka, October 16, 2019.

137. Y. Ni, Y. Yonezawa, N. Tanji, Flame retardant composition and flame retardant resin composition containing same, US Patent Application 2020/0048433, assigned to Adeka, February 5, 2020.

138. Y. Ni, Y. Inagaki, Y. Yonezawa, N. Tanji, Composition and flame-retardant resin composition, US Patent Application 2020/0148954, assigned to Adeka, May 14, 2020.

139. Y. Ni, Y. Yonezawa, N. Tanji, Composition and flame-retardant resin composition, US Patent Application 2020/0165448, assigned to Adeka, May 28, 2020.

140. H. Steenbakkers-Menting, R. van Giesen, M. van der Ven, R. Bekcx, Flame retardant long glass fibre reinforced polypropylene composition, US patent 10435540, assigned to Sabic, October 8, 2019.

141. Xiao, X., Hu, S., Zhai, J., Chen, T., Mai, Y., Thermal properties and combustion behaviors of flame-retarded glass fiber-reinforced polyamide 6 with piperazine pyrophosphate and aluminum hypophosphite, *J. Therm. Anal. Calorim.*, 125, 175, 2016.

142. D.H. Lee, S.W. Lee, S.-B. Kim, C.-G. Kim, S.-D. Ahn, J.-I. Ahn, Flame-retardant polyolefin resin containing piperazine-based metal salt blend, US Patent 9221961, assigned to Doobon, December 29, 2015.

143. J. Engelmann and D.Warting, Halogen-free flameproof polyester, European Patent 1423460, assigned to BASF, April 27, 2005.

144. T. Eckel, V. Taschner, D. Wittmann, E. Wenz, Flame-proof impact resistant-modified polycarbonate compositions, European Patent 2225317, assigned to Bayer, February 27, 2013.

145. A. Feldermann and B. Krauter, Flame retardant polyalkylene terphthalate/polycarbonate compositions, US Patent 8779039, assigned to Bayer, July 15, 2014.

146. S. Costanzi and M. Leonardi, Polyester compositions flame retarded with halogen-free additives, US Patent 7700680, assigned to Italmatch, April 20, 2010.

147. M. Roth, K. Uske, U. Wolf, Long-fiber-reinforced flame-retardant polyesters, US Patent 8962717, assigned to BASF, February 24, 2015.

148. Yang, W., Song, L., Hu, Y., Lu, H., Yuen, R.K.K., Investigations of thermal degradation behavior and fire performance of halogen-free flame retardant poly(1,4-butylene terephthalate) composites, *J. Appl. Polym. Sci.*, 122, 1480, 2011.

149. Yanga, W., Song, L., Hu, Y., Lu, H., Yuen, R.K.K., Enhancement of fire retardancy performance of glass-fibre reinforced poly(ethylene terephthalate) composites with the incorporation of aluminum hypophosphite and melamine cyanurate, *Compos.: Part B.*, 42, 1057, 2011.

150. Tang, G., Wang, X., Xing, W., Zhang, P., Wang, B., Hong, N., Yang, W., Hu, Y., Song, L., *Ind. Eng. Chem. Res.*, 51, 12009, 2012.

151. Chen, L., Luo, Y., Hu, Z., Lin, G.-P., Zhao, B., Wang, Y.-Z., An efficient halogen-free flame retardant for glass-fibre-reinforced poly(butylene terephthalate), *Polym. Degrad. Stab.*, 97, 158, 2012.

152. J. Li and G. Woodward, Thermoplastic polymer composition comprising an alkali metal hypophosphite salt, PCT Patent Application WO 2014/075289, assigned to Rhodia, May 22, 2014.

153. Ge, H., Tang, G., Hu, W.-Z., Wang, B.-B., Pan, Y., Song, L., Hu, Y., Aluminum hypophosphite microencapsulated to improve its safety and application to flame retardant polyamide 6, *Haz. J. Mater.*, ID 294186, 2015.

154. F. de Campo, A. Murillo, J. Li, T. Zhang, Process for stabilizing hypophosphites, US Patent 8940820, assigned to Rhodia, January 27, 2015.

155. Atabek, L.S. and Dogan, M., Flame retardant effect of zinc borate in polyamide 6 containing aluminum hypophosphite, *Polym. Degrad. Stab.*, 165, 101, 2019.

156. Atabek, L.S., Tayfun, U., Hancer, M., Dogan, M., The flame-retardant effect of calcium hypophosphite in various thermoplastic polymers, *Fire Mater.*, 43, 294, 2019.

157. M. Roth, Flame retardant compositions comprising sterically hindered amines, US Patent 8349923, assigned to BASF, January 8, 2013.

158. S. Costanzi, Flame retardant polymer compositions, US Patent 7619022, assigned to Italmatch, November 17, 2009.

159. Xu, M.-J., Wang, J., Ding, Y.-H., Li, B., Synergistic effects of aluminum hypophosphite on intumescent flame retardant polypropylene system, *Chinese J. Polym. Sci.*, 33, 318, 2015.

160. Zhu, P., Xu, M., Li, S., Zhang, Z., Li, B., Preparation and Investigation of efficient flame retardant TPE composites with piperazine pyrophosphate/aluminum diethylphosphinate system *J. Appl. Polym. Sci.*, ID 47711, 2020.

161. Xiao, X., Hu, S., Zhai, J., Chen, T., Mai, Y., Thermal properties and combustion behaviors of flame-retarded glass fiber-reinforced polyamide 6 with piperazine pyrophosphate and aluminum hypophosphite, *J. Therm. Anal. Calorim.*, 125, 175, 2016.

162. Y. Ni, Flame-retardant copolyetherester composition and articles comprising the same, PCT Patent Application WO 2013/085724, to Du Pont, June 13, 2013.

163. U. Zucchelli and M. Rosichelli, Environmental friendly flame retardant moulding compositions based on thermoplastic impact modified styrenic polymers, US Patent 10119019, assigned to Italmatch, November 6, 2018.

164. F. de Campo, A. Murillo, J. Li, T. Zhang, Flame retardant polymer compositions comprising stabilized hypophosphite salts, PCT Patent Application WO 2012/113145, assigned to Rhodia, August 30, 2012.

165. Y. Hirschsohn and E. Eden, Flame-retarded styrene-containing formulations, PCT Patent Application WO 2018/178985, assigned to Bromine Compounds, October 4, 2018.

166. Y. Hirschsohn, E. Eden, Y. Epstein Assor, S. Dichter, Flame-retarded polyester compositions, PCT Patent Application WO 2018/073818, assigned to Bromine Compounds, April 26, 2018.

167. S.R. Sandler, Polyester resins flame retarded by poly(metal phosphinate)s, US Patent 4180495, assigned to Pennwalt, December 25, 1979.

168. W. Herwig, H.-J. Kleiner, H.-D. Sabel, Flame-retarding agents and their use in the preparation of fire-proof thermoplastic polymers, European Patent Application 0006568, assigned to Hoechst, January 9, 1980.

169. H.-J. Kleiner and W. Budzinsky, Flameproofed polyester molding composition, US Patent 5780534, assigned to Ticona, July 14, 1998.

170. H.-J. Kleiner, W. Budzinsky, G. Kirsch, Low-flammability polyamide molding materials, US Patent 5773556, assigned to Ticona, June 30, 1998.

171. N. Weferling, H.-P. Schmitz, G. Kolbe, Process for the alkylation of phosphorus-containing compounds, US Patent 6248921, assigned to Clariant, June 19, 2001.

172. E. Jenewein, H.-J. Kleiner, W. Wanzke, W. Budzinsky, Synergistic flame protection agent combination for thermoplastic polymers, US Patent 6365071, assigned to Clariant, April 2, 2002.

173. E. Schlosser, B. Nass, W. Wanzke, Flame-retardant combination, US Patent 6255371, assigned to Clariant, July 3, 2001.

174. Dietz, M., Hoerold, S., Nass, B., Schacker, O., Schmitt, E., Wanzke, W., New environmentally friendly phosphorus based flame retardants for printed circuit boards as well as polyamides and polyesters in E&E applications, in: *Proc. Conf. Electronics Goes Green*, pp. 771–776, Berlin, September 2004.

175. Hoerold, S., Safety for thermoplastics, *Specialty Chem. Magazine*, p. 28, November 2008.

176. Naik, A.D., Fontaine, G., Samyn, F., Delva, X., Bourgeois, Y., Bourbigot, S., Melamine integrated metal phosphates as non-halogenated flame retardants: synergism with aluminium phosphinate for flame retardancy in glass fiber reinforced polyamide 66, *Polym. Degrad. Stab.*, 98, 2653, 2013.

177. W. Krause, H. Bauer, M. Sicken, S. Hoerold, W. Wanzke, Schlosser, Flame retardant-stabilizer combination for thermoplastic polymers, European Patent 2625220, assigned to Clariant, November 9, 2016.

178. Oyama, R., Shimomichi, H., Yamanaka, T., Development of aluminum phosphite and Its application, *Phos. Res. Bull.*, 32, 10, 2016.

179. Y. Saga and W.W., Zhang, Flame resistant semiaromatic polyamide resin composition and articles therefrom, European Patent 2414446, assigned to DuPont, February 13, 2013.

180. Tomiak, F., Schartel, B., Wolf, M., Drummer, D., Particle size related effects of multi-component flame-retardant systems in poly(butadiene terephthalate), *Polymers*, 12, 1315, 2020.

181. M. Bolourchi and S. Kochesfahani, Flame-retardant polymer composition, PCT Patent Application WO 2018/187638, assigned to Imerys Talc, October 11, 2018.

182. P.K. Borade, M. Elkovich, J. Pal, Flame retardant poly(arylene ether)/poly-amide compositions, methods, and articles, US Patent 7592382, assigned to Sabic, September 22, 2009.

183. Walz, R. and Baque, T., Metal Phosphinates in Industrial Applications, *Kunststoffe*, p. 113, December 2007.

184. Sullalti, S., Colonna, M., Berti, C., Fiorini, M., Karanam, S., Effect of phosphorus based flame retardants on UL94 and comparative tracking index properties of poly(butylene terephthalate), *Polym. Degrad. Stab.*, 97, 566, 2012.

185. W. Wanzke, Non halogen FR solutions for polyesters in E&E applications – the use of phosphinates, Paper presented at *AMI Conf. Fire Retardants in Plastics*, Denver, CO, June 2012.

186. Braun, U., Schartel, B., Fichera, M.A., Jaeger, C., Flame retardancy mechanisms of aluminium phosphinate in combination with melamine polyphosphate and zinc borate in glass-fibre reinforced polyamide 6, 6, *Polym. Degrad. Stab.*, 92, 1528, 2007.

187. Braun, U. and Schartel, B., Flame retardancy mechanisms of aluminium phosphinate in combination with melamine cyanurate in glass-fibre-reinforced poly(1,4-butylene terephthalate), *Macromol. Mater. Eng.*, 293, 206, 2008.

188. Seefeldt, H., Duemichen, E., Braun, U., Flame retardancy of glass fiber reinforced high temperature polyamide by use of aluminum diethylphosphinate: thermal and thermo-oxidative effects. *Polym. Int.*, 62, 1608, 2013.

189. Q. Yao, S.V. Levchik, G.R. Alessio, Phosphorus-containing flame retardant for thermoplastic polymers, US Patent 7807737, assigned to ICL-IP, October 5, 2010.

190. Zhao, B., Chen, L., Long, J.-W., Chen, H.-B., Wang, Y.-Z. Aluminum hypophosphite versus alkyl-substituted phosphinate in polyamide 6: flame retardance, thermal degradation, and pyrolysis behavior, *Ind. Eng. Chem. Res.*, 52, 2875, 2013.

191. Ding, Y., Stoliarov, S.I., Kraemer, R.H., Pyrolysis model development for a polymeric material containing multiple flame retardants: relationship between heat release rate and material composition, *Comb. Flame*, 202, 43, 2019.

192. Brehme, S., Koeppl, T., Schartel, B., Altstaedt, V., Competition in aluminium phosphinate-based halogen-free flame retardancy of poly(butylene terephthalate) and its glass-fibre composites, *e-Polymers*, 14, 193, 2014.

193. W. Qiu, Flame-retardant poly(arylene ether) composition and its use as a covering for coated wire, US Patent 7589281, assigned to Sabic, September 15, 2009.

194. Y. Araki, T. Sato, T. Hisasue, Flame-retardant resin composition, European Patent 2048198, assigned to Asahi Kasei, January 4, 2012.

195. S.H. Lim, N.J. Park, S.H. Lee, S.M. Lee, J.K. Choi, S.W. Na, Poly (arylene ether) flame retardant resin composition and non-crosslinked flame retardant cable, US patent 9631091, assigned to LG Chem, April 25, 2017.

196. E. Karayianni and J.-M. Philippoz, Flame retardant thermoplastic elastomer compositions, US Patent 7790790, assigned to Du Pont, September 7, 2010.

197. C.F. Fitie, Flame retardant composition and insulated wires for use in electronic equipment, PCT Patent Application, WO 2019/063554, assigned to DSM, April 4, 2019.

198. E. Karayianni, T. Li, Y. Ni, Fire-retardant copolyetherester composition and articles comprising the same, European Patent 2892956, to DuPont, June 14, 2017.

199. A. Vanhalle, H. Verbeke, S.M.L. Meynen, Flame retardant thermoplastic composition, US Patent 8872034, assigned to Huntsman, October 28, 2014.

200. S. Knop, M. Sicken, S. Hoerold, Flame retardant formulation, US Patent 7332534, assigned to Clariant, February 19, 2008.

201. S. Hoerold, Flame-retarding thermosetting compositions, US Patent 6420459, assigned to Clariant, July 16, 2002.

202. S. Knop, M. Sicken, S. Hoerold, Flame retardant duroplastic masses, European Patent Application 1403309, assigned to Clariant, March 31, 2004.

203. M. Sicken, S. Knop, S. Hoerold, H. Bauer, Flame retardant dispersion, US Patent 7273901, assigned to Clariant, September 25, 2007.

204. T.A. Koes and O. Hajjar, Low loss prepregs, compositions useful for the preparation thereof and uses therefor, US Patent 7364672, assigned to Arlon, April 29, 2008.

205. M. Sicken, S. Knop, S. Hoerold, H. Bayer, Flame retardant dispersion, European patent 1454949, assigned to Clariant, July 6, 2016.

206. H. Ushiyama and K. Watanabe, Epoxy resin composition, molded article, prepreg, fiber-reinforced composite material and structure, US Patent 10513577, assigned to Mitsubishi Chemical, December 24, 2019.

207. D. Bedner and W. Varnell, Flame retardant compositions with a phosphorated compound, US Patent 8129456, assigned to Isola, March 6, 2012.

208. T.-C. Liao, D.-R. Fung, Y.-T. Huang, H.-S. Chen, H.-Y. Chang, C.-L. Liu, Thermosetting resin composition and prepreg as well as hardened product using the same, US Patent 10023707, assigned to Nan Ya, July 17, 2018.

209. N. Yao, R. Wang, Z. Ma, Y. Zhang, B. Li, Z. Shang, M. Yuan, Resin composition and articles made therefrom, US Patent 10626251, assigned to Elite, April 21, 2020.

210. M. Sicken, E. Schlosser, W. Wanzke, D. Burkhardt, Fusible zinc phosphinate, European Patent Application 1454912, assigned to Clariant, September 8, 2004.

211. Didane, N., Giraud, S., Devaux, E., Lemort, G., Capon, G., Thermal and fire resistance of fibrous materials made by PET containing flame retardant agents, *Polym. Degrad. Stab.*, 97, 2545, 2012.

212. B. Messmore, P.E. Rollin jr., K.R. Samat, J.-C. Chang, Flame retardant poly(trimethylene) terephthalate compositions and articles made therefrom, European Patent 2582756, assigned to DuPont, March 28, 2018.
213. D.M. Sarzotti, T.E. Schmitt, A.W. Briggs, Flame retardant fibers, yarns, and fabrics made therefrom, US Patent 10640893, assigned to Invista, May 5, 2020.
214. B. Muessig, Wrapping tape having a film made of TPU, PCT Patent Application WO 2008/151897, assigned to Tesa, December 18, 2008.
215. Reuter, J., Greiner, L., Schoenberger, F., Doering, M., Synergistic flame retardant interplay of phosphorus containing flame retardants with aluminum trihydrate depending on the specific surface area in unsaturated polyester resin, *J. Appl. Polym. Sci.*, 136, 47270, 2019.
216. Buszard, D.L. and Dellar, R.J., The performance of flame retardants in rigid polyurethane foam formulations, in: *Fire and Cellular Polymers*, J.M. Buist, S.J. Grayson, W.D. Woolley (Eds.), pp. 265–288, Elsevier, London, 1984.
217. J.-G. Hansel, G. Jabs, J. Kaulen, H.-G. Adams, H.G. Froehlen, Process for preparing dimethyl propanephosphonate, US Patent 7119220, assigned to Lanxess, October 10, 2006.
218. J. Xu, Y.D. Qi, L. Lotti, X. Tai, Use of trialkyl phosphate as a smoke suppressant in polyurethane foam, European Patent 27560928, assigned to Dow, December 26, 2018.
219. L. Boehnke, D. Achten, H.-J. Lass, A.-C., Bijlard, Translucent polyurethane or polyisocyanurate foams, PCT Patent Application WO 2020/070140, assigned to Covestro, April 9, 2020.
220. Tebbe, H. and Sawaya, J., Comparison of different phosphorus based flame retardants in rigid polyurethane foam for the production of insulation materials, in: *Proc. CPI Tech. Conf. Polyurethanes 2011*, Nashville TN, September 2011.
221. Lorenzetti, A., Modesti, M., Besco, S., Hrelja, D., Donadi, S., Influence of phosphorus valency on thermal behaviour of flame retarded polyurethane foams, *Polym. Degrad. Stab.*, 96, 1455, 2011.
222. Chung, Y.J., Kim, Y., Kim, S., Flame retardant properties of polyurethane produced by the addition of phosphorous containing polyurethane oligomers (II), *J. Ind. Eng. Chem.*, 15, 888, 2009.
223. E.R., Larsen, Ecker, E.L., On the use of phosphorus as an FR-adjunct with halogenated unsaturated polyesters, *J. Fire Retardant Chem.*, 6, 182, 1979.
224. Hernangil, A., Rodriguez, M., Leon, L.M., Ballestero, J., Alonso, J.R., Experimental design of fire-retarded formulations. Low viscosity polyester resins, *J. Fire Sci.*, 17, 281, 1999.
225. T.J. Cebasek, M. Gruskiewicz, S.T. Searl, Flame and smoke spread retardant molding compounds and components molded from these compounds, US Patent 8487040, assigned to Premix, July 6, 2013.

226. Y.-N. Cheng, S.-J. Chang, Y.-C. Shen, S.-P. Juang, Process for preparing flame-retardant phosphorus-containing unsaturated polyester, US Patent 5571888, November 5, 1996.

227. Zhang, C., Liu, S.-M., Huang, J.-Y., Zhao, J.-Q., The synthesis and flame retardance of a high phosphorus-containing unsaturated polyester resin, *Chem. Lett.*, 39, 1270, 2010.

228. Mauerer, O., Organophosphorus compounds in respect of increasing fire safety requirements, *PU Magazine*, p. 234, April 2007.

229. J.-G. Hansel and O. Mauerer, Flame-retardant, curable molding materials, US Patent Application 2011/0028604, assigned to Lanxess, February 3, 2011.

230. Creyf H. and Fishbein J., Advance in Flexible Foam Technology, in: Fire and Cellular Polymers, J.M. Buist, S.J. Grason, W.d. Woolley (Eds.), pp. 279-288, Elsevier, London, 1984

231. Hirschler, M.M., Flame retardants and heat release: review of traditional studies on products and on groups of polymers, *Fire Mater.*, 39, 207, 2015.

232. Hirschler, M.M., Flame retardants and heat release: review of data on individual polymers, *Fire Mater.*, 39, 232, 2015.

233. Krasny, J.F., Parker, W.J., Babrauskas, V., *Fire Behavior of Upholstered Furniture and Mattresses*, Noyes Publications, Park Ridge, NJ, 2001.

234. Morgan, A.B., Revisiting flexible polyurethane foam flammability in furniture and bedding in the United States, *Fire Mater.,* 2020 https://doi.org/10.1002/fam.2848.

235. Kraemer, R.H., Zammarano, M., Linteris, G.T., Gedde, U.W., Gilman, J.W., Heat release and structural collapse of flexible polyurethane foam, *Polym. Degrad. Stab.*, 95, 1115, 2010.

236. Levchik, S.V., New developments in flame retardant polyurethanes, in: *Proc. BCC Conf. Recent Advances in Flame Retardancy of Polymeric Materials*, Stamford, CT, June 2003.

237. L.L. Bradford, E. Pinzoni, B. Williams, T. Halchak, Polyurethane foam containing flame retardant blend of non-oligomeric and oligomeric flame retardants, European Patent 1218433, assigned to Akzo Nobel, August 13, 2003.

238. J.-G. Hansel and H. Tebbe, Halogen-free poly(alkylene phosphate), European Patent 2687535, assigned to Lanxess, September 14, 2016.

239. J.-G. Hansel and H. Tebbe, Phosphoric acid ester compositions with reduced hygroscopicity, European Patent 2860211, assigned to Lanxess, July 10, 2019.

240. Luda, M.P., Bracco, P., Costa, L., Levchik, S.V., Discolouration in fire retardant flexible polyurethane foams. Part I. characterisation, *Polym. Degrad. Stab.*, 83, 215, 2004.

241. Levchik, S.V., Luda, M.P., Bracco, P., Nada, P., Costa, L., Discoloration in fire-retardant flexible polyurethane foams, *J. Cellular. Plast.*, 41, 235, 2005.

242. B. Williams and L. de Kleine, Blend of organophosphorus flame retardant, lactone stabilizer, and phosphate compatibilizer, US Patent 7122135, assigned to Supresta, October 17, 2006.

243. Wang, X.-L., Yang, K.-K., Wang, Y.-Z., Physical and chemical effects of diethyl N,N'-diethanolaminomethyl phosphate on flame retardancy of rigid polyurethane foam, *J. Appl. Polym. Sci.*, 82, 276, 2001.

244. C.L. Harowitz, Flame-resistant polyurethanes prepared from certain phorphorus compounds, US Patent 3525705, assigned to Mobil Oil, August 25, 1970.

245. J.K. Stowell, G. Francisco, E. Weil, Method of making hydroxymethylphosphonate, polyurethane foam-forming compositions, polyurethane foam and articles made therefrom, US Patent 9593220, assigned to ICL-IP America, May 23, 2017.

246. S. Snider, L. Wang, J. Asrar, Open cell spray fire-retardant foam, US Patent 10118985, assigned to Johns Manville, November 6, 2018.

247. S. Snider, L. Wang, J. Asrar, Wall insulation boards with non-halogenated fire retardant and insulated wall systems, US Patent 9523195, assigned to Johns Manville, December 20, 2016.

248. M. Sicken and H. Staendeke, Flame-resistant polyurethanes, US Patent 5985965, assigned to Clariant, November 16, 1999.

249. T. Dreier, R. Roers, M. Gossner, S. Meyer-Ahrens, Polyetherester as flame retardant additive for polyurethane flexible foams, European Patent 1555275, assigned to BASF, May 7, 2008.

250. C. Eilbacht and M. Sicken, Process for producing flame-retardant flexible polyurethane foams, US Patent 6380273, assigned to Clariant, April 30, 2002.

251. Tokuyasu, N., Fujimoto, K., Hamada, T., Halogen free flame retardant technology for polyurethane foam, in: *Proc. API Conf. Polyurethanes Expo*, Las Vegas, 2004, pp. 118–127, October 2004.

252. N. Tokuyasu and T. Matsumura, Phosphoric ester, process for preparing the same and use thereof, US Patent 6127464, assigned to Daihachi, October 3, 2000.

253. Wu, W. and Yang, C.Q., Correlation between limiting oxygen index and phosphorus/nitrogen content of cotton fabrics treated with a hydroxy-functional organophosphorus flame retarding agent and dimethyloldihydroxyethyleneurea, *J. Appl. Polym. Sci.*, 90, 1885, 2003.

254. Wu, W., Zhen, X., Yang, C.Q., Correlation between limiting oxygen index and phosphorus content of the cotton fabric treated with a hydroxy-functional organophosphorus flame retarding finish and melamine-formaldehyde, *J. Fire Sci.*, 22, 11, 2004.

255. Yang, H., Yang, C.Q., He, Q., The bonding of a hydroxy-functional organophosphorus oligomer to nylon fabric using the formaldehyde derivatives of urea and melamine as the bonding agents, *Polym. Degrad. Stab.*, 94, 1023, 2009.

256. Yang, C.Q. and Chen, Q., Flame retardant finishing of the polyester/cotton blend fabric using a cross-linkable hydroxy-functional organophosphorus oligomer, *Fire Mater.*, 43, 283, 2019.

257. Yang, H. and Yang, C.Q., Nonformaldehyde flame retardant finishing of the nomex/cotton blend fabric using a hydroxy-functional organophosphorus oligomer, *J. Fire Sci.*, 25, 425, 2007.

258. Padda, R. and Lenotte, G., General trends in textile flame retardants, *Specialty Chem. Magazine*, p. 43, September 2005.

259. Lam, Y.L., Kan, C.W., Yuen, C.W.M., Effect of titanium dioxide on the flame-retardant finishing of cotton fabric, *J. Appl. Polym. Sci.*, 121, 267, 2011.

260. Kapura, A., Chemistry of flame retardants: II. NMR and HPLC analysis of PYROVATEX CP, *J. Fire Sci.*, 12, 3, 1994.

261. Horrocks, A.R. and Roberts, D., Minimization of formaldehyde emission, in: *Proc. Conf. Ecotextile'98: Sustainable Development*, Woodhead Publishing Ltd., Cambridge, UK, 1999.

262. R. Malinowsky, M. Neunteufel, M. Crnoja-Cosic, C. Bisjak, D. Eichinger, C. Schrempf, Flame retardant lyocell filament, PCT Patent Application WO 2019/068927, assigned to Lenzing Aktiengesel-lschaft, April 11, 2019.

263. DeStio, P., Flame retardant polyesters. *Proc. BCC Conf. Recent Advances in Flame Retardancy of Polymeric Materials*, Stamford, CT, June 1991.

264. M.H. Schindzielorz and A. Kokeguchi, Flame resistant fabric for aviation air-bags, European Patent 3146100, assigned to Schroth Safety Products, August 14, 2019.

265. H.B. Chew, R.R. Joshi, A.G. Mack, A. Ibay, Flame retarded rigid polyure-thane foams and rigid polyurethane foam formulations, European Patent 2178955, assigned to Albemarle, November 9, 2011.

266. J.K. Stowell, A. Piotrowski, M. Nagridge, G. Symes, Flame retardant additive composition comprising cyclic phosphonate blend and bis-phosphate ester, and polyurethane foam containing the same, US Patent 10144872, assigned to ICL-IP America, December 4, 2018.

267. A. Lambert, A. Ponnouradjou, L. Leite, J.-E. Zanetto, Synthetic, flame-resistant yarns, fibres and filaments, US Patent 7758959, assigned to Rhodia, July 20, 2010.

268. V. Butz, Flame-retardant polyamide moulding materials, European Patent 2443192, assigned to Thor, April 17, 2013.

269. V. Butz, Flame-retardant composition comprising a phosphonic acid deriva-tive, US Patent 8349925, assigned to Thor, January 8, 2013.

270. T. Claessen, R. Willemse, A. Sigrist, U. Scholbe, J. Ebert, Flame-retardant polyolefin foam and its production, European Patent 2666626, assigned to Sekisui Alveo, December 31, 2014.

271. S.M. Andrews and T.F. Thomson, Flame retardant polyolefin articles, PCT Patent Application WO 2017/013028, assigned to BASF, January 26, 2017.

272. V. Butz, Materials having a fire-resistant coating containing a spirophospho-nate, PCT Patent Application WO2017/080633, assigned to Thor, May 18, 2017.

273. Goedderz, D., Weber, L., Markert, D., Schiesser, A., Fasel, C., Riedel, R., Altstaedt, V., Bethke, C., Fuhr, O., Puchtler, F., Breu, J., Doering, M., Flame retardant polyester by combination of organophosphorus compounds and an NOR radical forming agent, *J. Appl. Polym. Sci.*, 137, ID 47876, 2020.

274. K. Yamanaka, K. Koya, Y. Taketani, Flame-resistant resin composition and article molded therefrom, US Patent 7169837, assigned to Teijin, Janury 30, 2007.

275. K. Yamanaka, F. Kondo, K. Toyohara, Flame-retardant resin composition and molding derived therefrom, European Patent 2287253, assigned to Teijin, January 1, 2016.

276. K. Yamanaka and F. Kondo, Flame retardant resin composition and molded article thereof, US Patent 8859655, assigned Teijin, October 14, 2014.

277. K. Yamanaka, T. Miyake, M. Saito, M. Kinoshita, Flame-retardant resin composition and molded products thereof, European Patent 2511338, assigned to Teijin, April 27, 2016.

278. K. Yamanaka, T. Takeda, K. Imazato, Flame-retardant resin composition and molded article produced from same, European Patent 3249010, assigned to Teijin, June 3, 2020.

279. K. Yamanaka, K. Imazato, N. Takahashi, Organophosphorus compound, flame retardant, and flame-retardant product including same, European Patent 3514151, assigned to Teijin, May 13, 2020.

280. K. Yamanaka, T. Takeda, K. Kondo, M. Haruyoshi, Flame-retardant synthetic leather, European Patent 2860309, assigned to Teijin, May 10, 2017.

281. Wolf, R., Flame retardant viscose rayon containing a pyrophosphate, *Eng. Chem. Prod. Res. Dev.*, 20, 413, 1981.

282. G. Kroner, Fire-retardant cellulose fiber, use thereof, and method for the production thereof, European Patent 2473657, assigned to Lenzing, April 30, 2014.

283. W.R. Haaf and D.L. Reinhard, Flame retardant, non-dripping compositions of polyphenylene ether and acrylonitrile-butadiene-styrene, US Patent 4107232, assigned to General Electric, August 15, 1978.

284. K. Okada, Y. Maeda, R. Motoshige, M. Noro, Flame-retardant thermoplastic resin composition, US Patent 6071992, assigned to Techno Polymer, June 6, 2000.

285. Touchette, N.W., Flame-Retardant Plasticizers, in: *Handbook of Polyvinyl Chloride Formulating*, E.J. Wickson (Ed.), pp. 275–302, Wiley, New York, 1993.

286. F. Hupka, T. Hagen, J. Nordmann, J. Gaca, R. Kraemer, S. Menon, High temperature resistant foams having a high flame resistance, European patent 3259295, assigned to Covestro and BASF, December 18, 2019.

287. R. Nuetzel, Application device for a solid-filled PU foam, European Patent 3137391, assigned to Sika Technology, January 22, 2020.

288. H.M. Jung, J.C. Cho, C.K. Lee, Flame retardant resin composition for multilayer wiring board and multilayer wiring board including the same, US Patent 8309636, assigned to Samsung, November 13, 2012.

289. S. Elgimiabi, Thermal cycling resistant low density composition, European Patent 3489271, Assigned to 3M, May 27, 2020.

290. Heaps, J.M., Phosphate esters plasticizers for flexible PVC, *Plast. (London)*, 33, 366, 410, 1969.

291. D.R. Randell and W. Pickles, Polyvinyl chloride plasticized with mixed phosphate esters, US Patent 3919158, assigned to Ciba-Geigy, November 11, 1975.

292. Duke, A.J., Inhibition of human androgen receptor by bis(2-isopropylphenyl) phenyl phosphate, *Chimia*, 32, 457, 1978.

293. Nobile, E.R., Page, S.W., Lombardo, P., Characterization of four commercial flame retardant aryl phosphates, *Bull. Envir. Contam. Toxicol.*, 25, 755, 1980.

294. Shankwalkar, S.G. and Placek, D.G., Oxidation and weight loss characteristics of commercial phosphate esters, *Ind. Eng. Chem. Res.*, 31, 1810, 1992.

295. D.A. Bright, B. Williams, E. Pinzoni, E.D. Weil, Flame retardant composition and polyurethane foams containing same, PCT Patent Application WO2006/119369, assigned to Supresta, November 9, 2006.

296. R.S. Rose, D.L. Buszard, M.D. Philips, F.J. Liu, Higher alkylated triaryl phosphate ester flame retardants, European Patent EP 1421139, assigned to Pabu Services, October 21, 2009.

297. J.K. Stowell, L.A. De Kleine, B.A. Williams, Non-halogen flame retardant additives for use in rigid polyurethane foam, European Patent 1973965, assigned to Supresta, August 26, 2009.

298. A. Adedeji, G. Riding, B. Torrey, High performance plastic pallets, US Patent 7273014, assigned to Sabic, September 25, 2007.

299. K. Kosaka, X. Li, V. Mhetar, J. Tenenbaum, W. Yao, Flexible poly(arylene ether)composition and articles thereof, US patent 7517927, assigned to Sabic, April 14, 2009.

300. J.H. Shin, S.-J. Yang, Y.K. Chang, Flameproof thermoplastic resin compositions, US Patent 6593404, assigned to Samsung Cheil, July 15, 2003.

301. D.M. Singh and D.M. Singh, High heat resistant composition, US Patent 9447291, assigned to Akzo Nobel Coatings, September 20, 2016.

302. Moy, P.Y., Formulating vinyl for flame resistance, in: *Handbook of Vinyl Formulating*, R.F. Grossman (Ed.), pp. 287–304, Wiley Interscience, Hoboken, NJ, 2008.

303. Moy, P., Aryl phosphate ester fire-retardant additive for low-smoke vinyl applications, *J. Vinyl Add. Technol.*, 10, 187, 2004.

304. S. Cron, Polyurethane/polyurea-forming compositions, US Patent 6369156, assigned to Huntsman, April 9, 2002.

305. K. Call and R. Ruprecht, Low softener halogen free flame retardant styrenic block copolymer-based thermoplastic elastomer compositions, US Patent 9156978, assigned to Teknor Apex, October 13, 2015.

306. Bright, D.A., Dashevsky, S., Moy, P.Y., Williams, B., J. Vinyl., Resorcinol bis(diphenyl phosphate), a non-halogen flame-retardant additive, *Add. Technol.*, 3, 170, 1997.

307. J.C. Gosens, C.F. Pratt, H.B. Savenije, C.A.A. Claesen, Polymer mixture having aromatic polycarbonate, styrene-containing copolymer and/or graft polymer and a flame-retardant, articles formed therefrom, US Patent 5204394, assigned to GE, April 20, 1993.

308. N. Agarwal, S.K. Gaggar, D. Gupta, S. Gupta, R. Krishnamurthy, R. Kumaraswamy, R.S. Totad, S. Tyagi, Polymer compositions, method of manufacture, and articles formed therefrom, US Patent 7557154, assigned to Sabic, July 7, 2009.

309. T. Kitahara, K. Hosokawa, T. Shimizu, Polytetrafluoroethylene fine powders and their use, US patent 6503988, assigned to Daikin Industries, January 7, 2003.

310. T. Eckel, A. Seidel, J. Gonzalez-Blanco, D. Wittmann, Flame-resistant polycarbonate molding composition modified with a graft polymer, US Patent 7319116, assigned to Bayer, January 15, 2008.

311. R.J. Wroczynski, Flame resistant compositions of polycarbonate and monovinylidene aromatic compounds, European Patent 0909790, assigned to General Electric, January 4, 2004.

312. O. Kenichi, Phosphoric ester flame retardant, Japanese Patent 3043694, assigned to Asahi Chemical, May 22, 2000.

313. Levchik, S.V., Bright, D.A., Moy, P., Dashevsky, S., New developments in fire retardant non-halogen aromatic phosphates *J. Vinyl. Add. Technol.*, 6, 123, 2000.

314. E.W. Burkhardt, D.A. Bright, S. Levchik, S. Dashevsky, M. Buczek, Epoxy-stabilized polyphosphate compositions, US Patent 6717005, assigned to Akzo Nobel, April 6, 2004.

315. A. Seidel, T. Eckel, D. Wittmann, D. Kurzidim, Impact-modified polycarbonate blends, US Patent 7067567, assigned to Bayer, June 27, 2006.

316. A. Imada, Resin composition and resin molded article, US Patent 9562157, assigned to Fuji Xerox, February 7, 2017.

317. X. Li, Flame retardant thermoplastic molding composition, US Patent 8217101, assigned to Bayer, June 10, 2012.

318. I.J. Kim, K.H. Kwon, B.N. Jang, Scratch-resistant flameproof thermoplastic resin composition with improved compatibility, European Patent 2222786, assigned to Cheil, January 2, 2013.

319. H. Warth, S. Chen, Y. Wang, Y. Wang, H. Li, Blend of aromatic polycarbonate and polylactic acid, the method for preparing the same and the use thereof, US Patent 8242197, assigned to Bayer, August 14, 2012.

320. J. Liu, Heat resistant, flame retardant polylactic acid compounds, US Patent 9309403, assigned to PolyOne, April 12, 2016.

321. P.Y. Moy, A.I. Gregor, R.L. Pirelli, D.A. Bright, L. Bright, Oligomeric bisphosphate flame retardants and compositions containing the same, EP 2089402, assigned to ICL-IP America, August 5, 2015.

322. R.W. Avakian and L. Hu, Non-halogenated flame retardant polycarbonate compounds, PCT Patent Application WO 2014/018672, assigned to PolyOne, January 30, 2014.

323. T. Matsumura and Y. Tanaka, Crystalline powders of aromatic diphosphates, their preparation and use, European Patent 0509506, assigned to Daihachi, June 25, 1997.

324. M. Katayama, M. Ito, Y. Otsuka, Polycarbonate resin composition containing block copolymer, US Patent 6316579, assigned to Daicel, November 13, 2001.

325. K. Tabushi and B. Mori, Flame-retardant and electromagnetic interference attenuating thermoplastic resin composition, European Patent 1336645, assigned to Nippon, April 1, 2009.

326. H. Sakurai, G. Kokura, T. Kamimoto, Flame-retardant synthetic resin composition, European Patent 3037483, assigned to Adeka, September 6, 2017.

327. T. Kamimoto, Y. Tanaka, T. Tezuka, H. Muraze, Process for solidification of phosphoric ester flame retardants, European Patent 2292688, assigned to Adeka, May 11, 2016.

328. H.-J. Jeong, M.-H. Kim, S. Shin, Resin composition exhibiting good heat resistance and insulation properties, and product using same, European Patent 3351588, assigned to Lotte, February 19, 2020.

329. Pfaender, R. and Doering, M., Eco-friendly fire prevention. New developments in flame retardants, *Kunstoffe Int.*, p. 48, August 2014.

330. F. Sato, H. Sakurai, G. Kokura, N. Tanji, Flame retardant composition, flame retardant resin composition containing said flame retardant composition, and molded body of said flame retardant resin composition, European Patent Application 3690006, assigned to Adeka, August 5, 2020.

331. Pawlowski, K.H. and Schartel, B., Flame retardancy mechanisms of triphenyl phosphate, resorcinol bis(diphenyl phosphate) and bisphenol A bis(diphenyl phosphate) in polycarbonate/acrylonitrile-butadiene-styrene blends, *Polym. Int.*, 56, 1404, 2007.

332. Schartel, B., Pawlowski, K.H., Perret, B., Flame retardancy mechanisms in halogen-free PC/ABS blends, *Polym. Mater., Sci. Eng.*, 98, 245, 2008.

333. Murashko, E.A., Levchik, G.F., Levchik, S.V., Bright, D.A., Dashevsky, S., Fire retardant action of resorcinol bis(diphenyl phosphate) in PC-ABS blend. II. Reactions in the condensed phase, *J. Appl. Polym. Sci.*, 71, 1863, 1999.

334. Perret, B., Pawlowski, K.H., Schartel, B., Fire retardancy mechanisms of arylphosphates in polycarbonate (PC) and PC/acrylonitrile-butadiene-styrene: The key role of decomposition temperature, *J. Therm. Anal. Calorim.*, 97, 949, 2009.

335. Despinasse, M.-C. and Schartel, B., Aryl phosphate-aryl phosphate synergy In flame-retarded bisphenol A polycarbonate/acrylonitrile-butadiene-styrene, *Thermochim. Acta*, 563, 51, 2013.

336. Despinasse, M.-C. and Schartel, B., Influence of the structure of aryl phosphates on the flame retardancy of polycarbonate/acrylonitrile-butadiene-styrene. *Polym. Degrad. Stab.*, 97, 2571, 2012.
337. Schartel, B., Richter, K.H., Boehning, M., Synergistic use of talc in halogen-free flame retarded polycarbonate/acrylonitrile-butadiene-styrene blends, in: *Fire and Polymers VI: New Advances in Flame Retardant Chemistry and Science, ACS Symposium Series*, vol. 1118, pp. 15–36, Washington DC, 2012.
338. Wang, W., Zhao, G., Wu, X., Li, X., Wang, C., Investigation on phosphorus halogen-free flame-retardancy systems in short glass fiber-reinforced PC/ABS composites under rapid thermal cycle molding process condition, *Polym. Compos.*, 36, 1653, 2015.
339. Jachowicz, J., Kryszewski, M., Sobol, A., Thermal degradation of poly(2-methylphenylene oxide), poly(2,5-dimethylphenylene oxide) and poly(1,4-phenylene oxide), *Polymer*, 20, 995, 1979.
340. V. Abolins and F.F. Holub, Polyphenylene ether resin compositions having improved ductile impact strength, US Patent 4504613, assigned to General Electric, March 12, 1985.
341. Lin, S., Sun, S., He, Y., Wang, X., Wu, D., Effects of phosphate and polysiloxane on flame retardancy and impact toughening behavior of poly(2,6-dimethyl-1,4-phenylene oxide), *Polym. Eng. Sci.*, 52, 925, 2012.
342. G. Solenicki and K. Sharma, Rail interior compliant thermoplastic composite, European Patent Application 3659796, assigned to Sabic, June 3, 2020.
343. G. Chen, Y. Na, Y. Xu, Water pipe for mining operations, European Patent 3362517, assigned to Sabic, July 15, 2020.
344. Jiang, Z., Liu, S., Zhao, J., Chen, X., Flame-retarded mechanism of SEBS/PPO composites modified with mica and resorcinol bis(diphenyl phosphate), *Polym. Degrad. Stab.*, 98, 2765, 2013.
345. H. Kubo and S. Sato, Resin composition for wire and cable covering material, European Patent 1563008, assigned to General Electric, October 5, 2006.
346. H. Kubo, V.R. Mhetar, V. Rajamani, S. Sato, X. Tai, W. Yao, Flame retardant electrical wire, European Patent 1829056, assigned to General Electric, October 15, 2008.
347. K. Kosaka, X. Li, Y. Xiao, Flame retardant thermoplastic composition and articles comprising the same, US Patent 7799854, assigned to Sabic, September 21, 2010.
348. S.R. Klei, J.E. Pickett, J.J. Xu, Poly(arylene ether) composition, method, and article, US Patent 7576150, assigned to Sabic, August 18, 2009.
349. T. Harada, V. Rajamani, S. Sato, X. Tai, W. Yao, Thermoplastic composition, coated conductor, and methods for making and testing the same, US Patent 7504585, assigned to Sabic, March 17, 2009.
350. H. Gan, Poly(phenylene ether) composition and article, European Patent 3317350, assigned to Sabic, January 9, 2019.

351. Boscoletto, A.B., Checchin, M., Milan, L., Pannocchia, P., Tavan, M., Camino, G., Luda, M.P., Combustion and fire retardance of poly(2,6-dimethyl-1,4-phenylene ether) - high-impact polystyrene blends. II. Chemical aspects, *J. Appl. Polym. Sci.*, 67, 2231, 1998.

352. Boscoletto, A.B., Checchin, M., Milan, L., Camino, G., Costa, L., Luda, M.P., Mechanism of fire retardance in poly(2,6-dimethyl-1,4-phenylene ether) – high impact polystyrene, *Makromol. Chem., Macromol. Symp.*, 74, 35, 1993.

353. Liu, S., Jiang, L., Jiang, Z., Zhao, J., Fu, Y., The impact of resorcinol bis(diphenyl phosphate) and poly(phenylene ether) on flame retardancy of PC/PBT blends, *Polym. Adv. Technol.*, 22, 2392, 2011.

354. Takeda, K., Amemiya, F., Kinoshita, M., Takayama, S., Flame retardancy and rearrangement reaction of polyphenylene-ether/polystyrene alloy, *J. Appl. Polym. Sci.*, 64, 1175, 1997.

355. Murashko, E.A., Levchik, G.F., Levchik, S.V., Bright, D.A., Dashevsky, S., Fire retardant action of resorcinol bis(diphenyl phosphate) in a PPO/HIPS blend, *J. Fire Sci.*, 16, 233, 1998.

356. C. Hackl, F. Lehrich, G. Bittner, Self-extinguishing thermoplastic polyurethanes and their preparation, US Patent 5837760, assigned to Elastogran, November 17, 1998.

357. O.S. Henze, M. Hansen, K. Hackmann, D. Meier, C. Beckmann, O. Muehren, Halogen-free flame retardant TPU, European Patent 2247657, assigned to BASF, December 28, 2011.

358. J.M. Cogen, G. Chen, X. Tai, G.D. Brown, Color-stable, halogen-free flame retardant thermoplastic polyurethane compositions, European Patent 2454316, assigned to Dow, April 3, 2013.

359. J.G. Chen, B. Li, W.W. Ma, X. Tai, Y. Qi, K.P. Pang, D.H. Guo, Y. Sun, Thermoplastic composition with epoxidized novolac, European Patent 2445967, assigned to Dow, March 22, 2017.

360. S. Dermeik, K.-H. Lemmer, R. Braun, W. Nassl, Composition for treatment of fiber materials by exhaust method in particular, US Patent 8303835, assigned to Huntsman, November 6, 2012.

361. M. Takata, A. Fukumi, K. Kubo, M. Hosaka, Flame retardant composition, flame retardant fiber treated with flame retardant composition, and method for increasing amount of flame retardant component adhered onto fibers using said composition, US Patent 10301547, assigned to Adeka, May 28, 2019.

362. R.N. Bookins, Emulsification of hydrophobic organophosphorous compounds, US Patent 10190051, assigned to Alexium, January 29, 2019.

363. M.A. Strand and R.L. Piner, Flame retardant polyester compositions for calendaring, US Patent 7285587, assigned to Eastman, October 23, 2007.

364. J.C. Chang, Y. Liang, J.P. McKeown, M.A. Page, Flame retardant poly(trimethylene terephthalate) composition, PCT Patent Application WO 2010/045181, assigned to DuPont, April 22, 2010.

365. Levchik, S.V., Georlette, P., Bar-Yaakov, Y., New developments in flame retardancy of polyolefins, in: *Proc. ANTEC2011*, vol. 3, p. 2387, Boston, MA, May 1-5, 2011.

366. S. Fuchs, U. Keppler, T. Albert, Halogen-free flame-retardant mixtures for polyolefin foams, European Patent Application 3145988, assigned to BASF, March 29, 2017.

367. S. Gabriel, L. Krishnamurthy, M. Stachnik, M.G. Weinberg, Flame retardant flash spun sheets, US Patent 8883893, assigned to DuPont, November 11, 2014.

368. S.V. Levchik, S. Dashevsky, E.D. Weil, Q. Yao, Oligomeric, hydroxy-terminated phosphonates, US Patent 7449526, assigned to Supresta, November 11, 2008.

369. Wu, T., Piotrowski, A.M., Yao, Q., Curing of epoxy resin with poly(m-phenylene methylphosphonate), Levchik, S.V., *J. Appl. Polym. Sci.*, 101, 4011, 2006.

370. Levchik, S.V. and Wang, C.S., P-based flame retardants in halogen-free laminates, *OnBoard Technol.*, 18, April 2007.

371. M. Schmidt, L. Bottenbruch, D. Freitag, K. Reinking, H. Roehr, H.-D. Block, Aromatic polyphosphonates and a process for their production, US Patent 4331614, assigned to Bayer, May 25, 1982.

372. M. Vinciguerra, D. Freitag, N. Rice, R. Lusignea, Branched polyphosphonates, US Patent 7816486, assigned to Triton Systems, November 19, 2010.

373. D. Freitag, Poly(block-phosphonato-ester) and poly(block-phosphonato-carbonate) and methods of making same, US Patent 7645850, assigned to FRX Polymers, January 12, 2010.

374. D. Freitag, J.-P. Lens, M.A. Lebel, Polyphosphonate and copolyphosphonate additive mixtures, US Patent 8975367, assigned to FRX Polymers, March 10, 2015.

375. Y. Jeong, L. Kagumba, J.-P. Lens, Epoxy compositions, US Patent 10077354, assigned to FRX Polymers, September 18, 2018.

376. X. Zeng and N. Ren, Thermosetting resin composition and use thereof, US Patent 10029438, assigned to Shengyi, July 24, 2018.

377. L. Kagumba, C. Manzi-Nshuti, K. Wei, Halogen-free flame-retardant compositions for flexible polyurethane foams, PCT Patent Application WO 2019/204625, assigned to FRX Polymers, October 24, 2019.

378. F. Cao, G.S. Nestlerode, Q. Lu, Flame retardant thermoplastic polyurethane compositions, US Patent 9657172, assigned to Lubrizol, May 23, 2017.

379. M.-A. Lebel, L. Kagumba, P. Go, Phosphonate polymers, copolymers, and their respective oligomers as flame retardants for polyester fibers, US Patent 9290653, assigned to FRX Polymers, March 22, 2016.

380. S. Ashby, S. Mortlock, W. Macdonald, A. Lovatt, Polyester film comprising a polymeric phosphonate flame retardant, PCT Patent Application WO 2019/207314, assigned to DuPont and Teijin Films, October 31, 2019.

381. Wang, Y.-Z., Solubility parameters of poly(sulfonyidiphenylene phenylphos-phonate) and its miscibility with poly(ethylene terephthalate), *J. Polym. Sci., Polym. Phys.*, 41, 2296, 2003.

382. Weil, E.D. and Levchik, S.V., Phosphorus-containing polymers and oligomers, in: *Encyclopedia of Polymer Science and Technology*, Wiley, New York, 2010, published on- line http://mrw.interscience.wiley.com/emrw/9780471440260/home/.

383. Y.-J. Kim, Y.-S. Lee, S.-M. Kim, Method of preparing flame-retardant poly-ester fiber and flame-retardant polyester fiber, assigned to Kolon, US Patent 8388879, March 5, 2013.

384. K.-T. Park, Y.-J. Lee, J.-R. Kim, S.-Y. Hwang, Method for preparing a polyes-ter resin, US Patent 9328195, assigned to SK Chemicals, May 3, 2016.

385. J. Asrar, Polymer-bound non-halogen fire resistant compositions, US Patent 5750603, assigned to Solutia, May 12,1998.

386. H.J. Kleiner, Process for preparing 6-chloro-(6H)-dibenzo[c,e][1,2]oxaphos-phorin, US Patent 5481017, assigned to Hoechst, February 21, 1995.

387. H.J. Kleiner, Method of preparing 6H-dibenzo[c,e][1,2]oxaphosphorin-6-one, US Patent 5391798, assigned to Hoechst, January 2, 1996.

388. Freudenberger, V. and Jacob, F., Phosphorhaltige polyethylenterephthalate, *Angew. Makromol. Chem.*, 105, 203, 1982.

389. S. Endo, T. Kashihara, A. Osako, T. Shizuki, T. Ikegami, Phosphorus-containing compounds, US Patent 4127590, assigned to Toyobo, November 28, 1978.

390. Gyobu, S., Sato, M., Takeuchi, H., Development and properties of flame retar-dant fibers, in: *Proc. Spring FRCA Conference*, pp. 157–165, New Orleans, LA, March 2003.

391. Sato, M., Endo, S., Araki, Y., Matsuoka, G., Gyobu, S., Takeuchi, H., The flame-retardant polyester fiber: Improvement of hydrolysis resistance, *J. Appl. Polym. Sci.*, 78, 1134, 2000.

392. Yang, S.-C. and Kim, J.P., Flame-retardant polyesters. II. Polyester polymers, *J. Appl. Polym. Sci.*, 106, 1274, 2007.

393. Yang, S.-C. and Kim, J.P., Flame retardant polyesters. I. phosphorus flame retardants, *J. Appl. Polym. Sci.*, 106, 2870, 2007.

394. B. Just, H. Keller, S., Imeri, Flame-retardant material containing phospho-rous, European Patent 2284208, assigned to Schill & Seilacher, February 22, 2012.

395. Zhang, C., Huang, J.Y., Liu, S.M., Zhao, J.Q., The synthesis and properties of a reactive flame-retardant unsaturated polyester resin from a phospho-rus-containing diacid, *Polym. Adv. Technol.*, 22, 1768, 2011.

396. Levchik, S.V. and Weil, E.D., Thermal decomposition, combustion and fire-retardancy of epoxy resins – a review of the recent literature, *Polym. Int.*, 53, 1585, 2004.

397. R. Ultz and S. Sprenger, Flame-retardant epoxy resins and flame retarders for epoxy resins, European Patent 0806429, assigned to Schill & Sellacher, December 19, 2001.

398. C.-S. Wang and C.H. Lin, Epoxy resin rendered flame retardant by reaction with 9,10-dihydro-9-oxa-10-phosphaphenanthrene-10-oxide, US Patent 6291627, assigned to National Science Council of Taiwan, September 18, 2001.

399. Sprenger, S. and Utz, R., Reactive, organophosphorus flame retardants for epoxies, *J. Adv. Mater.*, 33, 24, 2001.

400. Doering, M., Phosphorus containing flame retardants in epoxy resins, in: *Flame Retardants*, London, 2008, Interscience Communication, pp. 121–131, 2008.

401. Y. Hirai, H. Suzuki, Y. Takeda, K. Oohori, S. Kamoshida, M. Kakitani, N., Abe, Resin composition, and use and method for preparing the same, US Patent 6720077, assigned to Hitachi, April 13, 2004.

402. Rakotomalala, M., Wagner, S., Doering, M., Recent developments in halogen free flame retardants for epoxy resins for electrical and electronic applications, *Materials*, 3, 4300, 2010.

403. T. Sagara, T. Takata, K. Ihara, H. Kakiuchi, K. Ishihara, C. Asano, M. Gunji, H. Sato, Epoxy resin compositions containing phosphorus, flame retardant resin sheet using said epoxy resin containing phosphorus, resin clad metal foil, prepreg and laminated board, multi layer board, US Patent 6524709, assigned to Matsushita, February 2003.

404. Vlad-Bubulac, T., Hamciuc, C., Petreus, O., Synthesis and properties of some phosphorus-containing polyesters, *High Perform. Polym.*, 18, 255, 2006.

405. Montero De Espinosa, L., Ronda, J.C., Galia, M., Cadiz, V., Meier, M.A.R., Fatty acid derived phosphorus-containing polyesters via acyclic diene metathesis polymerization, *J. Polym. Sci., Polym. Chem.*, 47, 5760, 2009.

406. T. Asana, K. Ogasawara, N. Ito, Epoxy resin composition, prepreg and multilayer printed-wiring board, European Patent 1103575, assigned to Matsushita, February 8, 2006.

407. T. Iwami, M. Matsumoto, T. Abe, T. Yonemoto, H. Fujiwara, Epoxy resin composition for prepreg, prepreg, and multilayer printed circuit board, European Patent 2554561, assigned to Panasonic, September 4, 2019.

408. D. Kishimoto and Y. Umeki, High melting point flame retardant crystal and method for manufacturing the same, epoxy resin composition containing the flame retardant, and prepreg and flame retardant laminate using the composition, US Patent 9371438, assigned to Sanko, June 21, 2016.

409. Y. Meng, K. Fang, Y. Xu, Flame-retardant polyphenylene ether resin composition, European Patent Application 3564316, assigned to Shengyi, November 6, 2019.

410. V.M. Chopdekar, A.R. Mellozi, A.T. Cornelson, Flame-retardant cyanate esters US Patent Application 2011/0054087, assigned to Lonza, March 3, 2011.

411. M. Yaginuma, S. Ito, Y. Ueno, Resin composition, prepreg and resin sheet, and metal foil-clad laminate, European Patent 269012 assigned to Mitsubishi Gas, June 17, 2020.

412. A.M. Piotrowski, J. Zilberman, M. Zhang, E. Gluz, S. Levchik, K.A. Suryadevara, Active ester curing agent compound for the thermosetting resins, flame retardant composition comprising same, and articles made therefrom, US Patent 10500823, assigned to ICL-IP America, December 10, 2019.

413. J. Gan, Phosphorus-containing compounds useful for making halogen-free, ignition-resistant polymers, US Patent 8143357, assigned to Dow, March 27, 2012.

414. R.-T. Wang, L.-C. Yu, Y.-T. Lin, Y.-J. Chen, W. Tian, Z. Ma, W. Lu, Halogen-free resin composition, copper clad laminate using the same, and printed circuit board using the same, US Patent 9650512, assigned to Elite, May 16, 2017.

415. Y.-T. Li, Halogen-free resin composition and its application for copper clad laminate and printed circuit board, US Patent 9005761, assigned to Elite, April 14, 2015.

416. Y.L. Angell, K.M. White, S.E. Angell, A.G. Mack, DOPO-derived flame retardant and epoxy resin composition, US Patent 8536256, assigned to Albemarle, September 17, 2013.

417. Q. Yao, A.G. Mack, J. Wang, Process for the preparation of DOPO-derived compounds and compositions thereof, US Patent 9012546, assigned to Albemarle, April 21, 2015.

418. G. He and T. Amla, Varnishes and prepregs and laminates made therefrom, European Patent 3055354, assigned to Isola, December 5, 2018.

419. W.F. Scholz and S. Paul, Circuit materials with improved fire retardant system and articles formed therefrom, US Patent 10104769, assigned to Rogers, October 16, 2018.

420. Frank, A.W., Daigle, D.J., Vail, S.L., Chemistry of hydroxymethyl phosphorus compounds: Part II. Phosphonium salts, *Text. Res. J.*, 52, 678, 1982.

421. Choudhury, A.K.R., *Flame Retardants for Textile Materials*, CRC Press, Boca Raton, 2021.

422. Horrocks, A.R. and Zhang, S., Enhancing polymer flame retardancy by reaction with phosphorylated polyols. Part 2. Cellulose treated with a phosphonium salt urea condensate (Proban CC) flame retardant. *Fire Mater.*, 26, 173, 2002.

423. LeBlanc, R.B., The durability of flame retardant-treated fabrics, *Text. Chem. Color.*, 29, 2, 19, 1997.

424. Y. Gao, C.Y., Hsieh, Z., Ma, Z., Wang, W., Tian, Polyphenylene oxide resin, method of preparing polyphenylene oxide resin, polyphenylene oxide prepolymer and resin composition, US Patent 9447238, assigned to Elite, September 20, 2016.

425. H. Saito, and H. Fujiwara, Curable composition, prepreg, metal foil with composition, metal-clad laminate and wiring board, US Patent Application 2018/0170005, assigned to Panasonic, June 21, 2018.

426. M. Taniguchi, Y. Tada, Y., Nishioka, Flame-retardant epoxy resin composition, molded article thereof, and electronic part, US Patents 6797750, assigned to Otsuka, September 28, 2004.

427. L.-C. Yu and T.-A. Lee, Halogen-free resin composition and copper clad laminate and printed circuit board using same, US Patent 9279051, assigned to Elite, March 8, 2016.

428. Y. He and S. Su, Halogen-free flame retardant resin composition and the use thereof, US Patent 9745464, assigned to Shengyi, August 29, 2017.

429. T. Sagara, H. Kakiuchi, K. Kashihara, Y. Kitai, H. Saito, D. Yokoyama, H. Fujiwara, Resin composition, resin varnish, prepreg, metal-clad laminate, and printed wiring board, US Patent 9528026, assigned to Panasonic, December 27, 2016.

430. K. Maruyama and R. Motoshige, Flame retardant thermoplastic resin composition, European Patent 0728811, assigned to Mitsubishi, September 17, 2003.

431. Yamamoto, Y. and Wursta, T., Phosphazene flame retardants as a green chemistry to replace halogen-type flame retardants. Paper presented at *ANTEC 2010*, Orlando, FL, May 2010.

432. Y. Tada and T. Inoue, Cyanato group-containing cyclic phosphazene compound and method for producing the same, US Patent 7767739 assigned to Fushimi Pharmaceutical, August 3, 2010.

433. S.-F. Liu and M.-H. Chen, Resin composition and uses of the same, US Patent 10196502, assigned to Taiwan Union Technology, February 5, 2019.

434. Mabey, W. and Mill, T., Critical review of hydrolysis of organic compounds in water under environmental conditions, *J. Phys. Chem. Ref. Data*, 7, 383, 1978.

435. Saeger, V.W., Hicks, O., Kaley, R.G., Michael, P.R., Mieure, J.P., Tucker, E.S., Environmental fate of selected phosphate esters, *Envir. Sci. Tech.*, 13, 840, 1979.

436. Boethling, R.S. and Cooper, J.C., Environmental fate and effects of triaryl and tri-alkyl/aryl phosphate esters, *Residue Rev.*, 94, 49, 1985.

437. Jurgens, S.S., Helmus, R., Waaijers, S.L., Uittenbogaard, D., Dunnebier, D., Vleugel, M., Kraak, M.H.S., de Voogt, P., Parsons, J.R., Mineralisation and primary biodegradation of aromatic organophosphorus flame retardants in activated sludge, *Chemosphere*, 111, 238, 2014.

438. Oros, D.R., Jarman, W.M., Lowe, T., David, N., Lowe, S., Davis, J.A., Surveillance for previously unmonitored organic contaminants in the San Francisco estuary, *Marine Pollut. Bull.*, 46, 1102, 2003.

439. Muir, D.C.G., Phosphate Esters, in: *The Handbook of Environmental Chemistry, Vol, 3, Part. C Anthropogenic Compounds*, O. Hutzinger (Ed.), pp. 41–66, Springer-Verlag, Berlin, 1984.

440. Watts, M.J. and Linden, K.G., Advanced oxidation kinetics of aqueous tri-alkyl phosphate flame retardants and plasticizers, *Environ. Sci. Technol.*, 43, 2937, 2009.

441. Liu, Y., Liggio, J., Harner, T., Jantunen, L., Shoeib, M., Li, S.-M., Heterogeneous OH initiated oxidation: A possible explanation for the persistence of organo-phosphate flame retardants in air, *Env. Sci. Technol.*, 48, 1041, 2014.

442. Ghisalba, O., Kuenzi, M., Ramos Tombo, G.M., Schar, H.-P., Microbial degradation and utilization of selected organophosphorus compounds, *Chimia*, 41, 206, 1987.

443. Frost, J.W., Loo, S., Li, D., Radical-based dephosphorylation and organo-phosphonate biodegradation, *J. Am. Chem. Soc.*, 109, 2166, 1987.

444. Quinn, J.P., Kulakova, A.N., Cooley, N.A., McGrath, J.W., New ways to break an old bond: The bacterial carbon-phosphorus hydrolases and their role in biogeochemical phosphorus cycling, *Environ. Microbiol.*, 9, 2392, 2007.

445. Lenoir, D., Kampke-Thiel, K., Kettrup, A., New Duroplastic Materials with Phosphorus Compounds as Flame Retardant: Fate and Behavior in Thermal Reactions, in: *Proc. BCC Conf. Recent Advances in Flame Retardancy of Polymeric Materials*, Stamford, CT, May 1996.

446. Lengsfeld, H., Altstadt, V., Sprenger, S., Utz, R., Flame-retardant curing, *Kunstoffe*, 91, 37, 2001.

447. Li, T.-Y., Bao, L.-J., Wu, C.-C., Liu, L.-Y., Wong, C.S., Zeng, E.Y., Organophosphate flame retardants emitted from thermal treatment and open burning of e-waste, *J. Hazard. Mater.*, 367, 390, 2019.

448. Ma, Y., Stubbings, W.A., Cline-Cole, R., Harrad, S., Human exposure to halogenated and organophosphate flame retardants through informal e-waste handling activities - A critical review, *Environ. Pollution*, 268, 115727, 2021.

449. Bacaloni, A., Callipo, L., Corradini, E., Gubbiotti, R., Samperi, R., Lagana, A., *Flame Retardants: Functions, Properties and Safety*, P.B. Merlani (Ed.), pp. 131–143, Nova Science Publishers, New York, 2010.

450. Martínez-Carballo, E., González-Barreiro, A., Sitka, S., Scharf, O., Gans, O., Determination of selected organophosphate esters in the aquatic environment of Austria, *Sci. Total Environ.*, 388, 290, 2007.

451. Yang, J., Zhao, Y., Li, M., Du, M., Li, X., Li, Y., A review of a class of emerging contaminants: The classification, distribution, intensity of consumption, synthesis routes, environmental effects and expectation of pollution abatement to organophosphate flame retardants (OPFRs), *Int. J. Mol. Sci.*, 20, 2874, 2019.

452. Marklund, A., Andersson, B., Haglund, P., Traffic as a source of organophosphorus flame retardants and plasticizers in snow, *Environ. Sci. Technol.*, 39, 3555, 2005.

453. Regnery, J. and Puttmann, W., Organophosphorus flame retardants and plasticizers in rain and snow from middle Germany, *Clean*, 37, 334, 2009.

454. Kim, U.-J. and Kannan, K., Occurrence and distribution of organophosphate flame retardants/ plasticizers in surface waters, tap water, and rainwater: Implications for human exposure, *Environ. Sci. Technol.*, 52, 5625, 2018.

455. Li, W., Wang, Y., Kannan, K., Occurrence, distribution and human exposure to 20 organophosphate esters in air, soil, pine needles, river water, and dust samples collected around an airport in New York State, United States, *Environ. Int.*, 131, ID 105054, 2019.

456. Salamova, A., Ma, Y., Venier, M., Hites, R.A., High levels of organophosphate flame retardants in the Great Lakes atmosphere, *Environ. Sci. Technol. Lett.*, 1, 8, 2014.

457. Bekele, T.G., Zhao, H., Wang, Q., Chen, J., Bioaccumulation and trophic transfer of emerging organophosphate flame retardants in the marine food webs of Laizhou Bay, North China, *Env. Sci. Technol.*, 53, ID 13417, 2019.

458. Yang, J., Zhao, Y., Li, M., Du, M., Li, X., Li, Y., A review of a class of emerging contaminants: The classification, distribution, intensity of consumption, synthesis routes, environmental effects and expectation of pollution abatement to organophosphate flame retardants (OPFRs), *Int. J. Mol. Sci.*, 20, 2874, 2019.

459. Saito, I., Onuki, A., Seto, H., Indoor organophosphate and polybrominated flame retardants in Tokyo, *Indoor Air*, 17, 28, 2007.

460. Yang, F., Ding, J., Huang, W., Xie, W., Liu, W., Particle size-specific distributions and preliminary exposure assessments of organophosphate flame retardants in office air particulate matter, *Environ. Sci. Technol.*, 48, 63, 2014.

461. Cequier, E., Ionas, A.C., Covaci, A., Marcee, R.M., Becher, G., Thomsen, C., Occurrence of a broad range of legacy and emerging flame retardants in indoor environments in Norway, *Environ. Sci. Technol.*, 48, 6827, 2014.

462. Gravel, S., Lavoué, J., Bakhiyi, B., Diamond, M.L., Jantunen, L.M., Lavoie, J., Roberge, B., Verner, M.-A., Zayed, J., Labrèche, F., Halogenated flame retardants and organophosphate esters in the air of electronic waste recycling facilities: Evidence of high concentrations and multiple exposures, *Environ. Int.*, 128, 244, 2019.

463. Estill, C.F., Slone, J., Mayer, A., Chen, I.-C., La Guardia, M.J., Worker exposure to flame retardants in manufacturing, construction and service industries, *Environ. Int.*, 135, 105349, 2020.

464. Saini, A., Thaysen, C., Jantunen, L., McQueen, R.H., Diamond, M.L., From clothing to laundry water: Investigating the fate of phthalates, brominated flame retardants, and organophosphate esters, *Environ. Sci. Technol.*, 50, 9289, 2016.

465. Van der Veen, I. and de Boer, J., Phosphorus flame retardants: properties, production, environmental occurrence, toxicity and analysis, *Chemosphere*, 88, 1119, 2012.

466. Chupeau, Z., Bonvallot, N., Mercier, F., Le Bot, B., Chevrier, C., Glorennec, P., Organophosphorus flame retardants: A global review of indoor

contamination and human exposure in Europe and epidemiological evidence, *Int. J. Environ. Res. Public Health*, 17, 6713, 2020.

467. Hou, R., Xu, Y., Wang, Z., Review of OPFRs in animals and humans: Absorption, bioaccumulation, metabolism, and internal exposure research, *Chemosphere*, 153, 78, 2016.

468. Kajiwara, N., Noma, Y., Takigami, H., Brominated and organophosphate flame retardants in selected consumer products on the Japanese market In 2008, *J. Hazard. Mater.*, 192, 1250, 2011.

469. Wensing, M., Uhde, E., Salthammer, T., Plastics additives in the indoor environment - flame retardants and plasticizers, *Sci. Total Environ.*, 339, 19, 2005.

470. Kemmlein, S., Hahn, O., Jann, O., Investigations on Emissions of Selected Flame Retardants. A Possible Route of Exposure?, in: *Advances in the Flame Retardancy of Polymeric Materials: Current perspectives, presented at FRPM'05*, B. Schartel (Ed.), pp. 249–262, Germany Books on Demand GmbH, Norderstedt, Germany, 2007.

471. Reemtsma, T., Quintana, J.B., Rodil, R., Garcia-Lopez, M., Rodriguez, I., Organophosphorus flame retardants and plasticizers in water and air I. Occurrence and fate, *Trends Anal. Chem.*, 27, 727, 2008.

472. Ospina, M., Jayatilaka, N.K., Wong, L.-Y., Restrepo, P., Calafat, A.M., Exposure to organophosphate flame retardant chemicals in the U.S. general population: Data from the 2013–2014 national health and nutrition examination survey, *Environ. Int.*, 110, 32, 2018.

473. Blum, A., Behhl, M., Birnbaum, L.S., Diamond, M.L., Phillips, A., Singla, V., Sipes, N.S., Stapleton, H.M., Venier, M., Organophosphate ester flame retardants: Are they a regrettable substitution for polybrominated diphenyl ethers?, *Environ. Sci. Technol. Lett.*, 6, 638, 2019.

474. Hartmann, P.C., Burgi, D., Giger, W., Organophosphate flame retardants and plasticizers in indoor air, *Chemosphere*, 57, 781, 2004.

475. Li, J., Zhao, L., Letcher, R.J., Zhang, Y., Jian, K., Zhang, J., Su, G., A review on organophosphate ester (OPE) flame retardants and plasticizers in foodstuffs: Levels, distribution, human dietary exposure, and future directions, *Environ. Int.*, 127, 35, 2019.

3

Mineral Filler Flame Retardants

Reiner Sauerwein

Nabaltec AG, Schwandorf, Germany

Abstract

This chapter gives an up-to-date overview on mineral filler flame retardants. The term filler has been chosen to differentiate this from chapters 6 and 7, which also deal with some inorganic flame retardants. A differentiation between mineral filler flame retardants and inorganic additive flame retardants is given in the introduction of this chapter. After a brief view on the market share of mineral filler flame retardants, the most relevant products are highlighted according to their chemical composition, manufacturing processes, physical characteristics and their impact on polymer material properties. The working principle of mineral filler flame retardants and correlation of most important filler parameters with flame retardant performance is a major part of this chapter. The second part gives an overview of the most important applications and developments. Processing specifics for highly filled polymers are discussed and exemplary composite formulations are given. Examples of applications of the highest commercial importance and new developments have been considered.

Keywords: Metal hydrate, functional filler, particle size, surface area, loading level, bulk density, compounding, coupling, endothermic decomposition, smoke suppression

3.1 Introduction

Mineral flame retardants by definition are inorganic compounds and functional fillers. They display their function of retarding ignition and flame spread by a chemo-physical process induced under the conditions of an ignition source or an initial fire. Inert fillers, to the contrary, do not act as

---

Email: sauerwein.reiner@nabaltec.de

Alexander B. Morgan (ed.) Non-Halogenated Flame Retardant Handbook 2nd edition, (101–168)
© 2022 Scrivener Publishing LLC

flame retardants, but thin the burnable organic mass of a composite material. Only in very rare cases does a mineral flame retardant also provide cost savings, for which chalk (calcium carbonate) is purposely used [1]. The term "filler", which is sometimes falsely used as a synonym for cost saving, should more correctly be differentiated from the term "additive" by the loading used in polymers. When used as single flame retardant solution, mineral fillers have to be used at minimum 40 weight (wt.)-percent (-%), most commonly at 55 – 65 wt.-% and sometimes even up to 80 wt.-%, depending on the polymer matrix and the flame retardant standard required. Nevertheless, there are exceptions. When used as co-flame retardant or synergist, minerals are found at much lower addition levels, sometimes as low as 1-2 wt.-%.

Flame retardant additives, on the other hand, are normally used at loadings ranging from 5 – 30 wt.-%. Due to higher physical densities of minerals, the higher wt.-% loading compared to organic flame retardant additives is very often equalized when it comes to volume percent loadings. Organic compounds, like most polymers, have physical densities in the range 1 - 1.2 g/cm^3, while most minerals have densities between 2 and 5 g/cm^3. Nevertheless, the incorporation of mineral fillers at typical wt.-% loadings requires specialized polymer processes, last but not least because of the relative low bulk density of some mineral powders.

Powder is the typical form in which flame retardant fillers are processed. Masterbatches are very rarely made from mineral flame retardant fillers. The relatively moderate thinning effect from masterbatch to end use loading does not justify the additional process step of masterbatch production. Fillers are supposed to disintegrate from their primary powder agglomerates into a well dispersed state when mixed with molten or liquid polymers, and do not dissolve in their organic matrices. Interfacial effects between the mineral surface and matrix are therefore crucial for all physical properties.

3.2   Industrial Importance of Mineral Flame Retardants

Mineral filler flame retardants or inorganic flame retardants are the most important group of fire retardants on a volume basis. A basic classification can be done by grouping mineral filler flame retardants into natural minerals and synthetic minerals. All mineral flame retardants are originally based on mineral ores, but the term natural mineral flame retardants is used for products which are produced by mechanical mineral ore separation and subsequent mechanical disintegration processes. To the contrary, synthetic mineral flame retardants are always manufactured by chemical production processes. For more details on natural vs. synthetic please refer to sections 3.2.2 and 3.2.3.

3.2.1   Market Share of Mineral FRs

Referring to market studies of Freedonia for 2015 [2] and Ceresena Research for 2018 [3], the total volume of the worldwide flame retardant market has reached 2.2 metric tons (MT). In both studies the two mineral flame retardant products aluminium tri-hydroxide (also alumina tri-hydrate, abbreviated ATH) and antimony trioxide (abbreviated ATO) are listed as their own categories. Other mineral flame retardants are accumulated together with other nonmineral flame retardants under "other". A differentiation between natural and synthetic flame retardants is not accessible via commonly available flame retardant market studies. The volume split between the flame retardant categories investigated in the two market studies is listed in Table 3.1.

In the context of this chapter within this handbook, ATO is not considered as mineral filler flame retardant. ATO is not a standalone flame

Table 3.1 Volumes and %-split of different flame retardant categories according to two market studies [2, 3]. TMT = Thousand Metric Tons.

Study - year	Freedonia – 2015		Ceresena - 2018	
	in TMT	in %	in TMT	in %
ATH	706	31.8	821	37.2
ATO	197	8.9	134	6.1
Brominated FR	438	19.7	391	17.7
Chlorinated FR	189	8.5	252	11.4
Phosphorous FR	322	14.5	323	14.7
Boron FR	165	7.4	--	--
Other FR	205	9.2	285	12.9
Total	2222	100	2206	100

retardant, but is a synergist which needs the presence of halogens, either being a part of the polymer matrix (such as polyvinylchloride) or as a halogenated flame retardant additive.

3.2.2   Synthetic Mineral FRs within the Industrial Chemical Process Chain

The two most important mineral flame retardant compounds also have significant industrial importance. Aluminium hydroxide (ATH) and magnesium di-hydroxide or magnesium hydroxide (MDH) are by far the most important metal hydroxide flame retardants, but even more importantly, they are large volume raw material chemicals used in many different industries.

Aluminium hydroxide is produced starting from the ore bauxite by the so-called Bayer process. Crushed bauxite is digested in sodium hydroxide liquor to form sodium aluminate from which the insoluble ingredients of the bauxite ore are separated, as so called red mud. The sodium aluminate is seeded to crystallize aluminium hydroxide. This product is called chemical grade ATH or wet hydrate, meaning the product has been filtered but not dried. Wet hydrates have remaining moisture levels of 5-10 % and median particle sizes of approximately 100 microns (μm). Most of this type of ATH is not used in chemical applications, but is calcined to alumina. Again, by far the majority of all alumina is used as smelter grade alumina in the production of aluminium metal by melt electrolysis. Figure 3.1 displays the mass balance of bauxite and alumina production for the year 2018. Out of estimated 130 million metric tons (MMT) of alumina and aluminium hydroxide (calculated as Al_2O_3) produced starting from estimated 274 MMT of bauxite, only 3.8 MMT (or 5.9 MMT as aluminium hydroxide) have been used in chemical applications [4].

Figure 3.2 shows how these 5.9 MMT of ATH have been used [4]. Only 15 % of all ATH, around 885 TMT, has been used in flame retardant applications. This volume is a bit higher than the volumes identified by the two market studies (see Table 3.1).

Magnesium hydroxide (MDH) is also heavily linked to its corresponding oxide industry. Magnesium oxide or magnesia markets are divided into dead burned magnesia (DBM), caustic calcined magnesia (CCM) and electro fused magnesia (EFM). In 2012 approximately 10 MMT of magnesia have been produced [6].

DBM and EFM are exclusively used in refractory applications. Only CCM, which is approximately 10 % of all magnesia, is used for the production of other magnesium chemicals. This is because of its relatively

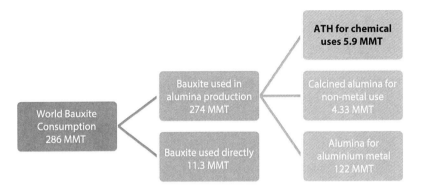

Figure 3.1 Mass balance for bauxite and alumina in 2018 (all figures as alumina, with the exemption of ATH for chemical uses) [4].

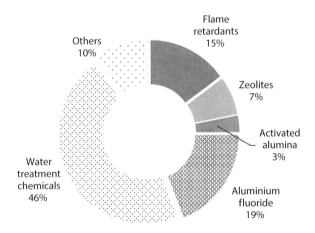

Figure 3.2 World consumption of ATH by markets, 2018 (%) [4].

high specific surface area, resulting in good reactivity. CCM as such is used in small volumes for rubber formulations as accelerator for curing reactions and as rheology modifier. Magnesium hydroxide (MDH) produced starting from CCM and used as flame retardant represents the most important polymer application. Other applications are outside the polymer field and include animal nutrition, fertilizers, waste water treatment, cement industry and pulp & paper. CCM is also used in toothpastes, pharmaceuticals and cosmetics and is the basis for other magnesium compounds [5].

Flame retardant market studies do not give any data for MDH, but categorize MDH under "other flame retardants". Freedonia identified 205 TMT of "other flame retardants" in 2015, while Ceresena Research identified 285 TMT for 2018 (see Table 3.1). In both studies the category "other flame retardants" also includes nitrogen based, inorganic phosphorous (red phosphorous, (Pred), and ammonium polyphosphate, (APP)). In case of the Ceresena study boron compounds are also included, while they are listed as their own category in the Freedonia study. Mineral filler flame retardants belonging to this category include MDH (synthetic and natural), other natural fillers, like huntite/hydromagnesites blends, and specialities, like zinc-hydroxy-stannates and boehmites (for details please refer to 3.3).

Because of the lack of public market figures, the author tried to elaborate the market volume of MDH based on market intelligence of his employer (Nabaltec AG). Based on this, MDH flame retardant market volume is estimated to be 60 TMT worldwide in 2018. Further the author assumes that synthetic MDH grades used worldwide are approximately 50 % of this volume, 30 TMT. When compared with ATH, total MDH flame retardant volume is roughly only 7-8 % of ATH.

3.2.3   Natural Mineral FRs

The volume of natural mineral flame retardants is very small compared to synthetic mineral flame retardants. Total volume of Huntite/Hydromagnesite (see section 3.3.1 for chemistry) blends and natural MDH (also called brucite) is believed to be in the range of 44 TMT worldwide for 2018 (own estimates). This is roughly 5 % of all mineral filler flame retardants.

These minerals are also used for other purposes. Huntite/hydromagnesite blends are used as rheology modifier in polymeric sealants or coatings. Due to their basic nature and layered structure these products may also be used as acid scavengers or catalyst carrier.

When dividing between ATH, MDH and "other mineral filler flame retardant", natural huntite/hydromagnesite blends are aggregated with synthetic zinc-stannates, zinc-hydroxy-stannates and boehmites and are estimated to be in the range of 18 TMT in total.

Considering the author's MDH and "other mineral filler flame retardant" estimates and taking the flame retardant market study of Ceresena Research as basis, the resulting market split for mineral filler flame retardants is as shown in Figure 3.3. The left diagram shows the split according to mineral flame retardant chemical type, while the diagram on the right shows the split between synthetic and natural mineral flame retardants.

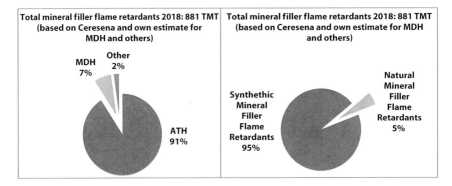

Figure 3.3 Volume split within mineral filler flame retardants based on flame retardant market study of Ceresena Research [3] and own estimates for MDH and other mineral flame retardants: right by mineral compounds, left differentiating between synthetic and natural mineral filler flame retardants.

3.3   Overview of Mineral Filler FRs

Mineral filler flame retardants can be described according to their chemical compositions, by the way they are produced, according to their physical properties and by the impact they have on the polymer properties of the composites in which they are used.

3.3.1   Mineral Filler Flame Retardants by Chemistry

All mineral flame retardants of industrial importance are either oxides, sulphides, hydroxides, carbonates, borates, stannates or mixed salts of these anions based on aluminium, magnesium and zinc. Table 3.2 lists the most important mineral flame retardants and some characteristic properties. As mentioned earlier, antimony trioxide (ATO) is not considered in this book, because it is a synergist to halogenated compounds only and not a standalone additive. The list also contains zinc borate. Zinc borate has several modifications and other inorganic boron compounds are also used, which are not mentioned here, because this handbook has a separate chapter on boron compounds (see chapter 7). Also hydrated layered silicates and organically modified variants are not discussed in this chapter, as they are part of chapter 10.

Metal hydroxides are by far the most important group of mineral flame retardants. The chemically incorrect term metal hydrate is more often used in industry and is therefore used synonymous in this chapter. Aluminium tri-hydrate, ATH, contains three hydroxyl groups per aluminium atom.

Table 3.2 List of mineral filler flame retardants and their most important properties. Tdeh. = onset temperature of dehydration, Refr. Ind. = Refractive Index.

Name, Abreviation	Chemical Formula	%-Loss on ignition (1000 °C)	Tdeh. (°C)	pH	Density g/cm³	Mohs Hardness	Refr. Ind.	Particle shape
Aluminium hydroxide, ATH	$Al(OH)_3$	34.6	200	9	2.4	3	1.58	spherical – platy
Magnesium hydroxide, MDH	$Mg(OH)_2$	31	330	10	2.4	3	1.57	spherical – platy
Boehmite, AOH	$AlOOH$	17	340	8	3.0	3-4	1.62	spherical – platy
Huntite, H	$Mg_3Ca(CO_3)_4$	35	450		2.7	1-2	1.22	Platy
Hydromagnesite, HM	$Mg_5(CO_3)_4(OH)_2 \cdot 4H_2O$	45	220		2.2	3-4	1.52	Platy
Zinc borate	$2ZnO \cdot 3B_2O_3 \cdot 3.5H_2O$	15.5	290	7-8	2.8		1.58	Platy
Zinc stannate	$ZnSnO_3$	2	560	9-10	3.9		1.9	Spherical
Zinc hydroxy stannate	$ZnSn(OH)_6$	17-19	200	9-10	3.8		1.9	Spherical

The other aluminium flame retardant compound of industrial relevance, boehmite, is an oxide hydroxide of aluminium (abbreviated AOH), having only one hydroxyl group per molecule and consequently a loss on ignition of only 17 % versus 34.6 % for ATH. Magnesium hydrate exists only as di-hydroxide (abbreviated MDH), having a loss on ignition of 31 % of its mass.

While all aluminium hydrates are produced via a chemical synthesis known as Bayer process, MDH flame retardants can either be produced via synthetic routes or are based on natural ores called brucite. The latter products are also called natural MDH.

Other mineral flame retardants of industrial importance are huntite and hydromagnesite blends. Both products occur as natural mixtures. Huntite is a magnesium calcium carbonate releasing carbon dioxide (CO_2) at temperature exceeding 450 °C. Hydromagnesite is a hydrated magnesium carbonate, releasing water above 220 °C. The total loss on ignition until 1100 °C is in the range of 41 – 43 % [6].

All zinc based flame retardants are synthetic chemicals. The most important zinc borate modification used as flame retardant is a hydrated version with a loss on ignition of 15.5 %. Dehydration starts at 290 °C. Zinc stannates are available in a non-hydrated version, stable up to 560 °C, while the hydrated version, zinc hydroxy stannate, starts giving off water at 200 °C [7].

Zinc borates and zinc stannates have also been applied as mineral coating on ATH and MDH carriers [8], but these products are so far commercially negligible.

3.3.2   Classification by Production Process

Mineral fillers in general can be distinguished by their production process. Grinding and air separation are mechanical processes which are applied to natural minerals as well as synthetic minerals. Precipitation is a common synthetic route for inorganic compounds. Surface treatment by organics is done to modify filler surface characteristics.

3.3.2.1   Crushing and Grinding

Crushing is applied as a very first mechanical operation when producing fillers starting from mineral lumps. It describes the disintegration of larger chunks of stones gained from mining operations. Grinding processes start either from fine lumps or coarse powders produced by mineral ore crushing or by crystallization processes respectively. Powdery fillers produced by grinding normally range from 10 μm up to 100 μm in median particle size, but some special grinding technologies will also result in finer products, some as fine as 1 μm. Even finer particle sizes can be achieved by so called wet grinding processes, where the mill charge is dispersed in water, more rarely in organic solvents.

Ground products have more irregular and splintery particle shapes than precipitated products. Depending on the grinding technology applied as well as filler chemistry and mineralogy, the filler particles shapes may be dominated by spherical aggregates, irregular plates or acicular, needle-like, geometries. The particle size distribution after grinding is very often not mono-modal, but bi- or even multi-modal. In general ground products have higher specific surface areas according to BET when compared with synthetic, precipitated powders of the same particle size.

Dry grinding technologies have the highest commercial importance and include rotating ball mills, vibration mills and rotary mills. In these mills the disintegration of large particles happens by mechanical shearing of the grinding media between grinding bodies (ball and vibration mills) or between a rotating and a static tool (rotary mills). Wet grinding is applied when a disintegrated slurry preparation is required. Wet grinding can also be used to produce very fine, submicron particles in powdery form, when a filtration and drying process follows the grinding operation. Therefore, wet grinding may also be used on precipitated products (see below).

3.3.2.2   Air Classification

Air separation or classification is applied as a further refining process after grinding. It uses differences in air drag, dependent on particle size, to separate according to particle size. Air classifiers are very often used to cut off particles larger than a defined top cut value. Some rotary mills have an integrated classifier, keeping particles larger than the desired top cut inside the mill. Air classifiers can also be combined with ball mills to form a continuous operating grinding and classifying facility.

3.3.2.3   Precipitation and Their Synthetic Processes

Synthetic minerals like ATH are always produced by a chemical process starting from chemicals which differ from the end product of the chemical process. In case of ATH, the aluminium hydroxide is precipitated starting from a sodium aluminate solution leaving sodium hydroxide behind. The equilibrium of this chemical reaction is dependent on concentration and temperature.

$$Na[Al(OH)_4] \rightarrow NaOH + Al(OH)_3$$

The particle size of the precipitated ATH is additionally determined by the seed concentration and, of course, the nature of the seed. Carbon dioxide, amorphous and crystalline ATH can be used as seed [9]. In the case of seed crystals the particle size and surface area has an influence. Chemical grade, or "wet hydrate" ATH produced in bauxite refineries is crystallized at approximately 100 µm. Such type of product is used as a raw material for ground ATH fillers used as flame retardants.

Fine precipitated products can either be produced from sodium aluminate out of the Bayer process directly (starting from bauxite) or from chemical grade ATH, which is "re-dissolved" in caustic soda. Nowadays most ATH flame retardant producers run specialized manufacturing sites, no longer processing bauxite ores in the classical Bayer process, but buying chemical grade ATH as a feedstock material [10]. Fine precipitated products are produced according to BET surface area by adjusting seed type and concentration and precipitation temperature. BET values ranging from 3-12 m^2/g are commonly produced on large volume scale. These products have median particle sizes from 0.7-2 µm. Table 3.3 displays the three most common BET-ranges produced by the leading fine precipitated ATH manufacturers. ATH grades with BET values higher than 12 m^2/g are industrially produced, but currently have much lower market importance.

Synthetic magnesium hydroxide is produced starting from caustic calcined magnesia (CCM). CCM itself can be produced from sea water brines or magnesium containing mineral ores like serpentine, magnesite or dolomite. When starting from mineral ore, magnesium chloride brine is leached out of the mineral with hydrochloric acid.

The magnesium chloride intermediate is converted to magnesium oxide by a spray roasting process [11]. The magnesia is subsequently hydrolysed

Table 3.3 Typical product range of fine precipitated ATH.

Specified BET range (m^2/g)	Typical values		
	D50 (µm)	BET (m^2/g)	Oil absorption (ml/100g)
3 - 5	1.5	3.5	22
5 - 8	1	6	28
10 - 13	0.9	11	37

to MDH (magnesium hydroxide). Particle shape and BET surface of MDH can be modified by a hydrothermal (pressurized suspension) refining process.

$$MgCl_2 + H_2O + \Delta T \rightarrow MgO + HCl$$

$$MgO + H_2O \rightarrow Mg(OH)_2$$

So called "seawater" MDH grades are produced by direct precipitation from magnesium chloride brine by the addition of lime.

$$MgCl_2 + Ca(OH)_2 \rightarrow Mg(OH)_2 + CaCl_2$$

Crystalline boehmite or aluminium oxide hydroxide (AOH) is produced under hydrothermal conditions. The reaction can be either started from an ATH suspension in water, or from an ATH suspension in sodium hydroxide in the presence of boehmite seed crystals. In water the conversion of the gibbsite crystals to crystalline boehmite is probably determined by a template effect of the gibbsite crystal, but the process conditions (hydrothermal pressure) and the reactor type used are also crucial for the resulting boehmite particle morphology and size [10, 12]. When sodium hydroxide is used, sodium aluminate is formed as an intermediate from which boehmite crystals precipitate (here called "autocatalytic hydrothermal crystallization"). Particle shape and size are mainly determined by the boehmite seed [13] Figure 3.4 displays reaction schemes for three synthetic routes.

Amorphous boehmites, also known as pseudo-boehmites, have high specific surface areas ranging from 100-300 m²/g. Pseudo-boehmites can

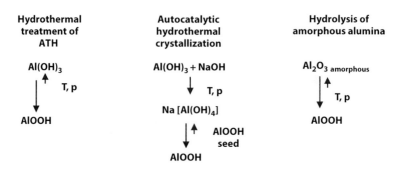

Figure 3.4 Three synthetic routes for the production of boehmites.

be produced by hydrolysis of amorphous aluminas or by hydrolysis of aluminium alcoholates. Their flame retardant usage is limited to synergistic dosings, because melt compounding and processing in liquid reactive resins at high loadings is difficult for such high BET fillers [14].

3.3.2.4   Surface Treatment

Surface treatment is an additional refining process to modify mineral fillers for better performance. Organic molecules are attached to the surface of the particulates to change their surface properties. For metal hydrates and other polar fillers, the surface treatment is done to change the filler surface from hydrophilic to a more hydrophobic behaviour, making it more compatible with non-polar polymer matrices. Hydrophobicity and lower moisture uptake is achieved by attaching organic molecules having a hydrophobic tail. The organic molecules can attach by a chemical reaction, like in the case with silanes, or by a combination of chemical bonding and hydrogen-bridges. Fatty acids, like stearic acid, belong to the latter group of molecules. Very often surface treatment agents possess a second functional group which is supposed to react with the polymer matrix when mixed. Bifunctional silanes are the most important products used for this purpose and vinyl-, amino and glycidyl-groups are of commercial relevance. Bifunctional silanes are also known as "coupling agents" because of their ability to chemically couple the filler to the matrix polymer [16, 17]. Table 3.4 lists the most important silane coupling agents according to the functional group and the most important polymers these couplings are used in. All these silanes react with hydroxyl groups on the filler surface via their silanol groups. The silanol groups themselves form during the coating process by fillers' surface moisture initiated hydrolysis of silano-alcoholates (methoxy or ethoxy). In competition to the surface bonding, silanol groups can also self-condensate, which should be avoided to guarantee an effective use of the silane.

In rare cases, surface treating of fillers can be done during grinding or precipitation. Such an integrated coating process is restricted to fatty acids or fatty acid salts. Normally it is an add-on process generating significant additional costs. There are several machinery setups available and many more very specifically designed operations may exist undisclosed at manufacturers' sites. The basic principle is probably the same with all processes: a fluidized powder bed of the filler is brought in contact with the surface coating agent. The organic additive is either applied as such or solubilised in an appropriate liquid. In case of solutions and when the reaction at the

Table 3.4 Sketch of silane coating, most important functional groups and the polymers they are applied in.

R $\sim\sim$ $\overset{OR_1}{\underset{R_1O}{\overset{|}{Si}}}-O\dot{R}_1$	Hydrolysable -OR$_1$	Functional Group R	Polymers
	-O-CH$_3$ -O-C$_2$H$_5$	-NH$_2$ (amino)	EVA, EVM, PA6, PA66, PBT, PVC
		-O-CH$_2$CHOCH$_2$ (glycidyl)	EP, PA6, PA66, PBT
		-CH=CH$_2$ (vinyl)	PE, (PP), EPDM,
		-O-CO-CH=CH$_2$ (acryl)	PP, PMA, PMMA, PVC
		-O-CO-C(CH$_3$)=CH$_2$ (methacryl)	PP, PMMA, PMA, PVC

filler surface releases low molecular weight compounds (like alcohols in the case of silanes), the process needs proper ventilation to remove volatiles.

Principally applicable to all flame retardant fillers having sufficient reactivity, synthetic metal hydroxides and especially fine precipitated products are preferred. For natural minerals and coarse ground products used in commodity applications, the additional cost for a separate surface treatment step very often cannot be justified by the performance gains. In general, surface treated fillers compete with so called *in situ* application of silanes or maleic anhydride grafted polyolefins during polymer processing.

3.3.3   Physical Characterisation of Mineral FRs

Mineral flame retardants are solid particulate materials. There basic flame retardant function is related to their chemical composition and chemical reactions they undergo in a case of an ignition or fire scenario. When discussing the working principle in detail in section 3.4, it will be shown that the fire retardant properties of filled compounds/composite additionally depends on some filler characteristics like fineness expressed as median particle size or specific surface area according to BET. It is crucial for the mineral flame retardant user to be aware of the most important properties which characterize particulate mineral fillers and influence their performance during processing and end application.

3.3.3.1   Particle Shape/Morphology/Aspect Ratio

All three terms are interrelated, but have slightly different meanings. Particle shape describes the geometric form of a particle. Spherical particles may be round or cubic and all three axes have very similar length. Acicular particles are needle-like, which means that the particles are much more elongated in one direction, while the other two axes are similar in length. Platey particles have two axes which are significantly longer than the third dimension, which describes the thickness of a flaky particle. Irregular particles have more complex and less symmetric shapes. Figure 3.5 displays sketches of the most important particle shapes.

The external shape of a natural mineral is very often manifested in its crystal modification. Some minerals of identical composition exist in different crystal modifications, but under normal conditions most compounds have only one thermodynamically stable modification.

The shape of particles is qualitatively determined by optical methods like optical microscopy or SEM (Scanning Electron Microscopy). To quantify particle shape it is possible to determine the aspect ratio, which is the ratio of the largest and the smallest dimensions. Ideal spherical particles have an aspect ratio of 1.

The term morphology of mineral particles is very close to "shape", but is commonly used to describe also how the surfaces of the particles are textured and how the individual particles may aggregate in a powder. The

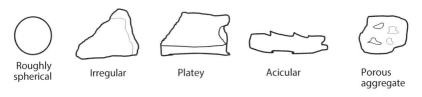

Roughly spherical   Irregular   Platey   Acicular   Porous aggregate

Some particle types likely to be found in common fillers

Figure 3.5 Typical particle shapes according to R. Rothon [18].

Figure 3.6 SEM of some alumimium hydrates: a) 4m²/g precipitated ATH with "porous edge surfaces", b) 4m²/g with smooth edge surfaces, c) loose agglomerates of submicron boehmite particles, d) submicron boehmite primary particles adhered to micron sizes aggregates.

surface texture can be very smooth, more uneven or even very porous. The primary crystal can occur as properly distinguished particle or as loosely attached agglomerates. Figure 3.6 shows some exemplary SEM pictures of aluminium hydroxides.

3.3.3.2 Particle Size Distribution

Particle size is measured by laser optical methods using optical particle counter technology (such as ISO 13320 or ASTM B822). Most manufacturers mention the median particles size D50 in their data sheets. But additionally, information on D10 and D90 is very useful to get an impression of the particle size distribution (PSD). D10 is a value representing the fine grains. 10 % of all particles are smaller than the D10 value. The D90 defines that 90 % of all particle are finer, so it provides a good indication of the

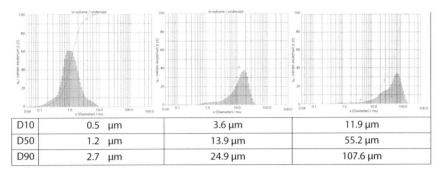

D10	0.5 µm	3.6 µm	11.9 µm
D50	1.2 µm	13.9 µm	55.2 µm
D90	2.7 µm	24.9 µm	107.6 µm

Figure 3.7 PSD curves of fine precipitated ATH (left), ground ATH (middle) and coarse ATH (logarithmic x-axis is equal) and corresponding D10, D50 and D90 values.

largest particles of the PSD. Some producers also give the D99 as a top cut value: only 1 % of particles are larger than the D99 value. The larger the difference between D10, D50 and D90, the broader is the PSD.

In general, a broader and coarser PSD is easier to process and especially beneficial for highly filled composites (see right PSD in Figure 3.7). Smaller particles can fill the free volumes built up by the larger particles, resulting in lower viscosity. Small particle size and PSD on the other hand is preferred for good physical properties, since small particles deteriorate mechanical properties less than large particulates. Smaller sized mineral flame retardant fillers with narrow PSD (Figure 3.7, left) are therefore preferred for thermoplastic and elastomeric compounds, in contrast to fibre reinforced thermosets based on reactive resins. In reactive resin processing the viscosity and rheology is extremely important for processability (see 3.6), while mechanical properties are dominated by the reinforcement.

3.3.3.3   Sieve Residue

Sieve residue determines coarse particle by a sieving method. For many functional filler end applications it is important to have a very low level of coarse particle or contaminants. ISO 66165, part 2 describes the procedure for sieving. A watery suspension of powder is prepared with the help of a dispersing agent. The suspension is poured onto the sieve and washed with water. For fine grains it is necessary to apply a vibration force. After oven drying the amount of residue is determined by a gravimetric method.

Sieve residue values for 325 mesh size or 45 micron are the most frequently used. Besides the absolute level, which is given as a percentage of the total mass, the nature of the residues is important for process control at the manufacturing site.

3.3.3.4   BET Surface Area

Specific surface area is a very important parameter for mineral filler characterization. The analysis method according to BET, named after the scientists Brunauer, Emmett and Teller, is based on a multilayer gas absorption of nitrogen (ISO 9277 or ASTM E2864).

The surface area of mineral fillers is very important because the properties of filled compounds are critically dependant on the interphase between the filler and the matrix. Besides the surfaces chemical composition, the size of the surface is most decisive. When properly dispersed, higher surface area per mass of filler means more interphase.

In comparison to precipitation processes, grinding and subsequent refining processes result in higher specific surface areas at the same or comparable particle sizes or vice versa. Table 3.5 compares precipitated and a ground ATH having the same BET value: the ground product is much coarser. This difference is because of the difference in particle shape and morphology. More irregular shapes, porosity and high aspect ratios result in higher BET values.

3.3.3.5   Oil Absorption

DIN EN ISO 787, part 5 gives a method for oil absorption measurement of fillers. A refined linseed oil is dispensed in small portions from a burette

Table 3.5 Fine precipitated versus grinded and classified ATH (magnification of the SEM for the precipitated product is half that of the grinded ATH).

SEM	ATH 7 m²/g precipitated	ATH 7 m²/g ground and classified
BET (m²/g)	7	7
D10 (µm)	0.5	1.2
D50 (µm)	1.1	7.0
D90 (µm)	2.7	17.0

and mixed with powder using palette knife or spatula until smooth consistency is obtained. The two corresponding ASTM standards (D281 and D1483) define the end point differently. Such differences in end point determination are the main reason why oil absorption is considered to be very much dependant on the operator. The test was originally developed as a formulation tool in the coatings industry. The result is a measure of the oil needed to wet the surface area of the filler and to fill the voids in and among the filler particles. Most often it is given in ml/100g. Beyond coating application, the value is especially useful for reactive resin applications, where flame retardant fillers are commonly used. Low oil absorption values will allow higher loadings at the same viscosity/rheology. Oil absorption is influenced by BET and PSD. Low BET and broad PSD favours low oil absorption.

For fillers used in PVC, oil absorption is preferably measured with phthalate plasticizers like DOP (di-octyl-phthalate), now more and more substituted by DINP (di-iso-nonyl-phthalate). In PVC and other thermoplastics, fillers with low oil absorption values are preferred for processing reasons and lower plasticizer demand. In elastomeric compounds higher oil absorption values correspond to higher tensile strength. When partially substituting reinforcing fillers like silica or carbon black, high surface area flame retardants with high oil absorption are therefore often welcomed in elastomers.

3.3.3.6   pH-Value/Specific Conductivity

The pH-value is measured according to DIN EN ISO 787, part 9 on a 10% suspension of filler in distilled water. The specific conductivity can be measured on suspensions of the same concentration (DIN EN ISO 787, part 14). Specific conductivity value is recorded in microSiemens per centimetre (μS/cm). In electrical applications, when the flame retardant compound fulfils insulation properties, a low specific conductivity of the flame retardant filler is crucial. This is especially true for wire & cable compounds used to insulate conductors, where mineral flame retardant fillers are used at high loading levels.

The pH-value is a basic property of the chemical compound. Fluctuations are very minor and there is practically no influence of powder parameters. But pH comes into play during formulation design. Considerations about potential pH-initiated reactions with other additives or the polymer matrix during processing or end application use should be part of formulator's functional specification. Often, such influences are only seen after ageing tests on composite materials.

3.3.3.7   Bulk Density and Powder Flowability

The physical density of a mineral flame retardant is a given constant. But bulk density and powder flowability are both correlated to fineness and influence powder handling in industrial processes. The higher the bulk density, the less air needs to be taken out in the mixing procedure, which is especially important for closed processes. Melt compounding output in batch mixers can be restricted with fluffy, low bulk density powders, allowing only limited batch sizes by mass. Low bulk density fillers very often correlate with poor powder flow, which is especially problematic in continuous compounding equipment using continuous operating feeders.

Bulk density can be given as tamped bulk density. Such a tamped bulk density can be useful for design of bag packaging sizes. For rating the handling performance, the loose bulk density is more advantageous (DIN EN ISO 23145-2).

To evaluate flow properties non-standardized funnel methods are broadly used - powder material is forced to flow through a funnel and the time needed for a fixed volume or mass is recorded. Powder rheometry generates a broader data base for comparison and is especially more accurate for non-free flowing powders [19]. A powder rheometer measures the flow properties of powders in terms of the energy needed to make them flow. A twisted blade is forced along a helical path down through a powder. The force on the blade is measured as it forces its way down. The basic flowability energy measured is the energy required to displace a constant volume of conditioned powder at a given flow pattern and flow rate. The Flow Rate Index (FRI) is the quotient of energy at tip speeds of 10 mm/s versus 100 mm/s. It is a measure of the extent to which the basic flowability energy is changed when the flow rate of the standard test is reduced by a factor of 10. Most powders have a value higher than 1 (more energy is needed when moving the powder more slowly). The stability index SI is a factor by which the measured energy changes during repeated testing. For ideal powders, SI = 1 is valid. Stability of powder flow is very important for industrial production, where powder handling equipment operates at fixed settings. Figure 3.8 shows a sketch of the equipment and a plot for different fine precipitated ATH of very similar particle size and BET surface area. Despite having very similar PSD and BET, distinct differences in stability index and flow rate index have been identified.

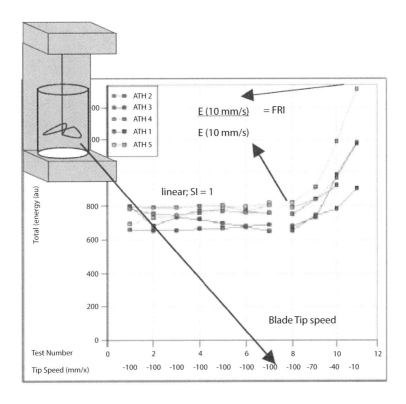

Material	Normalized Base Flow Energy (mJ/g)	Apparent bulk density (g/l)	Stability Index SI	Flow Rate Index (FRI)
A	11	430	1.05	1.50
B	14	380	1.17	1.73
C	9	490	1.19	2.21
D	10	530	1.18	1.86

Figure 3.8 Principle of powder rheometer and comparison of characteristic data for four fine precipitated ATH of 4 m²/g BET [20].

3.3.3.8   Thermal Stability/Loss on Ignition/Endothermic Heat

The thermal stability of a mineral flame retardant restricts its application during processing and/or end use. Processing or end use temperatures of the flame retardant polymer should always be well below the decomposition temperature of the flame retardant. If this is not guaranteed, the flame retardancy function cannot be fully utilised or will decrease during usage. Due to the release of volatile decomposition products, at least the optical appearance of the flame retardant plastic material will suffer, but most likely also mechanical properties. Additionally, the temperature range of decomposition between flame retardant and polymer matrix should match to maximise flame retardant effectivity.

To determine decomposition temperature and also mass loss during decomposition thermo gravimetric analysis according to ISO 11358/ASTM E1131 is used. The mass loss recorded can also be used to determine loss on ignition, when applying the temperature range defined under the relevant standards for loss on ignition (e.g. DIN 51006). Figure 3.9 displays TGA curves of the most important mineral filler flame retardants.

The loss on ignition is also reported as part of elemental or oxide analysis of a mineral. The volatile materials lost usually consists of "combined water" (hydrated water and hydroxy-compounds) and carbon dioxide from carbonates.

Endothermic heat uptake during decomposition is another important feature of mineral flame retardant fillers. It can be determined by differential scanning calorimetry (DSC). In general, endothermic heat expressed in kJ/kg correlates with flame retardant activity (see section 3.4).

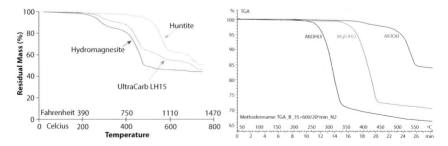

Figure 3.9 TGA curves for hydromagnesite, huntite and a commercial blend (left) [6] and ATH, MDH and AOH.

3.3.4   General Impact of Mineral FRs on Polymer Material Properties

Inorganic fillers are generally higher in density than organic polymers. Consequently filled polymer composites show higher densities than unfilled polymers. This may restrict some end applications where the low density of flame retardant polymer is crucial like in some transport applications. The impact on other important material properties is given as follows.

3.3.4.1   Optical Properties

The influence on compound colour is important, especially when colorants/pigments are added. The tone of the flame retardant filler may have an influence as such. Synthetic flame retardant fillers are generally whiter and very consistent in colour, while flame retardants based on mined minerals tend to off-white, sometimes also greenish white and consistency of the colour impression may fluctuate. In addition, sometimes absorption of organic dyes on filler surface indirectly impacts the colour impression. Last but not least the refractive index can impact the optical properties. In Table 3.2 the refractive indices of the mineral filler flame retardants are listed. In summary, even though most industrial mineral flame retardants are white powders, impact on pigmentation needs to be considered and potentially needs some adjustment compared to the unfilled reference.

3.3.4.2   Mechanical Properties

Increasing mineral filler load increases stiffness and material becomes increasingly brittle. When added at low to moderate levels, fillers can increase impact strength of filled polymers. But the loadings used to achieve flame retardancy (see also 3.4.1) are mostly in a range where impact strength or toughness is negatively affected. Formulation strategies which can improve impact properties based on modification of the polymer-filler interface are applicable. Despite such attempts, mineral flame retardant fillers are very rarely used as stand-alone solution to make engineering plastics fire resistant. But the development of combinations based on organic and mineral filler flame retardants for engineering plastics, used in electrical housings, is one research fields of highest activity (see also section 3.5.6 and Chapter 9 of this book).

High surface areas lead to a stiffening of compounds and composites, but are beneficial for flame retardant properties. Higher BET

Table 3.6 Synthetic boehmite (AOH) and MDH of low to moderate compared with natural MDH and a hydromagnesite/huntite with higher L/D-ratio in EVA compound.

Compound: EVA 38.3wt.-% Aminosilane 0.4wt.-% Mineral FR 61.3wt.-%	Synthetic AOH non surface treated	Synthetic MDH non surface treated	Natural MDH fatty acid treated	Natural Hydromagnesite/ Huntite blend fatty acid treated
BET (m^2/g)	6	7	9	9.5
L/D	1 – 2	4 - 6	7 – 10	10 – 20
Elongation at break (%)	221	129	95	83
Tensile strength (MPa)	16.8	14.3	11.7	11.0

fillers give higher LOI at the same loading (see also Figure 3.14 in 3.4.1). Likewise observed with all fillers, high aspect ratios of metal hydroxides may increase tensile strength, but very often limit elongation properties. When applied in cable compound formulations, ground ATH, brucite MDH and also huntite-hydromagnesite blends, all having relatively high L/D ratios of 5-10, give acceptable to good tensile properties, but struggle to fulfil required elongation properties. Table 3.6 compares synthethic boehmite and MDH with small L/D with natural MDH and a hydromagnesite/huntite blend having relatively high L/D ratios in EVA. Even though the two natural ground products were fatty acid treated, the drop in elongation compared to the synthetic fillers is striking.

3.3.4.3   Water Uptake and Chemical Resistance

Inorganic flame retardants, like all metal hydroxides can increase the water uptake of flame retardant polymeric materials. BET surface area has biggest influence. The higher the BET of the filler, the higher is the material water uptake. But it needs to be underlined that there are differences between products of same surface area too. Such differences can be related to the manufacturing process technology and become apparent when comparing offset products from different manufacturers. Even more relevant are differences related to surface treatments. This is shown in Figure 3.10 for 4 m^2/g fine precipitated ATH products loaded at 61.3 wt.-% in EVA. All

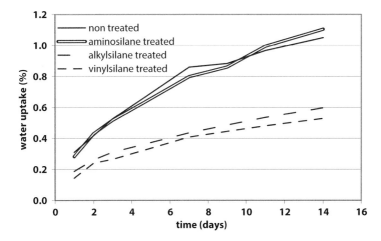

Figure 3.10 Water uptake of 4m^2/g ATH grades in EVA (61.3 wt.-% filler loading). Non coated and surface treated with different silanes are compared (specimens immersed in water at 70 °C for 14 days).

ATH originated from the same manufacturer (Nabaltec AG). The aminosilane treatment keeps the surface polar and consequently does not reduce the water uptake of the polymer compound, while vinylsilane and alkylsilane change the surface polarity to hydrophobic, resulting in significantly reduced water uptake.

The chemical nature and the production method influence the water uptake of mineral flame retardant compounds. In general it can be stated that metal hydroxides containing divalent earth alkaline metals, like magnesium and calcium, are more susceptible to water uptake than aluminium hydroxides. When comparing aluminium hydroxides, the slightly higher hydrophibicity of boehmites (AOH), due to a smaller number of hydroxide groups, also results in lower water uptake [21]. Ground fillers give higher water uptake than synthetic, precipitated products, presumably caused by the more irregular surface and surface defects generated during grinding.

In some applications chemical resistance of mineral flame retardant compounds is a required feature. Organic solvents, mineral oils and strong acids or bases are common test media. Organic substances do not attack mineral fillers, but the interphases generated in a filled polymer can cause issues. Nevertheless, the correct choice of polymer matrix is most important.

Because of its amphoteric nature, ATH shows good resistance against acids and basis. But in strong acids boehmite (AOH) is superior to ATH, making AOH a good candidate for such applications (e.g. battery cables, printed circuit boards). In Table 3.7, the solubilities of metal hydroxides in

Table 3.7 Comparison of metal hydrate solubility in battery acid (34 % H_2SO_4, D=1.25 g/cm³) at room temperature.

Product	Chemical composition	BET (m²/g)	Dissolved portion
Boehmite, AOH	AlOOH	3	0.2 %
Aluminium hydroxide, ATH	Al(OH)$_3$	3.5	8 %
Magnesium hydroxide, MDH	Mg(OH)$_2$ (1 % stearic acid)	8	48 %
Magnesium hydroxide, MDH	Mg(OH)$_2$	8	93 %

sulphuric acid are displayed. The dissolved portion after 24 h immersion is given. In such harsh conditions 8 % of the fine precipitated ATH dissolved, while the boehmite of comparable specific surface area was hardly dissolved at all. The non-surface treated ground MDH tested dissolved nearly completely, while a stearic acid surface treatment of this filler could reduce the dissolution to 48 %.

3.3.4.4 Thermal Properties

Mineral flame retardant composites show increased thermal conductivity and heat capacity. Metal hydrates are not excellent heat conductors, but are better than most polymers used as matrices. When used at high loadings, heat conductivity values of up to nearly 3 W/mK can be reached [22]. Figure 3.11 displays heat conductivity as a function of loading level in UP resin. ATH grades of different particle size have been used, but the influence of particle size and related differences of interphase sizes is negligible. Volume filling level is most decisive for thermal conductivity. Furthermore, the thermal expansion measured as CTE (coefficient of thermal expansion) of polymer composites are reduced when mineral filler flame retardants are present.

These thermal effects are purposely used for electrical encapsulations, where flame retardancy combined with thermal conductivity, low CTE and electrical insulation are frequently requested. E-mobility is a market of increasing importance for heat conductive components forming the battery case, the gap fillers in the battery stacks, various adhesives and gaskets

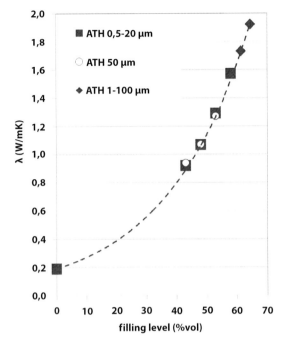

Figure 3.11 Heat conductivity as a function of ATH filling level in UP resin. Influence of particle size is minimal, loading level is important.

as well as the motor coil encapsulation. The thermal management of the electric motor and the lithium ion battery (LIB) plays a key role to obtain short charge and long life times of these key components of the electric drive systems.

3.3.4.5   Electrical Properties

Mineral flame retardants are non-conductive and are broadly used in insulation compounds. Nevertheless due to higher dielectrical leakage compared to polymers and because of traces of conductive materials originating from production processes, mineral filler flame retardants deteriorate the electrical insulation of polymers. The magnitude of such changes can significantly be limited when using fillers of very high chemical purity.

Low conductive ATH and MDH fillers, produced by precipitation followed by sophisticated washing processes, are nowadays standard products used in e.g. wire & cable applications.

3.3.4.6   Rheological Properties

Rheology is relevant during conversion processes of polymers. In thermoplastics, melt flow index (MFI) or melt volume rate (MVR) are used to generate material data and give a good indication of process-ability in, e.g., extrusion or injection moulding. Filled compounds have higher densities than neat polymer and therefore it makes more sense to compare melt flow on volume base (MVR given in $cm^3/10$ min). As matter of fact, MVR decreases significantly when a polymer is filled with mineral flame retardants, influencing processing rates like extrusion speed. But MVR is a single point measurement at defined temperature and weight load with practically no shear. Shear viscosity curves generated e.g. by capillary rheometers deliver important information for extrusion process configuration.

In thermoset production the reactive resin paste viscosity before cure is crucial. Different rheometers are used for evaluation. To compare filled with unfilled systems, relative viscosity values related to the unfilled resin paste are very often used.

Besides its relevance during processing, rheology of a flame retardant compound is also important in the case of fire. This becomes especially apparent when performing vertical flame tests. Polymers which flow too easily under the influence of heat may impact the fire performance of end products negatively. Melt dripping and feeding of flames by the polymer melt flowing into the burning area of test specimens may occur. Compound formulators need to find a good compromise between sufficiently good processing of mineral flame retardant compounds and sufficiently good fire retardant performance of the end products.

3.4   Working Principle of Hydrated Mineral Flame Retardants

The performance of metal hydrates as flame retardants is based on physical and chemical processes. The schematic drawing in Figure 3.12 shows the involved processes for a metal hydrate filled polymer in the case of a fire. In the presence of an ignition source - a flame or a hot object - the thermal decomposition of the metal hydroxide into the corresponding metal oxide and water takes place. During this process, energy is consumed from the ignition source, as the decomposition is an endothermic reaction. At the same time, the released water vapour cools the surface of the polymer and particularly dilutes the concentration of burnable gases in the surrounding area. The remaining metal oxide residue has a high

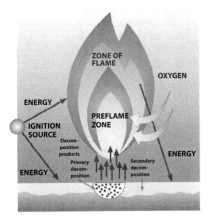

$$2\,Al(OH)_3 \xrightarrow[1075\,kJ/kg]{>200\,°C} Al_2O_3 + 3H_2O$$

$$Mg(OH)_2 \xrightarrow[1220\,kJ/kg]{>320\,°C} MgO + H_2O$$

$$2\,AlOOH \xrightarrow[700\,kJ/kg]{>340\,°C} Al_2O_3 + H_2O$$

Hydromagnesite

$$Mg_5(CO_3)_4(OH)_2.4H_2O \rightarrow 5MgO + 4CO_2 + 5H_2O$$

Huntite

$$Mg_3Ca(CO_3)_4 \rightarrow 3MgO + CaO + 4CO_2$$

HM/H commercial blend   >220°C ; 990kJ/kg

Figure 3.12 Scheme of the processes involved during burning of a metal hydroxide filled polymer (left) and decomposition reactions, their onset temperature and endothermic heat involved for ATH, MDH, AOH and HM/H blend.

internal surface where sooty particles, respectively polycyclic aromatic hydrocarbons, are absorbed. Additionally, the oxide layer acts as a barrier, disabling the further release of low molecular weight decomposition products as well as a heat barrier protecting the polymer against further decomposition.

When considering flame retardant fillers containing carbonates like Hydromagnesite/Huntite (HM/H), the mechanism stays basically the same, but includes the release of carbon dioxide which also dilutes the concentration of burnable gases released from the polymer matrix. Recent studies showed that huntite (H), which does not contribute much to the endothermic effect at the decomposition temperatures of most polymers, must have a positive overall effect. This was attributed to its ability to contribute and reinforce the inorganic and ash residues [23–25].

Such synergistic effect on ash residue stability has also been reported for boehmite AOH, when blended with ATH or MDH [26, 27].

The onset temperature of the above described mechanism is firstly determined by the decomposition temperature of the metal hydrate and secondly by the thermal stability of the polymer matrix. Ideally the thermal decomposition area of the inorganic filler and polymer overlap or are close to each other to give most effective flame retardant properties of the composite. The onset of mineral flame retardant decomposition may also restrict the processing and conversion conditions used during compound/composite production.

The hydrated forms of zinc stannates and borates also release water, but in addition to the process described above, a gas phase mechanism can be involved with these flame retardants. When used in halogenated polymers like PVC, borates and stannates are partially volatilised during combustion and may also exhibit vapor phase flame-inhibiting activity.

In halogen free systems, the charring ability of stannates and borates results in a complimentary action to the water release and endothermic effect of hydrated fillers. Both compounds appear to promote a thermally stable, cross linked or glassy char in the condensed phase, reducing toxic gas emissions and sometimes allowing lower overall filler loadings.

3.4.1   Filler Loading, Flammability and Flame Propagation

The "Limiting Oxygen Index" (LOI) in accordance with DIN EN ISO 4589-2 (ASTM D2863) is an indicative value for the assessment of flame retardancy and, in particular, the flammability of polymer materials. This describes the minimum concentration of oxygen in an oxygen/nitrogen mixture which is sufficient to support the combustion of a vertically oriented specimen. At lower oxygen concentration, the flame is extinguished. Hence, high LOI values indicate high flame retardancy or low flammability.

Figure 3.13 shows the LOI values for EVA-, Poly(ethylene-co-vinly acetate)-compounds filled with varying wt.-%-concentrations of three synthetic mineral flame retardants. Aluminium hydroxide (ATH), aluminium oxide hydroxide (AOH, boehmite) and magnesium hydroxide (MDH) all show a very similar dependency. Only when loading exceeds 50 wt.-% does the slope change from initially flat to steep. Because of the dominance of

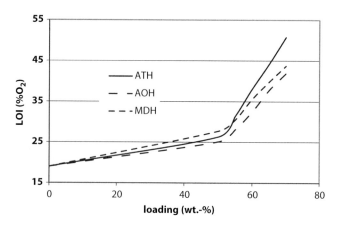

Figure 3.13 LOI in dependence of loading level in EVA (19 % VA-content).

the endothermic effect involved, a minimum filler loading is required to achieve decent fire retardant effects. Experience shows, that LOI values of at least 30 % oxygen are necessary to fulfil the basic flame retardancy requirements in a wide range of applications. In the EVA example given here, this corresponds to a filling level of around 55 wt.-% for ATH and MDH and 58 wt.-% for the boehmite (AOH). This higher loading for the monohydrate boehmite is linked to the lower loss on ignition compared to MDH and ATH.

However, the resulting oxygen index does not depend solely on the chemical composition of the metal hydroxide (and of course the polymer matrix), but also to a large extent on the fineness and the specific surface area of the filler additive. This relationship is shown for ATH, MDH and AOH in Figure 3.14 based on EVA (19 % VA-content) filled with 61.3 wt. % of the mineral flame retardants. The LOI values of the compounds produced via melt compounding were determined. The smaller the particles are and the higher their specific surface area is, the higher is the LOI value of the compound produced with it. This correlation is valid for all kinds of metal hydroxides.

Flame propagation is described by various flame retardant standards. The US standard UL 94 has asserted itself for polymer materials applied in electronics and has been adopted internationally (IEC 60695-1-10). In the most usual classification according to UL 94 V, the specimens are tested in vertical orientation. A standard Bunsen burner is used as the ignition source. The test bodies are repeatedly exposed to the flame and the after-burning time as well as any smouldering and dripping are used as rating

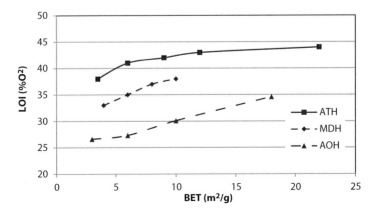

Figure 3.14 LOI in dependence of BET-surface area of metal hydroxide filler. 61.3 wt.-% filler in EVA (19 %VA).

Table 3.8 Typical loading levels to fulfill UL94 V0 classification at 3.2mm and corresponding LOI values for some selected polymers.

Polymer system	Filling level ATH [wt.-%]	LOI [%O$_2$]
Epoxy resins (EP)	55	39-45
Unsaturated polyester resins (UP)	58	35-50
Poly(ethylene-co-vinyl acetate) (EVA)	62	38-45
Polyethylene (PE)	65	32-36
Polypropylene (PP)	67	30-32

criteria. The descending classification is V0, V1, V2 and non classified - UL 94 V0 thus represents the highest level of fire protection. To reach the V0 classification in different polymer systems, different mineral flame retardant filling levels are required. It should be noted that UL 94 ratings are sample thickness dependent, with thinner specimens having a harder time obtaining the higher V ratings.

Experience shows that polymers which contain hetero-atoms, like oxygen, in their molecular structure, are inherently flame retarded to a certain degree. A certain level of flame retardancy can thus be reached using smaller amounts of flame retardants than in pure hydrocarbons. The facts described here are summarised in Table 3.8. Different polymers were filled with ATH. The filling level at which each polymer could be classified V0 according to UL 94 (specimen thickness 3.2 mm) and the LOI at this filling level were determined.

It is obvious that the polymers which contain hetero-atoms (ER UP, EVA) are classified V0 already at lower filling levels than the hydrocarbons PE and PP. The LOI values, also given in this table, are different, although all compounds are classified V0. There is no direct correlation between LOI values and UL 94 classifications. A correlation between these two values is limited to evaluation within one polymer system.

3.4.2   Smoke Suppression

All combustion is accompanied by smoke. Alongside the combustion conditions (intake of air, open fire or smouldering fire), the amount of smoke released depends to a large extent on the chemical composition of the materials involved. In the event of a fire, keeping the smoke density as low as possible can mean the difference between life and death. An escape route which remains visible for just a few more minutes can give many people the chance of saving their lives. Statistical analysis has shown that most victims die through inhalation of smoke and not as a result of the fire itself [28].

A common method for classifying materials in accordance with their smoke emissions in the event of fire is the smoke density determination in accordance with ASTM E662, also known by its older common name, the NBS (National Bureau of Standards) smoke chamber test. This method records the release of smoke over time on the basis of the optical density. The optical density is derived from the smoke released from the specimen into the chamber which weakens the transmission of a laser beam traversing the chamber. The more smoke that develops in the course of the test, the lower is the transmission and thus the higher is the optical density. As the smoke density depends to a great extent on the fire conditions, the test materials are subjected to a double test. In the "non-flaming" mode the specimens are only exposed to a radiant heater which causes smouldering, while in the "flaming" mode, the test is carried out using a small pilot flame. The maximum values of the smoke density are generally taken for comparison.

In Figure 3.15 smoke density curves over time are plotted for a UP resin filled with increasing ATH loading levels. As can be seen, with ATH the onset of smoke production in the initial stage of the test is retarded, the

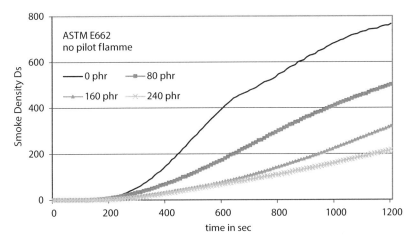

Figure 3.15 Smoke density over time of an UP resin loaded with increasing parts per hundred resin (phr) of ATH.

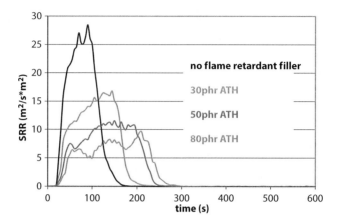

Figure 3.16 Smoke Rate Release over time measured by cone calorimeter at 50 kW/m². Material: plasticized PVC with increasing content of ATH.

smoke curves are flattened with increasing flame retardant filler content and the smoke values at 20 min are also significantly reduced.

Smoke can alternatively be measured during cone calorimetry (see 3.4.3). Smoke release rate (SRR) is plotted over time as shown in Figure 3.16, where the effect of increasing ATH loading for plasticized PVC is demonstrated. The addition of only 30 phr of ATH leads to a delay of smoke generation and the peak of smoke release rate is reduced. With increasing ATH loading all smoke values (total smoke, time to peak and peak of smoke release rate are further improved.

Most mineral flame retardant fillers, especially metal hydroxides, effectively reduce smoke generation. The metal oxides formed during their decompositions have high internal surfaces, where sooty particles, including polycyclic aromatic hydrocarbons, are easily absorbed. Zinc borates and zinc stannates promote the formation of stable chars, especially in the presence of metal hydroxides [29].

3.4.3   Heat Release

Fires spread through the dissipation of thermal energy which warms other materials in direct proximity until they ignite due to the heat itself or an igniting spark. The rate at which heat is released is a good indication whether a fire will grow and how quickly. Hence, materials which release a lot of heat will contribute a significant amount to the growth of a fire and will accelerate the sudden flashover of a fire.

Cone Calorimetry in accordance with ISO 5660/ASTM E1354 has established itself as a method to assess the heat emission of materials. Standardized specimens are exposed to a conical heat radiator which typically emits 20 - 100 kW/m² of thermal energy. Low-molecular weight products released by the decomposing polymer are ignited by an electric igniter and consequently set the specimen on fire. The amount of released heat is calculated using the oxygen consumption principle and plotted against the time (Heat Release Rate, HRR). In addition, the time to ignition, which should be as long as possible, the total amount of released heat (THR) and the maximum value of the heat release (PHRR) are used for the assessment. The smaller are the two latter values and the later the maximum is reached, the less the tested material will contribute to the propagation of flames.

The composition as well as the density of the combustion gases can also be analysed by Cone Calorimeter. The transmission rate of a laser beam in the chimney will provide the opacity of the combustion gases; the more smoke, the more the laser fails to reach the measuring diode. The content of carbon dioxide and toxic carbon monoxide in the gaseous combustion products is analysed using an IR-detector for those models so equipped with a multi-gas analyzer.

Figure 3.17 shows the heat release rate (HRR) versus time for plasticized PVC and increasing ATH content. Increasing the amount of ATH retards the time to ignition slightly, while the PHRR is substantially reduced and a considerable part of the heat is released at a later time during the measurement. Both effects are equivalent to a further improvement of the flame retardant properties and can, in the event of an actual fire scenario, delay or even prevent the fire from spreading to other objects (flashover).

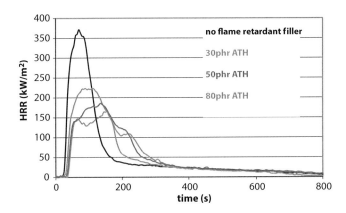

Figure 3.17 Heat Release Rate of plasticised PVC with increasing ATH load (at 50 kW/m²).

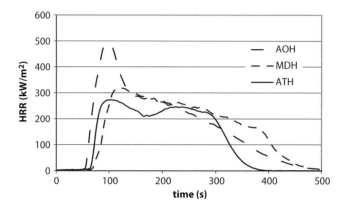

Figure 3.18 Heat Release Rate (HRR) of EVA (19 %VA) filled with 61.3 wt.-% of metal hydrate (at 50 kW/m²). AOH, MDH and ATH of comparable particle fineness are compared.

In Figure 3.18 ATH, MDH and AOH are compared in an EVA compound. All three metal hydrates are of similar fineness and surface area according to BET and have been used at the same loading.

All compounds resulted in approximately the same Total Heat Release, but as the curves in Figure 3.18 show, the heat is released differently. For ATH and MDH, which have very similar endothermic decomposition energies, time to ignition and HRR are relatively close together, boehmite (AOH) with its lower endothermic effect ignites earlier and gives higher HRR. The heat release process is finished earlier for ATH than for MDH and AOH based EVA compound. This corresponds with the decomposition temperature range of the metal hydrates, which is higher for the latter two metal hydroxides. But the high decomposition temperature of the boehmite (onset >340 °C) does not result in a longer time to ignition. Obviously, the lower endothermic decomposition energy of AOH is more decisive for the ignition process.

3.5   Thermoplastic and Elastomeric Applications

3.5.1   Compounding Technology

Sophisticated compounding technology is fundamental for the production of highly filled mineral flame retardant polymers at high consistency. Proper dispersion is a prerequisite for mineral flame retardants to display their full function and to keep other physical properties in the desired range. Special attention must be given to the process temperature limit preset by the decomposition temperature of the filler.

In the manufacture of filled elastomers as well as thermoplastics, the internal mixer (kneader) has proved most flexible. The addition of pre-weighed ingredients is done either in one shot or in a step process with multiple mixing cycles. The mixing procedure is controlled via temperature and/or torque monitoring. Another discontinuous kneader is the two roll mill, applied for elastomers and PVC. Elastomers require multiple mixing cycles starting with the melting of the polymer, while in case of PVC a dryblend of all ingredients is pre-manufactured on powder mixing equipment. This dryblend is often pre-compounded in a single screw or planetary kneader before using the two roll mill. Internal mixers and two roll mill compounders need a subsequent single screw extruder for the granulation process.

The two most important continuous processes applied for mineral flame retardant compounds are the co-kneader and the co-rotating twin screw extruder. They are considered to be superior in quality consistency, but most importantly continuous compounding enables higher output rates when a limited number of compound formulations are produced at high volumes. The co-rotating twin screw extruders used for mineral filled compounds commonly have an L/D ratio of at least 32 and two feeding ports. The second feed is preferably executed as a side screw feeder. The characteristic of the single screw co-kneader is an axial oscillation additional to the screw rotation. The compound is dispersed and conveyed through intermitting static kneading elements (bolts) and hence dispersed in a smooth oscillating manner. The co-kneader commonly has an L/D ratio of at least 11, but may be as high a 20 for special applications like silane cross-linkable compounds [30]. The mineral flame retardants are usually added via two feeding ports. Because of the oscillating screw, the co-kneader does not build up enough pressure to feed a melt granulator. A discharge screw arranged vertically to the kneading screw is used.

Non-intermeshing counter-rotating twin-screw technology is another continuous technique which can also be applied for mineral flame retardant compounds. Because of non-intermeshing screws and an L/D of only 6, this technology has some similarities with the discontinuous internal mixer principle. As with discontinuous and co-kneader compounding, a single screw hot feed extruder is needed for granulation.

In principle, screw geometries with low compression should be used for continuous compounding processes. Specific screw designs are based on the compound formulations, degree of filling and the mineral flame retardant used.

The relevance of powder flow properties, especially for continuous dosing and compounding, has already been discussed (see 3.3.3.7). To ensure constant filling levels, gravimetric dosing equipment is used for mineral flame retardant feeding. Vacuum degassing and/or ventilation openings ensure the removal of entrapped air and volatile organic substances.

3.5.2   Compound Formulation Principals

The basic aspects which need to be considered when designing a flame retardant compound based on mineral fillers have been given in chapter 3, 4. Additionally, in 3.3.3 and 3.3.4, powder parameters of mineral flame retardant fillers and their impact on mechanical, and other physical properties of polymer compounds and composites have been discussed. The focus here is to give a brief overview on compound formulation design strategy.

Target compound specifications are given by the end application and very often also define the polymer matrix required. The polymer or polymer blends used have the most significant influence on all physical properties, but mineral fillers, especially when used at high loadings, can significantly modify them. If specification allows, it is recommended to use polymers which accept inorganic fillers easily as blend components. In case of polyolefins, metallocene grades are known to be beneficial. In general polymer grades with relatively high MFI or low Mooney viscosity are recommended, due the scale down effect on melt flow properties caused by the filler load. The use of processing aids, like internal and external lubricants, is an additional measure, but not an alternative to the proper choice of polymer. Processing aids are low molecular weight organics and very often deteriorate flame retardant properties as well as tensile properties. They should be used at low levels and with great care.

With the exemption of filling masses used for, e.g., in cable bedding or aluminium composite panels, most compounds need to fulfil minimum mechanical requirements. Fine micron sized fillers give superior properties compared to coarser fillers. Therefore, most thermoplastic and elastomeric compounds are formulated based on fillers below 10 μm particle size. Additionally, coupling agents are needed to achieve good mechanical properties. Such coupling agents function as chemical mediators between the polar, hydroxide groups carrying surface of the mineral and the less polar or even non-polar polymer. Coupling agents can already be part of the mineral flame retardant, when the filler surface has been precoated with a bifunctional silane (see also 3.3.2.4). More often, the coupling agent is added as a separate ingredient to the compound formulation. In such case, the chemical bond between the organosilane and the filler surface is formed only during melt compounding. The coating of the mineral filler takes place "*in-situ*". The addition of coupling agent during melt compounding is therefore known as "*in-situ*" coupling process.

Alongside the use of bifunctional organosilanes, polymers grafted with maleic anhydride (in the case of PE shown as PE-g-MA), are the most widely used coupling agents (see Figure 3.19). While maleic anhydride grafted products are generally supplied in pellet form, organosilanes are liquid. If dosing of very small quantities of fluids is not favourable, they can also be used as a masterbatch on a porous polymer carrier. Other low molecular weight coupling agents include organotitanates, functionalized fatty acids and silicones. None of these latter products is used in large industrial scale, presumably because their use in the cost effective *in-situ* process is difficult.

A lot of development by polymer and additive manufacturers has assisted in a broader use of mineral flame retardants. Despite improved and more versatile coupling agent technology, modern polymer synthetic methods, like metallocene technology, provide polymers with a very high filler acceptance. Nevertheless, there are still some areas of applications where mineral

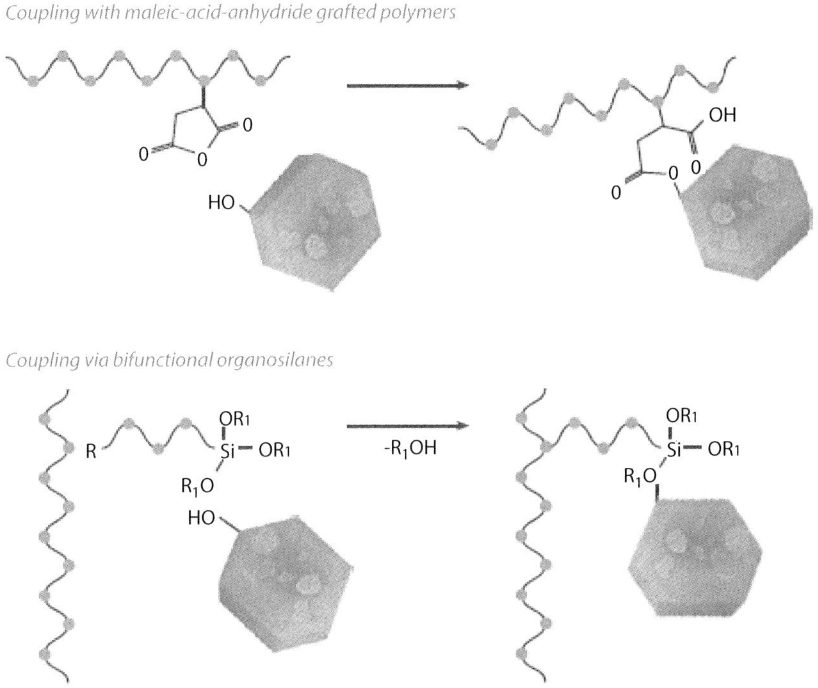

Coupling with maleic-acid-anhydride grafted polymers

Coupling via bifunctional organosilanes

Figure 3.19 Chemical working function of industrially most important coupling agents.

filler flame retardants as a stand-alone solution are extremely uncommon or impossible. Firstly engineering plastics produced by injection moulding of glass fibre reinforced compounds requiring very good impact properties and secondly, textile yarns, which do not allow high filler loadings because of the melt spinning process, are the most important applications to be mentioned here. Nevertheless combinations of mineral and organic flame retardants have already been established and are an area of ongoing development. Minerals are used at loading levels up to 10 wt.-% within such formulations (see 3.5.6).

3.5.3   Wire & Cable

Cables transmit energy and an ever growing amount of data. Large numbers of cables are installed to connect rooms horizontally and floors vertically in multi-story buildings. Cables are also found in elevators connecting these floors from the basement up to the top of the building. Transport vehicles like cars, trains, ships or airplanes and machines, industrial robots and computers are equipped and connected with energy and sensor cables. Low up to medium voltage cables, telephone and LAN cables are part of our everyday life.

Independent from the conductor/transmitter material used, all cables generally contain a substantial amount of inherently flammable polymer materials as insulation, sheathing, or bedding. Cables therefore represent a significant quantity of fuel for fires. Besides being a potential source of ignition due to overheating, arcing, short circuiting or other electrical faults, cables can tremendously contribute to the spread of fires, simply because they form an interpenetrating network.

Figure 3.20 shows a pyramid of commonly applied cable standards for buildings. The grouping of the standards was done according to the fire safety level of cables. Fire resistant or fire rated cables continue to operate in the presence of a fire and have to guarantee circuit integrity for a certain period of time and intensity of fire. Fire alarm and emergency lighting cables fall under this highest category and have to fulfil severe performance tests also under quenching water and mechanical stress.

Essential for all cables applied in buildings is their resistance to flame propagation, as this is the most important hazard originating from cables. Different standards and test regimes exist worldwide, but all of these tests have burning length and heat release criteria included.

The presence and intensity of smoke influences the escape and survival probability of fire victims and has therefore found entry in many standards. Low acid gas emission is a criteria frequently requested by insurance

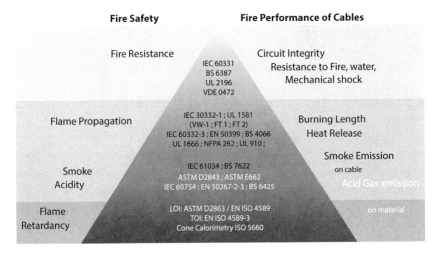

Figure 3.20 Pyramid of commonly applied cable standards for buildings according to fire safety terms (left) and fire performance categories.

companies. This is because high levels of acid gases emitted will cause significant consequential damage in conjunction with fire fighting water. Other than that, acidic gases are also irritants and may hinder fire victims to easily escape.

Terms like fire resistance and flame propagation are closely connected with real life scenarios in buildings. The correlating cable performance parameters verified in large scale tests are not only influenced by the polymer material, but also by the cable design. Smoke generation, acid gas emission and flame retardancy on the other hand side are material properties, determined by the choice of raw materials.

Flame retardant wire and cables (W&C) is by far the largest market for mineral filler flame retardants. According to Freedonia the total flame retardant demand for W&C in 2018 is around 587 TMT [31]. Based market company internal market intelligence the author estimates the ATH volume used in W&C insulation and sheathing to be around 400 TMT in 2018. It is believed that ATH will gain further market share against ATO within the next years. This will be realized by substitution of ATO in PVC and the further switch from PVC based compound technology to polyolefin based HFFR technology.

As of today, plasticised PVC (Poly(vinyl-chloride)) is still the most important polymeric material used in the cable industry. While showing a high degree of inherent flame retardancy, the presence of flammable plasticisers in soft PVC makes non-flame retardant PVC cables an important fuel source. Antimony trioxide (ATO) is a very effective flame retardant for flexible PVC when ignitability and flame propagation are considered. This is because ATO only works with halogens to assist in vapour phase combustion inhibition, as described earlier in this chapter. While ATO is an inorganic filler and is non-halogenated, it is not considered to be part of non-halogenated flame retardant technology because it only works in the presence of halogenated flame retardants and halogenated polymers.

Modern Low Smoke Flame Retardant (LSFR) PVC compounds are made by incorporating metal hydrate flame retardants like ATH or MDH. Zinc-hydroxy-stannates and Zinc-borates are additionally used in low quantities as synergistic flame retardants. Zinc Hydroxystannate acts as a synergist in PVC, acting in conjunction with chlorine to prevent the spread of flame and also to reduce smoke by char formation [32] (see Table 3.9).

Based on these flame retardants sheathing materials with low smoke generation (see also Figure 3.16 in section 3.4.2) and low hydrochloric (HCl) acid emission can be designed, also fulfilling the commonly requested ignition and flame propagation criteria. Table 3.10 lists the most important mineral flame retardants used in PVC and other halogenated polymers, their loading levels and main cable applications.

Highly flame retardant PVC compounds can also be used in one of the most severe construction situations found for fire retardant cables: plenum spaces. A plenum describes the space between the structural ceiling and a dropped ceiling. Plenum cables are laid in these plenum spaces of buildings. Two factors make this construction situation critical for fire propagation. Firstly, these spaces are actively ventilated as they are used for air

Table 3.9 Typical formulation and compound properties of LSFR-PVC compounds.

Component	phr	
PVC, K=70	100	*Typical properties*
DINP	50	Tensile Strength > 12 MPa
Ca/Zn stabiliser	5	Elongation at break > 200 %
PE-wax	0.5	LOI $> 32\%O_2$
ATH (4 m²/g)	75	
Zinc-borate or	5	
Zinc-hydroxy-stannate		

Table 3.10 Mineral flame retardant use levels in PVC and other halogenated polymers and main cable applications.

Filler flame retardant	Typical loadings	Polymers/ compounds	Main applications
Antimony trioxide (ATO)	1 – 3 wt.-% (3-10 phr) only for FR-PVC	FR-PVC • Flame Retardant PVC	Electrical cables - Low voltage - Medium voltage
Aluminium tri-hydroxide (ATH)	5 – 15 wt.% (10 – 30 phr) in FR-PVC (combined with ATO)	LSFR-PVC • Low Smoke Flame • Retardant PVC	
Magnesium di-hydroxide (MDH) - Synthetic - natural	20 – 50 wt.-% (50 – 100 phr) in LSFR-PVC	CPVC • Chlorinated PVC	Data cables - LAN cables - Telephone cables
Huntite/ Hydromagnesite	5 – 15 wt.-% (as synergist)	CR • Chlorinated Rubber	
Zinc-borates	2 – 5 wt.-% (as synergist)		
Zinc-hydroxy-stannates			- Plenum cables

circulation and heating / air conditioning systems. Secondly, over the lifetime of offices and functional buildings many new cables, especially new generations of data cables, are installed in such plenums, while abandoned cables are commonly not dismantled. So the amount of burnable mass increases over the years. In the US cables used in plenums are regulated under NFPA 90A and NFPA 262. Besides fluorinated ethylene polymers, only speciality PVC compounds containing large amounts of flame retardants can meet these severe requirements.

Halogen-free flame retardant (HFFR) or low-smoke free-of-halogen (LSFOH) polymer compounds have gained a significant market share, especially in Europe. Table 3.11 lists polymers and the corresponding mineral flame retardants, their loadings and main cable applications.

By far the most important compounds by volume used for HFFR wire and cables (W&C) are based on blends of EVA (poly(ethylene-co-vinyl

Table 3.11 Overview on HFFR compounds regarding typical mineral flame retardant loading, polymers used and application areas.

Mineral flame retardant	Typical loadings	Polymers	Applications
Aluminium tri-hydroxide (ATH) Magnesium di-hydroxide (MDH) - Synthetic - natural	58 – 67 wt.-% up to 80 wt.-% combined loading with chalk in bedding compounds	Polyolefins • Low-density polyethylene (LDPE) • Poly-ethylene vinyl- acetate copolymer (EVA) • Poly-ethylen-co-butene • Poly-ethylen-co-octene	Electrical cables - Low voltage - Medium voltage - PV cables - Emergency lighting
Huntite/ Hydromagnesite	5 – 30 wt.-% (used as partial replacement for ATH or MDH)		Control cables - Fire alarm cables
Boehmite (AOH) Zinc-borates Zinc-hydroxy-stannates	3 – 15 wt.-% (used as synergist)	Elastomers • Natural Rubber (NR) • Poly-Ethylene-Diene- • Rubbers (EPDM) • Poly-Styrene-Butadiene- • Rubbers (SBR) • Silicone rubbers (SiR)	Data cables - LAN cables - Telephone cables Control cables - lift cables - fire alarm cables
Aluminium tri-hydroxide (ATH)	20 – 55 wt.-% (used in combination with P- and or N-FR)		
		Thermoplastic Elastomers (TPE) Thermoplastic Poly urethanes (TPU)	

acetate)) and LLDPE (Linear Low Density Polyethylene) using fine precipitated ATH as the sole flame retardant filler at loadings of 60 – 65 wt.-% (see also Table 3.12). Coupling agents guarantee the required physical properties. These compounds are obtained by a standard extrusion processes without any cross-linking.

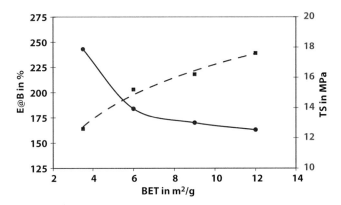

Figure 3.21 Tensile strength (TS, broken line) and elongation at break (E@B) in dependence of BET surface area of fine precipitated ATH for an EVA compound filled with 61.3 wt.-%.

Figure 3.21 displays the influence of BET surface area of fine precipitated ATH on tensile - elongation properties for an EVA compound (EVA with 26 % VA, aminosilane coupled, 61.3 wt.-% ATH). Tensile strength (TS) goes parallel with surface area, while elongation at break decreases. Consequently, ATH with higher BET are mainly used in elastomeric formulations, where tensile properties are more critical for filled systems. In thermoplastic compounds, elongation suffers more from filler load and hence, ATH grades of BET ranging from 4-6 m²/g are preferred.

Thermoset type or cross-linkable compounds are used whenever special requirements in regard to abrasion and chemical or temperature resistance are required, like, e.g., lift/elevator cables or cables used for photovoltaic modules. Elastomers based on EPDM (poly-ethylene-propylene-diene-copolymer) or EVM (EVA grades with high contents of vinyl-acetate) are an option. Cross-linking is induced by sulphur (EPDM) or peroxides. These compounds also contain ATH at loadings of 50 – 60 wt.-%, sometimes combined with zinc-borates as FR-synergist and other mineral fillers like silica, used to guarantee sufficient strength and hardness. A simple starting formulations based on a peroxide cross-linked EVM formulation is given in Table 3.13 (right).

Speciality cables for severe environments are based on elastomeric specialities like e.g. HNBR (hydrated nitrile butadiene rubber). HNBR can enable good media resistance against oil and mud and good mechanical low temperature resistance. Such requirements are typical for off shore cables. Additionally flame retardancy is also a must. In such applications, fine precipitated ATH grades with very high BET have proven their performance in HNBR/EVM blends [33].

Table 3.12 Exemplary basic HFFR compound formulations based on PE/EVA and compound properties.

Component	wt.-%	Component	wt.-%
LLDPE	15.8	LLDPE	9.66
EVA	19.0	EVA	29
PEgMA (coupling)	5	Vinylsilane /Peroxide (coupling) + Process aid	1.14
Stabiliser/Antioxidant	0.2	Stabiliser/Antioxidant	0.2
ATH, 4 m²/g	60	ATH, 4 m²/g	60
Characteristic compound data		*Characteristic compound data*	
Tensile Strength [MPa]	13.1 MPa	Tensile Strength [MPa]	11 MPa
Elongation at Break [%]	188 %	Elongation at Break [%]	260 %
MVR (cm3/10min, 160°C; 21.6kg)	4.3	MVR (cm3/10min, 160°C; 21.6kg)	9.4
LOI [%O$_2$]	35	LOI [%O$_2$]	37

Alternatively thermoplastic compounds may also be cross-linked by E-beam cure after wire / cable extrusion. Depending on the processing temperatures, sometimes MDH is preferred over ATH. E-beam cross-linking is a procedure often used for the production of photovoltaic cables, where resistance against sun light and high temperatures are most crucial requirements.

Thermoplastic base materials like TPU (Thermoplastic Poly-Urethane) give inherently good abrasion resistance. In this case the combination with phosphorous- or nitrogen-based flame retardants is an established technology. Melamine derivatives, organic phosphate and metal hydroxides, ATH or MDH, are used in different ratios depending on the manufacturers technology [34]. Table 3.13 (left) gives an example for HFFR TPU realized by a combination of an ATH treated with an alkyl silane and bisphenol-A bis(diphenyl phosphate) (abbreviated BDP). This compound gives very high LOI and despite the very high ATH loading excellent

Table 3.13 Exemplary basic HFFR compound formulations and properties based on TPU (left) and EVM (peroxide cross-linked).

Component	wt.-%	Component	phr
TPU	33.5	EVM (80 % VA)	100
ATH (4 m²/g, coated with alkyl silane)	60.0	ATH (6 m²/g)	160
BDP (Bispenol-A bis(diphenyl phosphate)	4.0	Zinc borate	10
Epoxyresin	2.0	Plasticiser	6
Antioxidant	0.5	Additives	9.5
		Cross-linking system	7
Characteristic compound data		*Characteristic compound data*	
Tensile Strength [MPa]	12 MPa	Tensile Strength [MPa]	7.2 MPa
Elongation at Break [%]	561 %	Elongation at Break [%]	298 %
MVR (cm3/10min, 160°C; 21.6kg)	4.3	H 23°C (Shore A)	79
LOI [%O_2]	56.7	LOI [%O_2]	37

elongation properties. Due to its low level of BDP phosphate ester this compound is very low in smoke emission [35].

Electrical cables are usually designed with a filling mass surrounding the individual insulated conductors, thus filling the empty space between the insulated conductors and the cable sheath. These bedding compounds are also mineral filled. In order to design flame retarded cables with lowest possible fire load, mineral filler flame retardants are also used here, partially or fully substituting calcium carbonate. The mechanical requirements for a filling mass are very low so that such compounds are designed with extremely high loadings of ground mineral flame retardant grades.

3.5.4 Other Construction Products

A very important application for mineral flame retardants is thermal insulation foams made of PVC/NBR (NBR = Nitrile Butadiene Rubber) blends. These elastomeric materials have to fulfil severe fire resistant requirements, especially when used in a linear product, e.g., for insulation of heating and plumbing pipe work in multi storey buildings.

A proper cell structure in combination with a low foam density is a must for thermal insulation foams. Consequently very fine metal hydrate flame retardant fillers are required to enable closed cell foam formation. The total loading is restricted by the foam density requirement. Mineral flame retardants like ATH are indispensable for reduced smoke release, but to fulfil requirements of construction product codes like EN 13501-1, class B or BS 476: part 6, class 0 a mix of flame retardants has to be used. Brominated flame retardants (Br-FR) like 1,2-bis-(pentabromophenyl)ethane (Decabromo-diphenylethane, DPDE) and antimony trioxide are used, although there are increasing regulations and requirements making it difficult to use certain Br FR compounds in insulation foams. Additionally chlorinated paraffins and phosphate esters used as plasticisers also contribute to reduced fire spread properties.

Current development work in industry focuses on the reduction or even elimination of brominated flame retardants. Because of ongoing discussion of the health and safety status of chlorinated-paraffin, formulators also should consider substitution of this plasticiser in future. Fully halogen free flame retardant flexible insulation foams based on EPDM and NBR/EVM blends are under development.

Flexible polyurethane (PUR) foams are used in construction or vehicle applications as thermal insulation or sound deadening and can be flame retardant by a post treatment process. The compressed foam is allowed to expand in an acrylic suspension of finely ground ATH. The ATH and the acrylic resin penetrate into the pores by capillary forces. After drying the resin and ATH adhere to the PUR.

Roofing membranes based on PVC, bitumen, or polyolefins very often require flame retardants. Mineral filler flame retardants, especially natural minerals are used in bitumen. Because bitumen is inherently flame retardant, the loadings required are often in the range of only 10 – 15 wt.-%. PVC based membranes are very often formulated in combination with chalk for cost saving reasons. Flame retardancy of polyolefin (PO) based roofing membranes is more critical and co-polymers with high filler acceptance and finer mineral flame retardant fillers at higher loadings are used. Sea water grades of MDH are most common in North America, while European manufacturers favour the use of precipitated or finely ground ATH.

Tarpaulins used for truck covers, temporary facade coverage or roofing for large tents are made of flame retardant PVC coated fabrics. Finely ground or precipitated ATH is used in these applications. The latter is preferred because of the surface finish requirements of the end product and process benefits because of proper fabric wetting by the PVC plastisol. For higher flame retardant requirements, more phosphate at the cost of phthalate plasticisers is used. Loadings of up to 50phr of ATH are realised in fabric coating. Loadings of only 30phr give more flexible fabrics with lower fire performance, but very often sufficient for automotive applications according to FMVSS 302. An exemplary formulation is displayed in Table 3.14.

Flooring based on PVC, EPDM or other elastomers used in public areas, like schools, hospitals or airports, are flame retardant (e.g. according to EN ISO 11925-2). Mineral filler flame retardant, ground and fine precipitated grades, are used in conjunction with liquid phosphate flame retardants and other minerals, like siliceous earth fillers. The latter fillers are used to improve other relevant physical properties like abrasion and tear resistance. The flammability of carpets used in public buildings and public transport vehicles can be controlled via the tuft or the backing. The tuft can be laminated with a flame retardant treatment or the fibres are produced based on flame retardant compounds. Mineral flame retardants are rarely used in the tuft, while the carpet backing made from latex can be heavily filled with mineral flame retardants. Coarse ATH grades are used in such carpet backings

Facades of multi storey buildings have to fulfil construction product standards, in which flame retardancy is one mandatory request. Besides fibreglass reinforced polymer composite panels based on cross-linked unsaturated polyester resins produced by SMC (see 3.6.1), aluminium composite panels (ACP) have gained significant market relevance as facade

Table 3.14 Basic formulation band for PVC plastisol used for coated fabrics. ATH loading and plasticiser composition dependant on specific performance requirements.

Component	Phr
PVC, K=57	100
Ca/Zn or Ba/Zn stabiliser	3
Phosphate plasticiser	0 – 45
Phthalate plasticiser	15 – 70
ATH	30 – 50

cladding. ACP is based on an inner and outer aluminium sheet separated by a polyolefinic filling mass. Most ACP is produced by a continuous T-die co-extrusion process, simultaneously applying polymer adhesive layers on both sides of the mineral filled compound sheet and subsequently laminating the polymer with aluminium sheets by roll milling. To fulfil relevant flame retardant construction product standards (EN 13501-1, class B, BS 476-6/7 class 0, ASTM-E 84), the polyolefin (PO) is filled with 67 – 80 wt.-% of mineral flame retardants. Due to the aluminium sheet structure mechanical properties of the filling compound are of minor importance. Ground ATH and natural, ground MDH (brucite) fillers are used.

Most extruded profiles, conduits and pipes are based on rigid PVC, which is inherently flame retardant. Such products normally do not need additional flame retardants. In some cases ATO is used at low loadings of 1-2 wt.-%. ATO can be substituted by zinc hydroxystannate at similar or slightly higher loading. Sometimes also metal hydroxides are used in rigid PVC, although they require higher addition levels risking deterioration of mechanical properties. But for profiles and sheets which contain fillers for other reasons, ATH or MDH is used. Flame retardant wood plastic composite (WPC) based on wood floor or other cellulose fillers is a growing application of this kind. Because of increasing requests for halogen free construction products, polyolefin based HFFR compounds gain importance in such rigid extrudates. The compound formulations have been developed in the style of HFFR cable compounds, sometimes using zinc borates or additionally some organic flame retardant synergist to keep mineral filler load at somewhat lower levels [36].

3.5.5   Special Applications

Conveyer belts installed in public buildings, like airports and in underground mines, have to fulfil flame retardant requirements. They are based on elastomers like natural rubber (NR) nitrile butadiene rubbers (NBR), Ethylene-diene-rubber (EPDM) and blends or are made of plasticized PVC. In both cases, combinations of mineral filler flame retardants and phosphate ester plasticizers are used. Table 3.15 displays an exemplary starting formulation for a conveyer belt compound based on PVC.

Infrastructure development projects in developing countries and upgrading of electrical distribution nets in industrialized countries, necessary because of increasingly decentralized renewable energy production, lead to an enormous demand for high voltage insulators in electrical distribution grids. Composite high voltage insulators are nowadays increasingly used for outdoor high voltage insulation. Low weight, higher

Table 3.15 Starting formulation for conveyer belts.

Component	phr
PVC, K=57	100
Ca/Zn or Ba/Zn stabiliser	3
Phosphate plasticiser	50
ATH/MDH/HM/M-blend	50
GCC (Ground Calcium Carbonate)	10
ATO	4

mechanical strength to weight ratio, resistance to vandalism and better performance in the presence of heavy pollution make them superior to traditional ceramic or glass insulators. Composite high voltage insulators consist of a glass fibre reinforced epoxy resin core which is covered by a silicone rubber shell. This shell is injection moulded onto the core and has the corona ring structure well known for high voltage insulators. Besides reinforcing silica filler, high end insulators contain fine precipitated ATH and are cured at elevated temperature (called HTV, high temperature vulcanizing silicone). HTV silicon as such shows good creep resistance because of the self-cleansing property of silicone surfaces. Creeping or current transport along the insulator surface initiated by corona discharges is supported by dirt particle and /or water films. The hydrophobic surface makes water droplets fall off, taking away dust and dirt. ATH is used to increase the corrosion resistance of the HTV silicon. The ATH particles act mainly in the bulk of the HTV silicone by avoiding spark trees to penetrate through the silicone. The finely dispersed ATH particles decompose to alumina when hit by an electrical current, taking up the electrical energy of the spark or charge. ATH is actively sacrificed to increase the lifetime of the silicone matrix. It has been shown that this mechanism is also relevant at the silicone surface during corona discharging, resulting in distinct differences between ATH containing and non ATH filled silicone rubbers [37].

Other than in case of a fire scenario, when all mineral flame retardant filler is consumed in a single event within a very short period of time, the ATH decomposition in a HTV silicone rubber is a grouping of occurrences over a long period of time, making HTV high voltage insulator to easily last for 40 years. Fine precipitated ATH is used at loadings ranging from 40-60 wt.-%. Most formulations use coated ATH with hydrophobic and/or vinyl groups. In any case, vinylsilane is added *in situ* to guarantee a properly cross-linked network structure of filler and silicone matrix.

3.5.6   Engineering Plastics for E&E Applications

Engineering plastics are used in applications where special performance is needed, e.g. high impact strength, high durability, low abrasion, high resistivity towards oil and other aggressive media or high thermal- and electrical resistance. To achieve these goals, compounders blend matrix resin with glass fibres and various additives. Most common as matrix polymers are poly-condensates like polyamide and polyesters. Injection molding is the dominating conversion process to produce switches, sockets, connectors or under-the-hood plastic parts, such as oil sumps, intake manifold covers and fuse boxes. Even supporting structures in cars are processed from glass fibre reinforced polyamides.

Especially electrical and electronic appliances made of engineering plastics have to fulfill flame retardant requirements, mostly according to UL-94 V-0 rating. While halogenated products are still dominant, non-halogenated flame retardants are gaining significant market share. Phosphorous- and/ or nitrogen-containing flame retardants have been developed offering excellent replacement solutions. Mineral flame retardants have not been used broadly so far. The main reason is deterioration of mechanical properties when used at loadings fulfilling UL 94 V-0. Additionally, the number of mineral flame retardants processable in polyamide and polyester matrices is limited by their temperature stability. Nevertheless, MDH is used as stand-alone solution for some commodity applications. But in the case of MDH, the comparatively high alkalinity is an additional drawback. In combination with moisture traces the alkaline surface of MDH catalyses the hydrolysis of ester and amide bonds leading to polymer chain scission, further impacting mechanical properties. Boehmite (AOH) is less alkaline but is also less effective as flame retardant because of less releasable water. AOH is not used as the sole flame retardant, but has proven to be an excellent synergist to phosphorous and nitrogen based flame retardants in polyamide and polyesters [38–40].

Besides its synergistic flame retardant performance, AOH acts as an acid scavenger. AOH can capture aggressive by-products formed during processing of phosphorous containing flame retardants, helping to avoid corrosion of compounding and injection moulding screws.

Figure 3.22 displays impact resistance data for PBT compounds filled with 20 wt.-% of glass fibre. When adding flame retardants, the charpy and notched charpy values suffer in general. When combining boehmite with

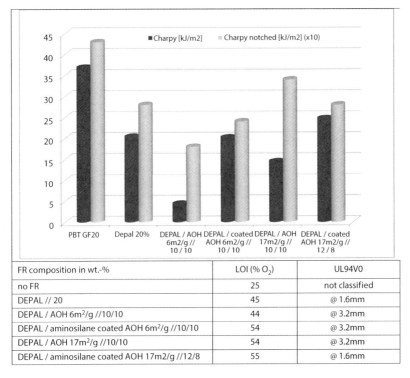

FR composition in wt.-%	LOI (% O$_2$)	UL94V0
no FR	25	not classified
DEPAL // 20	45	@ 1.6mm
DEPAL / AOH 6m^2/g //10/10	44	@ 3.2mm
DEPAL / aminosilane coated AOH 6m^2/g //10/10	54	@ 3.2mm
DEPAL / AOH 17m^2/g //10/10	54	@ 3.2mm
DEPAL / aminosilane coated AOH 17m^2/g //12/8	55	@ 1.6mm

Figure 3.22 Impact resistance, LOI and UL94V rating of PBT compound containing 20 wt.-% glass fibre (Charpy notched values are multiplied by 10).

metal phosphinate (aluminium-tris-(di-ethyl-phosphinate), abbreviated DEPAL), the proper ratio of the two flame retardant components and the choice of AOH filler influences the mechanical and flame retardant properties. While LOI generally increases when adding boehmite, the UL 94 classification may not necessarily. Best LOI and good physical properties are gained when combining DEPAL with an ultrafine AOH of 17 m^2/g specific surface area. This is further improved when applying a surface treated AOH. The ultrafine AOH with BET of 17 m^2/g and surface treated with aminosilane used at 8 wt.-% together with 12 wt.-% DEPAL resulted in the best combination of flame retardant and impact properties. The amino groups help to attach the AOH particles to the polymer backbone by chemical reaction with the ester groups of the PBT and amide formation or via hydrogen bridge building.

Another parameter influenced positively by a combination AOH mineral flame retardant filler and DEPAL phosphorous flame retardant is processability. Figure 3.23 (left) plots MVR values of non-flame retardant PBT

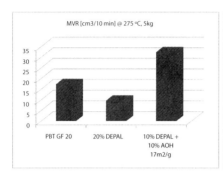 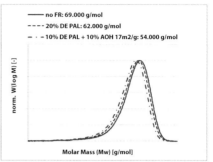

Figure 3.23 MVR of PBT compounds (20 % glass fibre) in dependence of FR-composition (left). Despite large MVR differences, no substantial change in molecular weight is detectable by GPC (right).

versus flame retardant compounds based on pure DEPAL and the AOH-DEPAL combination. The flame retardant combination shows the highest MVR, allowing lower injection pressure or faster filling of the mold. The GPC diagram on the right of Figure 3.23 demonstrates that scission of polymer chains are not responsible for such big differences. Addition of flame retardants does decrease the molecular weight and this decrease is more pronounced in the blend containing boehmite, but the moderate extent of molecular weight change does not correspond with the observed extent of MVR increase. It is assumed that the very fine boehmite particles (D50 = 400nm) promote the sliding of polymer chains and glass fibres against each other.

Other mineral flame retardants used in engineering plastics are zinc stannates. Zinc stannates are extremely high temperature stable and have proven as ATO replacement and effective smoke suppressant when combined with halogenated flame retardants [41].

3.6   Reactive Resins/Thermoset Applications

In the context of this chapter the term thermoset is restricted to cross-linked polymers based on liquid reactive resins. This is different than the definition and use of the expression in North America, where thermoset also includes cured elastomers and cross-linked thermoplastics.

For mineral flame retardancy in thermosetting systems, the differences in the inherent ignitability of the various resin types must be taken into account. Depending on the flame retardancy standard to be satisfied, phenolic resins can be used partially without any flame retardants due to their low inherent ignitability. Polyaromatic melamine and bis-maleidtriazine, which both contain nitrogen in their structure, show some inherent fire resistance, but do normally require additional flame retardants in most regulated applications. The large volume thermoset resins based on epoxy (EP), polyurethane (PUR) and unsaturated poly-esters (UP) require the addition of flame retardants. However, such differences are not necessarily manifested in simple material tests like LOI. Figure 3.24 illustrates the influence of glass fibre reinforcement on oxygen indices for some selected resins. LOI with and without a glass fibre reinforcement is plotted. Phenolic resin shows the biggest LOI increase between neat resin and reinforced composite. The LOI increase by glass fibre reinforcement is smaller for the other thermoset resins plotted, but with the exemption of EP the corresponding neat resins have higher LOI than phenolic resin.

As already shown for thermoplastic resins, the flame retardancy effect improves with increasing mineral flame retardant loading. A rough overview of the ATH loadings necessary to satisfy special flame retardant standards is shown for UP resin systems in Table 3.16. However, these loadings vary depending on the resin system and should therefore only be used as rough indication for other resins.

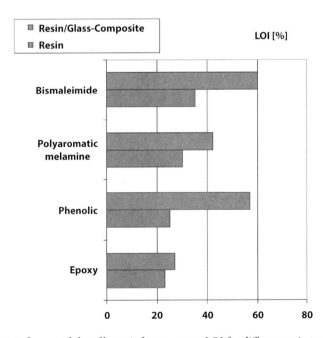

Figure 3.24 Influence of glass fibre reinforcement on LOI for different resin types [42].

Table 3.16 Loading levels of ATH required in UP resins to fulfil the listed flame retardant standards.

phr	Standard	Application
50	*Does not satisfy any standard*	
100	EN 13501-1, class D	Building sector
150	EN 45545-2, HL-1 UL94V-0	Railway E&E
250	EN 13501-1, class B	Building sector
350	EN 45545-2, HL-3	Railway
400	EN 13501-1, class B, s1 BS6853	Building sector Railway

Further, the ignitability of fibre reinforced plastics is also affected by composite thickness. The thinner a composite, the higher is its flammability. The type of fibre, namely glass, carbon, cellulose or aramid affects the fire behaviour of the end product in a different fashion and the fibres volume fraction in the composite determines the extent of this influence. For glass fibres, the flammability is reduced with increasing fibre content. Nevertheless, in the case of specific set ups of a flame test, glass fibres may have a negative effect. This is known for flammability test according to UL 94V, where vertical fibre orientation may cause a kind of chimney/candlewick effect in the specimen, resulting in lower ratings than expected.

3.6.1   Production Processes for Glass Fiber-Reinforced Polymer Composite

Glass fibre reinforced plastics (GFRP) are by far the biggest volume of all thermoset applications. As is the case for all reactive resin based system, the production processes for GFRP start with resin paste manufacturing.

3.6.1.1   Paste Production

The resin paste is produced by incorporation of mineral fillers, including filler flame retardants and other additives into liquid pre-polymers and monomers. Good dispersion of the paste is crucial for the storage stability of the resin paste, which is especially important when GFRP production is done at a different production site than resin paste manufacturing.

High shear mixers are commonly used, but good dispersion results can be achieved with standard dissolvers. In the case of paste formulations containing comparatively low loading, it may be advantageous to prepare a pre-blend with high flame retardant filler loading, which produces high shearing forces during mixing. This paste is subsequently diluted by adding resin until the required final loading is reached. This method can also be used to prevent agglomeration. In general it is advantageous to begin by adding the finest filler with highest BET surface in order to rapidly generate high shear rates. This improves the dispersion of all ingredients. However, depending on the system (fillers, resin, additives, etc.), fine-particle fillers may in turn result in agglomeration and hence in poor dispersion. Thus, in this case, it may make sense to first add coarser filler in order to increase the shear rates which will then aid the dispersion of the fine filler.

3.6.1.2   Hand Lamination/Hand-Lay-Up

Hand lamination or hand-lay-up describes the use of manual rollers to apply resin paste on glass fibre mats which are fixed on layer by layer on open moulds. This process is used for example for the production of rotor blades for wind power plants, in boat construction (hulls) and for the manufacture of components of passenger trains (wagon trim panelling, heads of train).

3.6.1.3   SMC and BMC

Bulk moulding compound (BMC) and Sheet Moulding Compound (SMC) are both used to describe GFRP raw materials and production processes. The matrix is almost exclusively based on UP resin.

SMC is a continuous process, where a carrier film is first laminated with resin paste before it passes underneath a chopper which cuts glass rovings onto the liquid resin layer. Another sheet is added on top which sandwiches the resin paste and glass fibres. The sheets are compacted by rollers and put onto a take-up roll. After pre-curing the carrier film is removed and the endless sheet is cut into defined sizes (see also sketch in Figure 3.25). SMC is molded under heat and pressure (compression moulding) to the required shape. When fully cured, the GFRP part is removed from the mould as the finished product.

BMC is manufactured by mixing strands of chopped glass fibres with polyester resin. This is done in special mixers. Due to the mixing the fibre glass length is reduced compared to SMC. After pre-cure the BMC is used for GFRP manufacturing. When processed discontinuously the cycle starts

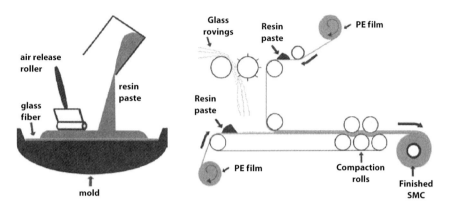

Figure 3.25 Sketch for the hand lamination (left) and SMC process.

by arranging the BMC mass in the mould. The compression molding cycle is finished when the GFRP part is fully cured and taken out of the mould. Due to its bulky morphology, BMC can also be processed continuously by injection molding.

3.6.1.4   Pultrusion

The continuous closed process of pultrusion is increasingly used for the manufacture of flame retardant profiles. A glass fibre reinforcement is pulled through an impregnation bath. The immersed reinforcement enters a heated extruder head where the required shape is given and curing starts. Subsequent heating and cooling zones control the reaction rate. Due to the continuous process and the particular rheological conditions of pultrusion, the requirements to be satisfied by the impregnating paste are very high.

3.6.1.5   RTM/RIM

Resin transfer moulding (RTM) and resin injection moulding (RIM) both describe discontinuous, closed curing processes, where the resin paste is brought in contact with a glass fibre mat reinforcement by using pressure difference. The paste is either pressed into (RIM) or pulled into (RTM) the mould by pumps. The reinforcement is arranged in the mould before closing and sealing. Rheological requirements for the paste are very high. Viscosity should be low enough to allow the total filling of the mould before curing starts. Viscosity adjustment via mould heating is limited by the curing reaction. Mineral filler use is additionally restricted by filtration effect which may occur at the fiber reinforcement. This needs to be considered when selecting the flame retardant filler during formulation design.

3.6.2   Formulation Principles

In contrast to thermoplastic or elastomeric compound formulations, mechanical properties for thermoset resins are of lesser importance. Fibre reinforced plastics (FRP) receive their mechanical strength mainly by the reinforcement. However, if the filler reduces the glass transition temperature (Tg) of the thermoset matrix, the upper thermal use temperature for the FRP may be reduced, as FRPs mechanically fail above their Tg. Non-reinforced thermoset applications are either thin coatings or cast applications, where the mechanical integrity is guaranteed by the carrier construction. Consequently, when designing flame retardant thermoset resin formulations, mineral filler loading is a minor issue when selectively considering properties of the cured composite. As already outlined, paste viscosity/rheology during paste processing is the most important parameter which needs to be controlled. By adding mineral flame retardant fillers, the viscosity of the reactive resin increases. Coarser fillers will give lower viscosity than fine fillers with higher BET surface areas. Because of the smaller particles filling the free volumes built up by the larger particles, a broad PSD (particle size distribution) is especially beneficial. Leading ATH manufacturers have therefore developed special ATH grades following this principle resulting in lower viscosity of filled resin pastes.

Figure 3.26 (left) displays relative viscosities of UP resin paste as a function of loading level for different ATH grades. Viscosity optimized grade give low to moderate viscosity increase up to high loading levels.

Sedimentation of mineral flame retardant filler is especially important when the non-cured resin paste is shipped or stored before further processing. Coarse particles tend to sediment more rapidly than fine particles, meaning that settling tendency and viscosity increase are inversely related to particle size. Depending on the process requirement, paste formulators need to design their optimized compromise between these two properties. Figure 3.26 (right) compares sedimentation over time for two different ATH grades. The viscosity optimized grade gives more sediment than the optimized grade, which was designed as a compromise between low viscosity and low sedimentation.

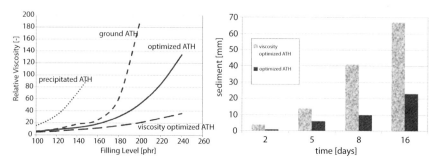

Figure 3.26 Relative viscosity of UP filled with different ATH and increasing filler load (left); sedimentation in mm sediment in dependence of time (right, UP resin, 175 phr loading level).

Last but not least the surface finish of a GFRP is influenced by mineral filler particles. Coarse fillers may give insufficient surface appearance. Good surface finish requires control of the mineral flame retardants top cut.

Mineral filler flame retardants' surfaces may absorb low molecular additive like curing agents, accelerators, or other additives. When demobilized on particle surfaces, such additives can partially be deactivated. The formulator needs to test such effects and potentially has to increase additive concentrations in reference to non-filled formulations. The basic nature of most mineral filler flame retardant additives can also influence the hardening/curing by chemically interaction. Many curing systems are sensitive to pH of resin paste.

3.6.3   Public Transport Applications of GFRP

Engineers designing transport vehicles in general request materials of ever decreasing specific weight to reduce fuel consumption. Plastics are favoured for their lower weight compared to metals. In cars the use of flame retardant GFRP is limited due to the relatively low requirements according to FMVSS 302, but regulations in mass transport vehicles like busses (UN/ECE R118), trains (EN 45545), ships (regulated by IMO, International Marine Organisation standards) and planes (regulated by FAA, Federal Aviation Agency standards) include partially quite severe flame retardant requirements. Especially railway and cruising ship interiors like cabin walls and doors and seat shells are made of mineral flame retardant GFRR But in case of railway, the traction unit of locomotives and subway trains or street

Table 3.17 UL 94 testing results for epoxy novolac formulations filled with boehmite. (* Average of 2 samples. 3rd sample was self extinguishing). According to Neumeyer *et al.* [44].

P-content of unfilled resin	0 wt-% boehmite	10 wt-% boehmite	30 wt-% boehmite	50 wt-% boehmite
0.5 wt-%	HB	HB	HB	HB
	18 ± 1 mm/min	12 ± 2 mm/min*	self extinguishing	self extinguishing
1.0 wt-%	HB	HB	HB	HB
	14 ± 1 mm/min	self extinguishing	self extinguishing	self extinguishing

cars is very often made of flame retardant GFRP too. Mostly unsaturated polyester (UP, also vinyl ester) are used as matrix material.

In airplanes weight consideration are extremely important. Highly filled, composites are therefore rarely found in aircraft interiors. Instead, phenolic matrices are preferred because of their inherent flame resistance [43], and carbon fibres are used instead of glass fibres. But phenolics have disadvantages such as shrinking and emission of volatiles during curing and restricted freedom of design. Industry and academia work intensively on alternatives. Some of these promising alternatives are based on combinations of organic non-halogenated and mineral filler flame retardants. Boehmite (AOH) combined with DOPO (9,10- Dihydro-9-oxa-10-phosphaphenanthrene-10-oxide) modified epoxy novolac resin has been reported [44, 45]. The AOH based composite showed improved flammability behaviour (see Table 3.17) and increased fracture toughness compared to resin only using DOPO as sole flame retardant. Furthermore and in contrast to DOPO, which decreases the crosslinking density, AOH does not negatively influence the glass transition temperature (Tg) of the epoxy matrix.

3.6.4 E&E Applications

Flame retardant GFRP produced by SMC is broadly used for enclosures of electrical equipment like, e.g., switchboards. The base material is UP resin, the loading needed for UL 94 V-0 rating is in the range of 120 – 160 phr depending on the glass fibre content of the composite.

Flame retardant epoxy resins are well established in electronic applications. Epoxy resins rated according to UL 94 V-0 can be found as cast

Table 3.18 Left: amine hardened epoxy cast resin satisfying UL94V0. Right: Dicyandiamide, fenuron hardened epoxy resins fulfilling UL 94 V-0 at 4mm. By combination with boehmite, the DOPO content can be significantly reduced.

Component	Parts	Flame retardant	Portion [Wt.-%]	T_g (DSC) [°C]
Epoxy resin	100	DOPO	11.2	158
Amine hardener TETA (Tri-ethanol-amine)	13.5	DOPO + AOH	2.9 + 30	168
ATH	150			

resins in electronic parts like capacitors and as laminating resins in prepreg laminates used to produce copper clad laminates (CCL) and finally printed circuit boards (PCB). Resins based on TBBPA (tetra-bromo-bisphenol-A) which is polymerized into the epoxy backbone still dominate, but halogen free flame retardants are gaining increasing market share.

Halogen free CCL use nitrogen or phosphorous flame retardants in combination with mineral filler flame retardants and sometimes also inert inorganic fillers. The organic non-halogenated flame retardants can be reactive type, meaning they chemically link to the epoxy backbone, or additive type, which means the flame retardant is dispersed or solubilised in the epoxy matrix. ATH and AOH particulate mineral flame retardant additives are used in combination. The high temperature stability and acid resistance of boehmite gives benefits in regard to process security during etching and soldering steps in PCB manufacturing. More importantly, reliability and service life of the end product is improved in the case of AOH [46–49]. Maximum reliability of PCBs is demanded in aviation, aerospace, and medical applications, where long service life of a minimum of 20 years is expected [50]. Table 3.18 (right) displays a comparison between an epoxy-formulation based on reactive DOPO and DOPO combined with AOH. Both fulfil UL 94 V-0, but the combination results in higher Tg and is commercially more attractive than the use of DOPO alone. Filler use in PCB generally improves CTE (coefficient of thermal expansion), making copper and epoxy resin more compatible. Heat dissipation is also improved with mineral fillers.

In cast resins, ATH has been an established flame retardant for many years. Table 3.18 (left) shows a typical loading for an UL 94 V-0 rated epoxy resin. The ATH is either a coarse grade with low oil absorption or a viscosity optimised type. Both ATH types enable low viscosity resin pastes with sufficiently good adhesion to the electronic components and casing being filled.

3.6.5   Construction and Industrial Applications

In construction applications UP is the most important matrix resin used for GFRP materials. Because of the more severe construction standards, mineral flame retardant loadings are higher than those used in electronic applications. Loadings of minimum 200 phr flame retardant filler on 100 phr resin are used. ATH is by far the dominant flame retardant filler in these applications. In Europe, building panels are often produced by compression moulding of SMC or BMC. In North America the pultrusion process is more common to produce FRP profiles and sheets. In each case ATH is used as flame retardant filler, but as discussed above in section 3.6.1, the particle size distribution needed for both processes is different and it is more common to use combinations of halogenated flame retardants and ATH in pultrusion because of potential processing issues with too high loading levels.

Flame retardant gratings made of GFR UP resin are used in industrial surroundings like chemical plants, where fire safety and chemical resistance is requested in parallel. They are also produced by pultrusion.

Artificial marble made of ATH filled and pigmented UP resin is widely used for interior applications. Flame retardancy is a side effect in here. The purpose of ATH use in artificial marble is its white colour and chemical resistance. In fact, artificial marble grades are selected by viscosity and especially colour criteria. Often artificial marble ATH grades are surface treated with acrylic functionalized silanes to further improve the chemical and thermo-shock resistance by improved filler-matrix bonding. This is important for working surfaces made of artificial marble, e.g., in the kitchen.

Gelcoats based on vinylesters are applied when the surface smoothness of GFRP is requested to be high quality for aesthetic reasons or practical reasons, like reduced aerodynamic resistance of wind mill blades. When the underlying GFRP construction is flame retardant, the gelcoat should also contribute to the reduced flammability. In such cases fine precipitated ATH grades are used at moderate loadings of 100 – 120 phr. The precipitated ATH is needed to guarantee top surface quality.

Epoxy cast resins for floors are used because of good abrasion properties. When applied on balconies of multi-storey building these coatings need to be flame retardant. In the case of pigmented coatings, ATH as mineral filler flame retardant can be used.

3.7   Conclusion, Trends and Challenges

Mineral filler flame retardants are mainly used in highly filled composites and such applications, especially in halogen free polymers, are a solid basis for their further dynamic market growth. As valid for all traditional, mature technologies, these applications are challenged by ever increasing cost pressure throughout their whole process chains. Manufactures of mineral flame retardant fillers have to answer this challenge by steadily improving specific energy consumption in production processes. This has gained even higher importance because of the recently agreed goals for CO_2-neutrality by 2050 [51]. LCA (life cycle analysis) data and/or CO_2-footprint need to be addressed by the mineral filler flame retardant manufacturers.

The increasing market relevance of electro mobility is pushing the use of mineral filler flame retardants. One of the biggest potential applications is construction material used in lithium-ion-battery pack and module assembly. The use mineral fillers is primarily driven by another functionality than flame retardancy. It is the ability to conduct heat, which boosts the use of mineral fillers in so called gap filler masses, tapes and adhesives used to connect battery cells into battery packs and modules. Flame retardancy given by metal hydrate fillers is a welcomed additional feature. The combination of these two functionalities is also important for cast resins used for electrical motor and power electronics (e.g. inverter and converters). Other electro mobility applications of high potential for mineral filler flame retardants are high temperature barriers for battery casings and GFRP used for charging stations.

Many academic development approaches in the area of mineral filler flame retardants focus on a reduction of loading. Some industrial approaches have been made too, especially by developing very fine flame retardant fillers [52, 53]. Other developments focus on the use of nano-composites in combination with mineral filler flame retardants and are discussed in chapters 9 and 10. But such products and combinations are very often not in line with the main challenge described above and are contrary to the goals of reduced specific energy demand and improved processing performance during polymer conversion. This in combination with parallel developments of new polymers with improved filler acceptance, improved coupling agent technology and innovative compounding techniques is why reduced filler loading has no commercial relevance in traditional application fields. Other than that, partial replacement of synthetic mineral filler flame retardants by cheaper, natural mineral fillers is a trend observed in this industry.

But nanocomposites and very fine submicron filler flame retardants will increasingly be used as add on additives when very stringent flame retardant requirements cannot be matched by increasing the loading of the base filler. Such developments have been seen in the course of the introduction of the Construction Product Regulation (305/2011) in the European Union. However, safety and environmental fate of nanoparticles continues to be an unknown, and as such, nanocomposite solutions have not made much progress entering the market.

Very fine mineral flame retardants like submicron metal hydrates, especially those with high temperature stability and chemical inertness will be more and more used in halogen free flame retardant blends. By combining mineral with organic flame retardants, polymer applications can be developed, which were technically not accessible for fillers and commercially not viable for organic halogen free flame retardants in the past.

References

1. Kulshreshtha, A., *Looking for the right mineral filler for plastics*, pp. 29–33, Industrial Minerals, London, United Kingdom Remark: Industrial Minerals is now Fatmarkets IM Head Office 8 Bouverie St London EC4Y 8AX United Kingdom,pp. 29–33, February 2013.
2. *World Flame Retardants, Industry Study #3258*, p. 43, The Freedonia Group Inc, Cleveland, Ohio, February 2015.
3. *Market Study Flame Retardants*, 5th Ed., Ceresena Research, Konstanz, German, p. 175ff, January 2018.
4. *Non Metallurgical Bauxite and Alumina Outlook to 2029*, 10th Ed., Roskill Information Services Ltd, London, United Kingdom, 2019.
5. Saxby, A., *Magnesia –the changing face of the industry*, Magmin 2013, Oslo, 13.-15. May, 2013.
6. Hollingbery, L., *Fire Retardants in Plastics 2013*, AMI (Apllied Market Information), 13.-14.June, 2013.
7. Wypych, G., *Handbook of Fillers*, ChemTec Publishing, 2nd Edition, p. 175, 1999.
8. Hornsby, P.R., Cusack, P.A., Cross, M., Töth., A., Zelei, B., Marosi, G., Zinc hydroxystannate-coated metal hydroxide fire retardants: Fire performance and substrate-coating interactions. *J. Mater. Sci.*, 38, 13, 2893, Jul 2003.
9. Bauxite and Alumina, *Global Industry Markets and Outlook*, 8th Edition, p. 20, Roskill Information Services Ltd, London, 2012.
10. Klimes, M. and Reimer, A., *Aluminium Hydroxide Preparation and application as a mineral flame retardant*, Die Bibliothek der Technik, Munich, Germany, vol. 370, 2015.

11. M. Grill, Process for the recovery of magnesium oxide of high purity, US 4,255,399, US Patent.

12. D., Prescher, *et al.*, Flammwidrige Kunststoffmischung und Verfahren zur Herstellung eines Füllstoffs DE Patent (German Patent), DE 19812279 C1.

13. A., Reimer, *et al.*, Feinkristalliner Böhmit und Verfahren zu dessen Herstellung, DE Patent (German Patent), DE 10 2006 012 268 A1.

14. Torno, O., Synthetic boehmite aluminas and hydrotalcites as performance fillers, in: *High Performance Fillers 2006*, iSmithers Rapra Publishing, Cologne, 21-22.March 2006, ISBN 1859575609.

15. Wypych, G., *Handbook of Fillers*, 2nd Edition, ChemTec Publishing, Toronto, Canada, pp. 176–177, 1999.

16. Wypych, G., *Handbook of Fillers*, 2nd Edition, pp. 305–325, ChemTec Publishing, 1999.

17. Rothon, R., *Particulate – Filled Polymer Composites*, Longman Group Ltd., London, United Kingdom, pp. 123–163, 1995.

18. Rothon, R., *Particulate – Filled Polymer Composites*, Longman Group Ltd London, United Kingdom, p. 14, 1995.

19. https://www.freemantech.co.uk/powder-testing/ft4-powder-rheometer-powder-flow-tester (accessed 02/09/21).

20. Sauerwein, R. and Klimes, M., Process Optimized Fine Precipitated Aluminium hydroxide offering outstanding Compound Properties, in: *IWCS-Focus 2004*, IWCS, Inc. & IWCS/Focus Conferences, Philadelphia, 14-17 November 2014.

21. Sauerwein, R., Newest developments in modified mineral hydrates under the aspect of PVC stabilization, fire retardancy and smoke suppression, in: *PVC Formulations 2010*, AMI (Apllied Market Information), Cologne, 17-18. March 2010.

22. https://nabaltec.de/fileadmin/user_upload/07_presse/Mediathek/White_Papers/Whitepaper_APYRAL_HC.pdf (accessed 02/02/21)

23. Hollingbery, L.A. and Hull, T.R., The fire retardant behavior of huntite and hydromagnesite – A review. *Polym. Degrad. Stab.*, 95, 2213–2225, 2010.

24. Hull, T.R., Witkowsky, A., Hollingbery, L.A., Fire retardant action of mineral fillers. *Polym. Degrad. Stab.*, 96, 1462–1469, 2011.

25. Hollingbery, L.A. and Hull, T.R., The fire retardant effects of huntite in natural mixtures with hydromagnesite. *Polym. Degradation Stability*, 97, 504–512, 2012.

26. Sauerwein, R., Mineral filler blends as flame retardants for processing temperatures up to 300°C, in: *Fire and Materials Conference 2001*, Interscience Communications, San Francisco, January 2001.

27. Sauerwein, R., Apymag AOH850 – a new mineral flame retardant for processing temperatures beyond 200°C, in: *Industrial Minerals 1999*, November 1999.

28. Babrauskas, V., Gann, R.G., Grayson, S.J., *Hazards of Combustion Products*, Interscience Comm. Ltd., London, 2008.

29. Horrocks, A.R., Smart, G., Nazaré, S., Kandola, B., Price, D., Quantification of zinc hydroxystannate and stannate synergies in halogen-containing flame retardant polymeric formulations. *J. Fire Sci.*, 28, 3, 217–248, 2010.

30. Niklas, A., *Hochleistungs HFFR Kabelcompounds – Neue Erkenntnisse in der Compoundierung*, VDI Fachtagung Kabelextrusion, Nuremberg, 4-5 December, 2012.

31. *World Flame Retardants, Industry Study #3258*, p. 76, The Freedonia Group Inc, Cleveland, Ohio, February 2015.

32. Horrocks, A.R., Smart, G., Kandola, B., Price, D., Zinc Stannates as Alternative Synergists in Selected Flame Retardant Systems. *J. Fire Sci.*, 27, 5, 495–521, 2009.

33. Roos, A. and La Rosa, M., High Performance Elastomers in Cables for Offshore and Arctic Regions, in: *NRC Nordic Rubber Conference 2009*, IWCS, Inc. & IWCS/Focus Conferences, Jönköping, 6-7 May 2009.

34. Muehren, O. and Westerdale, S., Thermoplastic Polyurethane (TPU) for High Performance Cable Applications: Current Applications and Future Developments, in: *Proceedings of the 61st IWCS Conference*, IWCS, Inc. & IWCS/Focus Conferences, Providence, 12-15 November 2012.

35. Sauerwein, R., Töpfer, O., Englmann, T., Luks, A., New aluminium hydrates as flame retardant fillers for TPU, in: *Proceedings of the 61st IWCS Conference*, IWCS, Inc. & IWCS/Focus Conferences, Providence, 12-15 November 2012.

36. Stark, N.M., White, R.H., Mueller, S.A., Osswald, T.A., Evaluation of various fire retardants for use in wood flour polyethylene composites. *Polym. Degrad. Stab.*, 95, 9, 1903–1910, September 2010.

37. Bi, M., Gubanski, S.M., Hillborg, H., Seifert, J.M., Ma, B., Effects of long term corona and humidity exposure of silicone based housing materials, *ELECTRA*, 267, pp. 4–15, April 2013.

38. Berneck, J., New flame retardant systems for connectors, in: *Cables 2010*, Cologne, 10-11. March 2010.

39. Töpfer, O., More than preventing fire – mineral based flame retardants in Engineering Plastics, in: *Vision in Plastics 1/2011*, CHEMManager, Wiley-VCH Verlag GmbH & Co. KGaA, Darmstadt, Germany, pp. 24–25, 2011.

40. Töpfer, O., Boehmite as halogen free co-flame retardant for Engineering Plastics, in: *High Performance Engineering Plastics*, Chatsworth Associates Pte Ltd, Singapore, 10-12. April 2011.

41. Cusack, P.A., Heer, M.A., Mon, A.W., Zinc hydroxystannate: A combined flame retardant and smoke suppressant for halogenated polyesters. *Polym. Degrad. Stab.*, ECCM http://www.escm.eu.org/eccm15/start.html, 32, 2, 177–190, 1991.

42. Kandola, B.K., Flammability and fire resistance of composites, in: *FRPM 2007*, Bolton, 4-5 July 2007.

43. Mouritz, A.P. and Gibson, A.G., *Fire Properties of Polymer Composite Materials*, Springer, Dordrecht, 2006.

44. Neumeyer, T., Bonotto, G., Kraemer, J., Altstaedt, V., Doering, M., Fire behaviour and mechanical properties of an epoxy hot-melt resin for aircraft interiors, in: *ECCM15,15TH EUROPEAN CONFERENCE ON COMPOSITE MATERIALS*, Venice, Italy, 24-28 June 2012.

45. Neumeyer, T., Bonotto, G., Kraemer, J., Altstaedt, V., Doering, M., Fire behaviour and mechanical properties of an epoxy hot-melt resin for aircraft interiors. *Compos. Interfaces*, Vol.6, p.443-455, 2013, https://doi.org/10.1080/15685543.2013.807153.

46. Ihmels, C.W., Boehmites as Flame Retardant Fillers for Highly Temperature-Resistant Base Laminates, in: *EIPC / CPCA International Symposium*, EIPC, Shanghai, China, March 20, 2007.

47. Dietz, M. and Ihmels, C.W., *Halogen-Free Flame Retardant Systems for EP-Based PWB's*, Electronic Goes Green, Berlin, September 9, 2008.

48. Ihmels, C.W., Thermally Stable Boehmites as Halogen Free Flame Retardant Fillers Allowing the Manufacture of Green E&E Products with Highest Quality and Reliability, in: *EIPC Summer Conference*, EIPC, Nürnberg, June 7/8, 2010.

49. Töpfer, O., Boehmite as Halogen Free Flame Retardant Filler Utilized in Cost Effective Production of Highly Reliable Base Laminates, in: *EIPC Summer Conference*, EIPC, Milan, September 13 & 14, 2012.

50. Birch, B. and Reid, P., *EIPC Reliability Workshop*, EIPC, Nürnberg, June 6, 2010.

51. https://ec.europa.eu/info/strategy/priorities-2019-2024/european-green-deal_en (accessed 02/09/21)

52. Sauerwein, R., Application of Submicron Metal Hydrate Fillers in Flame Retardant Cable, in: *55th IWCS-Focus, 2006*, IWCS, Inc. & IWCS/Focus Conferences, Rhode Island, 12-15. November 2006.

53. Luks, A. and Sauerwein, R., Halogen free and flame retardant elastomeric cable compounds with submicron sized fillers, in: *57th IWCS-Focus, 2008*, IWCS, Inc. & IWCS/Focus Conferences, Rhode Island, 9-12. November 2008.

4

Intumescence-Based Flame Retardant

Serge Bourbigot[1,2]

[1]University of Lille, CNRS, INRAE, Centrale Lille, UMR 8207 - UMET - Unité Matériaux et Transformations, F-59000 Lille, France
[2]Institut Universitaire de France (IUF)

Abstract

This chapter reviews the applications of intumescence in the field of fire protection (resistance to fire) and fire retardancy (reaction to fire) of material. It starts describing the basics of intumescence and gives the main intumescent products available on the market associated with their typical application's fields. It covers the reaction to fire of intumescent polymers and textiles including the latest developments of flame retardants as well as synergists. Resistance to fire is also considered using intumescent coating at the large and reduced scale. Mechanisms of action are always discussed considering the chemical, physical and thermal aspects.

Keywords: Intumescence, fire protection, flame retardant, review

4.1 Introduction

The word "intumescence" comes from Latin "intumescere" which means "to swell up". From the dictionary, intumescence is defined as the act or process of swelling or enlarging. Some definition also mentions the action of heat to expand the body. French Writers like Jules Verne or François-René de Chateaubriand used the word 'intumescence' with different meanings: a hilly area on volcano (J. Verne in The Mysterious Island, 1874) or the waves' swell (F.R. de Chateaubriand in Mémoires d'Outre-Tombe, 1848). Taken 'as is', those definitions are not directly linked to the flame

Email: serge.bourbigot@centralelille.fr

Alexander B. Morgan (ed.) Non-Halogenated Flame Retardant Handbook 2nd edition, (169–238)
© 2022 Scrivener Publishing LLC

retardancy of polymeric materials or to the fire protection. Nevertheless, the above definitions describe well the behavior of an intumescent material used for fire retardancy. When heating beyond a critical temperature, it begins to swell and then to expand. The result of this process is a foamed cellular charred layer on the surface which protects the underlying material from the action of the heat flux or the flame. Visually, the swelling and the expansion looks like 'black waves' swollen at the surface of the material and the final char exhibits hemispheric shape with a roughed or smooth surface.

We can consider intumescence was first reported in the scientific literature by J.L. Gay-Lussac in 1821 in the case of the flame retardancy of textiles [1]. The word 'intumescence' was not mentioned but we suspect an intumescent phenomenon (similar experiments were done by this author) when woven fabrics in hemp and flax were coated by a mixture of borate and ammonium phosphate: here there are the three ingredients of intumescence (it will be commented further in the paper): the acid source, namely borate and phosphate, the char former, namely flax or hemp, and the blowing agent, namely the evolving ammonia from the phosphate. In 1934, a German patent (extended in the US in 1938) claimed the fire protection of wood using a mixture of diammonium phosphate and formaldehyde [2]. It is reported the formation of a swollen char layer upon heating, protecting wood but the word 'intumescence' is not used in the text. Only in 1940, the word 'intumescence' was used in an US patent where it was claimed the making of silicate-based intumescent coating to protect combustible [3]. However, a US patent of 1912 [4] also mentioned the word 'intumescence' claiming the making of intumescent materials based on the same technology using alkali silicates. It was not for fire protection but to create heat insulator. The first comprehensive paper was published in the early 70s by H. Vandersall and gives the fundamentals of intumescence [5]. In this latter paper, only coatings were considered and it is only in the 80s that G. Camino applied with success this concept to bulk polymers (mainly thermoplastics) [6]. This basic work was useful for the development of products which were launched by different companies in the 90s.

We may think intumescence is an old concept since the first comprehensive paper was published in the early 70s but a brief review of the literature shows that intumescence is still largely employed to make flame retardant (FR) polymers and FR paints and that some recent developments are very promising for fire protection [7–9] and for flame retardancy [10–12]. This concept of intumescence appears as an attractive topic as demonstrated by the number of papers or patents dealing with this topic which is still

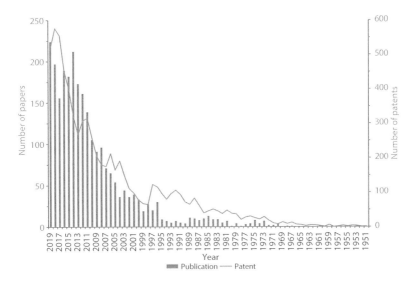

Figure 4.1 Number of publications (all types) and patents (extracted from the database of US patent office, Japan patent office, European patent office, World intellectual property organization and UK intellectual property office) devoted to intumescence from 1951 to 2019 (Scopus database with the keywords combination: ['intumescence' È 'intumescent'] Ç ['flame retardant' È 'flame retardancy' È 'fire retardant' È 'fire retardancy' È 'fire protection']).

growing since the 50's (Figure 4.1). The number of papers (all types) and patents was the highest between 2013 and 2019 evidencing the interest for this concept from the scientific community and the industry. It is then the purpose of this chapter to review the developments done in the reaction and resistance to fire with the intumescence concept.

This chapter is organized in four parts. The two first sections survey briefly intumescence to provide the reader the basic understanding on the mechanisms of action by intumescence and to give the main intumescent products available on the market associated with their typical application's fields. Reaction to fire of intumescent polymers and textiles, i.e. the contribution of the material to fire growth, is examined in the third section. The latest developments of flame retardants as well as synergists are considered, and mechanism of action is commented based on published work and on our own experience. The fourth section is devoted to the resistance to fire of materials using intumescent paint or coating. Resistance to fire is defined as the ability of materials to resist the passage of fire and/or gaseous products of combustion, and is capable of meeting specified performance criteria to those ends (the reader should make the distinction between reaction and resistance to fire). A short description of the normalized tests is first given. It also explains why bench-scale testing must be developed and correlation small/large scale tests is shown. Mechanism of action are discussed considering the chemical, physical and thermal aspects.

4.2 Fundamentals of Intumescence

Flame retarding polymers or textiles by intumescence are essentially a special case of a condensed phase mechanism [13–17]. Intumescent systems interrupt the self-sustained combustion of the polymer at its earliest stage, i.e. the thermal decomposition with the evolution of gaseous fuels (Figure 4.2). The intumescence process results from a combination of charring and foaming at the surface of the substrate. So, the charred layer acts as a physical barrier which slows down heat and mass transfer between gas and condensed phase. The formation of an effective char occurs via a semi-liquid phase, which coincides with gas formation and expansion of the surface. Gases released from the decomposition of the intumescent material, and in particular of the blowing agent, have to be trapped and to diffuse slowly in the highly viscous melt degraded material in order to create a layer with appropriate morphological properties. The viscosity of the degraded matrix in the blowing phase is therefore a critical factor. Another significant aspect of intumescent formulations is the mechanical strength of the intumescent

Figure 4.2 Intumescent polylactide (PLA) during a cone calorimeter experiment. Note the small flames on the side of the intumescent 'cake' showing how the intumescent coating smother the fire.

char. In the conditions of a fire, char destruction can proceed not only by means of ablation and heterogeneous surface burning but also by means of an external influence such as wind, mechanical action of the fire or convective air flows. The mechanical stability of the intumescent char depends both on the structure and porosity of the foamed intumescent material. If the structure of the char (morphology, distribution of voids inside the char) is appropriate, the thermal conductivity of intumescent chars can be very low and it limits efficiently heat transfer from the heat source to the substrate. Finally, the construction of the whole intumescent structure is controlled by its chemical composition and its dynamics (kinetics).

As mentioned above, the intumescence mechanism is that the intumescent material reacts upon heating to produce gases that are partially trapped in a viscoelastic matrix (rheological aspect). The matrix expands as gases are produced (from the blowing agent and/or from the decomposition products of the polymeric matrix) and at the same time, cross-linking reactions and charring cause the matrix to harden thereby producing a coherent highly porous char (chemical aspects and formulation). The porosity of the char is generally extremely high that the resulting structure has extremely low thermal conductivity. The most important parameters of the expanded char affecting its thermal insulation performance are its heat conductivity and its ability to swell rapidly (thermal aspects). The thermal properties are fully dependent of the rheology and of the chemistry of the intumescent and so, the appropriate values cannot be reached without considering them.

a. Chemistry of intumescence

The usual chemistry of intumescence involves composition containing an inorganic acid or a material yielding acidic species, of a char former and of a component that decomposes to enable the expansion of the system (blowing agent). Typical examples of components used in intumescent systems are reported in Table 4.1.

The following sequences of events have been proposed to describe the development of the intumescent phenomenon:
– the inorganic acid is released at temperatures depending on its source and other components
– the acid esterifies the carbon rich components at temperatures slightly above the acid release temperature
– the mixture of materials melts prior or during esterification
– the ester decomposes via dehydration resulting in the formation of a carbon-inorganic residue

Table 4.1 Examples of components of intumescent systems.

Acid source	Char former
Phosphoric	Starch
Sulfuric	Dextrins
Boric	Sorbitol, mannitol, xylitol
Ammonium salts	Pentaerythritol, monomer, dimer,
Phosphates, polyphosphates	trimer
Borates, polyborates	Phenol-formaldehyde resins
Sulfates	Methylol melamine
Phosphates of amine or amide	Char former polymers (PA-6,
Products of reaction of urea or	PA-6.6, PU, PC, PVC …)
Guanidyl urea with phosphoric	
acids	Blowing agents
Melamine phosphate	Urea
Product of reaction of ammonia	Urea-formaldehyde resins
with P_2O_5	Dicyandiamide
Organophosphorus compounds	Melamine
Tricresyl phosphate	
Alkyl phosphates	
Haloalkyl phosphates	

- released gases from the above reactions and decomposition products cause the carbonizing material to foam
- as the reaction nears completion, gelation and finally solidification occurs, the resulting solid is in the form of multicellular foam.

A typical example is the case of polypropylene (PP)-ammonium polyphosphate (APP)/pentaerythritol (PER) system [18, 19]. The reaction of the acidic species (APP and its decomposition products into orthophosphates and phosphoric acid) with the char former agent (PER) takes place in a first stage (T < 280°C) with formation of esters mixtures. The carbonization process takes then place at about 280°C (via the formation of double bonds followed by Diels Alder reaction and a free radical process increasing the size of the polyaromatic structure [20]). In a second step, the blowing agent decomposes to yield gaseous products (i.e., evolved ammonia from the decomposition of APP) which cause the char to swell (280 < T < 350°C). The intumescent material decomposes then at higher temperatures and loses its foamed character at about 430°C. Concurrently, the heat conductivity of the char decreases between 280°C and 430°C and the insulation of the substrate is enhanced [21].

The method to develop the intumescence phenomenon is mainly based on series of chemical reactions occurring at the right time. Another way to make intumescence in a polymeric matrix is the physical expansion. In other words, the rapid sublimation of a molecule (and/or the decomposing products) creates the expansion of the top degraded layer of the polymer to make an intumescent coating. This mechanism occurs when using expandable graphite (EG). EG is a synthesized intercalation compound of graphite that expands or exfoliates when heated. A wide variety of chemical species can be used to intercalate graphite materials (e.g. sulfate, nitrate, various organic acids …) [22]. A typical example is the incorporation of 10 wt% EG in polypropylene (PP) [23]. PP/EG was evaluated by cone calorimetry at an external heat flux of 35 kW/m² and compared to neat PP (Figure 4.3 – (a)). When incorporated in PP, the peak of heat release rate (pHRR), is decreased by about 70% and total heat release rate (THR) is also significantly reduced (by 35%). The formation of an expanded charred layer is observed at the surface of PP/EG evidencing an intumescent phenomenon. Not as conventional intumescent char which exhibits flat expanded carbonaceous layer, PP/EG shows

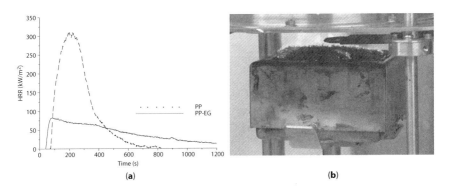

(a)   (b)

Figure 4.3 (a) HRR curves as a function of time of intumescent PP (external heat flux = 35 kW/m²) and (b) intumescent coating formed from PP/EG during the cone calorimetry experiment (note the formation of an entangled network made of graphite worms).

an expanded 'hairy' char during the experiment (Figure 4.3 – (b)). The intercalation compounds contained in EG decompose rapidly into gaseous products, which blast off the graphite flakes. Those flakes make then worms to form an entangled network at the surface of the material. This network acts as a protective layer.

Alkali silicates can be also used to make intumescent coating [3, 24, 25]. The mechanism of intumescence is the rapid evolution of water at temperature higher than 100°C creating a silicate-based foamed material (chemical reaction in Scheme 1) (Figure 4.4). This foamy material exhibits low heat conductivity and is appropriate for fire protection [24]. In addition to this, the evolution of water is strongly endothermic (4280 J/g) and hence it created an additionnal cooling effect and dilution [26].

b. Rheology of intumescence

Upon heating the intumescent material yields a viscoelastic char with the appropriate rheological properties (viscosity) to be expanded by the increasing internal pressure coming from the accumulation of the decomposition products and/or of the blowing agent. It must occur at the right time and at the right temperature to be efficient by controlling the kinetics of the decomposition reactions (dynamic of the

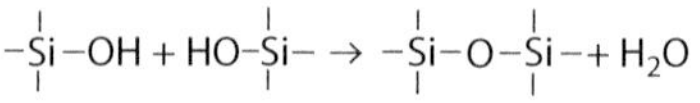

$$-\overset{|}{\underset{|}{Si}}-OH + HO-\overset{|}{\underset{|}{Si}}- \rightarrow -\overset{|}{\underset{|}{Si}}-O-\overset{|}{\underset{|}{Si}}- + H_2O$$

Scheme 4.1 Chemical reaction occurring during the expansion of silicates.

Figure 4.4 Intumescent silicate-based coating prepared in a furnace at high temperature.

intumescence). This phenomenon is relatively well understood but its measurement have been reported only late 90's using a homemade device in a calibrated tube [27] and a rheometer [28].

A typical example is the rheological measurements using a rheometer in a parallel plate configuration which were done to measure viscosity and expansion of an epoxy-based intumescent coating as a function of temperature [29]. It is clearly shown that expansion occurs when the viscosity is the lowest and when the temperature of charring is reached (about 250°C) (Figure 4.5-a). It is noteworthy that the viscosity must be high enough to accommodate internal stresses due to the internal pressure coming from the decomposition gases (for comparison the viscosity at 250°C of the virgin epoxy is twice lower than that of the intumescent coating). Expansion is also measured in the rheometer (convective heating in a furnace) and it gives useful indications on which

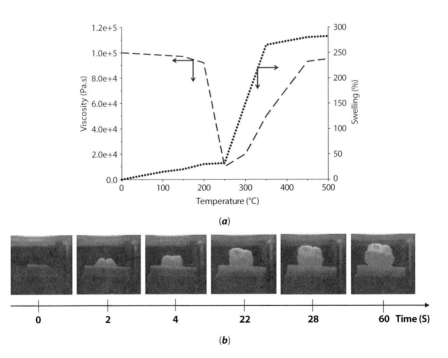

Figure 4.5 (a) viscosity and swelling as a function of temperature of an epoxy-based intumescent coating; (b) formation of an intumescent char as a function of time during cone calorimetry (external heat flux at 50 kW/m²) until the full expansion.

temperatures intumescence takes place. Another method can be used to follow the expansion '*in situ*' during cone calorimetry experiment (the heat source is a radiative heat flux). It is done using an infrared camera combined with image analysis and it provides the expansion as a function of time (Figure 4.5-b). Those protocols have been applied to numerous intumescent coatings and the same trends were observed for each system evidencing the general behavior of intumescent material [30–32].

c. Thermal behavior of intumescence

The main purpose of an intumescence coating is to create an insulative material at the surface of the substrate to limit heat transfer and so, to slow down the decomposition of the substrate and/or the diffusion of heat. If the structure of the char (morphology, distribution of voids inside the char) is appropriate [33] (Figure 4.6), the thermal conductivity of intumescent chars is very low and hence, it limits efficiently heat transfer from the heat source to the substrate.

Heat conductivity (k) is one of the most influencing parameters determining the efficiency of an intumescent barrier. In a general way [34], intumescent coating reacts around 250°C forming char and starting to swell. At higher temperatures (up to 350°C), the porosity of the intumescent char increases as well as its volume. Finally, between 400 and 600°C, intumescent char degrades slowly and its expansion remains constant. Experimental results of k were determined with the transient plane source (TPS) method [35] up

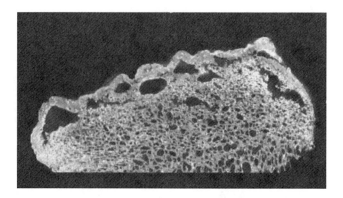

Figure 4.6 Internal structure of an intumescent char observed by X-ray tomography.

to 800°C for an epoxy-based intumescent paint [31]. At low temperature (T<200°C), k increases but when the material starts reacting and forming the char, k decreases dramatically (k drops down from 0.45 to 0.1 W/m.K). It is attributed to the formation of an expanded foamed char exhibiting low heat conductivity. At higher temperatures (T>375°C), k increases considerably up to 500°C due to the shrinkage and the partial destruction of the char. At T>500°C, k increases smoothly. From this experiment, it is then evidenced that intumescent coating exhibits low heat conductivity limiting heat transfer. This example can be extended to other intumescent systems [36, 37] and this trend can be generalized. It makes sense because the general mechanism of intumescence is based on the formation of multicellular structure from a bulk material at a given temperature and hence, the structure changes lead to a specific k-changes as a function of temperature.

4.3   Intumescence on the Market

According to researchandmarkets.com, the market of halogen-free flame-retardants is projected to witness $3,871.09 million in 2025 owing to the increasing government regulations for environmental safety and the flourishing electrical & electronics and automotive industries. This market is expected to increase at a highest growth rate between 2020 and 2025. Electrical and electronics (E&E) industry is the most significant end-use industry in the global halogen-free flame-retardant market. Phosphorus segment (phosphorus is one of the main ingredients to make intumescent systems) held the major revenue share of 57% in 2020, followed by minerals segment. Its global market is likely to log the fastest 2019-2026 CAGR (compounded annual growth rate) of 5.6% and reach a projected US$3.1 billion by 2026. Some of the key players operating in the global phosphorus-based flame retardants market are Clariant AG (Switzerland), Lanxess (Germany), Adeka (Japan), Huber Engineered Materials (USA), Israel Chemicals Limited (Israel), BASF SE (Germany), Italmatch Chemicals (Italy), Chang Chun Group (Taiwan), Rin Kagaku Kogyo Co (Japan), Jiangsu Liside (China), Shandong Ruixing (China), Shouguang Weidong Chemical Co (China), Shandong Moris Tech Co (China), Zhejiang Wansheng Co (China).

Typically, a phosphorus-based FR is designed to develop its activity just before the starting decomposition of the specific polymer it is used for. It can offer a partial gas phase contribution to the flame extinguishing effect, which is comparable to bromine- or chlorine containing FRs. However, the main feature is mostly char-forming activity like intumescence. The phosphorus-based FRs can be used in numerous polymers and resins such as polyamides, polyesters, polyolefins, epoxy and styrenics. The main area of application for the compounded materials is injection-molded E&E parts. Besides the electrical market, a very important market is flame retardant fabrics for public buildings and public transportation seating. Other compounds as phosphorus can be used in intumescent systems (Table 4.2) but they are not widely employed except boron-based compounds for non-durable textile applications and in intumescent coating for fire protection, and also expandable graphite and silicates for applications including intumescent seals, foams and strips. Table 4.2 gathers (it is not our intention to make an exhaustive list) intumescent compounds as typical examples of product available on the market. It shows that four main groups of additives are available and used industrially: phosphate, borate and expandable graphite.

4.4   Reaction to Fire of Intumescent Materials

a. Evaluation of the reaction to fire
   This section intends to remind the reader the basics of fire testing of polymeric materials and to explain why certain physical parameters were considered to evaluate fire. Various organizations (e.g. ASTM, UL, CEN, ISO, IMO etc.) throughout the world wrote fire standards. Their purpose was to create various types of fire standards, guidance documents and test methods addressing most of the major fire properties. It is interesting that fire standards throughout the world are all tending in the same direction: away from simple ranking tests and geared towards test results that can be used as input into mathematical fire models or fire hazard assessment and research, and real-scale tests to be used for validation of the small-scale tests and models [38, 39].
   The growth of fire follows distinct phases starting with ignition and early growth (it includes fire spread), to the fully developed stage, leading through to the eventual decay of the fire. These different stages are associated to fire scenarios

Table 4.2 Examples of intumescent compounds available on the market.

Type	Brandname	Supplier	Examples of application
Ammonium polyphosphate[1]	AP420 series	Clariant (Switzerland)	Intumescent gelcoats (Epoxy, UP or PUR based)
Ammonium polyphosphate combined with char former and nitrogen synergist[2]	AP740/750/760 series		FR polyolefins for E&E and building/transportation Applications – FR for lightweight composites
Aluminum phosphinate salt + melamine polyphosphate[3]	OP1310 series		FR polyamides for E&E applications
Ammonium polyphosphate[1] (crystal forms I and II)	FR CROS 580/750 series	Budenheim (Germany)	FR for PU, EP and paints
Melamine phosphate	Budit 310 series (310 to 312)		Unsaturated polyester resins, particularly for the SMC-process
Melamine borate	Budit 313		In combination with melamine phosphate for phenolic resin bound nonwoven based on cotton fibers

(Continued)

Table 4.2 Examples of intumescent compounds available on the market. (*Continued*)

Type	Brandname	Supplier	Examples of application
Coated ammonium polyphosphate[2]	Budit 660 series		FR polyolefins for E&E and building/ transportation applications
Melamine polyphosphate	Melapur 200	BASF (Germany)	PA66/Glass fibers, epoxies, synergistic blends with other flame retardants
Hydrated sodium silicate	Palusol series		Intumescent seals which prevent the passage of smoke, heat and flames
Cresyl diphenyl phosphate[4]	Disflamoll DPK	Lanxess (Germany)	For thermosets like phenolic, polyester and epoxy resins, PC/ABS, HIPS/PPE blends, TPU compounds, PUR- foams (rigid and flexible) and rubbers
APP-based mixture containing char former and blowing agent	FlameOff	FlameOff (USA)	Intumescent powder for polyolefins

(*Continued*)

Table 4.2 Examples of intumescent compounds available on the market. (*Continued*)

Type	Brandname	Supplier	Examples of application
Piperazine pyrophosphate containing char former and blowing agent	ADK Stab FP 2000 series	Adeka (Japan)	Intumescent powder for polyolefins
Cresyl diphenyl phosphate[4]	CDP	Shandong Ruixing (China)	For thermosets, PVC and natural rubbers
Expandable graphite	GrafGuard	GrafTech (USA)	FR additive in plastics (processing up to 200°C), foams, putties and coatings

[1]Typical acid source (dehydrating agent) used in numerous intumescent formulations.
[2]This product contains all intumescent ingredients, i.e. acid source, char former, blowing agent and a synergist.
[3]In some specific grade, zinc borate is added as synergist and processing aid.
[4]Intumescence is observed when used in char forming polymer (e.g. TPU, epoxy …).

mimicked by the standards depending on the application field (E&E, building, transportation etc.). They are briefly commented hereafter giving some fire tests as examples:

i. *Ignition*: All fires start with an ignition event, where a source of heat comes into contact with a fuel (here materials) in the presence of oxygen. Different standards can determine the ignition temperature or the ignition time on samples of small size (barrel or plaque of few centimeters and of few grams). The glow wire test (EN 60335-1) is one of them and it is used to evaluate the flammability of material used within an appliance mimicking over-current or short circuit failures. It is an indirect method (no direct contact with the flame but with a heating glow wire element) and the ignition temperature can be determined. The well-known UL-94 test is a direct method where the flame is applied directly to a vertically or horizontally mounted specimen under controlled conditions. Test results from applying these methods provide a way to compare the materials' tendency to resist ignition, self-extinguish flames (ignition should occur) and to not propagate fire via dripping. The cone calorimeter (ISO 5660) is another test method which permits to vary the incident heat flux and to measure the time to ignition (among many other parameters, see below). It simulates a developing fire scenario.

ii.   *Flame spread*: An important parameter is the measurement of the rate at which a flame front tends to propagate along a flammable material. Usually, the tests measure the propagation of a flame away from the source of ignition across the surface of a material or assembly and evaluate the potential for spreading flames in the event of a fire. UL-94 test in its two configurations (horizontal and vertical burning) can estimate the flame spread. It provides preliminary indications on the flammability of polymeric materials. FMVSS 302 (or ISO 3795) is an horizontal burning test which measures the flame spread of samples used in passengers vehicles. The LIFT (Lateral Ignition and Flame Spread Test) apparatus (ISO 5658-2) provides ignitability and flame spread information on, typically, vertically oriented samples. It can generate fundamental data for fire modeling.

iii. *Developing fire (or heat release rates measurement)*: The rate of heat release is arguably the most important fire parameter, since it controls the rate of growth in the fire, including heat and ultimately the amount of smoke and toxic gas generated. It is measured by cone calorimetry setting heat flux resulting from an electrical heater rod, tightly wound into the shape of truncated cone. It is the so-called cone calorimeter and it yields heat release rate information on small samples. The results of the test include the heat release rate (HRR) versus time at a specified external heat flux. Other tests can also provide HRR measurement such as Ohio State University apparatus (or OSU in short) used for cabin aircraft materials and EN 30399 to evaluate the burning behavior of bunched cables.

iv. Extinction: Limiting oxygen index (LOI, ISO 4589 part 2) serves as a measure of the ease of extinction of the material. It determines the minimum oxygen concentration required to support candle-like downward flaming combustion.

All the measurements described above can be applied on intumescent materials. Their behavior during the test does not create issues in general. However, it can be problematic with samples exhibiting high expansion during a cone experiment. The char layer formed upon heating can hit the electrical heater and, in some cases, can plug the extraction chimney (Figure 4.7). A way to avoid this is to adjust the distance between the cone heater and the sample holder but if the distance is too long then the homogeneity of the heat flux at the surface of the sample is lost. Another way is to use a grid as recommended in the standard but in this case, the benefit of the expansion to create a heat barrier is lost.

b. Intumescent organic polymeric materials
i. Organic-based systems

In the 80's, intumescent systems (APP-based formulations) were used to flame retard polymer and in particular thermoplastics [40–42]. However, such systems present some drawbacks (water solubility, migration of the additives throughout the polymer) that can be prevented by synthesizing novel phosphorus- and/or nitrogen-containing systems. The last 40 years have seen the improvement of the intumescent concept applied to polymers in terms of performance, processability and durability. A survey of the recent literature shows that numerous new intumescent compounds have been synthesized[17]. Basic concepts are used to combine in a single molecule the three main ingredients of intumescence, i.e. the acid source, the char former and the blowing agent (see

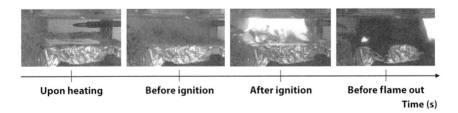

| Upon heating | Before ignition | After ignition | Before flame out |

Time (s)

Figure 4.7 Snapshots as a function of time of a burning intumescent epoxy-based polymer during a cone experiment.

the pioneering paper of Halpern [42]) and to develop new systems containing complex organic molecules (e.g. hyperbranched molecules [43]).

The use of polyols such as pentaerythritol, mannitol or sorbitol as 'conventional' char formers in intumescent formulations for thermoplastics is associated with migration and water solubility issues. Moreover, these additives are often not compatible with the polymeric matrix and the mechanical properties of the formulations are then very poor. Those problems can be solved (at least partially) by the synthesis of additives that concentrate the three intumescent flame retardant elements in it as suggested by the pioneering work of Halpern [42]. b-MAP (4) (melamine salt of 3,9-dihydroxy-2,4,8,10-tetraoxa-3,9-diphosphaspiro[5,5]-undecane-3,9-dioxide) and Melabis (5) (melamine salt of bis(1-oxo-2,6,7-trioxa-1-phosphabicyclo[2.2.2]octan-4-yl-methanol)phosphate) were synthesized from pentaerythritol (2), melamine (3) and phosphoryl trichloride (1) (Figure 4.8). They were found to be more effective to fire retard polypropylene than standard halogen-antimony fire retardant.

Based on this work and searching for environmental friendly process, Fontaine *et al.* [44] suggested the synthesis

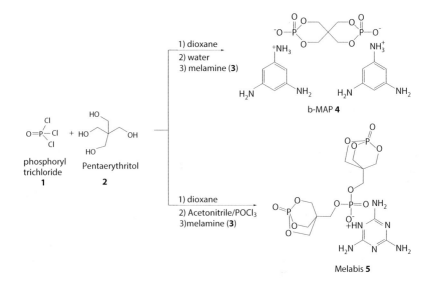

Figure 4.8 Melabis and b-MAP synthesis.

of PEPA (1-oxo-4-hydroxymethyl-2,6,7-trioxa-l-phospha-bicyclo[2.2.2]octane) and bis(PEPA)phosphate-melamine salt derivatives prepared by a novel and safe protocol (one step – one pot process). These new salts were incorporated in PP and exhibit high fire performance via an intumescence mechanism (LOI of 32 vol.-% and V-0 rating at UL-94 on bars of 3.2 mm thick). Moreover, Liu *et al.* [45] proposed to use catalytic action of phosphotungstic acid in the synthesis of melamine salts of pentaerythritol phosphate (called MPP in their paper but it is the same molecule as b-MAP) in order to solve the problems of conventional preparation methods (the use of $POCl_3$; the high temperature of the reaction and thus the high energy consumption…) (Figure 4.9). It was shown that the acid catalysis can enhance the conversion degree of the reaction and decrease the reaction temperature while keeping a satisfactory conversion, thus greatly controlled the energy-consuming in the preparation process of MPP (or b-MAP). The results also indicated that as compared to non-catalyzed MPP, catalyzed MPP flame retardant system remarkably improved the flame retardant properties reinforcing and stabilizing the char layer, and maintaining acceptable mechanical properties.

Other authors also suggested the use of organic compound '3 in 1' containing the three main ingredients of

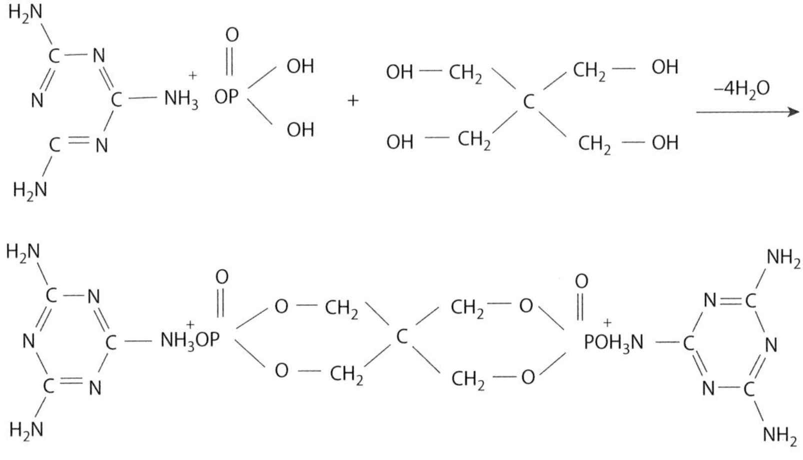

Figure 4.9 Synthesis of melamine salts of pentaerythritol phosphate (MPP or b-MAP) (from Ref. [45]).

intumescence. A clever application of a well-known natural product, namely deoxyribose nucleic acid or in short DNA [46], was reported by Alongi et al. in thermoplastics [47, 48]. They showed that DNA could compete with conventional FRs when evaluated by cone calorimetry.

Wang et al. reported [49] the synthesis of a novel intumescent flame retardant, poly(2,2-dimethylpropylene spirocyclic pentaerythritol bisphosphonate) to yield polyethylene terephthalate (PET) with both excellent flame retardancy and anti-dripping properties. A novel phosphorus-nitrogen containing intumescent flame retardant was prepared via the reaction of a caged bicyclic phosphorus (PEPA) compound and 4,4-diamino diphenyl methane (DDM) in two steps (Figure 4.10) [50]. Incorporated in PBT and in combination with thermoplastic polyurethane (TPU) as additional char former, the intumescent PBT exhibits V-0 rating at the UL-94 test (3.2 mm) and an enhanced thermal stability. It is suggested that P-N bonds detected in the charred structure might play a role in its efficiency.

To avoid dissolution or extraction of the FR, FR oligomer can be used instead of monomer. Using this approach, Ma et al. [51] reported the synthesis of phosphate–polyester copolymers from spirocyclic pentaerythritol di(phosphate acid monochloride)s. It was shown that LOI of the copolymer increases with increasing phosphate content to reach a maximum of 30 vol.-%. As the polyol-based char formers need to be substituted, Li et al. [52] reported the synthesis of a novel char former for intumescent system based on triazines and their derivatives. It is a macromolecular triazine

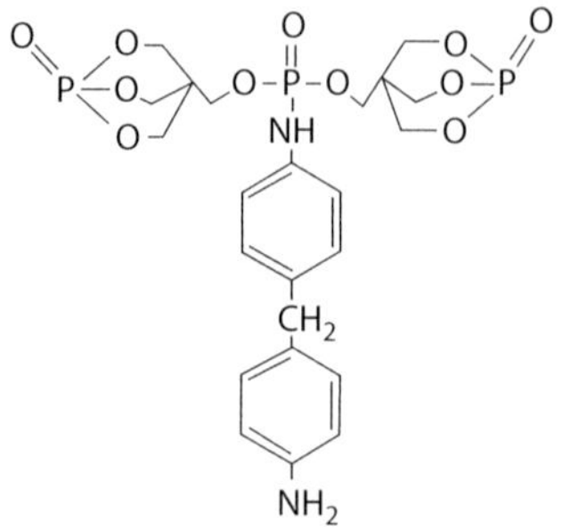

Figure 4.10 Phosphorus-nitrogen intumescent flame retardant.

derivative containing hydroxyethylamino, triazine rings and ethylenediamino groups (Figure 4.11). They showed that the new char former in an intumescent formulation containing APP and a zeolite as synergist can achieve low flammability at only 18 wt.-% loading in PP (LOI of 30 vol.-% and V-0 rating (3.2 mm) at the UL-94 test). No tentative of mechanism is commented but we suspect that it should be closed to that we described in a previous paper [53].

Schartel's group followed a promising approach developing novel hyperbranched molecules as potential FRs [54, 55]. They synthesized halogen-free aromatic and aliphatic hyperbranched polyphosphoesters (hbPPEs), which were prepared by olefin metathesis polymerization and investigated them in epoxy resins. The aliphatic version of hbPPEs exhibits a gas phase activity and they are not intumescent. On the contrary, the aromatic version of hbPPEs increases the char yield and enhances the thermal stability of the epoxy-based material. Intumescence is not mentioned in the paper [54] but the authors observed the formation of 'voluminous char layer'. So, it can be suspected the formation of an intumescent material (it is also confirmed by the pictures of the cone residues). hbPPEs were then modified by sulfur moieties in their structure [55]. The authors reported that the presence of sulfur increased thermal stability of the flame retardants and introduced additional condensed phase action. Once again, the formation of char

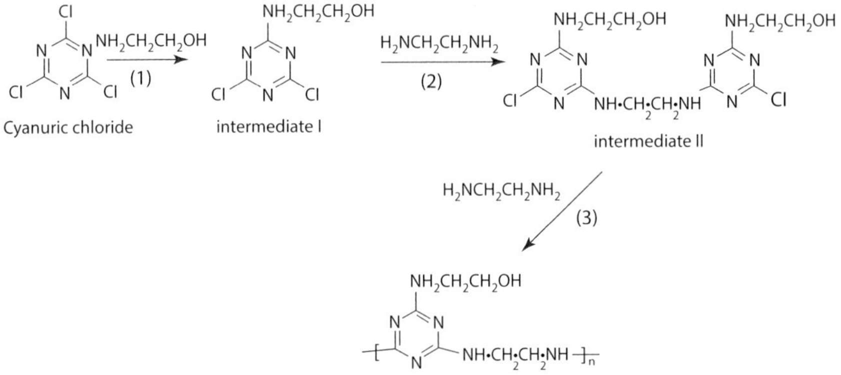

Figure 4.11 Synthesis of macromolecular triazines derivatives as char former for intumescent systems (from Ref [52]).

acting as protective layer was formed during cone testing. According to the pictures showed in the paper, it can be said an intumescent char was created upon heating. Other authors synthesized hyperbranched molecules as char former in intumescent systems. Examples are molecules containing s-triazine, diphenylmethane and urea groups [56] and hyperbranched and phosphorus-containing triazine derivative [57]. They are very efficient in combination with the conventional APP in polypropylene (PP) as the pHRR is reduced by about 80% in the two cases.

ii. Inorganic-based systems

The organic intumescent systems represent a large part of the studies dealing with intumescence while the processing of intumescent mineral systems is even older [4]. However, it is rarely commented in the literature. Mineral intumescent systems are based on alkali silicates. The swelling of the material upon heating or on contact with a flame is due to an endothermic process and is associated with the emission of water vapor that is ionically hydrated in the silicate system [26]. The solid foam formed is rigid and consists of hydrated silica. The structure remains solid until it reaches its glass-softening point. Since only water vapor is released, toxic fumes that may be released from organic-based systems are eliminated. But intumescent alkali silicates have serious limitation, in particular they are sensitive to carbon dioxide and water which is present in the atmosphere causing the silicate coating to gradually lose its intumescence, to become brittle, and to lose its adhesion.

Hermansson *et al.* [53] showed that incorporation of chalk filler and silicone can greatly improve the FR properties of ethylene-butyl acrylate (EBA) formulations. A filling of 30 wt.% of chalk filler and 5 wt.% silicone was employed. Compared to the pure polymer, an increase of LOI from 18 to 30 vol.-%, and a decrease in the pHRR from 1300 to 330 kW/m^2 were measured. The improvements were assigned to an intumescent process. The suggested mechanism is as follows [58]: (i) ester pyrolysis occurs when decomposition starts at 300°C producing carboxylate ions on the copolymer backbone, (ii) volatile decomposition products of EBA copolymer lead to foaming of the melt, which is stabilized by the formation of carboxylate ions and (iii) calcium ions from the chalk (Figure 4.12). Other copolymers (EBA blended

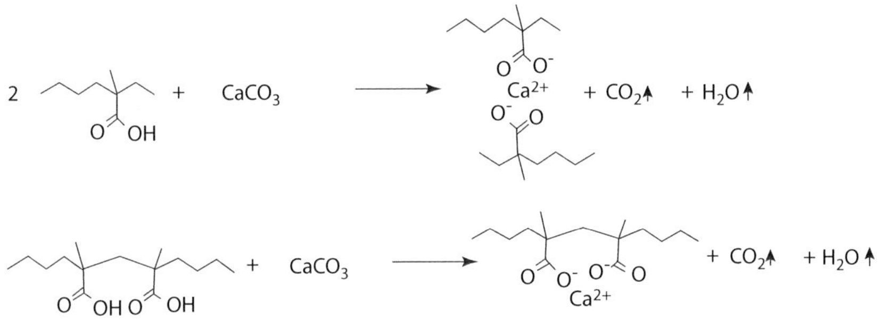

Figure 4.12 Reaction scheme for the calcium salt formation in EBA copolymer containing chalk (from Ref [59]).

with PP and poly(ethylene-co-methacrylic acid) (EMAA)) have been investigated by Krämer *et al.* [59] and they found that EMAA formulation has the most effective intumescent process with a low heat release rate and good char stability.

Polyhedral silsesquioxanes (POSS) are not known as intumescent FR but as nanoparticles for making polymer nanocomposites. POSS can provide low flammability to polymer via a mechanism in condensed phase [60]. In a previous work [61], it was shown that the incorporation of POSS in thermoplastic polyurethane (TPU) used as coating on woven PET fabrics permitted 50% reduction in pHRR. The suggested mechanism is char formation at the surface of the material which can act as an insulative barrier. It was also reported [62] that the incorporation of 10 wt.-% FQ-POSS (Poly(vinylsilsesquioxane) with the brand name Fire Quench from Hybrid Plastics – USA) in TPU (in bulk polymer) permits to decrease by 80% the pHRR (Figure 4.13-a) without any significant enhancement of LOI (22 vs. 23 vol.-%) and UL-94 (V-2 at 3.2 mm in the two cases). Visually, the mechanism of protection occurs via an intumescent phenomenon (Figure 4.13-b) and it was evidenced that the mechanism consists in the following steps: (i) charring of TPU and POSS taking place at similar temperatures with the formation of a 'viscous' paste, (ii) expansion of a ceramified char (char reinforced by a silicon network) from the evolution of the degraded products and partial sublimation of POSS, and (iii) strong reduction of heat transfer to the substrate (measured experimentally) by the formation of the intumescent layer.

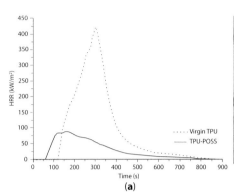

(a)   (b)

Figure 4.13 HRR as a function of time of pure TPU and TPU/FQ-POSS composite (external heat flux= 35 kW/m²) (a) and intumescent char residue at the end of the cone experiment (b) (from Ref [62]).

The incorporation of different types of POSS at low loading (about 5 wt%) in epoxy resins decreases their flammability via an intumescent mechanism. Gerard *et al.* [31] reports the formation of highly expanded intumescent coating during cone calorimetry but the multi-layered structure does permit a high reduction of pHRR (only -10%). With another type of POSS (reactive POSS vs not reactive) able to be chemically incorporated in the epoxy network, Raimondo *et al.* [63] showed that pHRR is decreased by 40%. It is assigned to the formation of an intumescent coating exhibiting a multi-layered structure. According to the pictures published by the authors, the char created with the reactive POSS looks more cohesive than that formed with the non-reactive POSS. With another reactive POSSs in an aeronautic epoxy resin (specific chemistry), Laik *et al.* [64] showed up to 65% reduction of pHRR. An intumescent mechanism was reported and it was evidenced the formation of a foamy and cohesive char. This foamy structure containing well distributed voids explains the superior performance measured by cone calorimetry. In those papers, the mechanism was not clearly detailed but it is reasonable to assume the formation of ceramified char by the interactions between the degrading epoxy resin and POSS. It is followed by the expansion of the char (viscous charred material) because of the evolving decomposition products of the epoxy resin and of the organic parts of POSS.

Silicates can be used as intumescent FRs in some polymers [65] like unsaturated polyester resins and silicones. More specific silicate systems with bound water were incorporated in polyvinyl chloride (PVC) and polyethylene (PE) and the resulting materials were evaluated by cone calorimetry [66]. The loading was quite high (50 wt%) but the purpose was to make a comparison with the conventional aluminum trihydroxide (ATH). It was shown that the silicate system is more efficient than ATH and the pHRR was decreased by 70% for the two polymers. The reported mechanism is intumescence combined with the endothermal dehydration of the silicates and the dilution effect of the water release.

In an unconventional way, it was reported that calcium carbonate ($CaCO_3$) was an intumescent filler in PE [67, 68]. To get this effect, stearic acid must be added to well disperse the particles of $CaCO_3$. Upon heating, a layer of $CaCO_3$ appears at the surface of the sample and thanks to its good dispersion, the decomposition gases of the polymer are trapped in the cohesive structure and the increasing internal pressure creates the expansion of the structure. The gases' release through the sample was slowed down thanks to the high tortuosity of the intumescent structure, which causes the reduction of the HRR during cone calorimetry (reduction by 80% compared to virgin PE).

Expandable graphite represents another class of inorganic intumescent systems as mentioned in the section 2 of this chapter. Expandable graphite is a graphite intercalation compound (GIC). It appears in the literature in 1840 [69]. It is a layered crystal consisting of sheets of carbon atoms tightly bound to each other. Chemicals (such as sulfuric acid) may be inserted between the carbon layers. Upon heating EG expands and generates a voluminous insulative layer thus providing fire performance of interest to the polymeric matrix. EG has been used advantageously in polyurethane coating to develop fire protective coating for polymeric substrates [70]. The intercalation compound can also be more sophisticated than the convention sulfuric acid. In combination with $KMnO_4$ and H_2SO_4, hydrated sodium silicate was employed [71, 72]. It was shown that V-0 rating was achieved in ethylene vinyl acetate copolymer (EVA). The mechanism is an intumescent phenomenon with the formation of stable char.

c. Intumescent textile

In this part, we should distinguish between natural and synthetic fibers because different methods are usually involved to provide them flame retardancy by intumescence. Very often the authors describe the mechanism of action of their materials as 'mechanism via charring enhancement' or something similar (see [73] as general review on flame retardant for textiles and references therein). However, based on the chemical nature of the flame retardant used and by the described mechanism, we can sometimes suspect an intumescent behavior. A typical example is that mentioned in the introduction about the paper by J.L. Gay-Lussac in 1821 [1]. It deals with the flame retardancy of curtains used in theater where fire hazard was high because of candles enlightening the scene and the room. Because the ingredients used, we can then suspect an intumescent behavior as protection mechanism against fire. In 2017, other authors repeated the same formulations on cotton and polyester and they got acceptable performance [74]. No comments were made on the mechanism of action but the formulation was similar.

Horrocks *et al.* ([75] and references therein) undertook considerable work on intumescents applied to textile structures, in particular substantive fiber treatments for cellulose. Based on the work of Halpern [42] on cyclic organophosphorus molecules, they developed a phosphorylation process for cotton fibers achieving intumescent cotton fabric with considerable durability. Char enhancement is as high as 60 wt.-% compared to the virgin cotton, which is associated with very low flammability. Coating (or back-coating) on fabric is another way to provide flame retardancy to cotton. Horrocks' group [76] used intumescent back-coatings based on APP as the main FR combined with metal ions as synergist. Metal ions promote thermal decomposition of APP at lower temperatures than in their absence, and this enables flame retardant activity to commence at lower temperatures in the polymer matrix thereby enhancing flame retardant efficiency. Giraud *et al.* [77, 78] developed the concept of microcapsules of ammonium phosphate embedded in polyurethane and polyurea shells to make an intrinsic intumescent system compatible in normal polyurethane

(PU) coating for textiles. The advantage of this concept is to reduce the water solubility of the phosphate and to produce textile back-coatings with good durability. The flame retarding behavior of these coated cotton fabrics was evaluated by cone calorimetry evidencing a significant FR effect. Development of intumescent char at the surface of the fabric was observed confirming the expected mechanism.

Layer-by-layer (LbL) assembly has been used as a surface treatment to impart flame resistance to cotton fabric. It is a simple versatile method to incorporate various polymers, colloids, or molecules into a thin film that is typically no thicker than 1 μm [79, 80]. So, it gives the opportunity to create intumescent systems selecting the right combination of ingredients. It was first evidenced by the Grunlan's group using LbL assembly of poly(sodium phosphate) (PSP), which acts as the acid source and is negatively charged in water, and poly(allylamine) (PAAm), which is used as the blowing agent [81]. From 5 to 20 bi-layers (BL) were applied on cotton fabric and the coated fabrics were subjected to vertical flame testing (VFT). The uncoated control fabric was completely consumed by direct flame, while coated fabric with a 5 BL coating was preserved as a complete piece. For the 20 BL-coated fabric, the flame extinguished right after ignition while the 10 BL-coated fabric exhibits an intermediate behavior (Figure 4.14). It is evidenced an intumescent

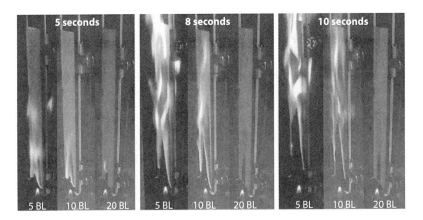

Figure 4.14 Cotton fabric coated with intumescent bi-layers subjected to vertical flame test; images during the flame testing recorded at 5, 8, and 10 s (from Ref [81]).

behavior protecting the fibers against the combustion. Other ingredients including chitosan (CH, acid source), APP (acid source and blowing agent) and branched polyethylenimine (BPEI) were used to coat cotton [82]. It is revealed an intumescent behavior during the VFT and the mechanism was completely elucidated. A combination of both condensed (formation of aromatic char) and gas phase (release of water and highly flammable gases) mechanisms impart the FR behavior promoting a kind of "microintumescence" phenomenon.

The application of DNA as all-in-one intumescent system (see the previous section) was first reported to impart flame retardancy to cotton fabric [16, 83, 84]. By combining DNA with chitosan, the authors made LbL assemblies on cotton fabric. The system is able to extinguish flame in horizontal flammability tests and strongly reduce the heat released during combustion. It was confirmed that DNA is an intrinsically intumescent material evidenced by charring and small expansion in the residues. It was reported that DNA layers promote the char formation of chitosan counterparts, by releasing phosphoric and polyphosphoric acid. The obtained char is thermally stable and it imparts flame retardant character to cotton.

Within the area of natural fibers, wool has the highest inherent non-flammability. It exhibits a relatively high LOI of about 25 vol.-% and low flame temperature of about 680°C [85]. The inherent flame retardant activity of the fiber can be associated with char-forming reactions which may be enhanced by a number of flame retardants. Based on their fundamental work to enhance char formation, Horrocks *et al.* [86] offer intumescent formulations based on melamine phosphate to flame retard wool. From TGA and SEM characterization, they proposed a comprehensive model on the mechanism of protection via an intumescent process. It involves the formation of cross-linked char by P-N and P-O bonds resistant to oxidation. In another work, they used spirocyclic pentaerythritol phosphoryl chloride (SPDPC) phosphorylated wool to achieve intumescent wool which exhibits large char expansion and good flame retardancy [87]. More recently, a water-soluble polyelectrolyte complex (PEC) consisting of bio-resourced phytic acid and chitosan was prepared to impart flame retardancy to wool fabric [88]. The technology is close to LbL process but it is faster. The coated wool exhibits an intumescent behavior upon testing using VFT and LOI. LOI jumps from 24% to 33% and the char length is short with the coating.

Polyester fibers are the main synthetic fibers used in the industrial manufacturing sector and can be found in several areas of application. Several flame retardants have also been designed for polyester extrusion (bisphenol-S-oligomer derivatives from Toyobo or phosphinate salts such as OP950, from Clariant). According to our previous work on PET [89], phosphinate salts provide flame retardancy to PET via an intumescent process but it is never clearly commented in the literature when it is applied to textile. Chen *et al.* [90] proposed the use of a novel anti-dripping flame retardant, poly(2-hydroxy propylene spirocyclic pentaerythritol bisphosphonate) (PPPBP) to impart flame retardancy and dripping resistance to PET fabrics. Flammability of PET fabrics treated with PPPBP was investigated by VFT which showed a significant enhancement of the flame retardancy (producing a non-ignitable fabric) and either a significant reduction of melt dripping at low levels or absence of dripping at higher levels of PPPBP. The same authors investigated in details the mechanism of flame retardancy of their FR PET fabrics [91]. They showed that it is a condensed-phase mechanism via char promotion in which PPPBP produces phosphoric or polyphosphoric acid upon thermal decomposition leading to the formation of phosphorus-containing complexes at higher temperatures. They suggest that the high yields of char are protected from thermo-oxidation by the presence of phosphoric acid contained in the charred residue and because of the high thermal stability of C=C groups in the char. In their paper, the authors do not report any intumescent behavior but we suspect because of the type of the compound used and according to the described mechanism that an intumescent process should be involved.

Like polyester, polyamides are synthetic fibers made from semicrystalline polymers which find use in a variety of applications in textiles almost similar as those of polyester. Horrocks *et al.* [75] has investigated the effect of adding selected intumescent flame retardants based on APP, melamine phosphate, pentaerythritol phosphate, cyclic phosphonate, and similar formulations into nylon 6 and 6.6 in the presence and absence of nanoclay. They found that in nylon 6.6 all of the systems comprising the nanoclay demonstrated significant synergistic behavior except for melamine phosphate because of the agglomeration of the clay. Recently, it was reported that enzymatically modified nylon 6.6 fabric, coated via LbL deposition of an intumescent APP/CH coating, with and without urea or thiourea provide self-extinguishing behavior [92] (horizontal test). This result is not exceptional in terms of fire retardancy but it is encouraging for further development.

Polypropylene (PP) is one of the fastest growing fibers for technical end-uses where high tensile strength coupled with low-cost are essential features. While polypropylene fibers may be treated with flame retardant finishes and back-coatings (intumescent or not) in textile form with varying and limited success [93], the ideal flame retardant solution for achieving fibers with good overall performance demands that the property is inherent within the fiber. Presently, the use of phosphorus-based, halogen-free flame retardants in PP fibers is prevented by the need to have at least 15–20 wt.-% additive. Since the latter are char-promoting while all halogen-based systems are essentially non-char-forming in polypropylene, the way forward for a halogen-free, char-forming flame retardant conferring acceptable levels of retardancy at low additive levels, 10 wt.-% will require either completely new FR chemistry or the development of a suitably synergistic combination. Based on this, Zhang *et al.* [93] reported that the flammability of polypropylene is reduced by the addition of small amounts of clay in conjunction with a conventional phosphorus-containing FR and a hindered amine. The authors suspect a P-N synergism to exist and the LOI value for the best formulation is 22 vol.-% compared to 19 vol.-% for neat PP, with only 6 wt.-% total loading. Microcapsules containing phosphate embedded in different polymeric shells (polyurethane, melamine formaldehyde and poly(hexamethylene adipate glycol) - melamine formaldehyde) were padded on PP fabrics [94]. The resulting fabrics were evaluated by cone calorimetry and it was shown that pHRR was decreased by up to 50%. The mechanism involved is the formation of an expanded charred layer protecting the PP fabric.

d. Intumescent inorganic polymeric materials

Few papers report the fire behavior of intumescent inorganic polymers. A possible explanation is that inorganic polymers are inherently flame retardant and are used "as is" [95]. The particular case of inorganic-organic polymers is either they are already flame retardant or they are used in applications which do not require high level of flame retardancy. The only published work devoted to intumescent inorganic polymer concerns polyphosphazene.

While polyphosphazene exhibits some level of flame retardancy, its performance is not enough to fulfill the requirements of the FAA for aircraft interiors. In order to make ultra-fire-resistant elastomers, the use of expandable graphite was investigated [96]. In a cone calorimeter experiment, the addition of the expandable graphite to the polyurethane reduces its heat release rate to a level approximating that of the pure polyphosphazene rubber. The addition of the expandable graphite to the polyphosphazene rubber reduces its heat release rate to a level approximating that of the fire-resistant engineering plastics or of thermoset resins currently used in aircraft interiors ($<100 \, kW/m^2$). The post fire test photos (not shown) show that the polyurethane rubber is completely consumed in the fire test. In contrast, the polyphosphazene leaves a char residue equal to 35% of the original weight. The polyurethane and polyphosphazene formulated with the expandable graphite leave a light friable char on the order of 20–50 times the original sample volume. The expanded graphite char insulates the underlying polymer from burning. The high thermal efficiency of the expanded char results in a peak heat release rate that is 5 and 7 times lower than that for the virgin polyphosphazene and polyurethane polymers, respectively.

e. Synergy in intumescents

Interesting developments have occurred which involve unexpected "catalytic" effects in various intumescent systems. Performance in terms of LOI, UL-94 or cone calorimetry was enhanced dramatically adding small amount of an additional compound leading to a synergistic effect. In the following, we use the definition of synergy as "a synergistic effect occurs when the combined effects of two chemicals are much greater than the sum of the effects of each agent give alone".

In the literature, numerous synergists (micro- and nanofillers) have been used in conventional "three-based ingredients" intumescent formulations. It covers the boron compounds (zinc borates [44], borophosphate [97], borosiloxane [98]), phosphorus compounds (phosphazene [99], $ZrPO_4$ [100]), silicon compounds (silica [101], silicone [102], silicalite [103]), aluminosilicate (mordenite [104], zeolite [103], montmorillonite [105]), rare earth oxides [106, 107] (La_2O_3, Nd_2O_3), metal oxides (Al_2O_3 [101], MnO_2 [108], ZnO [109], Ni_2O_3 [110], Bi_2O_3 [111], TiO_2 [112], ZrO_2 [113], Fe_2O_3 [114]) and others (carbon nanotubes [115], silsesquioxanes [116], layered double hydroxides [117], Cu [118], Fe [119], talc [120], sepiolite [121], zinc and nickel salts [122]). The presence of this additional filler can modify the chemical (reactivity of the filler versus the ingredients of the intumescent system) [105] and physical (expansion, char strength and thermophysical properties) [37] behavior of the intumescent char when undergoing flame or heat flux leading to enhanced performance. Those aspects are rarely described for the systems and papers only report the description of the enhanced performance with few physical and chemical characterizations. Our understanding is the chemical reaction between the fillers and the acid source (mainly phosphate derivatives) yields to phospho-X compounds (e.g. phosphosilicate, zinc phosphate, borophosphate etc.) reinforcing the structure and/or the action of the fillers (or its reaction products) as nucleating agent permits the formation of a homogeneous foamed structure with appropriate thermophysical properties (lower heat conductivity, lower emissivity at the surface etc.). Note in the particular case of using phosphinates or phosphonates in intumescent formulations, no reaction occurs between the filler and the other ingredients and in this case, the mechanism of action is mainly due to a physical effect [123]. Some examples are reported in the following to show how large can be the improvement and the mechanism involved in the synergy.

The first paper reporting synergy in intumescent systems was written by Scharf et al. [112]. They used metal oxides (TiO_2 and SnO_2) and they found a dramatic increase of LOI when incorporating a small amount of oxides in APP/PER system. The explanation was a chemical and physical inter-action stabilizing the char and modifying its morphology. It was not clearly evidenced but it is correct because APP reacts with TiO_2 to yield titanium phosphate upon heating. This last molecule is stable at high temperature and it can prevent the char from oxidation (so it explains why high LOI can be measured). Concurrently, Bourbigot *et al.* did all types of characterization to explain this phenomenon and to conceptualize the aspects of synergy in intumescent systems [37, 124]. Based on the same type of mechanism, they showed that adding small amount of minerals such as zeolites [103], natural clays [104] and zinc borates [44, 125] in intumescent systems, the FR performance (LOI, UL-94 and cone calorimetry) can be dramatically enhanced. It was confirmed by Levchik *et al.* [108, 126] in the same years. They reported that the use of small amount of talc and manganese dioxide combined with APP in PA-6 could promote charring and enhance insulative properties of the intumescent coating leading to a significant improvement of the FR performance.

Large synergistic effects are observed when incorporating nanofillers in intumescent formulation and most of the recent works on synergy are devoted to this. The basic mechanism remains similar to that described above but the incorporation of nanofiller can bring additional properties (e.g. mechanical properties). It is noteworthy that the nanofiller is a crucial ingredient (reactant) in the development of intumescence forming additional species stabilizing the structure (it is also true with microfiller) and modifying the rheological behavior (more specific to nanofiller when the dispersion is really at the nanoscale). The nanofiller is incorporated at amount as low as 1 wt.-% (sometimes less as in the case of the incorporation of nanoparticles of copper at amount as low as 0.1 wt.-% in epoxy resin containing APP [127]) and it permits the formation of active species selecting chemical reactions in the condensed phase and yielding to char with the dynamic properties of interest.

A typical example of using nanofillers as synergists in an intumescent system was given by Muller *et al.* [37]. The polymeric matrix was a polyurethane (PU) synthesized in the lab by conventional methods. It was known that APP in PU provides an intumescent because PU is the char former and

APP is the acid source and the blowing agent. The nanofillers considered were MgO, SiO₂ and octamethyl polyhedral
silsesquioxane (OMPOSS). The incorporation of 30 wt.%
APP in PU decreases by 62% the pHRR compared to virgin
PU when measured by cone calorimetry (Figure 4.15). The
time to ignition is similar for all the materials and it occurs
at about 45 seconds. The substitution of 2 wt.% of APP by
nanoparticles shows significant effects on the reaction to fire
of the intumescent PU. PU/APP and PU/APP-SiO₂ exhibits
similar HRR curves as a function of time. The incorporation of OMPOSS in PU/APP brings an additional decrease
to pHRR: decrease by 71% vs. 62% compared to virgin PU.
A larger synergistic effect is observed with the incorporation
of MgO since pHRR is decreased by 83% (PU/APP-MgO)
instead of 62% (PU/APP) compared to virgin PU. The HRR
curves are stretched over time (pHRR occurs at 200s compared to 100s for PU-APP) and even if the time to ignition
of PU/APP-MgO is not significantly modified, HRR remains
quite low (pHRR<100 kW/m²) and its slope increases significantly only from 100s. In all cases, the reduction of HRR
is assigned to an intumescence phenomenon protecting the
material (visual observation). The char of PU/APP exhibits an expansion of 1100% associated to a residual mass of
44wt% while those containing the nanoparticles exhibit an

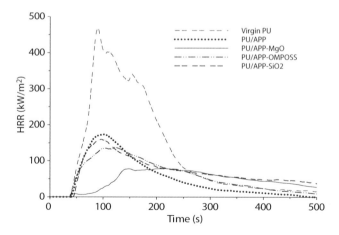

Figure 4.15 HRR curves as a function of time for PU, PU/APP, PU/APP-MgO, PU/APP-
SiO₂ and PU/APP-OMPOSS (external heat flux = 50 kW/m²).

expansion of 1200% (SiO_2), 1300% (OMPOSS) and 2200% (MgO) associated to a residual mass of 44wt% (SiO_2), 42wt% (OMPOSS) and 52wt% (MgO).

The nanofiller of MgO gave the largest synergistic effect and the mechanism of action was investigated. It was shown that a reaction between APP and MgO occurred yielding to the formation of magnesium phosphates (evidenced by solid state NMR of [31] P and by XRD). Those latter stabilizes phosphates at high temperature which can prevent the oxidation of the intumescent char [128] and hence, the protection is more efficient. The char morphology was also examined since it governs all the thermo-physical properties (heat conductivity for example) leading to an efficient thermal barrier. PU-APP char morphology exhibits an irregular surface but it is more homogeneous inside the char. The inner structure is composed of big bubbles separated with thick porous walls (Figure 4.16). On the contrary, PU/APP-MgO char is highly expanded and it exhibits a regular surface. Inside the char, two distinct structures are observed: at the bottom, there are many bubbles of various sizes separated by thin walls; on the top, only one or two huge bubbles are observed (Figure 4.17). Compared to the PU/APP char structure, the walls of PU/APP-MgO char are generally thinner than those of PU/APP char. In particular the crust is very thin at the top of the char where huge bubbles are observed while the walls between bubbles are thicker at the bottom of the char where many cells are observed. The incorporation of MgO modifies the morphology of the char obtained with PU/APP. The expansion, the size, number and repartition of cells, the walls thickness, the aspect and size

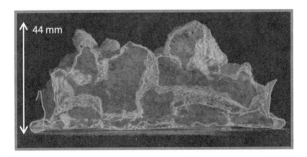

Figure 4.16 X-ray tomography picture of the inner structure of the char formed by PU/APP after cone calorimetry.

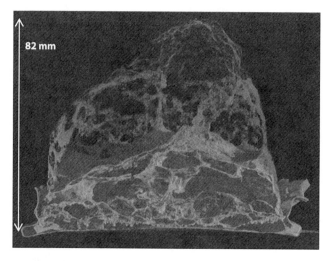

Figure 4.17 X-ray tomography picture of the inner structure of the char formed by PU/APP-MgO after cone calorimetry.

of the crust are indeed affected by the presence of MgO. It is observed two different parts in the char structure with many unorganized cells at the bottom and huge voids at the top. The morphology exhibited by the intumescent PU formulations is a result of a swelling and blowing mechanism leading to the formation of cells of different size. The fillers incorporated in PU participate to this bubbles development to a certain extent depending on the fillers or combination of filler used. Upon thermal stress, the fillers degrade and/or react leading to the release of gases which are "used to build the intumescent structure". The increasing internal pressure from decomposition gases allows the char to expand. It was evidenced that the incorporation of MgO affects the final char structure which is much more complex than that without MgO. The synergy mechanism between APP and MgO in PU is consequently partly due to the char morphology and the development of intumescence.

4.5   Resistance to Fire of Intumescent Materials

Intumescent coatings have had a large increase in use, as a method of passive fire protection, over the past few years. Their main purpose is to protect construction materials such as steel and wood in case of fire. Intumescent coatings swell when subjected to high temperatures and they must resist to fire and avoid its passage and/or that of gaseous products of combustion. The goal is to keep the integrity and the functionality of the substrate as long as possible. That is why the main parameters considered to measure the resistance to fire of a material is the 'time-temperature curve' and the determination of a 'time to failure' (time to reach a given temperature, temperature depending on the specification). In this part, evaluation of the resistance to fire of materials are presented and commented in the first section. The second section is devoted to the development of bench scale tests and to recent protocols developed to measure key parameters of intumescence. The performance and the behavior of intumescence coatings are then surveyed in the third section.

a. Evaluation of the resistance to fire

In the event of a fire in buildings, vehicles or means of transportation, the fire resistance time of the materials and products is the most critical variable and has a direct influence on the occupants' safety. Fire-resistant products should allow the structures to retain their minimum functions during the time required to evacuate the people, even despite the extreme conditions of heat and pressure to which they are subjected. The products must be evaluated according to the applicable regulation.

After about a century, the standard fire resistance test is the predominant means to characterize the response of structural elements in fires. In 1908 the American Society for Testing Materials (ASTM) published a standard test on the need to develop a common approach to evaluating the fire safety of construction materials. In 1918 a time-temperature curve was established as part of the standard based on the maximum temperatures experienced in real fires at the time (known as ASTM E119 [129]). The curve was not based on the response of building components to a real fire but rather what the authors have described as a worst-case time-temperature relationship to be expected during a fire. This approach is still in use at this time even if some authors pointed out many shortcomings and showed it is sometimes unrealistic (see e.g. the review of Bisby et al. [130]).

Required levels of protection are specified in terms of time and temperature on the basis of one or more criteria, which may include statutory requirements, design considerations and insurance cost implications. It can vary from a few minutes to several hours but it usually takes the form of 15 minutes increments. The duration is established by a time rating which is determined by testing in accordance with an approved standard. The standards depend on the kind of fire the material should resist. There are three main categories of fires: the cellulosic or wood fire, the hydrocarbon fire and the jet fire. However, a number of different fire test curves have been proposed as shown in Figure 4.18 [131–133].

The cellulosic fire curve (ASTM E119) simulates the rate of temperature increase observed in a residential or commercial building fire where the main sources of combustion fuel are cellulosic in nature, such as wood, paper, furniture and common building materials. In the development of new fireproofing products, this test method is currently used primarily to evaluate the relative performance of fireproofing on full scale standard sizes and shapes of structural steel under controlled conditions. Although this fire curve is still used, it is noteworthy that the burning rates for certain materials e.g. petrol gas, chemicals, etc., are well in excess of the rate at which, for instance, timber burns. As such, there was a need for an alternative fire test for the evaluation of structures and materials used within the petrochemical industry and therefore the hydrocarbon test curve was developed.

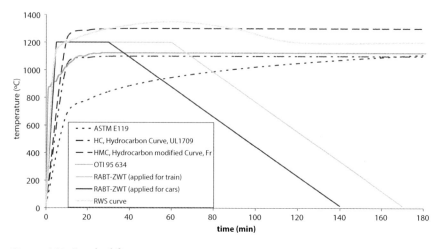

Figure 4.18 Standard fire test curves.

The hydrocarbon test curve should duplicate or be indicative of the rapid temperature rise seen when a hydrocarbon fuel such as oil or natural gas burns: the temperature rises rapidly to 900°C within 4 minutes and significantly higher overall temperatures are reached (between 1100°C and 1200°C). This hydrocarbon fire test curve, developed by the Mobil Oil Company in the early 1970's and adopted by a number of organisations and in particular Underwriters laboratories (UL 1709 "Rapid Temperature Rise"), UK Dept. Of Energy, BSI, ISO and the Norwegian Petroleum Directorate is now a common test method for high risk environments such as petrochemical complexes and offshore platforms. Derived from the above-mentioned hydrocarbon curve, the French regulators asked for a modified version, the so-called hydrocarbon modified curve (HCM). The maximum temperature of the HCM curve is 1300°C instead of the 1100°C standard HC curve. However, the temperature gradient in the first few minutes of the HCM fire is as severe as all hydrocarbon-based fires.

The RABT (Richtlinien für die Ausstattung und den Betrieb von strassen Tunnels) curves were developed in Germany [134]. In the RABT curve, the temperature rise is very rapid up to 1200°C within 5 minutes, faster than the Hydrocarbon curve which rises only to 1150°C in 10 minutes. The duration of the 1200°C exposure is shorter than other curves with the temperature drop off starting to occur at 30 minutes. This test curve can be adapted to meet specific requirements. In testing to this exposure, the heat rise is very rapid, but is only held for a period of 30 minutes: it is similar to the sort of temperature rise that would be expected from a simple truck fire, but with a cooling down period of 110 minutes.

The RWS (Rijks Water Staat) curve [134] was developed by the Ministry of Transport in the Netherlands. This curve assumes that in a worst-case scenario, a fuel, oil or petrol tanker fire with a fire load of 300MW could occur, lasting up to 120 minutes. The RWS curve was based on the results of testing carried out by TNO Centre for Fire Research in the Netherlands in 1979. The difference between the RWS and the hydrocarbon curve is that the latter is based on the temperatures that are expected from a fire occurring within a relatively open space, where some dissipation of the heat occurs, whereas the RWS curve is based on temperature found in a fire occurring in an enclosed area, such as a tunnel, where there is little or no chance of heat dissipating into the surrounding atmosphere. The RWS curve simulates the initial rapid growth of a fire using a petroleum tanker as the source, and the gradual drop in temperatures to be expected as the fuel load is burnt off.

All the tests of fire resistance are not based on time-temperature curve. Some of them involve jet fires to mimic a specific fire scenario. The properties of jet fires depend on the fuel composition, release conditions, release rate, release geometry, direction and ambient wind conditions. Jet fires represent a significant element of the risk associated with major accidents on offshore installations or in plants [135]. The jet fire scenario assumes a flammable or explosive atmosphere when tank, vessel or pipeline containing pressurized and/or liquefied gas fail and hence, gas leakage entrains air with the gas effluent stream. The gas dilution starts at the leaking point because of the high pressure in the container. Therefore, a turbulent-free jet is formed, wherein the gas concentration and velocity along its axis depend on the orifice diameter. The heat fluxes released from these fires are very high, ranging from 200 to 400 kW/m² depending on the type of fuel. This range of heat fluxes is considered by the standard ISO 22899-1:2007 according to the Health & Safety Executive, Offshore Technology Report OTI 95 634: 1995. The applicability of this test was critically reviewed by Bradley *et al.* [136] and they found that a single heat flux was inadequate means of specifying the severity of the hazard but instead, a detailed description of the nature of the release should be provided.

In transportation such as aviation, jet fires can occur because of the large quantities of highly flammable fuel which can ignite in case of crash or during flight. So, the threat is particularly high compared to other situations. Materials and composites used in the construction of transport category aircraft must comply with national and international regulations with most countries having adopted the U.S. Federal Aviation Regulations (FAR). The fire resistance of components, equipment and structure located in 'fire zones' (e.g. compartments containing main engines and auxiliary power units) is conventionally evaluated according to the international standard ISO 2685 [137]. The test is established on a pass/fail basis thereby an element is considered to pass its fire test if it is capable to sustain its design function after the standard flame exposure for a specific lapse of time. It uses a specific burner fed by propane or kerosene with a flame temperature of 1100°C and delivering a heat flux of 116 kW/m². In addition to this standard and to increase survivability, the Federal Aviation Administration (FAA) has developed a medium-scale laboratory test to analyze the burnthrough resistance of aircraft skin components using an impinging high heat flux ($\approx$ 180 kW/m²) jet fuel flame from an oil burner to simulate conditions observed in post-crash aviation fires [138].

The standards listed above are the most commonly used standards but they are not exhaustive. They give an overview of fire testing in the field of resistance to fire and they show the conditions can be very severe and much more than those involved in the case of reaction to fire. It is therefore the challenge of intumescence to provide the protection of interest in those severe conditions. Intumescent materials are classified as either thick or thin film intumescent coatings. Intumescent thick films are usually based on epoxy, vinyl or other elastomeric resins. These films are particularly efficient in the case of hydrocarbon and jet fires. They are hard and durable and can provide excellent protection against corrosion. This is due to their very high adhesion to the substrate and resistance to impact, abrasion and vibration damage. High tensile and compressive strengths can be obtained and weather resistance is excellent. On the other hand, thin intumescent films are used for protection from cellulosic type fires. They typically use thermoplastic acrylic or polyvinylacetate resin, and they respond rapidly and intumesce quickly when exposed to a cellulosic type fire environment. One-hour protection can be achieved with between 1 and 3 mm of product. Thin film intumescents are often referred to as "fire retardant paints" rather than "fireproofing" materials due to their inferior fire resistance compared to thick films intumescents. They are mostly used inside buildings because of their poor durability.

b. Testing and characterization of intumescence at reduced scale

Resistance to fire of intumescent coating is typically measured by a curve 'temperature as a function of time' or with the impingement of burner flame into the material (see above). In all cases, it requires large scale equipment and those tests are very expensive and time consuming. Even if due to the complexity of fire phenomenon, full-scale tests remain the main and the most credible tool for investigating fire-related issues, the cost of those tests significantly increases with scale and it prevents the fast development of new fire-resistant materials and of new structures. This section examines the scale reduction of different tests to mimic fire scenarios at the reduced scale. Using selected representative examples, it is shown how to evaluate fire resistant intumescent material.

The protection of metallic materials against fire has become a very important issue in the construction and petrochemical industries, as well as in the marine and military fields. Structural steel loses a significant part of its load-carrying ability when its temperature exceeds 500°C. Prevention of the structural collapse of a building is crucial to ensure the safe evacuation of people and it is a prime requirement of building regulations in many countries. A small-scale furnace test (internal volume of 40 dm^3) was built to evaluate the fire performance of steel protected by intumescent coatings in cellulosic and hydrocarbon fires. This test was designed to mimic the ISO834 (or ASTME119) and the UL1709 normalized temperature/time curves, respectively related to cellulosic fire and to hydrocarbon fire using propane burners [139]. Samples of 10 x 10 cm^2 are tested with this equipment and the temperature as a function time is recorded on the backside of the steel plate. Intumescent paints based on silicone as polymeric matrix were applied on steel plates and were evaluated within the small-scale furnace according to ISO 834 temperature curve (Figure 4.19). The silicone-based intumescent paint was compared to a commercial epoxy-based intumescent paint. The failure temperature was taken at 500°C. Uncoated steel plate reaches 500°C at 1000s while the coated steel with intumescent coatings reaches the failure temperature at times

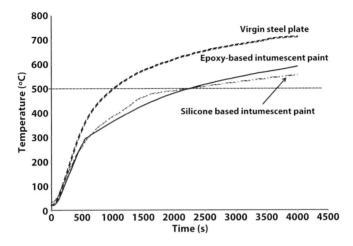

Figure 4.19 Temperature as function of time on the backside of steel plate protected or not by an intumescent paint when undergoing ISO 834 heating in the small furnace.

higher than 2400s. It shows that intumescence is an efficient way to protect steel since the time to failure temperature is twice longer than that with virgin steel. The silicone paint was compared to the commercial paint and it appears that its behavior is similar as epoxy-based paint.

Similar approach was followed by Schartel *et al.* modifying a lab muffle furnace to comply with the time/temperature curve described in EN 1363-1 [140]. The furnace is completely instrumented with in particular, a high temperature endoscope permitting to observe the foaming/expansion process of the intumescent coating (Figure 4.20). This specific instrumentation gives new insights on the formation of the intumescent coating which allows the ranking of the intumescent materials and also, the development of new ones exhibiting higher performance. They also evaluate different intumescent coatings under four different but similar shaped heating curves with different maximum temperatures (standard time-temperature curve, hydrocarbon curve and two self-designed curves with reduced temperature) [141]. They found that the time/temperature curves affected the inner structures and the residue surface leading to difference thermal performance. They suggested that the coating should be appropriately selected according to those curves.

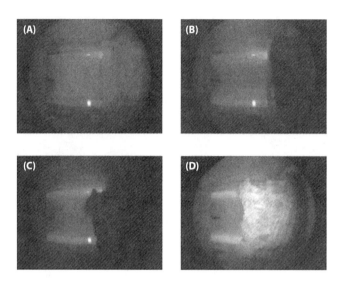

Figure 4.20 (a–d), pictures as a function of time of the foaming specimens recorded by the high-temperature endoscope inside the furnace (from Ref [140]).

A novel fire testing method, named the Heat-Transfer Rate Inducing System (H-TRIS), was presented by Maluk *et al.* [142] (Figure 4.21-a). The method directly controls the thermal boundary conditions imposed on a test specimen by controlling a specified time-history of incident radiant heat flux at its exposed surface. It was used to examine the effectiveness of thin intumescent coatings (Figure 4.21-b) [143]. It was found the onset of swelling is directly influenced by

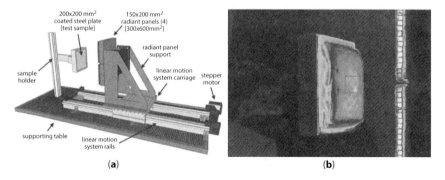

Figure 4.21 (a) scheme of the H-TRIS bench-scale test and (b) analysis of an intumescent coating using infrared thermography undergoing radiative heat flux on H-TRIS test (from Ref [143]).

the heating conditions at the exposed surface and the original applied dry film thickness.

An horizontal burner test was developed to mimic the burnthrough taking place in ISO 2685 (Figure 4.22 (a)). ISO 2685 is used to evaluate the fire resistance of components located in 'fire zones' of aircraft [144]. The test consists of impacting a 100x100mm² sample by a propane flame at 1100°C and with a calibrated heat flux of 116 kW/m². A commercial epoxy-based intumescent coating was evaluated for protecting steel when the flame directly impinges the intumescent paint. The comparison of the temperatures as a function of time on the backside of steel plate with and without paint shows the efficiency of the intumescent protection (Figure 4.22 (b)): in the steady state, the temperature only reaches 290°C with the intumescent protection while the temperature is 450°C without protection.

Infrared images (Figure 4.23 (a) and (b)) show the development of the intumescent coating and the flame contour during the test. Note the infrared images were obtained using specific filters to observe the flame shape (Figure 4.23 (a)) and the surface of the sample through the flame (Figure 4.23 (b)). The burner flame engulfs the intumescent coating (Figure 4.23 (a)) and the expansion of the intumescent coating is limited. It does not form a regular expanded char but the expansion is a distribution of small bubbles at the surface of the material (Figure 4.23 (b)). It is probably

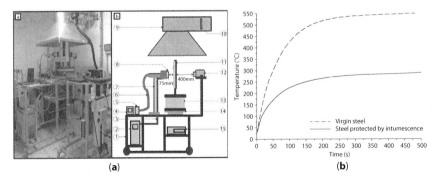

(a)   (b)

Figure 4.22 (a) Picture and scheme of the horizontal burner test bench (from Ref [144]) and (b) Temperature as a function time at the burnthrough test of steel protected by intumescent coating.

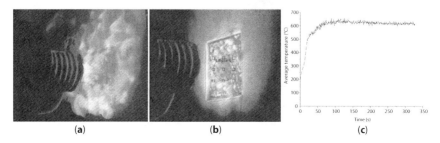

Figure 4.23 Intumescent coating on steel plate during the burnthrough test (a) infrared image of flame contour and (b) infrared image of the intumescent surface through the flame - (c) Average temperature at the surface of the intumescent coating on the zone defined by the ellipse.

due to the velocity of the gas jet limiting the expansion. A crude assumption without any further measurement on the emissivity of the surface (here the emissivity is assumed to be one) permits to estimate the surface temperature as a function of time of the intumescent coating during the test (Figure 4.23 (c)). It is an average temperature measured in an elliptic zone as shown in Figure 4.23 (b). The temperature increases sharply when applying the burner flame and reaches a pseudo steady state temperature of 615°C. The temperature gradient between the front side and the back-side is therefore of 325°C on a small thickness. It evidences the efficiency of the intumescence for fire protection and the approach permits to get quantitative data on temperature gradient involved in the intumescent char.

Jetfire scenario involves extremely high heat flux (up to 450 kW/m²) and it is generally not investigated at the reduced scale. In our lab, we developed bench-scale test mimicking jetfire for testing intumescent paint on steel plate (Figure 4.24-a) [145]. A powerful premixed propane/air burner provides a flame impinging the sample to be tested. Heat fluxes up to 450kW/m² can be reached by adjusting the propane to air ratio of the burner as well as their mass flow. The sample holder is shaped like a box dimensioned in order to avoid backfiring of the flame. Opposite to the burner, two boxes are used respectively to perform the calibration of the burner (the heat flux and temperature of the flame are measured using respectively a water-cooled heat gauge and

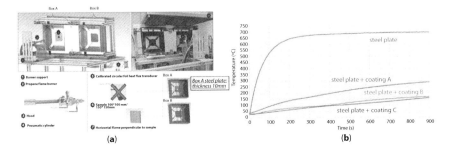

Figure 4.24 (a) Schematic description of the jetfire bench-scale test (note heat gauge is substituted by steel plate for day to day calibration) and (b) time/temperature curves of intumescent coating steel plates compared to virgin one.

thermocouples) and to evaluate the behavior of materials. Three intumescent coatings based on different chemistries were evaluated: (i) coating A (silicone-based coating containing expandable graphite), (ii) coating B (water-based coating) and (iii) coating C (epoxy-based coating). They were applied on steel plate and submitted to a heat flux of 300 kW/m². Time-temperature curves were recorded during the experiments and the results are shown in Figure 4.24-b. The profile obtained with the uncoated plate is compared to those obtained for coated plates whose basis weight was 1g/cm², leading to thicknesses of 6 mm, 7 mm and 5 mm for coatings A, B and C, respectively. All the coated plates show an improved behavior compared to the uncoated one thanks to the formation of an intumescent char. After 15 minutes test, the plates coated with A, B and C coatings reach respectively 287°C, 169°C and 154°C compared to 697°C for the uncoated one. It evidences that intumescent coating can resist to high velocity jet and at high heat flux.

Schartel *et al.* [146] also build up a bench scale at very high heat flux using a powerful oxygen-propane burner. The heat flux was not estimated but the temperature of the flame was extremely high in the range of 1200-1800°C. It was shown it was useful for evaluating intumescent coating in extreme conditions.

c. Fire protective intumescent materials

Few recent scientific papers are devoted to the resistance to fire involving new developed intumescent coatings while existing a growing market for this application and while academics are active in understanding the mechanisms of protection (see the previous section). However, we should cite the excellent review of E. Weil published in 2011 [147] reporting recent developments for intumescent coatings and their application for fire protection. An updated version was then done by Mariappan in 2016 [148].

A rapid survey of the literature shows that the used formulations are always based on the well-known triplet APP/PER-Melamine as it was reviewed by Weil *et al.* in 1994 [149]. The combination APP/PER-Melamine offers the possibility to substitute the char former (PER) by its dimer and trimer. Comparison in temperature development, foaming ratios, and rheological behavior was performed by Andersson *et al.* [150] between formulations containing PER, di–PER, and tri–PER. A simulated fire test developed by the authors (the coated face is upside down and exposed to Bunsen burner and an infrared pyrometer records the temperature on the backside), in which the temperature increase during intumescence was studied. They showed that the formulations containing PER were considerably more efficient in keeping a low temperature throughout the process. A more rapid temperature development was displayed when using di–PER and tri– PER as char former. Rheometer tests indicate that PER formulations enter the intumescent process at a lower temperature and stays in it for a longer time than di–PER and tri–PER formulations. No explanation is given why PER is more efficient than the two others. It might be proposed it is because of the higher reactivity of PER to APP (reaction of esterification between APP and PER occurs at 190°C [151]) occurring at lower temperature permitting a faster development of the intumescent coating.

Other char formers in substitution to PER and in combination with APP, can be found in epoxy-based commercial formulations such as Chartek (Akzo Nobel) and Pittchar (PPG) (see [147] and references therein). They are complex formulations involving numerous ingredients but the triplet acid source/char former/blowing agent is still there in majority. The char former was tris(hydroxyethyl) isocyanurate (THEIC) in Chartek and bisphenol-A derivative in Pittchar. Those two formulations are used nowdays as thick coatings for steel protection in off-shore oil drilling platforms and petrochemical chemical installations. It is noteworthy the presence of boron molecules namely boric acid (Chartek) or zinc borate (Pittchar). They play a particular role reinforcing the char forming borophosphate-type molecules upon heating [152]. In epoxy-based formulations, the role of char former can be the matrix itself [31]. Another interesting substitution of the char former is the use of ketone thermoplastic resin in epoxy-based intumescent coating [147].

Zeng *et al.* [153] report the mapping performance of intumescent coatings as a function of concentration of ingredients. The char former was the polymeric matrix (namely epoxy) and they examined the effects of APP, melamine, TiO_2, $CaCO_3$, and vitreous silicate fiber on the fire-resistance performance of zinc borate (ZB) containing and ZB-free hydrocarbon intumescent coatings with exposure to the hydrocarbon fire testing curve UL 1709. As previously reported, they found the coating with 25 wt.% APP or 5 wt.% MEL obtained the best performance. TiO_2 and $CaCO_3$ and the fiber do not significantly modify the performance of the ZB-containing intumescent coatings, but the time to failure (here 550°C) and char appearance were strongly affected for the ZB-free coatings. Increasing the content of any of the three inorganics led to reduced expansion of the ZB-free coatings, but improved the chars with more mechanically stable structures. The longest time to failure were found with the ZB-free formulations containing 1.5 wt.% TiO_2, 2.5 wt.% $CaCO_3$ and 5 wt% vitreous silicate fiber.

The use of additional ingredients acting potentially as synergists (micro- and nanoparticles) was also investigated in intumescent coatings. Multi-walled carbon nanotube (MWCNT) was found to be effective as flame retardant [154] and Beheshti *et al.* [155] evaluated it in intumescent coating as a potential synergist. Using a burner test, they found it is ineffective for improving the protection in acrylic based polymer containing APP/PER/MEL. Li *et al.* [156] suggested combining EG and/or molybdenum disilicide ($MoSi_2$) in an intumescent system based on APP/PER-Melamine. The results show that incorporating EG and $MoSi_2$, the time to failure (time required to reach 500°C in this case) is prolonged (heat insulation test similar as that described in the previous paragraph) and char formation rate is enhanced.

The largest improvement is achieved combining the two ingredients MoSi$_2$ and EG in an appropriate ratio. The suggested mechanism of action is revealed using scanning electron microscopy (SEM) images and thermogravimetry showing that the synergistic effect is obtained through a ceramic-like layer produced by MoSi$_2$ covered on the surface of an "open-cellular" structural char and a better resistance to the thermooxidation. EG was also used as additional blowing agent in benzoxazine-based coating containing APP and boric acid to make an efficient intumescent coating [157]. In another paper, Wang et al. [158] report the use of organo-modified layered double hydroxide (OLDH) as nanofiller combined with an intumescent system in acrylate resin. The intumescent paints were evaluated on steel plate measuring the temperature on the backside of the plate as a function of time when undergoing temperature (standard ISO-834). It was found that incorporating 1.5% OLDH the fire-resistant time (time to reach 300°C) jumps at 100 min compared to 60 min without OLDH (virgin coating). It is also noteworthy that the thickness of char layer of the formulation with 1.5% OLDH is similar to that of the formulation without. It confirms our results suggesting that the highest expansion is not necessary to get the best performance. The improvement of the performance is partially explained by the char strength. The morphology of char layer exhibits close holes but the diameters of holes of the char containing OLDH are much smaller (10-30 µm) than those of the char without OLDH. Wang suggests that small holes reinforce the char strength and avoid the formation of cracks at the surface of the char. Indeed, the formation of close cells in the char structure like in foam evenly dispersed, reduces heat transfer and then increases the efficiency of the char. When the cells are too big, char strength is reduced and cracks can appear.

In addition to the oxides as synergists, Ahmad et al. [159] investigated the effect of nano-sized boron nitride reinforced with alumina in epoxy-based intumescent coatings in terms of improvement of char expansion and of the morphology. Epoxy plays the role of char former and the formulation contains boric acid and EG in addition with APP and MEL. They showed that the formulations reinforced with nano-alumina effectively protected steel and kept its temperature well below the time to failure (measured at the burner test). Nano-alumina decreases the expansion but forms an additional protection of alumina and of alumino-phosphate at the surface of the char layer. It is also believed that the reduction of pore size in the char structure increases the insulation performance of the layer.

Environmental concerns promote the development of bio-based materials. Bio-fillers as additional ingredients in intumescent formulations were suggested in the recent literature. Nasir *et al.* [160] used rice husk ash (RHA) and chicken eggshell (CES) from waste product as bio-filler in intumescent coating. The formulation was based on APP/PER/MEL in acrylic polymer and were evaluated with a home-made burner test. When applied on steel plate, it was found that RHA acts as a synergist providing the lowest temperature in the steady state. Its mechanism of action is not clear and it is only assigned to better insulation due to the high expanded cross-linked and foamed char layer. Lignin is a natural product and is abundant. Baldissera *et al.* [161] used it as char former in intumescent formulation in combination with polyaniline. They found that the new formulations gave similar protection at the burner test as those containing oil-based char formers. In addition to bio-based molecules, recycled materials might be interesting to solve our environmental issues. Liu *et al.* [162] used recycled powders containing flame-retardants from the precipitate of industrial effluents (the composition is not clearly defined in the paper) and apply them in an intumescent coating. The intumescent coating was based on APP/PER/MEL with styrene-acrylic polymer as binder in which the recycled powder was incorporated. The formulations were evaluated with a torch test and it was shown the benefit of recycled powder as it enhances the char strength and the heat insulation of the intumescent layer.

Instead of the triplet APP/PER/MEL, other ingredients can be employed to make intumescent coating. It includes diglycidyl methylphosphonate, triglycidyl phosphate, bis[2-(methacryloyloxy) ethyl] phosphate or dihydroxa-phosphaphenanthrene oxide (DOPO), or preferably an adduct of DOPO with an epoxy-novolac, in intumescent coatings especially for steel [163]. The purpose was to achieve a substantial reduction in the film thickness of intumescent coating required to protect a given substrate from the effects of fire (cellulosic and hydrocarbon fires). The mechanism was not described in the patent but we can suspect an additional effect in the gas phase thanks to the action of DOPO enhancing the performance at low thickness.

Silicones have the properties to exhibit low thermal conductivity, to be water and heat resistant, to evolve few toxic gases during their decomposition and to have high durability: they are therefore good candidate to make fire protective coating. Gardelle *et al.* showed that intumescent silicone based-coating is very efficient to protect steel in the case of hydrocarbon fire [139] and cellulosic fire [164] scenarios. Expandable graphite (EG) was used as main ingredient in a silicone-based coating and provides good performance protecting steel (Figure 4.25). The performance increases as a function of EG loading in the coating and at 25wt% EG, the silicone coating exhibits higher performance (time to reach 500°C) than conventional commercial intumescent paint. This is explained by a high swelling velocity (18%/s), a high expansion (3400%), an impressive cohesion of the char and a low heat conductivity at high temperature (0.35 W/(K.m) at 500°C). This type of intumescent silicone based-coatings

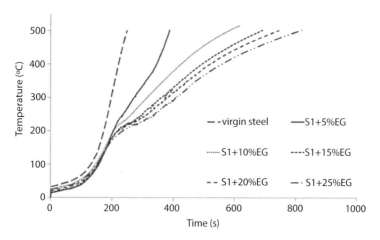

Figure 4.25 Temperature as a function of time measured on the backside of steel covered by an intumescent silicone-based coating at different loading of EG (hydrocarbon fire scenario according to UL1709 done in a furnace at a reduced scale, plaque of 10x10 cm²) (from Ref [139]).

were also evaluated on carbon fiber reinforced polymer (CFRP) using a bench developed in the laboratory mimicking a jet fuel fire occurring at high heat flux (200 kW/m²) [165]. A large intumescence was observed and the composite does not pierce after 15 min. EG was also incorporated thermoset resins for protecting wood composite and was successful at the Steiner tunnel (ASTM E84) [166].

Alkali silicates can be sued to make intumescent inorganic coating for fire protection [24]. The expansion of the material upon heating over 100°C is due to the rapid release of water vapor [26]. It makes a foamy material which is rigid and is constituted of hydrated silica. It acts as thermal insulator and provides the fire protection of interest. Veinot et al. [26] evaluated the performance of different formulations of silicate-based coatings applied on aluminum according to the time/temperature curve of ASTM E-119 in a tubular furnace at the reduced scale. They showed that they could keep the temperature of the substrate below 300°C for 15 min with the appropriate thickness and the right amount of an acidic curing agent. This agent was not described in the paper but it is reported it has a beneficial effect on the intumescent structure which is solid and thick almost like cement in texture. Silicate-based coating was also applied on wood and gives satisfactory performance [25]. Additional magnesium-based ingredients were incorporated in the formulations providing higher expansion and more efficient structure limiting the diffusion combustion gases. An interesting version of silicate-based coating was to make silicate-polymer hydrogels [167]. The resulting coating is transparent and hence, it can be used as varnish or for window protection. The hydrogel was incorporated between two glass panes and was evaluated in a furnace of fire resistance. It was shown that the hydrogel-based materials meet the requirements for fire tests in the construction. It was revealed the formation of an intumescent protective coating and that the temperature on the pane was kept below 180°C for 30 min for the optimized formulation of hydrogel.

A novel way to make intumescent coating is to use aluminosilicate composites or in other words, geopolymers. This approach was evaluated by Schartel et al. [168] to protect steel plates according to cellulosic fire scenario (ISO 834).

Two aluminosilicate formulations were prepared based either on microsilica and alumina or on metakaolin and microsilica containing or not flame retardants including boron trioxide, borax, ATH and MDH. It was observed an intumescent behavior during fire testing (up to 300% expansion) which is an unusual behavior for geopolymers (Figure 4.26). The best fire insulation was obtained with 10% borax in the formulation based on microsilica and alumina (original layer thickness of 6 mm). The fire protection can reach up to 30 min (temperature of failure of 500°C).

Other geopolymers were prepared consisting of metakaolin-based alkaline aluminosilicate with a swelling agent and calcium carbonate as intumescent coating for the fire protection of structural elements [169]. An intumescent behavior was observed and the resulting material exhibits low heat conductivity and good mechanical stability. In another work [170], rice husk ash was incorporated in a metakaolin-based geopolymer to make fire resistant coating for structural insulated panel. The optimized geopolymer-based coating permits to resist 2h at the torch test before reaching 100°C. Relatively high degree of expansion was observed but some cracks and holes can be distinguished on the residues after fire testing. A foam version of geopolymers also exists, which is made with metakaolin and by acid- or alkali-activation [171]. Geopolymer (GP) foams based on mixtures of alkali-activated metakaolin and silica fume were prepared in the lab [172]. Hydrogen peroxide (H_2O_2) was used as a porogen agent and a cationic surfactant (cetyltrimethyl ammonium bromide) as foam stabilizer (Figure 4.27 – (a)). The foam was coated on steel plate and was evaluated at the

Figure 4.26 Intumescent coating of geopolymer containing borax after fire testing (scale bar in cm) (from Ref [168]).

burnthrough test mimicking ISO 2685 at the reduced scale (heat flux of 116 kW/m²) [144]. The temperature reached on the backside of the steel pate is much lower for the geopolymer foam compared to the virgin steel (Figure 4.27 – (b)). In the steady state, the temperature difference is about 250°C. It is also noteworthy that a plateau can be distinguished at about 100°C. It is assigned to H_2O molecules released from the foam, because of structure changes (amorphization) and vaporizing almost instantaneously.

An alternative approach to make novel intumescent coating is to develop novel design. Polymer metal laminates consisting of alternating thin aluminum foils and thin epoxy resin layers provide protection against fire due to the delamination between layers during burning (radiant panel at 50 kW/m²) [173]. However, it is not enough when considering other and more severe scenarios and new concept called 'Intumescent Polymer Metal Laminates' or IPML was developed [174]: the thin epoxy resin layers were substituted by selected intumescent coatings. When evaluating at the burnthrough test, it was shown the slopes of the time/temperature curves was dramatically decreased at the beginning of the test compared to the unmodified intumescent coating but the temperature in the steady state was not lower. It was attributed to the delamination of the aluminum foils (Figure 4.28). The concept was then further enhanced combining the use of two different

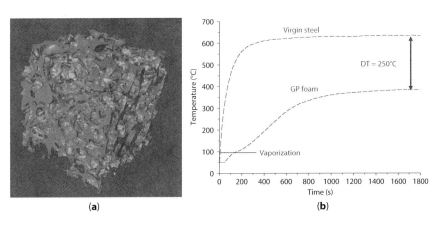

(a)   (b)

Figure 4.27 (a) X-ray tomography of GP foam showing the internal foamy structure and (b) temperature as a function of time of GP foam compared to virgin steel plate undergoing burnthrough test.

Before fire testing

After 15 min fire testing

Figure 4.28 IPML before and after fire testing at the burnthrough test, delamination of the aluminum foils can be clearly seen (adapted from Ref [174]).

intumescent paints and a metal laminate structure [175]. The design with two aluminum foils and the overlay of both intumescent coatings reveals efficient fire protection with stabilization at low temperatures after 30 min of fire exposure at the burnthrough test. It was shown this type of design exhibits much higher performance than that of the conventional intumescent coating.

4.6   Conclusion and Future Trends

This chapter has discussed the developments of intumescent fire retarded materials in terms of reaction and resistance to fire. Research work in intumescence is still very active even if the first papers on this topic appeared at the beginning of the 20th century and the concept of intumescence remains attractive. The quick overview of the mechanisms of action reveals that the formation of an expanded charred insulative layer acting as thermal shield is involved. The chemistry is very often based on the well-known triplet APP/PER/MEL. Numerous other char formers were suggested as well as other blowing agent but the basic mechanism remains similar. New molecules (commercial molecules and new concepts like all-in-one molecules) appeared based on this. Micro- and/or nanofillers can be also combined with the conventional ingredients of intumescence to create synergistic effects. Reaction can take place between the additional filler and some ingredients of the intumescent formulation (e.g. the phosphate) in order to thermally stabilize the charred structure (chemical stabilization and prevention of oxidation). Physical interactions are also observed (e.g. additional 'boost' of expansion). Those interactions permit the reinforcement of the char strength and to avoid the formation of cracks. The development rate and the quality of the layer are of the primary importance and research work should be focused on this.

However, other chemistries cannot be ignored namely those based on a physical expansion. Silicate-based material can create an inorganic intumescent layer by the vaporization of water molecules. It is an old concept which is used in many applications. Few recent papers are devoted to this but it should be explored further because of the advantages of having inorganic materials (e.g. thermal stability and resistance to oxidation of the intumescent layer). Expandable graphite provides an intumescence based on physical expansion (i.e. sublimation of the insertion molecules): it makes an insulative protective layer formed by entangled worms of graphite while the graphite is highly thermally conductive. The mechanism is not completely elucidated and it opens the door to novel way of research. Inorganic chemistry can be still considered with geopolymers giving the opportunity of novel chemistries to design new intumescent materials exhibiting high performance in extreme fire scenarios. Finally, the design of intumescent multi-materials should be also considered as it gives an exceptional opportunity to make materials of high performance and it paves the way for novel concepts.

Environment is a main concern in our society. It can be considered making new intumescent materials with positive life cycle analysis (LCA). Bio-based and recycled ingredients and/or polymers offer this opportunity based on our knowledge on the diverse chemistries of intumescence. The resulting materials could exhibit higher performance or at least, similar one and hence, it is an alternative to be explored to make fire-safe and environmentally friendly intumescent materials.

References

1. Gay Lussac, L.J., Note sur la propriété qu'ont les matières salines de rendre les tissus incombustibles. *Ann. Chim. Phys.*, *18*, 211–218, 1821.
2. H. Tramm, C. Clar, P. Kühnel, W. Schuff, Fireproofing of wood, USA, 1-2, United States Patent Office 1938.

3. Q. Sverre, Silicate-containing product and process for making it, United States Patent Office, 1940.

4. W. Arthur, Intumescent material and process of making the same, in: U.S.P. Office (Ed.) United State Patent Office, General Electric, USA, 1912-10-15, 3, 1912.

5. Vandersall, H.J., Intumescent coating systems, their development and chemistry. *J. Fire Flammabl.*, 2, 97–140, 1971.

6. Camino, G., Costa, L., Trossarelli, L., Study of the mechanism of intumescence in fire retardant polymers: Part I-Thermal degradation of ammonium polyphosphate-pentaerythritol mixtures. *Polym. Degrad. Stab.*, 6, 4, 243–252, 1984.

7. Gardelle, B., Duquesne, S., Vandereecken, P., Bellayer, S., Bourbigot, S., Resistance to fire of intumescent silicone based coating: The role of organoclay. *Prog. Org. Coat.*, 76, 11, 1633–1641, 2013.

8. Zeng, Y., Weinell, C.E., Dam-Johansen, K., Ring, L., Kiil, S., Exposure of hydrocarbon intumescent coatings to the UL1709 heating curve and furnace rheology: Effects of zinc borate on char properties. *Prog. Org. Coat.*, 135, 321–330, 2019.

9. Nicoară, A.I., Bădănoiu, A.I., Voicu, G., Dinu, C., Ionescu, A., Intumescent coatings based on alkali-activated borosilicate inorganic polymers. *J. Coat. Technol. Res.*, 17, 3, 681–692, 2020.

10. Bourbigot, S., Sarazin, J., Samyn, F., Jimenez, M., Intumescent ethylene-vinyl acetate copolymer: Reaction to fire and mechanistic aspects. *Polym. Degrad. Stab.*, 161, 235–244, 2019.

11. Tawiah, B., Zhou, Y., Yuen, R.K.K., Sun, J., Fei, B., Microporous boron based intumescent macrocycle flame retardant for poly(lactic acid) with excellent UV protection. *Chem. Eng. J.*, 402, 2020.

12. Liu, X., Meng, X., Sun, J., Tang, W., Chen, S., Peng, X., Gu, X., Fei, B., Bourbigot, S., Zhang, S., Improving the flame retardant properties of polyester-cotton blend fabrics by introducing an intumescent coating via layer by layer assembly'. *J. Appl. Polym. Sci.*, 137, 41, 2020.

13. Camino, G., Costa, L., Trossarelli, L., Study of the mechanism of intumescence in fire retardant polymers: Part V-Mechanism of formation of gaseous products in the thermal degradation of ammonium polyphosphate. *Polym. Degrad. Stab.*, 12, 3, 203–211, 1985.

14. Bourbigot, S., Le Bras, M., Delobel, R., Fire degradation of an intumescent flame retardant polypropylene. *J. Fire Sci.*, 13, 1, 3–22, 1995.

15. Zhang, S. and Horrocks, A.R., Substantive Intumescence from Phosphorylated 1,3-Propanediol Derivatives Substituted on to Cellulose. *J. Appl. Polym. Sci.*, 90, 12, 3165–3172, 2003.

16. Alongi, J., Carletto, R.A., Di Blasio, A., Cuttica, F., Carosio, F., Bosco, F., Malucelli, G., Intrinsic intumescent-like flame retardant properties of DNA-treated cotton fabrics. *Carbohydr. Polym.*, 96, 1, 296–304, 2013.

17. Alongi, J., Han, Z., Bourbigot, S., Intumescence: Tradition versus novelty. A comprehensive review. *Prog. Polym. Sci.*, *51*, 28–73, 2015.
18. Camino, G., Costa, L., Trossarelli, L., Study of the mechanism of intumescence in fire retardant polymers: Part II-Mechanism of action in polypropylene-ammonium polyphosphate-pentaerythritol mixtures. *Polym. Degrad. Stab.*, *7*, *1*, 25–31, 1984.
19. Camino, G., Costa, L., Luda di Cortemiglia, M.P., Overview of fire retardant mechanisms. *Polym. Degrad. Stab.*, *33*, *2*, 131–154, 1991.
20. Bourbigot, S., Le Bras, M., Delobel, R., Carbonization mechanisms resulting from intumescence association with the ammonium polyphosphate-pentaerythritol fire retardant system. *Carbon*, *31*, *8*, 1219–1230, 1993.
21. Bourbigot, S., Duquesne, S., Leroy, J.M., Modeling of heat transfer of a polypropylene-based intumescent system during combustion. *J. Fire Sci.*, *17*, *1*, 42–56, 1999.
22. Focke, W.W., Badenhorst, H., Mhike, W., Kruger, H.J., Lombaard, D., Characterization of commercial expandable graphite fire retardants. *Thermochim. Acta*, *584*, 8–16, 2014.
23. Bourbigot, S., Sarazin, J., Bensabath, T., Samyn, F., Jimenez, M., Intumescent polypropylene: Reaction to fire and mechanistic aspects. *Fire Saf. J.*, *105*, 261–269, 2019.
24. Bulewicz, E.M., Pelc, A., Kozlowski, R., Miciukiewicz, A., Intumescent silicate-based materials: Mechanism of swelling in contact with fire. *Fire Mater.*, *9*, *4*, 171–175, 1985.
25. Kazmina, O., Lebedeva, E., Mitina, N., Kuzmenko, A., Fire-proof silicate coatings with magnesium-containing fire retardant. *J. Coat. Technol. Res.*, *15*, *3*, 543–554, 2018.
26. Veinot, D.E., Langille, K.B., Nguyen, D., Bernt, J.O., Soluble silicate-based coatings for fire protection. *Fire Technol.*, *25*, *3*, 230–240, 1989.
27. Zubkova, N.S., Butylkina, N.G., Chekanova, S.E., Tyuganova, M.A., Khalturinskii, N.A., Reshetnikov, I.S., Naganovskii, Y.K., Rheological and fireproofing characteristics of polyethylene modified with a microencapsulated fire retardant. *Fibre Chem.*, *30*, *1*, 11–13, 1998.
28. Bugajny, M., Le Bras, M., Bourbigot, S., New approach to the dynamic properties of an intumescent material. *Fire Mater.*, *23*, *1*, 49–51, 1999.
29. Jimenez, M., Duquesne, S., Bourbigot, S., Characterization of the performance of an intumescent fire protective coating. *Surf. Coat. Technol.*, *201*, *3-4*, 979–987, 2006.
30. Duquesne, S., Magnet, S., Jama, C., Delobel, R., Thermoplastic resins for thin film intumescent coatings - Towards a better understanding of their effect on intumescence efficiency. *Polym. Degrad. Stab.*, *88*, *1*, 63–69, 2005.
31. Gérard, C., Fontaine, G., Bellayer, S., Bourbigot, S., Reaction to fire of an intumescent epoxy resin: Protection mechanisms and synergy. *Polym. Degrad. Stab.*, *97*, *8*, 1366–1386, 2012.

32. Bodzay, B., Bocz, K., Bárkai, Z., Marosi, G., Influence of rheological additives on char formation and fire resistance of intumescent coatings. *Polym. Degrad. Stab.*, 96, 3, 355–362, 2011.

33. Okyay, G., Naik, A.D., Samyn, F., Jimenez, M., Bourbigot, S., Fractal conceptualization of intumescent fire barriers, toward simulations of virtual morphologies. *Sci. Rep.*, 9, 1, 2019.

34. Bourbigot, S., Le Bras, M., Duquesne, S., Rochery, M., Recent Advances for Intumescent Polymers. *Macromol. Mater. Eng.*, 289, 6, 499–511, 2004.

35. Gustavsson, M.K. and Gustafsson, S.E., *Different ways of evaluating thermal transport properties from measurements with the transient plane source (hot disk) method*, pp. 397–402, ITCC29, Birmingham, AL, 2008.

36. Gardelle, B., Duquesne, S., Rerat, V., Bourbigot, S., Thermal degradation and fire performance of intumescent silicone-based coatings. *Polym. Adv. Technol.*, 24, 1, 62–69, 2013.

37. Muller, M., Bourbigot, S., Duquesne, S., Klein, R., Giannini, G., Lindsay, C., Vlassenbroeck, J., Investigation of the synergy in intumescent polyurethane by 3D computed tomography. *Polym. Degrad. Stab.*, 98, 9, 1638–1647, 2013.

38. Hirschler, M.M. Survey of american test methods associated with fire performance of materials or products. *Polym. Degrad. Stab.*, 54, 2-3 SPEC. ISS., 333–343, 1996.

39. Troitzsch, J.H., The globalisation of fire testing and its impact on polymers and flame retardants. *Polym. Degrad. Stab.*, 88, 1, 146–149, 2005.

40. Brady, D.G., Moberly, C.W., Norell, J.R., Waltrs, H., Intumescence: a novel effective approach to flame retarding polypropylene. *J. Fire Retardant Chem.*, 4, 150–164, 1977.

41. Montaudo, G., Scamporrino, E., Vitalini, D., Intumescent flame retardants for polymers. II. The Polypropylene-Ammonium Polyphosphate-Polyurea system. *J. Polym. Sci. Part A-1, Polym. Chem.*, 21, 12, 3361–3371, 1983.

42. Halpern, Y., Mott, D.M., Niswander, R.H., Fire Retardancy of Thermoplastic Materials by Intumescence. *Ind. Eng. Chem. Prod. Res. Dev.*, 23, 2, 233–238, 1984.

43. Yan, J. and Xu, M., Design, synthesis and application of a highly efficient mono-component intumescent flame retardant for non-charring polyethylene composites. *Polym. Bull.*, 78, 2, 643-662, February 2021.

44. Fontaine, G., Bourbigot, S., Duquesne, S., Neutralized flame retardant phosphorus agent: Facile synthesis, reaction to fire in PP and synergy with zinc borate. *Polym. Degrad. Stab.*, 93, 1, 68–76, 2008.

45. Liu, Y. and Wang, Q., Catalytic action of phospho-tungstic acid in the synthesis of melamine salts of pentaerythritol phosphate and their synergistic effects in flame retarded polypropylene. *Polym. Degrad. Stab.*, 91, 10, 2513–2519, 2006.

46. Alongi, J., Di Blasio, A., Milnes, J., Malucelli, G., Bourbigot, S., Kandola, B., Camino, G., Thermal degradation of DNA, an all-in-one natural intumescent flame retardant. *Polym. Degrad. Stab.*, 113, 110–118, 2015.

47. Alongi, J., Cuttica, F., Bourbigot, S., Malucelli, G., Thermal and flame retardant properties of ethylene vinyl acetate copolymers containing deoxyribose nucleic acid or ammonium polyphosphate. *J. Therm. Anal. Calorim.*, *122*, 2, 705–715, 2015.

48. Alongi, J., Cuttica, F., Carosio, F., DNA Coatings from Byproducts: A Panacea for the Flame Retardancy of EVA, PP, ABS, PET, and PA6? *ACS Sustain. Chem. Eng.*, *4*, 6, 3544–3551, 2016.

49. Wang, D.Y., Ge, X.G., Wang, Y.Z., Wang, C., M.H., Qu, Zhou, Q., A novel phosphorus-containing poly(ethylene terephthalate) nanocomposite with both flame retardancy and anti-dripping effects. *Macromol. Mater. Eng.*, *291*, 6, 638–645, 2006.

50. Gao, F., Tong, L., Fang, Z., Effect of a novel phosphorous-nitrogen containing intumescent flame retardant on the fire retardancy and the thermal behaviour of poly(butylene terephthalate). *Polym. Degrad. Stab.*, *91*, 6, 1295–1299, 2006.

51. Ma, Z., Zhao, W., Liu, Y., Shi, J., Synthesis and properties of intumescent, phosphorus-containing, flame-retardant polyesters. *J. Appl. Polym. Sci.*, *63*, *12*, 1511–1515, 1997.

52. Li, B. and Xu, M., Effect of a novel charring-foaming agent on flame retardancy and thermal degradation of intumescent flame retardant polypropylene. *Polym. Degrad. Stab.*, *91*, 6, 1380–1386, 2006.

53. Bourbigot, S. and Duquesne, S., Fire retardant polymers: Recent developments and opportunities. *J. Mater. Chem. A*, *17*, 22, 2283–2300, 2007.

54. Markwart, J.C., Battig, A., Velencoso, M.M., Pollok, D., Schartel, B., Wurm, F.R., Aromatic vs. aliphatic hyperbranched polyphosphoesters as flame retardants in epoxy resins. *Molecules*, *24*, 21, 2019.

55. Battig, A., Markwart, J.C., Wurm, F.R., Schartel, B., Sulfur's role in the flame retardancy of thio-ether–linked hyperbranched polyphosphoesters in epoxy resins. *Eur. Polym. J.*, *122*, 2020.

56. Yan, H., Zhao, Z., Ge, W., Zhang, N., Jin, Q., Hyperbranched Polyurea as Charring Agent for Simultaneously Improving Flame Retardancy and Mechanical Properties of Ammonium Polyphosphate/Polypropylene Composites. *Ind. Eng. Chem. Res.*, *56*, 30, 8408–8415, 2017.

57. Zhu, C., He, M., Cui, J., Tai, Q., Song, L., Hu, Y., Synthesis of a novel hyperbranched and phosphorus-containing charring-foaming agent and its application in polypropylene. *Polym. Adv. Technol.*, *29*, 9, 2449–2456, 2018.

58. Lundgren, A., Hjertberg, T., Sultan, B.A.,Influence of the structure of acrylate groups on the flame retardant behavior of ethylene acrylate copolymers modified with chalk and silicone elastomer. *J. Fire Sci.*, *25*, 4, 287–319, 2007.

59. Krämer, R.H., Blomqvist, P., Hees, P.V., Gedde, U.W., On the intumescence of ethylene-acrylate copolymers blended with chalk and silicone. *Polym. Degrad. Stab.*, *92*, 10, 1899–1910, 2007.

60. Gupta, S.K., Schwab, J.J., Lee, A., Fu, B.X., Hsiao, B.S., POSS™ reinforced fire retarding EVE resins, in: *47th International SAMPE Symposium and Exhibition*, pp. 1517–1526, Long Beach, CA, 2002.
61. Devaux, E., Rochery, M., Bourbigot, S., Polyurethane/clay and polyurethane/POSS nanocomposites as flame retarded coating for polyester and cotton fabrics. *Fire Mater.*, 26, 4-5, 149–154, 2002.
62. Bourbigot, S., Turf, T., Bellayer, S., Duquesne, S., Polyhedral oligomeric silsesquioxane as flame retardant for thermoplastic polyurethane. *Polym. Degrad. Stab.*, 94, 8, 1230–1237, 2009.
63. Raimondo, M., Russo, S., Guadagno, L., Longo, P., Chirico, S., Mariconda, A., Bonnaud, L., Murariu, O., Dubois, P., Effect of incorporation of POSS compounds and phosphorous hardeners on thermal and fire resistance of nanofilled aeronautic resins. *RSC Adv.*, 5, 15, 10974–10986, 2015.
64. Laik, S., Galy, J., Gérard, J.F., Monti, M., Camino, G., Fire behaviour and morphology of epoxy matrices designed for composite materials processed by infusion. *Polym. Degrad. Stab.*, 127, 44–55, 2016.
65. Horacek, H. and Pieh, S., Importance of intumescent systems for fire protection of plastic materials. *Polym. Int.*, 49, 10, 1106–1114, 2000.
66. Liu, T.M., Baker, W.E., Langille, K.B., Nguyen, D.T., Bernt, J.O., New Silicate-Based Powders for Fire Protection of Thermoplastics. *J. Vinyl Addit. Technol.*, 4, 4, 246–258, 1998.
67. Bellayer, S., Tavard, E., Duquesne, S., Piechaczyk, A., Bourbigot, S., Mechanism of intumescence of a polyethylene/calcium carbonate/stearic acid system'. *Polym. Degrad. Stab.*, 94, 5, 797–803, 2009.
68. Bellayer, S.P., Tavard, E., Duquesne, S., Piechaczyk, A., Bourbigot, S., Natural mineral fire retardant fillers for polyethylene. *Fire Mater.*, 35, 3, 183–192, 2011.
69. Schafhaeutl, C., Ueber die Verbindungen des Kohlenstoffes mit Silicium, Eisen und anderen Metallen, welche die verschiedenen Gallungen von Roheisen, Stahl und Schmiedeeisen bilden. *J. Prakt. Chem.*, 21, 1, 129–157, 1840.
70. Duquesne, S., Le Bras, M., Bourbigot, S., Delobel, R., Vezin, H., Camino, G., Eling, B., Lindsay, C., Roels, T., Expandable graphite: A fire retardant additive for polyurethane coatings. *Fire Mater.*, 27, 3, 103–117, 2003.
71. Pang, X.Y., Tian, Y., Weng, M.Q., Preparation of expandable graphite with silicate assistant intercalation and its effect on flame retardancy of ethylene vinyl acetate composite. *Polym. Polym. Compos.*, 36, 8, 1407–1416, 2015.
72. Pang, X.Y., Tian, Y., Zhai, Z.X., Duan, M.W., Preparation of expandable graphite intercalated by sulfuric acid and sodium silicate and its flame retardancy application for ethylene vinyl acetate copolymer. *Asian J. Chem.*, 26, 14, 4297–4302, 2014.
73. Weil, E.D. and Levchik, S.V., Flame retardants in commercial use or development for textiles. *J. Fire Sci.*, 26, 3, 243–281, 2008.

74. Buyukakinci, B.Y. and Yilmaz, A., Investigation of boric acid and sodium borate effect on flame retardancy of cotton and polyester fabrics. *Asian J. Chem.*, *29*, *4*, 893–895, 2017.

75. Horrocks, A.R., Kandola, B.K., Davies, P.J., Zhang, S., Padbury, S.A., Developments in flame retardant textiles - A review. *Polym. Degrad. Stab.*, *88*, *1*, 3–12, 2005.

76. Davies, P.J., Horrocks, A.R., Alderson, A., The sensitisation of thermal decomposition of ammonium polyphosphate by selected metal ions and their potential for improved cotton fabric flame retardancy. *Polym. Degrad. Stab.*, *88*, *1*, 114–122, 2005.

77. Giraud, S., Bourbigot, S., Rochery, M., Vroman, I., Tighzert, L., Delobel, R., Microencapsulation of phosphate: Application to flame retarded coated cotton. *Polym. Degrad. Stab.*, *77*, *2*, 285–297, 2002.

78. Giraud, S., Bourbigot, S., Rochery, M., Vroman, I., Tighzert, L., Delobel, R., Poutch, F., Flame retarded polyurea with microencapsulated ammonium phosphate for textile coating. *Polym. Degrad. Stab.*, *88*, *1*, 106–113, 2005.

79. Decher, G. and Hong, J.D., Buildup of ultrathin multilayer films by a self-assembly process, 1 consecutive adsorption of anionic and cationic bipolar amphiphiles on charged surfaces. *Makromol. Chem., Macromol. Symp.*, *46*, *1*, 321–327, 1991.

80. Decher, G., Lvov, Y., Schmitt, J., Proof of multilayer structural organization in self-assembled polycation-polyanion molecular films. *Thin Solid Films*, *244*, *1-2*, 772–777, 1994.

81. Li, Y.C., Mannen, S., Morgan, A.B., Chang, S., Yang, Y.H., Condon, B., Grunlan, J.C., Intumescent all-polymer multilayer nanocoating capable of extinguishing flame on fabric. *Adv. Mater.*, *23*, *34*, 3926–3931, 2011.

82. Jimenez, M., Guin, T., Bellayer, S., Dupretz, R., Bourbigot, S., Grunlan, J.C., Microintumescent mechanism of flame-retardant water-based chitosan-ammonium polyphosphate multilayer nanocoating on cotton fabric. *J. Appl. Polym. Sci.*, *133*, *32*, 2016.

83. Alongi, J., Carletto, R.A., Di Blasio, A., Carosio, F., Bosco, F., Malucelli, G., DNA: A novel, green, natural flame retardant and suppressant for cotton. *J. Mater. Chem. A*, *1*, *15*, 4779–4785, 2013.

84. Carosio, F., Di Blasio, A., Alongi, J., Malucelli, G., Green DNA-based flame retardant coatings assembled through Layer by Layer. *Polym. (U.K.)*, *54*, *19*, 5148–5153, 2013.

85. Horrocks, A.R., High performance textiles for heat and fire protection, in: *High Performance Textiles and Their Applications*, pp. 144–175, Amsterdam, Elsevier Ltd, 2014.

86. Horrocks, A.R. and Davies, P.J., Char formation in flame-retarded wool fibres. Part 1. Effect of intumescent on thermogravimetric behaviour. *Fire Mater.*, *24*, *3*, 151–157, 2000.

87. Horrocks, A.R. and Zhang, S., Char formation in polyamides (nylons 6 and 6.6) and wool keratin phosphorylated by polyol phosphoryl chlorides. *Text. Res. J.*, *74*, *5*, 433–441, 2004.

88. Cheng, X.W., Guan, J.P., Yang, X.H., Tang, R.C., Yao, F., A bio-resourced phytic acid/chitosan polyelectrolyte complex for the flame retardant treatment of wool fabric. *J. Clean. Prod.*, *223*, 342–349, 2019.

89. Vannier, A., Duquesne, S., Bourbigot, S., Alongi, J., Camino, G., Delobel, R., Investigation of the thermal degradation of PET, zinc phosphinate, OMPOSS and their blends-Identification of the formed species. *Thermochim. Acta*, *495*, *1-2*, 155–166, 2009.

90. Chen, D.Q., Wang, Y.Z., Hu, X.P., Wang, D.Y., Qu, M.H., Yang, B., Flame-retardant and anti-dripping effects of a novel char-forming flame retardant for the treatment of poly(ethylene terephthalate) fabrics. *Polym. Degrad. Stab.*, *88*, *2*, 349–356, 2005.

91. Liu, W., Chen, D.Q., Wang, Y.Z., Wang, D.Y., Qu, M.H., Char-forming mechanism of a novel polymeric flame retardant with char agent. *Polym. Degrad. Stab.*, *92*, *6*, 1046–1052, 2007.

92. Jordanov, I., Kolibaba, T.J., Lazar, S., Magovac, E., Bischof, S., Grunlan, J.C., Flame suppression of polyamide through combined enzymatic modification and addition of urea to multilayer nanocoating. *J. Mater. Sci.*, *55*, *30*, 15056–15067, 2020.

93. Zhang, S. and Horrocks, A.R., A review of flame retardant polypropylene fibres. *Prog. Polym. Sci.*, *28*, *11*, 1517–1538, 2003.

94. Vroman, I., Giraud, S., Salaün, F., Bourbigot, S., Polypropylene fabrics padded with microencapsulated ammonium phosphate: Effect of the shell structure on the thermal stability and fire performance. *Polym. Degrad. Stab.*, *95*, *9*, 1716–1720, 2010.

95. Ramgobin, A., Fontaine, G., Bourbigot, S., Thermal Degradation and Fire Behavior of High Performance Polymers. *Polym. Rev.*, *59*, *1*, 55–123, 2019.

96. Lyon, R.E., Speitel, L., Walters, R.N., Crowley, S., Fire-resistant elastomers. *Fire Mater.*, *27*, *4*, 195–208, 2003.

97. Dočan, M., Ylmaz, A., Bayraml, E., Synergistic effect of boron containing substances on flame retardancy and thermal stability of intumescent polypropylene composites. *Polym. Degrad. Stab.*, *95*, *12*, 2584–2588, 2010.

98. Anna, P., Marosi, G., Csontos, I., Bourbigot, S., Le Bras, M., Delobel, R., Influence of modified rheology on the efficiency of intumescent flame retardant systems. *Polym. Degrad. Stab.*, *74*, *3*, 423–426, 2001.

99. Yang, S., Wang, J., Huo, S., Wang, M., Wang, J., Zhang, B., Synergistic flame-retardant effect of expandable graphite and phosphorus-containing compounds for epoxy resin: Strong bonding of different carbon residues. *Polym. Degrad. Stab.*, *128*, 89–98, 2016.

100. Khanal, S., Lu, Y., Dang, L., Ali, M., Xu, S., Effects of α-zirconium phosphate and zirconium organophosphonate on the thermal, mechanical and flame

retardant properties of intumescent flame retardant high density polyethylene composites. *RSC Adv.*, *10*, *51*, 30990–31002, 2020.

101. Wei, P., Hao, J., Du, J., Han, Z., Wang, J., An investigation on synergism of an intumescent flame retardant based on silica and alumina. *J. Fire Sci.*, *21*, *1*, 17–28, 2003.

102. Wang, C., Wei, P., Qian, Y., Liu, J., The synthesis of a novel flame retardant and its synergistic efficiency in polypropylene/ammonium polyphosphate system. *Polym. Adv. Technol.*, *22*, *7*, 1108–1114, 2011.

103. Bourbigot, S., Le Bras, M., Bréant, P., Trémillon, J.M., Delobel, R., Zeolites: New synergistic agents for intumescent fire retardant thermoplastic formulations - Criteria for the choice of the zeolite. *Fire Mater.*, *20*, *3*, 145–154, 1996.

104. Le Bras, M. and Bourbigot, S., Mineral fillers in intumescent fire retardant formulations - Criteria for the choice of a natural clay filler for the ammonium polyphosphate/pentaerythritol/polypropylene system. *Fire Mater.*, *20*, *1*, 39–49, 1996.

105. Bourbigot, S., Le Bras, M., Dabrowski, F., J.W., Gilman, T., Kashiwagi, PA-6 clay nanocomposite hybrid as char forming agent in intumescent formulations. *Fire Mater.*, *24*, *4*, 201–208, 2000.

106. Ren, Q., Wan, C., Zhang, Y., Li, J., An investigation into synergistic effects of rare earth oxides on intumescent flame retardancy of polypropylene/ poly (octylene-co-ethylene) blends. *Polym. Adv. Technol.*, *22*, *10*, 1414–1421, 2011.

107. Zhang, H., Lu, X., Zhang, Y., Synergistic effects of rare earth oxides on intumescent flame retardancy of Nylon 1010/ethylene-vinyl-acetate rubber thermoplastic elastomers. *J. Polym. Res.*, *22*, *2*, 1–10, 2015.

108. Levchik, S.V., Levchik, G.F., Camino, G., Costa, L., Lesnikovich, A.I., Mechanism of action of phosphorus-based flame retardants in nylon 6. III. Ammonium polyphosphate/manganese dioxide. *Fire Mater.*, *20*, *4*, 183–190, 1996.

109. Hu, S., Zheng, H., Shen, S., Han, H., Jing, H., Li, S., Synergistic effects of zinc oxide on intumescent flame-retarded polypropylene composites, in: *6th East Asian Symposium on Functional Ion Application Technology and 2010 International Forum on Ecological Environment Functional Materials and Industry*, pp. 279–284, Shanghai, 2011.

110. Liu, G., Zhou, Y., Hao, J., Thermal degradation mechanism of Ni2O3 synergistic intumescent flame retardant polypropylenes. *Gaofenzi Cailiao Kexue Yu Gongcheng/Polym. Mater. Sci. Eng.*, *29*, *9*, 98–102+106, 2013.

111. Xu, Z., Jia, H., Yan, L., Chu, Z., Zhou, H., Synergistic effect of bismuth oxide and mono-component intumescent flame retardant on the flammability and smoke suppression properties of epoxy resins. *Polym. Adv. Technol.*, *31*, *1*, 25–35, 2020.

112. Scharf, D., Nalepa, R., Heflin, R., Wusu, T., Studies on flame retardant intumescent char: Part I. *Fire Saf. J.*, *19*, *1*, 103–117, 1992.

113. Wang, X., Wu, L., Li, J., A study on the performance of intumescent flame-retarded polypropylene with nano-ZrO2. *J. Fire Sci.*, 29, 3, 227–242, 2011.

114. Lu, H. and Wilkie, C.A., Study on intumescent flame retarded polystyrene composites with improved flame retardancy. *Polym. Degrad. Stab.*, 95, 12, 2388–2395, 2010.

115. Bourbigot, S., Duquesne, S., Fontaine, G., Bellayer, S., Turf, T., Samyn, F., Characterization and Reaction to Fire of Polymer Nanocomposites with and without Conventional Flame Retardants. *Mol. Cryst. Liq. Cryst.*, 486, 2008.

116. Vannier, A., Duquesne, S., Bourbigot, S., Castrovinci, A., Camino, G., Delobel, R., The use of POSS as synergist in intumescent recycled poly(ethylene terephthalate). *Polym. Degrad. Stab.*, 93, 4, 818–826, 2008.

117. Ding, P., Tang, S., Yang, H., Shi, L., PP/LDH nanocomposites via melt-intercalation: Synergistic flame retardant effects, properties, and applications in automobile industries, in: *2009 Advanced Polymer Processing International Forum, APPF2009*, pp. 427–432, Qingdao, 2010.

118. Lewin, M. and Endo M., Catalysis of intumescent flame retardancy of polypropylene by metallic compounds. *Polym. Adv. Technol.*, 14, 1, 3–11, 2003.

119. Chen, X.L., Jiao, C.M., Wang, Y., Synergistic effects of iron powder on intumescent flame retardant polypropylene system. *Express Polym. Lett.*, 3, 6, 359–365, 2009.

120. Duquesne, S., Samyn, F., Bourbigot, S., Amigouet, P., Jouffret, F., Shen, K.K., Influence of talc on the fire retardant properties of highly filled intumescent polypropylene composites. *Polym. Adv. Technol.*, 19, 6, 620–627, 2008.

121. Huang, N.H., Chen, Z.J., Wang, J.Q., Wei, P., Synergistic effects of sepiolite on intumescent flame retardant polypropylene. *Express Polym. Lett.*, 4, 12, 743–752, 2010.

122. Wu, N., Ding, C., Yang, R., Effects of zinc and nickel salts in intumescent flame-retardant polypropylene. *Polym. Degrad. Stab.*, 95, 12, 2589–2595, 2010.

123. Samyn, F. and Bourbigot, S., Protection mechanism of a flame-retarded polyamide 6 nanocomposite. *J. Fire Sci.*, 32, 3, 241–256, 2013.

124. Bourbigot, S., Le Bras, M., Delobel, R., Trémillon, J.M., Synergistic effect of zeolite in an intumescence process: Study of the interactions between the polymer and the additives. *J. Chem. Soc - Faraday Trans.*, 92, 18, 3435–3444, 1996.

125. Samyn, F., Bourbigot, S., Duquesne, S., Delobel, R., Effect of zinc borate on the thermal degradation of ammonium polyphosphate. *Thermochim. Acta*, 456, 2, 134–144, 2007.

126. Levchik, S.V., Camino, G., Costa, L., Levchik, G.F., Mechanism of action of phosphorus-based flame retardants in nylon 6. I. Ammonium polyphosphate. *Fire Mater.*, 19, 1, 1–10, 1995.

127. Antonov, A., Yablokova, M., Costa, L., Balabanovich, A., Levchik, G., Levchik, S., The effect of nanometals on the flammability and thermooxidative

degradation of polymer materials. *Mol. Cryst. Liq. Cryst. Sci. Technol., Sect. A: Mol. Cryst. Liq. Cryst.*, *353*, 203–210, 2000.

128. McKee, D.W., Spiro, C.L., Lamby, E.J., The inhibition of graphite oxidation by phosphorus additives. *Carbon*, *22*, 3, 285–290, 1984.

129. ASTM E119-20, *Standard Test Methods for Fire Tests of Building Construction and Materials*, ASTM International, West Conshohocken, PA, 2020.

130. Bisby, L., Gales, J., Maluk, C., A contemporary review of large-scale non-standard structural fire testing. *J. Fire Sci.*, *2*, *1*, 1, 2013.

131. Schmid, J., Brandon, D., Werther, N., Klippel, M., Technical note - Thermal exposure of wood in standard fire resistance tests. *Fire Saf. J.*, *107*, 179–185, 2019.

132. Blagojević, M.D. and Pešić, D.J., A new curve for temperature-time relationship in compartment fire. *Therm. Sci.*, *15*, 2, 339–352, 2011.

133. Barnett, C.R., Replacing international temperature-time curves with BFD curve. *Fire Saf. J.*, *42*, 4, 321–327, 2007.

134. Khorasani, N.E., Billittier, J., Stavridis, A., Structural performance of a railway tunnel under different fire scenarios, in: *2018 Joint Rail Conference, JRC 2018*, American Society of Mechanical Engineers (ASME), 2018.

135. White, G.C. and Shirvill, L.C., Fire testing a review of past, current and future methods, in: *Proceedings of the 14th International Conference on Offshore Mechanics and Arctic Engineering. Part 5 (of 5)*, pp. 379–388, ASME, New York, NY, USA, 1995.

136. Bradley, I., Willoughby, D., Royle, M., A review of the applicability of the jet fire resistance test of passive fire protection materials to a range of release scenarios. *Process Saf. Environ. Prot.*, *122*, 185–191, 2019.

137. Sikoutris, D.E., Vlachos, D.E., Kostopoulos, V., Jagger, S., Ledin, S., Fire burnthrough response of CFRP aerostructures. Numerical investigation and experimental verification. *Appl. Compos. Mater.*, *19*, 2, 141–159, 2012.

138. Dierdorf, D., Menchini, C., Sellers, R., J, C., *Design and Analysis of Alternative High Heat Flux Sources for Materials Fire Testing - PREPRINT'*, p. 5, Atlantic City, NJ, 2007.

139. Gardelle, B., Duquesne, S., Vandereecken, P., Bourbigot, S., Characterization of the carbonization process of expandable graphite/silicone formulations in a simulated fire. *Polym. Degrad. Stab.*, *98*, 5, 1052–1063, 2013.

140. Morys, M., Illerhaus, B., Sturm, H., Schartel, B., Revealing the inner secrets of intumescence: Advanced standard time temperature oven (STT Mufu+)—μ-computed tomography approach. *Fire Mater.*, *41*, 8, 927–939, 2017.

141. Morys, M., Häßler, D., Krüger, S., Schartel, B., Hothan, S., Beyond the standard time-temperature curve: Assessment of intumescent coatings under standard and deviant temperature curves. *Fire Saf. J.*, *112*, 2020.

142. Maluk, C., Bisby, L., Krajcovic, M., Torero, J.L., A Heat-Transfer Rate Inducing System (H-TRIS) Test Method. *Fire Saf. J.*, *105*, 307–319, 2019.

143. Lucherini, A. and Maluk, C., Assessing the onset of swelling for thin intumescent coatings under a range of heating conditions. *Fire Saf. J.*, *106*, 1–12, 2019.

144. Tranchard, P., Samyn, F., Duquesne, S., Thomas, M., Estèbe, B., Montès, J.L., Bourbigot, S., Fire behaviour of carbon fibre epoxy composite for aircraft: Novel test bench and experimental study. *J. Fire Sci.*, *33*, 3, 247–266, 2015.

145. Adanménou, R., *Mesures à échelle réduite de paramètres pertinents issus de scénarios feu*, University of Lille, Lille, 2020.

146. Krüger, S., Gluth, G.J.G., Watolla, M.B., Morys, M., Häßler, D., Schartel, B., New ways: Reactive fire protection coatings for extreme conditions'. *Bautechnik*, *93*, 8, 531–542, 2016.

147. Weil, E.D., Fire-protective and flame-retardant coatings - A state-of-the-art review. *J. Fire Sci.*, *29*, 3, 259–296, 2011.

148. Mariappan, T., Recent developments of intumescent fire protection coatings for structural steel: A review. *J. Fire Sci.*, *34*, 2, 120–163, 2016.

149. Weil, E. and McSwigan, B., Melamine phosphates and pyrophosphates in flame-retardant coatings: Old products with new potential. *J. Coat. Technol.*, *66*, *839*, 75–82, 1994.

150. Andersson, A., Landmark, S., Maurer, F.H.J., Evaluation and characterization of ammoniumpolyphosphate-pentaerythritol- based systems for intumescent coatings. *J. Appl. Polym. Sci.*, *104*, 2, 748–753, 2007.

151. Delobel, R., Le Bras, M., Ouassou, N., Alistiqsa, F., Thermal Behaviours of Ammonium Polyphosphate-Pentaerythritol and Ammonium Pyrophosphate-Pentaerythritol Intumescent Additives in Polypropylene Formulations'. *J. Fire Sci.*, *8*, *2*, 85–108, 1990.

152. Jimenez, M., Duquesne, S., Bourbigot, S., Intumescent fire protective coating: Toward a better understanding of their mechanism of action. *Thermochim. Acta*, *449*, *1-2*, 16–26, 2006.

153. Zeng, Y., Weinell, C.E., Dam-Johansen, K., Ring, L., Kiil, S., Effects of coating ingredients on the thermal properties and morphological structures of hydrocarbon intumescent coating chars. *Prog. Org. Coat.*, *143*, 2020.

154. Kashiwagi, T., Grulke, E., Hilding, J., Groth, K., Harris, R., Butler, K., Shields, J., Kharchenko, S., Douglas, J., Thermal and flammability properties of polypropylene/carbon nanotube nanocomposites. *Polymer*, *45*, 12, 4227–4239, 2004.

155. Beheshti, A. and Heris, S.Z., Is MWCNT a good synergistic candidate in APP-PER-MEL intumescent coating for steel structure? *Prog. Org. Coat.*, *90*, 252–257, 2016.

156. Li, G., Liang, G., He, T., Yang, Q., Song, X., Effects of EG and MoSi2 on thermal degradation of intumescent coating. *Polym. Degrad. Stab.*, *92*, 4, 569–579, 2007.

157. Beraldo, C.H.M., da S. Silveira, M.R., Baldissera, A.F., Ferreira, C.A., A new benzoxazine-based intumescent coating for passive protection against fire. *Prog. Org. Coat.*, *137*, 2019.

158. Wang, Z.Y., Han, E.H., Ke, W., Fire-resistant effect of nanoclay on intumescent nanocomposite coatings. *J. Appl. Polym. Sci.*, *103*, 3, 1681–1689, 2007.

159. Ahmad, F., Zulkurnain, E.S.B., Ullah, S., Al-Sehemi, A.G., Raza, M.R., Improved fire resistance of boron nitride/epoxy intumescent coating upon minor addition of nano-alumina. *Mater. Chem. Phys.*, *256*, 2020.

160. Nasir, K.M., Sulong, N.H.R., Johan, M.R., Afifi, A.M., Synergistic effect of industrial- and bio-fillers waterborne intumescent hybrid coatings on flame retardancy, physical and mechanical properties. *Prog. Org. Coat.*, *149*, 2020.

161. Baldissera, A.F., Silveira, M.R., Dornelles, A.C., Ferreira, C.A., Assessment of lignin as a carbon source in intumescent coatings containing polyaniline. *J. Coat. Technol. Res.*, *17*, 5, 1297–1307, 2020.

162. Liu, S., Wang, C., Hu, Q., Huo, S., Zhang, Q., Liu, Z., Intumescent fire retardant coating with recycled powder from industrial effluent optimized using response surface methodology. *Prog. Org. Coat.*, *140*, 2020.

163. Bradford, S., Hallam, S., Taylor, A.P., Intumescent coating composition, British Patent Application Patent No. 2,451,233, UK, 2007.

164. Gardelle, B., Duquesne, S., Vandereecken, P., Bourbigot, S., Resistance to fire of silicone-based coatings: Fire protection of steel against cellulosic fire'. *J. Fire Sci.*, *32*, 4, 374–387, 2014.

165. Bourbigot, S., Gardelle, B., Duquesne, S., Intumescent silicone-based coatings for the fire protection of carbon fiber reinforced composites, in: *11th IAFSS Symposium*, International Association for Fire Safety Science (IAFSS), Christchuch, New Zealand, 2014.

166. Qureshi, S.P. and Krassowski, D.W., Intumescent resin system for improving fire resistance of composites, in: *Proceedings of the 1997 29th International SAMPE Technical Conference*, pp. 625–634, Orlando, FL, USA, 1997.

167. Mastalska-Popławska, J., Izak, P., Wójcik, Ł., Stempkowska, A., Góral, Z., Krzyżak, A.T., Habina, I., Synthesis and characterization of cross-linked poly(sodium acrylate)/sodium silicate hydrogels. *Polym. Eng. Sci.*, *59*, 6, 1279–1287, 2019.

168. Watolla, M.B., Gluth, G.J.G., Sturm, P., Rickard, W.D.A., Krüger, S., Schartel, B., Intumescent geopolymer-bound coatings for fire protection of steel. *J. Ceram. Sci. Technol.*, *8*, 3, 351–364, 2017.

169. Sotiriadis, K., Guzii, S.G., Mácová, P., Viani, A., Dvořák, K., Drdácký, M., Thermal Behavior of an Intumescent Alkaline Aluminosilicate Composite Material for Fire Protection of Structural Elements. *J. Mater. Civ. Eng.*, *31*, 6, 2019.

170. Abdul Rashid, M.K., Ramli Sulong, N.H., Alengaram, U.J., Fire resistance performance of composite coating with geopolymer-based bio-fillers for lightweight panel application. *J. Appl. Polym. Sci.*, *137*, 47, 2020.

171. Shuai, Q., Xu, Z., Yao, Z., Chen, X., Jiang, Z., Peng, X., An, R., Li, Y., Jiang, X., Li, H., Fire resistance of phosphoric acid-based geopolymer foams fabricated from metakaolin and hydrogen peroxide. *Mater. Lett.*, *263*, 2020.

172. J. Sarazin, C. A. Davy, S. Bourbigot, G. Tricot, J. Hosdez, D. Lambertin, and G. Fontaine: 'Flame resistance of geopolymer foam coatings for the fire protection of steel', *Composites Part B: Engineering*, 222, 109045, 2021.
173. Christke, S., Gibson, A.G., Grigoriou, K., Mouritz, A.P., Multi-layer polymer metal laminates for the fire protection of lightweight structures. *Mater. Des.*, 97, 349–356, 2016.
174. Geoffroy, L., Samyn, F., Jimenez, M., Bourbigot, S., Intumescent polymer metal laminates for fire protection. *Polymers*, 10, 9, 2018.
175. Geoffroy, L., Samyn, F., Jimenez, M., Bourbigot, S., Bilayer Intumescent Paint Metal Laminates: A Novel Design for a High-Performance Fire Barrier. *Ind. Eng. Chem. Res.*, 59, 7, 2988–2997, 2020.

5

Nitrogen-Based Flame Retardants

Alexander B. Morgan[1†] **and Martin Klatt**[2*]

¹University of Dayton Research Institute, Dayton, OH, USA
²BASF SE, Ludwigshafen, Germany

Abstract

Nitrogen-based flame retardants cover a wide range of vapor-phase and condensed phase mechanisms of flame retardant activity which is depending upon the specific chemistry of the flame retardant, as well as the chemistry of the polymer that the flame retardant is added into. Nitrogen based flame retardants are very often used as synergists for phosphorus based flame retardants but are also effective alone and should be treated as a separate class of flame retardants.

This chapter describes nitrogen based flame retardants and their use and application in different polymer systems. Besides commonly known ammonium and melamine based flame retardants, special members of this family are N-alkoxy hindered amines, azoalkanes, and phosphorus-nitrogen compounds such as phosphazenes, phospham, and other P-N compounds. Recently polymeric nitrogen compounds based on cyanuric acid have been developed and will be discussed as well.

Keywords: Melamine, melamine cyanurate, melamine polyphosphate, N-alkoxy hindered amines, azoalkanes, polymeric cyanuric acid derivatives, phosphazenes

5.1 Introduction

With continued pressure on the elimination of halogenated flame retardants from polymer applications (please see Chapter 1 of this book for more information), there is a continued increase in the use of non-halogenated flame retardant chemistries. One of the classes which has continued to

---

*Corresponding author: alexander.morgan@udri.udayton.edu

†Original Chapter Author for 1st Edition. This chapter has been updated with new content by Alexander Morgan.

Alexander B. Morgan (ed.) Non-Halogenated Flame Retardant Handbook 2nd edition, (239–270)
© 2022 Scrivener Publishing LLC

grow in use are nitrogen-based flame retardants. While exact sales numbers on the growth of this class are not available, in general, nitrogen-based flame retardants are growing, behind such classes as phosphorus and mineral fillers. Currently, inorganic metal hydroxides, mainly aluminum hydroxide (ATH), continue to be the largest class of flame retardants by volume due to the fact that large volumes of this particular flame retardant are required for flame retardant efficacy [1–3].

Since 2014, market growth for non-halogenated flame retardants, and nitrogen-based flame retardants, has continued to grow, although exact numbers are not currently available since there are multiple vendors of this flame retardant class worldwide, and exact sales are often considered to be trade secrets of the companies selling these flame retardant additives [4]. Still, growth has been robust, with the main nitrogen-based flame retardant driving sales being melamine. While nitrogen-based flame retardants can be used alone, they are often combined with other flame retardants, namely phosphorus [5].

Nitrogen-based flame retardants can be broken down into sub groups, including ammonium-based, melamine and melamine-based, radical generators, and organophosphorus-nitrogen compounds. Each of these groups will be discussed below, with additional breakdown in groups to highlight specific chemistries.

5.2   Main Types of Nitrogen-Based Flame Retardants

Nitrogen based flame retardants (NFR) can be divided into two groups with respect to their mode, or mechanism, of flame retardant action. By far the most dominant and well established group is derived from either ammonia or melamine, although some derivatives of urea and guanidine are known [6–8]. The primary mode of action of these NFRs based on ammonium or melamine is endothermic decomposition with release of large amounts of non-flammable gases such as nitrogen and ammonia, making them active as predominantly vapor-phase flame retardants. The compounds are often modified further to be salts (often generated by reactions between inorganic acids and melamine/ammonia) to impart higher thermal stability and lower volatility in the final application. Typical products and applications include melamine in polyurethane flexible foams, melamine phosphates or ammonium polyphosphate-pentaerythritol systems in polyolefins, melamine, melamine phosphates or dicyandiamide in intumescent paints, guanidine phosphates in textiles, ammonium salts in select textiles, and guanidine sulfamate in wallpapers.

The second group of NFR is characterized by a strong interaction with the matrix polymer during thermal decomposition of the polymer, namely in the condensed phase where the NFRs do not volatilize and instead leads either to decomposition of the matrix polymer to enable dripping away from the flame, or to intensive charring which resists further thermal decomposition. Since the mode of action requires a specific interaction with the matrix polymer during decomposition, use of this type of NFR is restricted to limited numbers of base polymers, for example N-alkoxy hindered amines in polyolefins, [9] melamine cyanurate for polyamides, [10, 11] and phosphazenes in epoxy resins, with the latter being heavily researched in China [12–22].

While NFRs can be grouped into vapor phase and condensed phase mechanisms of flame retrardancy, NFR can show different effects depending upon the polymer/flammable material they are combined with. As will be discussed below, there are cases where mostly vapor-phase active NFRs can show very strong condensed phase char formation, and where mostly condensed-phased active "char formers" can act more as vapor-phase flame retardants when in the right polymer. It is important to consider not only the chemistry of the flame retardant, but also the chemistry of polymer that the flame retardant is added into, as the reactions between the two materials (flame retardant and flammable material) will drive the exact mechanism of flame retardancy [23, 24].

5.3   Ammonia-Based Flame Retardants

Ammonia based flame retardants are the largest group of NFR by volume. Due to its being a gas down to -33°C, ammonia can only be used as salt when added to other solid materials as a flame retardant. In all cases, ammonia based flame retardants undergo endothermic decomposition with release of ammonia, leaving behind an acid which causes charring. There are only a few commercial ammonia based flame retardants used, of which the most prominent is ammonium polyphosphate (APP) followed by ammonium pentaborate and ammonium sulfamate. Ammonium phosphate (in its monobasic, dibasic, and tribasic forms) often gets used as a "dry powder" extinguishing agent for fires, rather than added directly to materials. The exception to this is when it is added directly to cellulosic fibers for use as flame retardant barriers for mattress applications [25, 26]. Ammonium sulfate will also be employed to flame retard cellulosic fibers for use in flame retardant barriers, but is not used as a dry extinguishing powder [27].

5.3.1   Ammonium Polyphosphate

One of the most commonly used nitrogen containing flame retardants is ammonium polyphosphate (APP). APP (CAS No 68333-79-9) is an inorganic salt of polyphosphoric acid and ammonia containing both linear and branched chains. As its chemical formula is $[NH_4PO_3]_n(OH)_2$ the monomer consists of an orthophosphate anion neutralized by an ammonium cation leaving two bonds free to polymerize. In case of branching some monomers have no ammonium anion and instead link to three other monomers. APP is an intumescent flame retardant (see Chapter 4 for more detail) and finds application in thermoplastics, thermosets, foams and coatings.

There are two main families of ammonium polyphosphate: Crystal phase I APP (APP I) and Crystal phase II APP (APP II) (see Figures 5.1 and 5.2). Crystal phase I APP (APP I) is characterized by a variable linear chain length, possessing a lower decomposition temperature (~150°C) and higher water solubility than Crystal Phase II APP. In APP I, n (number of phosphate units) is generally lower than 100.

As shown in Figure 5.2, the APP II structure can be cross linked or branched. The molecular weight is much higher than APP I with "n" greater than 1000. APP II has higher thermal stability (decomposition starts at approximately 300°C) and lower water solubility than APP I.

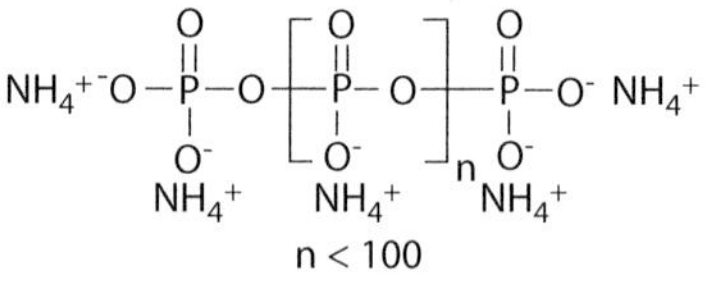

Figure 5.1 Linear APP.

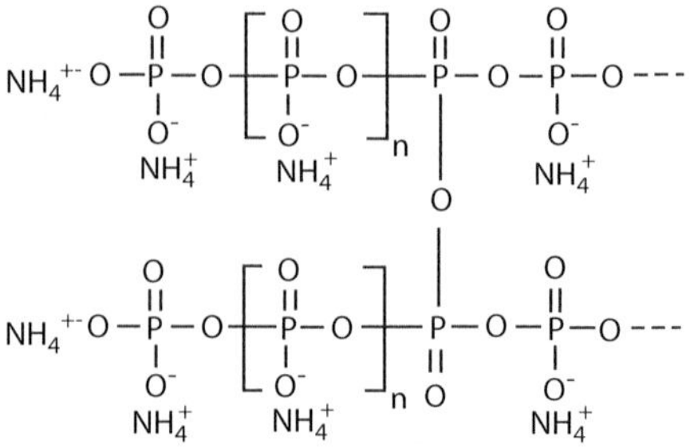

Figure 5.2 Branched APP, n > 1000.

When plastics or other materials containing APP are exposed to accidental fire or heat, the flame retardant starts to decompose. The decomposition products are polymeric phosphoric acid and ammonia. As aforementioned, the endothermic decomposition removes heat and the ammonia and water subsequently evolved dilute the combustible gases. Polyphosphoric acid reacts with hydroxyl groups of the matrix polymer or of an added synergist to form an unstable phosphate ester. In the next step, the dehydration of the phosphate ester follows. A foamed char layer forms on the surface facing the heat source (charring). The foamed char layer then acts as insulation, preventing further decomposition of the polymer. Addition of synergistic products like pentaerythritol derivatives, carbohydrates and foaming agents (melamine etc.) significantly improve the flame retardant performance of APP. The mechanism of decomposition has been investigated in order to establish a model for intumescent systems [28–30] and a reaction scheme (Figure 5.3) can be found for example in reference [31].

The addition of polyhydric alcohols such as pentaerythritol or starch as char forming agents increases the efficiency and allows lower loadings. The blowing effect can be greatly increased by using additional gas-forming compounds such as melamine. Nonetheless, 20% to 30% of such a synergistic APP mixture is needed to achieve acceptable flame retardancy in polyolefins. Figure 5.4 shows a polypropylene specimen containing APP and synergists after the burning test. The intumescent char can be seen at the tip of the specimen which was exposed to the external flame. Mixtures of APP with synergists are available under different trade names. Figure 5.5 shows the dependency of LOI on the loading level of a typical synergistic mixture (Exolit® AP 760) in polypropylene. LOI increases nearly linearly with loading. In addition it can be seen that LOI is higher for thicker materials and the slope of the line tends to be greater for the larger thicknesses.

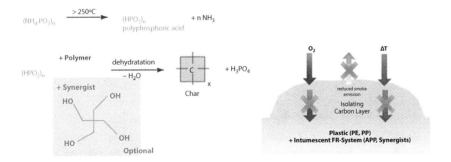

Figure 5.3 Reaction scheme for intumescence and char formation by APP.

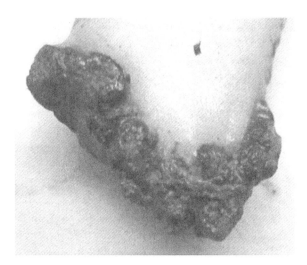

Figure 5.4 UL 94 testing with APP flame retarded polypropylene (left).

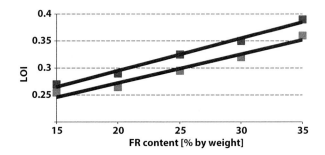

Figure 5.5 Dependency of LOI on loading level of Exolit® AP 760 in polypropylene (data courtesy of Clariant GmbH); red box 1.6 mm thickness, blue box 3.2 mm thickness (right).

Since the mechanism is based on solid phase charring reactions, there is a strong reduction of smoke density for intumescent flame retardants such as APP compared to halogenated flame retardants. Intumescent flame retardants such as APP and charring synergists protect the polymer from burning but do not interfere with the flame in the gas phase. As a result, a reduced amount of combustible gas is released leading to some non sustainable flame. As there is no interaction with the gas phase reaction in the flame, the smoke density is comparable to the virgin material. Figure 5.6 shows the smoke density of polypropylene with and without APP (Exolit® AP 760) in comparison to two different halogenated flame retardants.

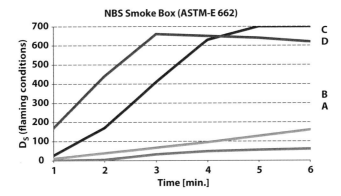

Figure 5.6 Smoke density (D_S) of burning PP (V-0, 1.6 mm) with different flame retardants. (A = without flame retardant, B = containing 26% Exolit® AP 760, C = 45% chlorinated cycloaliphate (65% Cl) + Sb_2O_3 + $ZnBO_3$, D = 33% Decabromodiphenylether + Sb_2O_3 (data courtesy of Clariant GmbH)).

APP has been successfully combined with many other flame retardants, and continues to be used in a wide range of applications and formulations, far too many to be covered in this chapter, but many of its applications and combinations are covered in other chapters of this book. For additional details on APP, please see Chapters 2 (Phosphorus Based Flame Retardants) and 4 (Intumescent Flame Retardants) of this book.

While APP is very versatile in providing flame retardant benefit, it does have two well-known drawbacks that can limit its use in some applications. The first is use temperature. Specifically, the upper temperature at which APP can be safely used before it activates and begins to thermally decompose. This is relevant when considering thermoplastic compounding such that the APP does not react during melt compounding resulting in damage to the extruder and a mess to clean up! APP, depending upon the grade and crystal phase available, will have an upper use temperature of ~150 – 200 °C, with the 200 °C being a limit only available to specialized grades of APP. Users of this chemical are strongly advised to work with their vendor on use temperature limits, and, to verify this use temperature with thermal analysis prior to melt compounding. The other drawback of APP is water absorption, which limits the use of APP to applications that are not water-sensitive. Electrical and electronic applications which are water sensitive often cannot use APP due to the ability of APP to absorb water over time. There are specialized grades of APP available which reduce the amount of water absorbed and users of this chemical are encouraged to discuss with vendors of APP on what is available and how well these specialized grades and resist water absorption.

5.3.2   Other Ammonia Salts

There are only a few other commercial flame retardants based on ammonium. Ammonium pentaborate, ammonium sulfamate, and ammonium sulfate have been described in literature. Ammonium pentaborate ($NH_4B_5O_8$) is a white crystalline alkaline salt product resulting from the controlled reaction of ammonia, water and boric acid. It is used when a readily soluble alkali borate is needed or when alkali metals cannot be used. Borates change the oxidation reactions in the combustion of cellulosic materials and cause the formation of carbon residue. The charred material forms a barrier to combustion, and retains the decomposition products. Ammonium pentaborate solutions can be applied to paper by spraying or dipping to yield a fire-retarded product. It can also be used as a component in flame proofing formulations for cellulosic materials (such as cotton textiles). Ammonium pentaborate has also been used as a flame retardant in polymers such as epoxy, thermoplastic polyurethane (TPU), urethane foam, etc. It should be noted that the water solubility of the ammonium pentaborate prevents its use as a launder durable flame retardant for textiles, and it is also not resistant to humidity/water exposure that would wash the ammonium pentaborate off paper or any other substrate it was applied to.

Myers *et al.* [32] claim that ammonium pentaborate creates a glassy multicellular structure which protects the underlying TPU. Levchik *et al.* [33] investigated the use of ammonium pentaborate in polyamide 6 (PA6). In this work ammonium *pentaborate* was shown to reduce thermal stability of PA6, thereby increasing formation of carbonated residue during combustion. In terms of its fire retardant effect ammonium *pentaborate* is similar to ammonium polyphosphate but leading to a greater reduction in thermal stability. This salt has also found use as a component of multi-flame retardant solutions in rigid polyurethane foam [34].

There have been investigations [35–38] on the flame resistance of PA6 treated with ammonium sulfamate ($NH_4SO_3NH_2$) and diammonium imidobisulfonate ($NH(SO_3NH_2)_2$). It was shown that only 2% ammonium sulfamate was sufficient to achieve a UL94 V0 in PA6 when combined with small amounts (1%) of an additional char forming synergist such as dipentaerythritol. They claimed that a condensed phase sulfating reaction of the primary PA6 amino groups and the OH groups of the dipentaerythritol leads to the formation of a protective char layer.

Wilkie [39] reported a reduction in the rate of heat release (HRR) in a cone calorimeter measurement of a polystyrene sample containing as little as 5% of ammonium sulfamate as flame retardant. He found a reduction of HRR by about 50%. An increase of the amount of ammonium sulfamate to 10% did not result in any further improvement of flame retardancy.

Finally, ammonium sulfate has been used as a flame retardant treatment for cellulosic fiber burn barriers in mattresses and has been known to be useful for this since the 1970s [40]. It has not shown effectiveness in any other materials besides cellulosic (wood, cotton) materials.

Similar to the APP discussion above, these ammonium salts have use temperatures above which they will thermally decompose which will limit what polymers/materials they can be used with. Additionally, the ammonium salts have issues with water absorption (as well as water solubility) that would severely limit their use in electrical/electronic applications and any other water-sensitive end-use application. Otherwise, these ammonium salts do work to provide flame retardant benefit when used in the right end-use application that will not have water/humidity exposure issues.

5.4 Melamine-Based Flame Retardants

Melamine and its condensation products and salts are widely used as flame retardants. In most applications they are combined with P-containing flame retardants, similar to the ammonium based flame retardants mentioned above. Combinations with metal hydrates and halogenated flame retardants are known as well. In all cases, melamine undergoes endothermic decomposition leading to cooling of the polymer matrix and the release of non-combustible gases such as water, CO_2 and ammonia [41]. These dilute the combustible decomposition products of the matrix polymer further destabilizing the flame. While this endothermic reaction is mostly independent of the matrix polymer, melamine cannot be broadly applied as flame retardant, and it can be more effective in select polymer classes. When considering melamine derivative salts, the choice of the derivative mainly depends on the processing temperature of the matrix polymer and its sensitivity towards hydrolysis. When melamine salts are used, the acid very often contributes to the flame retardant effect. Since this is typically achieved by charring or other interactions with the matrix polymer, the selection of the right acid too make into a melamine salt is crucial for its efficiency. The thermal stability of these melamine salts also strongly depends on the acid source, leading to a range of chemical solutions involving melamine salts as flame retardants.

5.4.1   Melamine as Flame Retardant

Melamine (2,4,6-triamino-1,3,5 triazine, CAS# 106-78-1) is a white crystalline powder with a melting point of approximately 354°C and a density of 1.573 grams/cc [42]. At about 200°C, melamine undergoes sublimation and thereby dilutes the fuel gases and oxygen near the combustion source. The energy of sublimation is about 29 kcal/mole. On the other hand, its decomposition is strongly endothermic in the order of ~470 kcal/mole [43] and melamine acts as a heat sink in fire situation. Melamine is only slightly soluble in cold water but shows much higher solubility in hot water [44].

Since melamine can be dispersed in polyols, its largest application as flame retardant is in flexible polyurethane foams used for upholstered furniture or mattresses which need to pass the requirements of the Crib V tests in UK [45]. Due to its limited efficiency, a high amount of about 20 – 30 parts by weight polyol of melamine should be used in addition with a gas phase active flame retardant such as trischloropropylphosphate (TCPP) to meet the required fire performance.

Figure 5.7 shows the burning behavior of a high resilient flexible PU foam with and without melamine [46]. In addition the polyol used was tested as well to show its potential contribution to heat release. It can be seen that the heat release rate of the melamine treated foam is reduced by a factor of 2. Ignition of the two PU foams occurs almost immediately upon

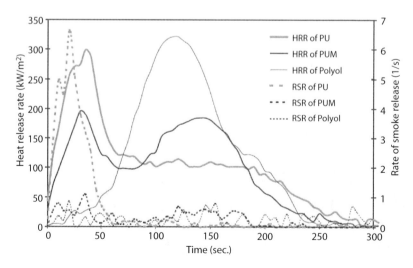

Figure 5.7 Heat release rate and rate of smoke release of polyurethane foam with (PUM) and without melamine (PU) and the corresponding polyol; external heat flux 35kw/m².

exposure to a low intensity heat flux ($35kW/m^2$), while the polyol shows a markedly delayed ignition (~25s) under the same conditions. The time to peak heat release for both foams (PU and PUM) is comparable at about 40s. The polyol itself reaches its maximum heat release at about 125s. This indicates that rapid ignition of the PU foam is not caused by the polyol but by the other component of the PU foam, the isocyanate [47, 48].

The cone calorimeter experiment also clearly demonstrates the reduction of rate of smoke release from 6.5/s for PU to below 1/s for the melamine treated foam (PUM). Again, the polyol does not show a significant smoke release rate. The authors concluded that reaction of melamine with isocyanate plays a role in both the reduction of the rate of heat release and of smoke release. They proposed that an amino group of melamine attacks the isocyanate forming urea (Scheme 5.1) which is not readily volatile and therefore, cannot enter the gas phase and fuel the flame. Reduced fuel leads to a reduced heat release rate. In addition, reduced sooting is observed, because less of the highly aromatic isocyanate is emitted during combustion.

This thorough investigation of the mode of action of melamine in PU foams shows that the reactive amino groups of melamine can play an active role in flame retardation. Therefore, the general assumption that melamine's effectiveness is only due to endothermic decomposition, intumescence and dilution of combustible gases by inert gases is only partially

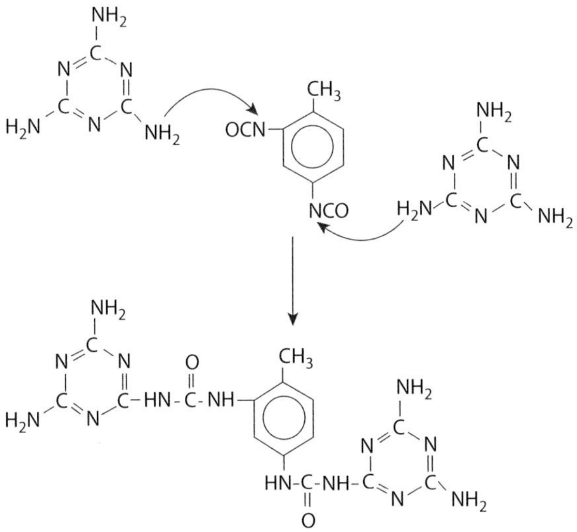

Scheme 5.1 Reaction between melamine and toluene diisocyanate.

true. In most cases, this might be the dominant mechanism but a detailed investigation is needed to fully understand the complete picture. Most likely, the ability of melamine to react with chemical groups in the polymer structure during thermal decomposition of said polymer would likely affect the mechanism of flame retardancy. In a polymer like polyurethane where the NH_2 groups can interact with the urethane linkage, one would expect to see some change in thermal decomposition chemistry. But for a polymer where this reaction in unlikely to happen (such as a polyolefin), one should expect melamine to not be as effective at changing condensed phase phenomena or polymer decomposition chemistry.

5.4.2   Melamine Salts

Due to its low sublimation temperature, melamine itself cannot be used in polymers which are processed at temperatures above 200°C, including polypropylene (select grades), polyamide and polyesters. In order to increase the thermal stability, salts of melamine have to be applied, the most common being melamine cyanurate and melamine phosphate/ (poly)-phosphate. In Figure 5.8 the thermal stability of melamine and some of its salts is shown.

While melamine phosphate is even slightly less thermally stable than melamine, the salts with cyanuric acid, melamine cyanurate, and with polyphosphoric acid, melamine polyphosphate, are distinctively more stable. Melamine cyanurate starts to decompose at about 300°C and melamine polyphosphate shows an onset temperature of around 350°C.

Although they are quite similar in structure, their mode of action is completely different as will be discussed in the following sections of this chapter.

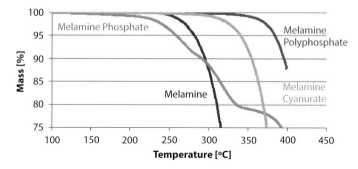

Figure 5.8 Thermogravimetric analysis of melamine and some of its salts.

5.4.3   Melamine Cyanurate

Melamine cyanurate (CAS Nr. 37640-57-6), also known as melamine-cyanuric acid adduct or melamine-cyanuric acid complex, is a crystalline complex formed from a 1:1 mixture of melamine and cyanuric acid. The substance is not a salt despite its name. The complex is held together by an extensive two-dimensional network of hydrogen bonds between the two compounds, reminiscent to that seen in DNA base pairing as illustrated in Figure 5.9.

This structure is responsible for its higher thermal stability relative to pure melamine and its insolubility in common solvents. In some application, this is advantageous for isolation properties of the polymer compound. Furthermore, there is only limited likelihood of migration of MC out of the matrix polymer.

Melamine cyanurate can be used in polymers with processing temperatures up to 300°C such as polyamides, thermoplastic polyurethanes and polyesters. Above 320°C, it undergoes endothermic decomposition to melamine and cyanuric acid which further decompose at higher temperatures to ammonia, water and CO_2.

Since the 1970s MC has been used as flame retardant in polyamide 6 and some co-polyamides such as polyamide 6,6 (PA6,6 or blends of PA66 and PA6). As it starts to decompose at around 300°C it cannot be used it PA66, which is typically processed at temperatures up to 330°C.

The mode of action of MC has been extensively investigated [49–52]. During decomposition, MC acts as a heat sink. The vaporized melamine

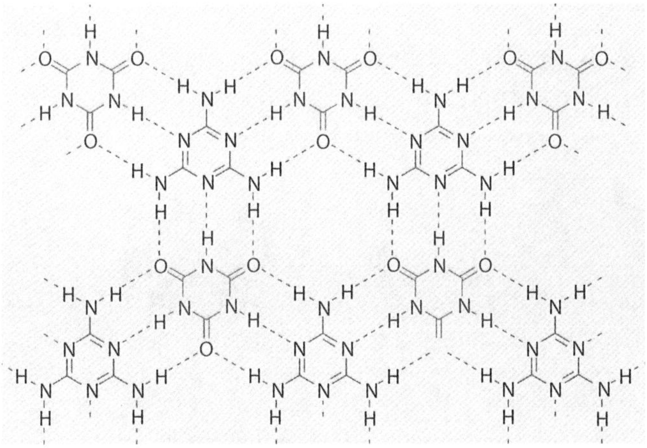

Figure 5.9 Structure of melamine cyanurate.

and further decomposition products are inert gases diluting the oxygen and the fuel gases present at the point of combustion. However, when used in polyamides, MC has some additional effects it imparts to provide flame retardancy.

In polyamides, MC is applied as a stand-alone flame retardant. In fact MC is one of the most efficient flame retardant in unreinforced PA6 and PA66. Only about 6 - 10% of MC is needed to achieve a flammability rating of UL94 V0 at thicknesses down to 0.4mm. The reason for this outstanding performance is due to the so-called "run-away" mechanism [53]. Effectively, the MC is causing rapid decrease in melt viscosity for the polyamide, making the solid material turn to liquid rapidly and allowing it to drip away from the heat source. Once the polymer is away from the heat source, it cannot pyrolyze/burn further. This dripping effect is not permissible in all fire safety applications, especially in cases where the dripping material may fall upon people escaping from a compartment.

When a flame is applied to polyamide containing MC, the decomposition of MC starts almost immediately (see picture 1 in Figure 5.10). Small bubbles are formed by the decomposition of MC indicating the creation of decomposition products from the MC and the polyamide. No char at all is formed under these conditions. In addition, the whole specimen starts to form a spit from which dripping begins typically during the second flaming period. These drops are so small that they cease burning by the heat sink/diluent effect of MC before reaching the bottom, thus resulting in a UL94 V0 classification.

The combination of the heat sink/diluent effect of MC with the runaway mechanism is very specific to polyamide. This is due to interaction of cyanuric acid with polyamide causing degradation and polymer breakdown which leads to a strongly reduced viscosity of the polymer melt [48–51]. It should be noted that while the melamine is serving as a flame retardant within the polymer drips, that this sort of phenomena where the polymer is decomposed allowing it to drip away from the heat source does

Figure 5.10 UL94 V0 test of polyamide containing 6% MC during first and second and after second application of flame.

result in an increase in heat release for the polymer because it is effectively decomposing and pyrolyzing faster than the non-flame retarded material. So for some applications where the material cannot drip away from the heat source, using flame retardants like MC in polyamides can actually make flammability worse. Users of MC are advised to look at their specific fire test requirements to ensure that it is fully appropriate for the end use fire risk scenario.

The situation is different for glass fiber reinforced polyamides. The glass fibers form entanglements which reduce the flowability of the melt. Therefore, much larger droplets are formed under the UL94 testing conditions required. As these larger droplets contain too much energy, the heat sink/diluent effect of MC is not sufficient to extinguish the flame before landing.

However, if short reinforcing fibers are used, there is a possibility to achieve at least a UL94 V2 rating with MC in polyamide compounds. Such is the case for milled glass fibers or fibrous fillers like Wollastonite [54].

As shown above, MC serves as a heat sink and source of inert gas and can be used alone as a flame retardant in polyamide where the runaway mechanism occurs. However, MC alone is not very efficient for most other polymer systems. Therefore, it is often combined with phosphorus based flame retardants. Typical examples are the combination of MC with aluminum diethylphosphinate in polyesters or MC with organic phosphates in thermoplastic polyurethanes.

Hackl *et al.* [55] disclose the use of MC with organic phosphates for thermoplastic polyurethane. They claim that a combination of 25% MC with 7.5% of an organic phosphate ester is sufficient to reach a UL94 V0. Tabuani *et al.* [56] presented a detailed study on the synergism of MC and nanoclay in TPU. While neither nanoclay nor MC alone is effective as flame retardant in TPU the combination of the two shows clear evidence of synergism as shown by thermogravimetric analysis and combustion tests, namely LOI and cone calorimeter. However, MC and organoclay tends to be antagonistic in vertical flame spread behavior in polyamides since the organoclay acts as an anti-drip flame retardant by increasing polymer viscosity, in competition with the MC mechanism of enabling the polyamide to melt back/drip away from the flame [57].

The combination of MC and organic phosphinates in polyesters has been investigated by Braun *et al.* [58]. They claimed that the main flame retardancy action comes from the gas phase active phosphinate while MC contributes through its heat sink/dilution effect. Inorganic phosphinates are also used together with MC in polyesters [59]. However, the efficiency of this system is not good enough to achieve an UL94 V0 at thicknesses below 1 mm. Addition of crosslinking agents led to better UL performance.

5.4.4   Melamine Polyphosphate

Melamine phosphates are substances combining the synergistic effect of melamine with P-containing components. Their thermal stability increases as follows: Melamine phosphate < Melamine pyrophosphate < Melamine polyphosphate.

Melamine (mono)-phosphate is a salt of melamine and phosphoric acid (Figure 5.11). Above ~200°C it is converted to melamine pyrophosphate and finally, when heated above 260°C, melamine polyphosphate. These transformations involve the release of water and lead to a heat sink effect. Above 350°C, melamine polyphosphate (MPP) undergoes endothermic decomposition. The released polyphosphoric acid coats and therefore, shields the condensed combustible polymer. Intensive charring also occurs. This char formed on the surface reduces the amount of oxygen present at the combustion source and protects the underlying polymer. The released melamine acts as blowing agent creating a protective foamed char.

Figure 5.12 shows the intumescent charring effect for a mixture of MPP in PA6 during a UL94 V test procedure. The intumescent char is almost immediately formed upon first application of the flame and grows during the second flaming. No dripping is observed.

Due to its higher thermal stability MPP is mostly used in thermoplastic and thermoset applications. Commercial products are available from BASF (Melapur® 200 range) and Budenheim (Budit® 3114). Although there are numerous publications on the use of neat MPP in polyamide, its efficiency is rather low. In highly glass filled polyamides, low amounts of MPP can provide an UL94 V0 rating. A virtually linear dependence of MPP loading

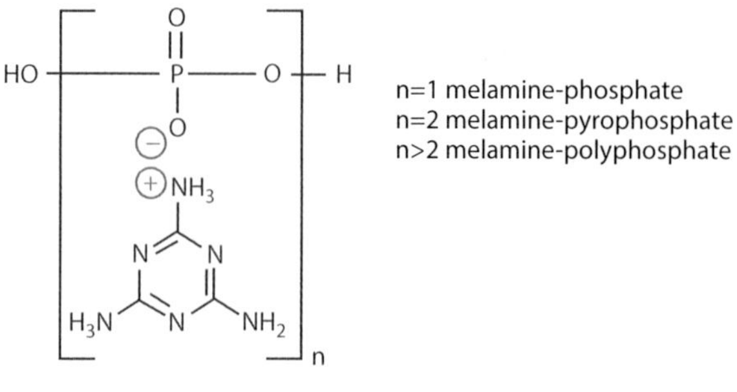

n=1 melamine-phosphate
n=2 melamine-pyrophosphate
n>2 melamine-polyphosphate

Figure 5.11 Structure of melamine phosphates.

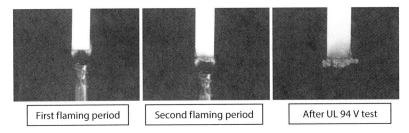

| First flaming period | Second flaming period | After UL 94 V test |

Figure 5.12 Polyamide compound with MPP under UL94 V test conditions.

and the amount of glass fibers required to reach V0 has been claimed [60] (see Figure 5.13).

Most reported applications of MPP describe its use as a synergist for P-containing flame retardants. Yang *et al.* [61] showed a synergistic effect of MPP with aluminum hypophosphite in polybutylene terephthalate. Synergistic mixtures of MPP with aluminum diethylphosphinate (DEPAL) are available from Clariant. They can be applied in both polyesters and polyamides. The mode of action of MPP and DEPAL in polyamide and polyesters has been extensively investigated by Braun *et al.* [62]. They suggest that the charring and intumescence of MPP together with gas phase interaction of DEPAL leads to the flame retardancy effect.

Recently, a new class of melamine polyphosphates was developed by Floridienne Chimie and was sold to JM Huber. These contain stoichiometric amounts of metal (zinc or magnesium, see Figure 5.14) and show similar synergistic effect as MPP. There is another version of this chemistry that contains an additional phosphorus compound, but the exact structure of this additive (Safire 3000) has not been provided.

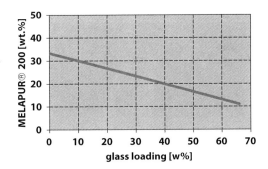

Figure 5.13 Concentration of MPP to get UL94 V-0 in PA66 depending on the glass fiber content.

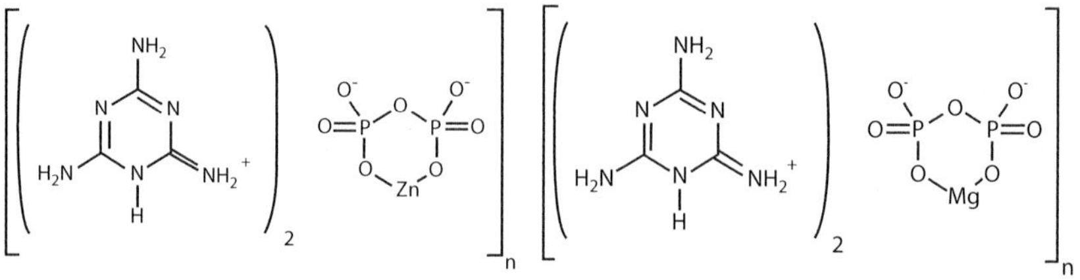

Figure 5.14 Metal modified MPP products by Floridienne Chimie (Safire® 400, Safire® 600).

5.4.5   Melamine Condensates and Its Salts

While melamine itself is rather volatile and sublimes at around 220°C, its condensation products are much more stable. Melam, melem and melon (Scheme 5.2) are formed under thermal treatment of melamine. An extensive overview of the melamine condensation products and their salts can be found in the review by Schwarzer [63].

Generally melam is generated by heating of melamine or its salts at temperatures around 300°C.

A detailed laboratory method for synthesis of melam is given in reference [64]. More recently, a commercial synthesis process for melam and its salts by thermal treatment of melamine in presence of an acid, preferred toluene sulfonic acid, was disclosed [65].

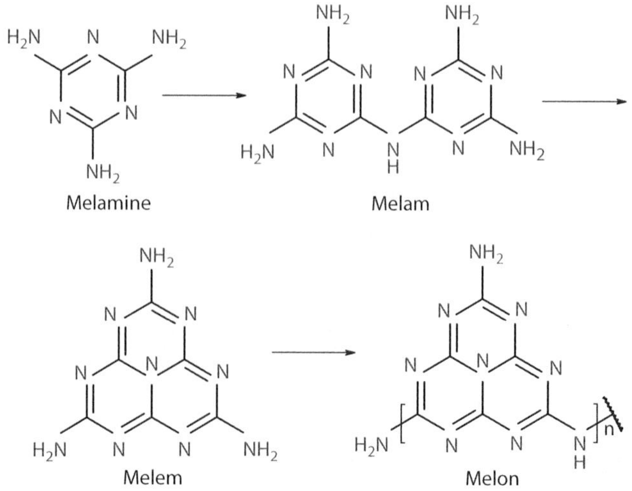

Scheme 5.2 Melamine and its condensation products.

Melam, melem and their salts are used as flame retardants for thermo-plastics, especially polyamide. DSM reported the use of about 7% melam as a flame retardant in PA 6, PA 66 and PA 46 [66].

Some patents describe the use of melam and melem salts of phosphoric acid, polyphosphoric acid or methane sulfonic acid as flame retardants. These salts are directly produced via heating of melamine, neutralized with the respective acid, at temperatures >300°C [67]. A mixture of melamine and polyphosphoric acid, for example, was calcinated at 300°C – 400°C for a several hours to yield melem polyphosphate. Depending on the exact cal-cination conditions (time and temperature), predominately melam poly-phosphate is formed [68].

Due to a general trend towards more temperature stable polyamides in electrical and electronic applications, a shift towards the thermally more stable condensation products of melamine and their salts is expected.

5.5   Nitrogen-Based Radical Generators

There is a class of nitrogen based flame retardants which has been specifically developed for use in polyolefins. These are radical generators which lead to extensive dripping and withdrawal of the polyolefin from the flame. Similar to melamine cyanurate, their effectiveness is due to the run-away effect. Of all the various chemical structures described in literature, N-alkoxy hin-dered amines [69] (NOR type) and azoalkanes [70] are the most prevalent. Commercially, the NOR type flame retardants find application in polyolefin films and fibers. Figure 5.15 shows the structure of Flamestab® NOR® 116.

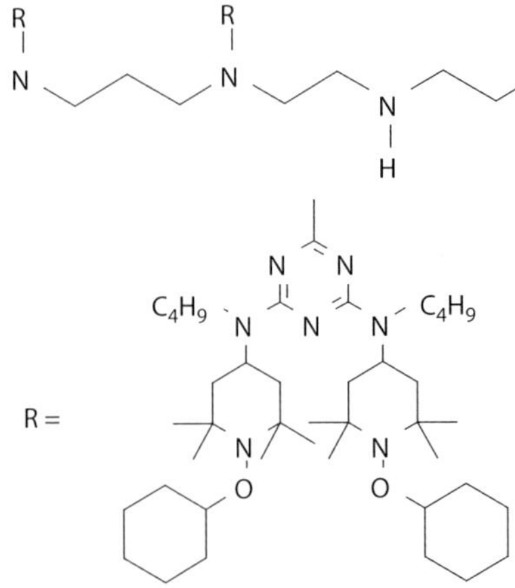

Figure 5.15 Hindered N-alkoxy amine stabilizer Flamestab® NOR® 116.

The use of Flamestab® NOR® 116 in polyolefins has been extensively investigated [71].

Film samples with and without Flamestab® NOR® 116 burnt according to the NFPA 701 1989 test are shown in Figure 5.16. It can be seen that the film without additive burns up to the clamp and a char layer is formed. The film containing 1% Flamestab® NOR® 116 displayed significantly reduced after flame and passed the test. There is almost no char visible.

In Scheme 5.3, the different modes of action of N-alkoxy hindered amines are shown. The upper reaction scheme illustrates the stabilizing effect, which dominates at temperatures below 150°C. Free radicals are consumed in the so-called Denisov Cycle [72]. At higher temperatures during combustion, the hindered amines decompose through scission of the N-O or the O-R bond and yield active radicals which catalyze the radically induced decomposition of the polyolefin. This leads to the formation of smaller breakdown products. However, these oligomeric products are not readily volatile and they just drip away without catching fire, similar to a burning candle. Wax will melt and then drip without catching fire.

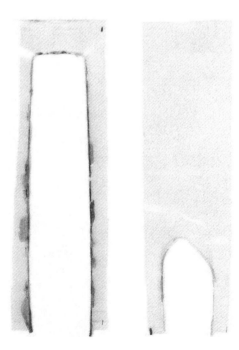

Figure 5.16 Polypropylene film samples with and without NOR-116 after burning in the NFPA 701 1989 test.

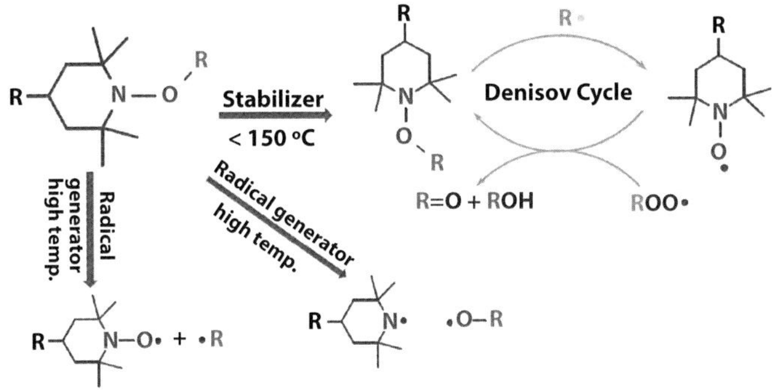

Scheme 5.3 Mode of action of N-alkoxy hindered amines.

Results from the NFPA 701 tests conducted on polypropylene compression molded films are shown in Table 5.1. Increasing the concentration of Flamestab® NOR® 116 to 5 and 10% decreased the flame retardancy performance as evidenced by the burn test. There seems to be an inverse relationship between the concentration of Flamestab® NOR® 116 and its flame retardancy performance. Since the additive system is organic, perhaps the incorporation of excess combustible materials results in decreased performance.

However, with increasing thickness, the efficiency of the run-away effect decreases. Test results for polypropylene molded plaques are given in Table 5.2. It can be seen that the first flame always extinguishes in a very short time when Flamestab® NOR® 116 is present, whereas the molded plaque without additive burns up to the clamp. Nonetheless, the NOR stabilized material also burns up to the clamp during the second flaming.

Similar to the discussion of melamine cyanurate above, use of the NOR materials results in rapid depolymerization of the polyolefin, which means

Table 5.1 NFPA 701 (1989) vertical burn test results in polypropylene compression molded (10 mil) films.

Additive	After flame (s)	Burn drips (s)	Char length (")	Rating
Blank	28	> 2	10.0	Fail
1% NOR-116	< 2	0	3.4	Pass
5% NOR-116	24	> 2	9.3	Fail
10% NOR-116	21	> 2	10.0	Fail

Table 5.2 UL-94 V test results in polypropylene compression molded specimens.

NOR-2 (%)	1st After flame (s)	2nd After flame (s)	Burning drips	Burn to clamp	Rating
-	172	-	Y	Y	Fail
1.0	4	106	Y	Y	Fail

an increase in heat release for this material as the polyolefin pyrolyzes at a faster rate than the polyolefin which does not have the NOR present. For a high heat release material like polyolefins, NOR can make flammability much worse in certain fire risk scenarios where a high heat release material generates even higher heat release in the presence of NOR, and therefore the user of this chemistry is strongly reminded to pay attention to their end-use application and fire risk scenarios before using this material.

Due to the above issues, NOR compounds are often used in combination with a phosphorus based synergist. Recently, the combination of methane phosphonates (cyclic dimethylspirophosphonate – Figure 5.17, ethylenedi-amine methane phosphonate and melamine methane phosphonate) with N-alkoxy hindered amines has been claimed as an efficient flame retardant for polyolefins of thickness higher than 100μm [73].

Along with the above mentioned N-O compounds, there have been some other nitrogen based radical generators which have shown flame retardant effects, again via the radicals changing polymer molecular weight and melt viscosity to allow the polymers to drip away from the flame. This have included azo and triazine compounds, [74–76] which form radicals due to the thermal decomposition of the azo/triazine groups with the subsequent release of N_2 gas. These materials are not currently commercial, and appear to be limited to laboratory studies. Still, this class of materials does appear to have some utility as a flame retardant, assuming that the end-use fire test will allow for the material to drip away from the flame.

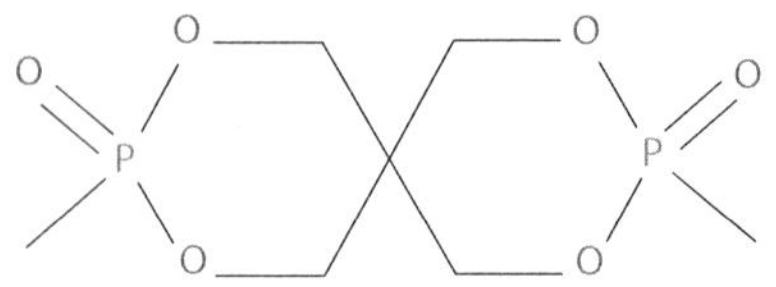

Figure 5.17 Cyclic spirophosphonate used as synergist for N-alkoxy hindered amines.

5.6   Phosphazenes, Phospham and Phosphoroxynitride

Phosphazenes, phospham and phosphoroxynitride are compounds, in which a phosphorus atom is covalently linked to two nitrogen atoms by a double and a single bond. In case of phosphazene individual pentavalent compounds are also known. Phospham and phosphoroxynitride are characterized by a highly cross-linked P-N structure, which contains varying amounts of hydrogen. In case of phosphoroxynitride, the phosphorus has a double bond to oxygen.

A comprehensive overview on phosphazenes is available since this chemistry has been around for since the 19th century [77]. Phosphazene derivatives are typically prepared via cyclic chlorophosphazene obtained through direct reaction of PCl_5 and ammonium chloride [78] (equation 5.1) in chlorobenzene. Sublimation affords mainly the trimer and tetramer $(PNCl_2)_{3/4}$ or the pure trimer, depending on the conditions. These cyclic phosphorus compounds were already described by Liebig in 1832 [79, 80] in his study of the reaction of PCl_5 and NH_3

$$PCl_5 + NH_4Cl \rightarrow 1/n\ (NPCl_2)_n + 4\ HCl \qquad (5.1)$$

The use of phosphazenes as flame retardant has been described by Allan [81]. Hexaphenoxy-phosphazene finds commercial use in epoxy-resins in printed circuit boards, where its solubility in methyl tert-butyl ether is advantageous. An 18% loading is needed to achieve a UL94 V0 rating [82]. Clariant has reported a synergy between hexaphenoxyphosphazene and diethyl phosphinic acid aluminum salt [83].

Levchik et al. [84] investigated the use of phosphazenes, phospham and phosphoroxynitride in polybutyleneterephthalate (PBT). In Figure 5.18, a thermogravimetric analysis of different phosphazenes in PBT is shown in comparison with pure PBT. All phosphazene derivatives lead to a destabilization of PBT as seen by the lowering of the onset temperature of up to 50°C. The LOI of PBT was increased from 22 to about 26 with a loading of about 20% phosphazene. A 20% loading of any of the phosphazenes investigated led to an UL94 V2 classification. Tris-(phenylene-1-amino-2-oxy)-tricyclophosphazene ($[PN(O)(NH)Ph]_3$ caused an increase in LOI increased to almost 30.

The use of phospham and phosphoroxynitride (PON) in PBT was also investigated. Figure 5.19 shows the thermogravimetric analyses of both phospham and PON, alone, and their mixtures with PBT.

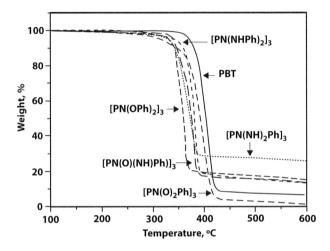

Figure 5.18 Thermogravimetric analysis of various phosphazenes.

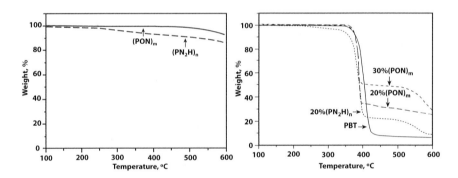

Figure 5.19 TGA of phospham, PON and mixtures with PBT.

Both Phospham and PON are very stable substances. Even at temperatures above 500°C phospham shows a mere weight loss of 10%. Phospham destabilizes the decomposition of PBT but it does not generate additional residue. PNO on the other hand, leads to an increase in residue yield, higher than that predicted from the independent decomposition of the resin and additive. For example, a loading of 30% PNO yields a residue of about 50%, whereas the calculation of both TGA curves would result in only 33% residue. Flame retardancy performances of these mixtures were nonetheless discouraging. Neither affected in any increase in UL 94 rating. In case of 20% phospham, the LOI was even reduced to 18. The authors concluded that the residue induced by PON is not protective but rather detrimental leading to an increase in combustible product emission.

The use of phospham in nylons has been described [85]. It was found that addition of phospham to PA46 increases LOI from 21 to 28 [86]. Addition of Novolak was beneficial for flame retardancy. More details on this class of P-N compounds can be found in Chapter 2 of this book.

5.7 Cyanuric-Acid Based Flame Retardants

Another class of nitrogen-based flame retardants has been developed based on cyanuric acid. It is known that cyanuric chloride reacts stepwise with alcohols and amines to form the corresponding amides or esters. Reaction with one mole of morpholine leads to a bifunctional cyanuric chloride derivative which can be polymerized with bifunctional amines or esters. MCA Technologies produces a polymer with morpholine and piperazine with the following formula [87] marketed under the trade name ppm-Triazine HF™ (Figure 5.20). These products are characterized by a molar mass of about 2800 D (Dalton) and a nitrogen content of about 31%.

Some work has been done on interactions and synergies with other flame retardants. In general, these triazine polymers induce charring and, as such, support other flame retardants such as APP or phosphinates.

Interesting results were obtained with a mixture of ppm-Triazine HF™ and APP in a 1:3 ratio in polypropylene, marketed as an intumescent system. They found an increase in LOI from 19 for pure polypropylene to up to 31 for a 20% loading and thus achieved an UL94 V0 rating. Cone calorimeter data showed a drastic reduction in heat release rate by a factor of 5 with a 10% loading and by a factor of 10 with a 20% loading. The total smoke emission is also significantly reduced in case of a 20% loading. The 10% loading does not show such a big improvement in smoke emission and this is mirrored in the burning behavior (Figure 5.21).

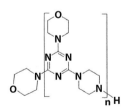

Figure 5.20 Polymer of piperazine and cyanuric acid as described by MCA Technologies.

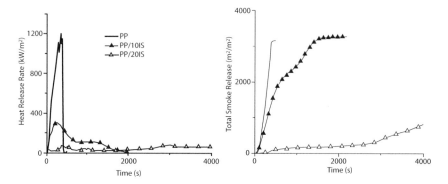

Figure 5.21 Heat release rate and smoke release of polypropylene with 10% and 20% loading of an intumescent APP ppm-Triazine HF™ (3:1) mixture (IS) (Data courtesy of MCA Technologies).

5.8   Summary and Conclusion

Nitrogen based flame retardants are a special class of substances with a broad application spectrum. Most of them decompose endothermically into inert volatiles under combustion conditions suppressing the propagation of a fire by intumescence and heat removal. These compounds are often accompanied with phosphorus acid generators in order to intensify charring. Here nitrogen based flame retardants are rather passive supporters of the intumescent mechanism by developing the foaming gases and additionally working as a heat sink due to endothermic decomposition. Additionally, nitrogen based compounds can either assist in the polymer dripping away from the heat source (NOR or melamine cyanurate) or assist in char formation (triazine) depending upon the chemistry of the polymer these additives are incorporated into. Otherwise, the more common approach is to consider combinations of nitrogen and phosphorus to yield the desired flame retardant effect in a polymer. Nitrogen-based flame retardants are a useful non-halogenated chemistry to consider, and they likely will continue to be used where their use is appropriate and they provide a meaningful fire safety benefit. As discussed in the first chapter of this book, all flame retardant chemistries are under regulatory scrutiny, including nitrogen-based flame retardants. When it is possible for the flame retardant to react into the polymer, this approach should be utilized. Other than melamine in polyurethane foam, most nitrogen-based flame retardants are non-reactive additives, which suggests their use may be limited in some areas (crystalline thermoplastics), but still very allowable in others (thermoset polymers). However, the regulations are in a state of flux and this class of flame retardant, which to date has had some favorable persistence, bioaccumulation, and toxicity profiles (such as select ammonium salts) may free itself from scrutiny and would end up with more use in the future.

References

1. Townsend Solutions Estimate.
2. Hull, T.R., Witkowski, A., Hollingbery, L., Fire retardant action of mineral fillers. *Polym. Degrad. Stab.*, 96, 1462–1469, 2011.
3. Please also see Chapter 3 of this book.
4. https://www.researchandmarkets.com/reports/5003998/flame-retardant-chemicals-a-global-market?utm_source=dynamic&utm_medium=CI&utm_code=zfnf83&utm_campaign=1362654+-+Global+Overview+for+Flame+Retardant+Chemicals+(2019+to+2026)+-+Featuring+3M%2c+Adeka+and+Clariant+AG+Among+Others&utm_exec=jamu273cid (cited 07/20/20).
5. Weil, E.D., Patel, N., Huang, C.H., Zhu, W., Phosphorus-nitrogen synergism, antagonism and other interactions, in: *Proceedings of the 2nd Beijing International Symposium/Exhibition on Flame Retardants*, 1993b.
6. Jordanov, I., Magovac, E., Fahami, A., Lazar, S., Kolibaba, T., Smith, R.J., Bischof, S., Grunlan, J.C., Flame retardant polyester fabric from nitrogen-rich low molecular weight additives within intumescent nanocoating. *Polym. Degrad. Stab.*, 170, 108998, 2019.
7. Lecoeur, E., Vroman, I., Bourbigot, S., Delobel, R., Optimization of monoguanidine hydrogen phosphate and aminopropylethoxysilane based flame retardant formulations for cotton. *Polym. Degrad. Stab.*, 91, 1909–1914, 2006.
8. Grześkowiak, W. Ł., Effectiveness of new wood fire retardants using a cone calorimeter. *J. Fire Sci.*, 35, 565–576, 2017.
9. Business Information BASF SE.
10. Gijsman, P., Steenbakkers, R., Furst, C., Kersjes, J., Differences in the flame retardant mechanism of melamine cyanurate in polyamide 6 and polyamide 66. *Polym. Degrad. Stab.*, 78, 219–224, 2002.
11. Diniz, A.T.S., Huth, C., Schartel, B., Dripping and decomposition under fire: Melamine cyanurate vs. glass fibres in polyamide 6. *Polym. Degrad. Stab.*, 171, 109048, 2020.
12. Business Information Otzuka Chemicals Co, Ltd.
13. Ding, J., Shi, W., Thermal degradation and flame retardancy of hexaacrylated/hexaethoxyl cyclophosphazene and their blends with epoxy acrylate. *Polym. Degrad. Stab.*, 84, 159–166, 2004.
14. Ding, J., Liang, H., Shi, W., Shen, X., Photopolymerization and Properties of UV-Curable Flame-Retardant Resins with Hexaacrylated Cyclophosphazene Compared with Its Cured Powder. *J. App. Polym. Sci.*, 97, 1776–1782, 2005.

15. Liu, R., Wang, X., Synthesis, characterization, thermal properties and flame retardancy of a novel nonflammable phosphazene-based epoxy resin. *Polym. Degrad. Stab.*, 94, 617–624, 2009.

16. Qian, L.-J., Ye, L.-J., Xu, G.-Z. *et al.*, The non-halogen flame retardant epoxy resin based on a novel compound with phosphaphenanthrene and cyclotriphosphazene double functional groups. *Polym. Degrad. Stab.*, 96, 1118–1124, 2011.

17. Qian, L., Ye, L., Qiu, Y., Qu, S., Thermal degradation behavior of the compound containing phosphaphenanthrene and phosphazene groups and its flame retardant mechanism on epoxy resin. *Polymer*, 52, 5486–5493, 2011.

18. Liu, H., Wang, X., Wu, D., Novel cyclotriphosphazene-based epoxy compound and its application in halogen-free epoxy thermosetting systems: Synthesis, curing behaviors, and flame retardancy. *Polym. Degrad. Stab.*, 103, 96–112, 2014.

19. Liu, H., Wang, X., Wu, D., Synthesis of a novel linear polyphosphazene-based epoxy resin and its application in halogen-free flame-resistant thermosetting systems. *Polym. Degrad. Stab.*, 118, 45–58, 2015.

20. Liang, W.-J., Zhao, B., Zhang, C.-Y., Jian, R.-K. *et al.*, Enhanced flame retardancy of DGEBA epoxy resin with a novel bisphenol-A bridged cyclotriphosphazene. *Polym. Degrad. Stab.*, 144, 292–303, 2017.

21. Guo, X., Wang, H., Ma, D., He, J., Lei, Z., Synthesis of a novel, multifunctional inorganic curing agent and its effect on the flame-retardant and mechanical properties of intrinsically flame retardant epoxy resin. *J. App. Polym. Sci.*, 2018.

22. Schmidt, C., Ciesielski, M., Greiner, L., Doring, M., Novel organophosphorus flame retardants and their synergistic application in novolac epoxy resin. *Polym. Degrad. Stab.*, 158, 190–201, 2018.

23. Schartel, B., Wilkie, C.A., Camino, G., Recommendations on the scientific approach to polymer flame retardancy: Part 1 – Scientific terms and methods. *J. Fire Sci.*, 34, 447–467, 2016.

24. Schartel, B., Wilkie, C.A., Camino, G., Recommendations on the scientific approach to polymer flame retardancy: Part 2 – Concepts. *J. Fire Sci.*, 35, 3–20, 2017.

25. Rochery, M., Vroman, I., Tighzert, L., Delobel, R., Poutch, F., Giraud, S., Bourbigot, S., Flame retarded polyurea with microencapsulated ammonium phosphate for textile coating. *Polym. Degrad. Stab.*, 88, 106–113, 2005.

26. Li, S., Zhong, L., Huang, S., Wang, D. *et al.*, A novel flame retardant with reactive ammonium phosphate groups and polymerizing ability for preparing durable flame retardant and stiff cotton fabric. *Polym. Degrad. Stab.*, 164, 145–156, 2019.

27. George, C.W. and Susott, R.A., Effects of Ammonium Phosphate and Sulfate on the Pyrolysis and Combustion of Cellulose, in: *Research Paper INT-90. Intermountain Forest and Range Experiment Station*, USDA Forest Service, Ogden, UT USA, April 1971.

28. Camino, G., Grassie, N., McNeill, I.C., Influence of the fire retardant, ammonium polyphosphate, on the thermal degradation of poly(methyl methacrylate). *J. Polym. Sci., Polym. Chem. Ed.*, 16, 95, 1978.

29. Camino, G., Costa, L., Trossarelli, L., Study of the mechanism of intumescence in fire retardant polymers: Part III—Effect of urea on the ammonium polyphosphate-pentaerythritol system. *Poly. Deg. Stab.*, 6, 243, 1984.

30. Camino, G., Costa, L., Trossarelli, L., Study of the mechanism of intumescence in fire retardant polymers: Part VI—Mechanism of ester formation in ammonium polyphosphate-pentaerythritol mixtures. *Poly. Deg. Stab.*, 12, 203, 1985.

31. http://www.specialchem4polymers.com/tc/ammonium-polyphosphate/?id=mechanism

32. Myers, R.E., Dickens, E.D., Jr., Licursi, E., Evans, R.E., Ammonium pentaborate: an intumescent flame retardant for thermoplastic polyurethanes. *J. Fire Sci.*, 3, 6, 432–49, 1985.

33. Levchik, G.F., Levchik, S.V., Selevich, A.F., Lesnikovich, A.I., Influence of *ammonium* pentaborate on combustion and thermal decomposition of polyamide 6, *Vestsi Akad. Navuk BSSR, Ser. Khim. Navuk*, 3, 34–9, 1995.

34. Akdogan, E., Erdem, M., Erdem Ureyen, M., Kaya, M., Synergistic effects of expandable graphite and ammonium pentaborate octahydrate on the flame-retardant, thermal insulation, and mechanical properties of rigid polyurethane foam. *Polym. Composites*, 41, 1749–1762, 2020.

35. Lewin, M., Brozek, J., Martens, M., The system polyamide/*sulfamate*/dipentaerythritol: flame retardancy and chemical reactions. *Polym. Advanced Technol.*, 13, 10–12, 1091–1102, 2002.

36. Lewin, M., Zhang, J., Pearce, E., Gilman, J., Flammability of polyamide 6 using the *sulfamate* system and organo-layered silicate. *Polym. Advanced Technol.*, 18, 9, 737–745, 2007.

37. Coquelle, M., Duquesne, S., Casetta, M. *et al.*, Investigation of the decomposition pathway of polyamide 6/ammonium sulfamate fibers. *Polym. Degrad. Stab.*, 106, 150–157, 2014.

38. Dahiya, J.B., Kandola, B.K., Sitpalan, A., Horrocks, A.R., Effects of nanoparticles on the flame retardancy of the ammonium sulphamate-dipentaerythritol flame-retardant system in polyamide 6. *Polym. Adv. Technol.*, 24, 398–406, 2013.

39. Wilkie, A. and He, J., Some new sulfur-containing materials as putative fire retardants, in: *24th Annual Conference on Recent Advances in Flame Retardancy of Polymeric Materials*, BCC Research, 2013.

40. LeVan, S.L., Chemistry of Fire Retardancy, in: *The Chemistry of Solid Wood*, R. Rowell (Ed.), Chapter 14, pp. 531–574, American Chemical Society, Washington, DC USA, 1984.

41. Costa, L. and Camino, G., Thermal Behaviour of Melamine. *J. Therm. Anal.*, 34, 423–429, 1988.

42. Thieme Chemistry (Hrsg.), RÖMPP Online - Version 3.5, Georg Thieme Verlag KG, Stuttgart, 2009.

43. Crews, G.M., Ripperger, W., Kersebohm, D.B., Güthner, T., Mertschenk, B., Melamine and Guanamines, Ullmann's Encyclopedia of Industrial Chemistry, Weinheim, Germany, 2006.

44. https://chem.nlm.nih.gov/chemidplus/rn/9003-08-1 (accessed 06/16/2021)

45. The Furniture and Furnishings (Fire) (Safety) Regulations 1988, http://www.legislation.gov.uk/uksi/1988/1324/contents/made (accessed 04/14/20).

46. Price, D., Liu, Y., Milnes, G.J., Hull, R., Kandola, B.K., Horrocks, A.R., An investigation into the mechanism of flame retardancy and smoke suppression by melamine in flexible polyurethane foam. *Fire Mater.*, 26, 201–206, 2002.

47. Kramer, R.H., Zammarano, M., Linteris, G.T., Gedde, U.W., Gilman, J.W., Heat release and structural collapse of flexible polyurethane foam. *Polym. Degrad. Stab.*, 95, 1115–1122, 2010.

48. Pitts, W.M., Role of two stage pyrolysis in fire growth on flexible polyurethane foam slabs. *Fire Mater*, 38, 232–238, 2014.

49. Levchik, S.V., Balabanovich, A.I., Levchik, G.F., Costa, L., Effect of melamine and its salts on combustion and thermal decomposition of polyamide 6. *Fire Mater.*, 21, 2, 75–83, 1997.

50. Kersies, J., Jr., Furst, Ch., Interflam '99, Flame retardant mechanistic aspects of melamine cyanurate in polyamide 6 and 66. *Proceedings of the International Conference, 8th*, Edinburgh, United Kingdom, June 29-July 1, 1999, vol. 2, pp. 1211–1216, 1999.

51. Levchik, S.V., Balabanovich, A.I., Levchik, G.F., Costa, L., Mechanistic study of fire- retardant action of melamine and its salts in nylon 6, in: *Recent Advances in Flame Retardancy of Polymeric Materials*, vol. 7, pp. 64–76, 1997, Volume Date 1996.

52. Casu, A., Camino, G., De Giorgi, M., Flath, D., Morone, V., Zenoni, R., Fire-retardant mechanistic aspects of melamine cyanurate in polyamide copolymer. *Polym. Degrad. Stab.*, 58, 3, 297–302, 1997.

53. Klatt M., Fire Retardancy of Thermoplastics, in: *Fire Resistance in Plastics conference*, p. 18, 2012.

54. Klatt, M., Non-halogenated solutions for thermoplastics and thermosets for E&E Industry, in: *PINFA conference*, Green Electronics, 2013.

55. Self-extinguishing polyurethanes retaining their mechanical properties - contain an org. phosphate and/or phosphate as flame retardant, opt. together with a melamine derivative, WP1994295744, BASF patent application (expired).

56. Tabuani, D., Bellucci, F., Terenzi, A., Camino, G., Flame retarded Thermoplastic Polyurethane (TPU) for cable jacketing application. *Polym. Degrad. Stab.*, 97, 12, 2594–2601, 2012.

57. Hu, Y., Wang, S., Ling, Z., Zhuang, Y., Chen, Z., Fan, W., Preparation and Combustion Properties of Flame Retardant Nylon-6/Montmorillonite Nanocomposite. *Macromol. Mater. Eng.*, 288, 272–276, 2003.

58. Braun, U. and Schartel, B., Flame retardancy mechanisms of aluminum phosphinate in combination with melamine cyanurate in glass-

fibre-reinforced poly(1,4-butylene terephthalate). *Macromol. Mater. Eng.*, 293, 3, 206–217, 2008.

59. Halogen-free flame retardant polyester composition, WO2013045965 A. Filed April 4, 2013, https://patentscope.wipo.int/search/en/detail.jsf?docId=WO2013045965 (accessed 06/16/21).

60. Herbst H., Fire Resistance in Plastics Conference, Applied Market Information (AMI), Ltd., Cologne, Germany, Nov 29-Dec 1, 2011.

61. Yang, W., Song, L., Hu, Y., Lu, H., Yuen, R.K.K., Investigations of thermal degradation behavior and fire performance of halogen-free flame retardant poly(1,4-butylene terephthalate) composites. *J. Appl. Polym. Sci.*, 122, 1480–1488, 2011.

62. Braun, U., Schartel, B., Fichera, M., Jaeger, Chr., Flame retardancy mechanisms of aluminum phosphinate in combination with melamine polyphosphate and zinc borate in glass-fibre reinforced polyamide 6,6. *Polym. Degrad. Stab.*, 92, 8, 1528–1545, 2007.

63. Schwarzer, A., Saplinova, T., Kroke, E., Tris-s-triazines (s-heptazines) – From a mystery molecule to industrially relevant carbon nitride materials. *Coord. Chem. Rev.*, 257, 13–14, 2032–2062, 2013.

64. Gavrilova, N.K., Gal'perin, V.A., Finkel'shtein, A.I., Koryakin, A.G., Synthesis of melam from melamine. *Zh. Org. Khim.*, 13, 3, 669–70, 1977.

65. Melamine condensates prodn. for flame-retardant mouldings - by heating melamine (salt) in presence of organic acid or ammonia or melamine salt of organic acid under suitable reaction conditions, WO9616948., Patent application (inactive) https://patents.google.com/patent/WO1996016948A1/zh (accessed 06/16/21).

66. Melam-based flame retardant polyamide composition, WO9617013 A, Patent application (inactive) https://pubchem.ncbi.nlm.nih.gov/patent/WO-9617013-A1 (accessed 06/16/21).

67. Methanesulfonic acid melam as a flame-retardant for polyamide resins, comprises reacting melamine with methanesulfonic acid and baking product at a specific temperature, WP2002121816, BASF internal patent application, expired.

68. Resin composition comprising polyamide resin, EP0994156 A, https://patents.google.com/patent/EP0994156A1/en (accessed 06/16/21).

69. Kaprinidis, N., Earhart, N., Zingg, J., Overview of recent advances in flame retardant compositions UV stable flame retardant systems; fully formulated antimony free flame retardant products for polyolefins, in: *Proceedings of International Wire and Cable Symposium*, 51st, pp. 594–596, 2002.

70. Aubert, M., Nicolas, R., Pawelec, W., Wilen, C.-E., Roth, M., Pfaendner, R., Azoalkanes-novel flame retardants and their structure-property relationship. *Polym. Adv. Technol.*, 22, 11, 1529–1538, 2011.

71. Arnaboldi, P., Davis, L., Samuels, S.-B., Zenner, J. M., Vulic, I. A Revolutionary Light Stabilizer System for Polyolefins and Other Resins. *Polym. Polym. Compos.*, 10, 93–99, 2002.

72. Hodgson, J.L. and Coote, M.L., Clarifying the Mechanism of the Denisov Cycle: How do Hindered Amine Light Stabilizers Protect Polymer Coatings from Photo-oxidative Degradation. *Macromolecules*, 43, 10, 4573–4583, 2010.

73. Flame-retardant composition comprising a phosphonic acid derivative, WO2010026230 A, https://patentscope.wipo.int/search/en/detail.jsf?docId=WO2010026230 (accessed 06/16/21).

74. Aubert, M., Nicolas, R.C., Pawelec, W., Wilen, C.-E., Roth, M., Pfaendner, R., Azoalkanes – novel flame retardants and their structure-property relationship. *Polym. Adv. Technol.*, 22, 1529–1538, 2011.

75. Tirri, T., Aubert, M., Wilen, C.-E., Pfaendner, R., Hoppe, H., Novel tetrapotassium azo diphosphate (INAZO) as flame retardant for polyurethane adhesives. *Polym. Degrad. Stab.*, 97, 375–382, 2012.

76. Pawelec, W., Aubert, M., Pfaendner, R., Hoppe, H., Wilen, C.-E., Triazene compounds as a novel and effective class of flame retardants for polypropylene. *Polym. Degrad. Stab.*, 97, 948–954, 2012.

77. M. Gleria and R. De Jaeger (Eds.), *Phosphazenes: A Worldwide Insight*, Nova Science Publishers Inc, Hauppauge, NY USA, 2004.

78. Holleman, A.F. and Wiberg, E., *Inorganic Chemistry*, Academic Press, Cambridge, MA USA, Nov 5, 2001.

79. Stokes H.N., On the chloronitrides of phosphorus. *Am. Chem. J.*, 17, 275, 1895.

80. Liebig-Wöhler, *Briefwechsel*, vol. 1, p. 63, *Ann. Chem. (Liebig)*, 11, 146, 1834.

81. Allen, Chr. W., The use of phosphazenes as fire resistant materials. *J. Fire Sci.*, 11, 4, 320–328, July 1993.

82. Business brochure, Otzuka Chemical Co., Ltd, Japan.

83. Rakotomala M., Flame retardant polyesters for injection moulding- improving material properties, in: *AMI conference in Cologne; Fire Resistance in Plastics*, 2013.

84. Levchik, G.F., Grigoriev, Y.V., Balabanovich, A.I., Levchik, S.V., Klatt, M., Phosphorus–nitrogen containing fire retardants for poly(butylene terephthalate). *Polym. Int.*, 49, 1095–1100, 2000.

85. Weil, E.D. and Patel, N.G., Phospham – A stable phosphorus-rich Flame Retardant. *J. Fire Sci.*, 18, 1–7, 1994.

86. Flame retardant thermoplastic containing phospham, EP 0417839. https://patents.google.com/patent/EP0417839A1/en (accessed 06/16/21).

87. MCA Technologies – Commercial Publications.

6

Silicon-Based Flame Retardants

Alexander B. Morgan[1][*] and Mert Kilinc[2][†]

[1]University of Dayton Research Institute, Dayton, OH
[2]Philips, Groningen, The Netherlands

Abstract

Silicon based materials, in particular oxides, tend to be inherently flame retardant materials due to the fact that they can form ceramics/glasses when they are exposed to heat and flames. This has led to the use of polymeric materials like silicones to be used as flame retardant materials, and, has led to use of silicon-based materials to be added to polymers to impart fire protection properties. This chapter will cover the chemistry of these materials, how they impart fire protection/flame retardancy, and how they get used today. Current research trends in this class of materials organized by chemical type will also be discussed.

Keywords: Silicone, silica, silicate, flame retardants, glasses, ceramics, flame retardancy mechanism

6.1 Introduction

Silicon is a tetravalent metalloid and the 14th element in the periodic table. Although it does not occur naturally in free form, in its combined form (typically as an oxide) it accounts for about 25% of the earth's crust. Silicon compounds are unique materials both in terms of the chemistry and in their wide range of useful applications. Silicon can be combined with organic compounds to provide unique properties that function over a wide temperature range, making the silicon based products less temperature sensitive than most other carbon-based materials. These properties can be attributed to the strength

---

*Corresponding author: alexander.morgan@udri.udayton.edu
†Original Chapter Author for 1st Edition. This chapter has been updated with new content by Alexander Morgan

Alexander B. Morgan (ed.) Non-Halogenated Flame Retardant Handbook 2nd edition, (271–308)
© 2022 Scrivener Publishing LLC

and flexibility of the Si-O bond (bond energy of Si-O is 452 kJ/mol and bond length is 1.63 Å [1]), its partial ionic character and the low interactive forces between the non-polar methyl groups, characteristics that are directly related to the comparatively long Si-O and Si-C bonds (bond energy of Si-C is 360 kJ/mol) [1]. The length of the Si-O (1.63 Å)[1] and Si-C (1.88 Å)[1] bonds also allows an unusual freedom of rotation, which enables the molecules containing this chemistry to adopt the lowest energy configuration at interfaces, providing a surface tension that is substantially lower than that of organic polymers.

One of the most basic technical errors made by people referring to materials is confusing silicon with silicone. The former, silicon, refers to the elemental material, (Si); the latter refers to materials in which silicon is bonded to oxygen. Silicon is the elemental refined material from which all silicone chemistry finds its roots. Since it is not at all common in the elemental form in nature, the first step in the chemistry is to produce silicon from quartz.

Silicon is obtained by the thermal reduction of quartz (SiO_2) with carbon (C). The reaction is conducted at very high temperatures and therefore is commonly carried out where there is abundant inexpensive power, like near hydrothermal power plants. The reaction is as follows:

$$SiO_2 + C \xrightarrow{1700°C} Si + CO_2 \qquad (6.1)$$

By analogy with ketones, the name silicone was given in 1901 by Kipping to describe new compounds of the generic formula R_2SiO. The name silicone was adopted by the industry and most of the time refers to polymers where R=methyl. The methyl groups along the chain can be substituted by many other groups, e.g., phenyl, vinyl, trifluoropropyl, etc. The simultaneous presence of "organic" groups attached to an "inorganic" backbone gives silicones a combination of unique properties and allows their use in fields as different as aerospace (low and high temperature flexibility), electronics (high electrical resistance), medical (excellent biocompatibility) and construction (resistance to weathering).

6.2   Basics of Silicon Chemistry

Silicon is in the same family of elements as carbon in the periodic table, group 4A. Group 4A elements have four valence electrons in their highest-energy orbitals (ns^2np^2). In their most stable state, silicon and carbon will both covalently bond to four other atoms; silicon based chemicals exhibit significant physical and chemical differences compared to analogous

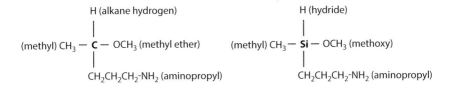

Figure 6.1 Carbon vs. silicon chemistry.

carbon based chemicals. Silicon is more electropositive than carbon, does not form stable double bonds, and is capable of very special and useful chemical reactions. Silicon based chemicals include several types of mono-meric and polymeric materials.

Monomeric silicon chemicals are known as silanes. A silane structure and an analogous carbon-based structure are shown in Figure 6.1.

The four substituents have been chosen to demonstrate differences and similarities between silicon and carbon based chemicals. A silane that contains at least one carbon-silicon bond (CH_3-Si-) structure is known as an organosilane. The carbon-silicon bond is very stable, very non-polar and gives rise to low surface energy, non-polar, hydrophobic effects. Similar effects can be obtained from carbon based compounds, although these effects are often enhanced with silanes. The silicon hydride (-Si-H) structure is very reactive. It reacts with water to yield reactive silanol (-Si-OH) species and, additionally, will add across carbon-carbon double bonds to form new carbon-silicon-based materials. The methoxy group on the carbon compound gives stable methyl ether, while its attachment to silicon gives a very reactive and hydrolysable methoxysilyl struc-ture. The organofunctional group, the aminopropyl substituent, will act chemically the same in the organosilicon compound as it does in the carbon-based compound. The distance of the amine, or other organo-functional group, from silicon will determine whether the silicon atom affects the chemistry of the organofunctional group. If the organic spacer group is a propyl linkage (e.g., $-CH_2CH_2CH_2-$), then the organic reactiv-ity in the organofunctional silane will be similar to the organic analogs in carbon chemistry. Certain reactive silanes, particularly vinyl silanes (-Si-CH=CH$_2$) and silicon hydrides (-Si-H), are useful reactive groups in silicon chemistry, even though the reactive group is attached directly to the silicon atom.

Attachment of chlorine, nitrogen, methoxy, ethoxy or acetoxy directly to silicon yields chlorosilanes, silyamines (silazenes), alkoxysilanes and acyloxysilanes, respectively; these are very reactive and exhibit unique inorganic reactivity. Such molecules will react readily with water, even moisture absorbed on a surface, to form silanols. These silanols can then react with other silanols to form a siloxane bond (-Si-O-Si-), a very stable structure; or in the presence of metal hydroxyl groups on the surface of glass, minerals or metals, silanols will from very stable –Si-O-metal bonds to the surface.

Chloro-, alkoxy-, and acetoxy- silanes and silazenes (-Si-NH-Si-) will react readily with active hydrogen on any organic chemical (e.g., alcohol, carboxylic acid, amine, phenol or thiol) via a process called silylation [2].

$$R_3SiCl + R^IOH \rightarrow R_3Si\text{-}OR^I + HCl \qquad (6.2)$$

Silylation is very useful in organic synthesis to protect functional groups while other chemical manipulations are being performed.

6.3   Industrial Applications of Silicones

Silicones in industry usually refer to linear polydimethylsiloxanes (PDMS). A combination of properties such as their backbone flexibility, low surface tension, low intermolecular interactions, and thermal stability explain many of their applications. But the name silicone also is used for more complex structures, where some of the methyl groups have been replaced by other functional groups, from branched polymers to resinous materials and even cross-linked elastomers. This allows for modifying some of the silicones properties to specific needs. Silicones are widely used in food industry, paper industry, textile industry, household cleaning applications, coatings, construction industry, electronics, plastics industry, beauty care and medical applications.

In food-related processes, silicones are very much associated with foam control agents because of the low surface tension displayed by polydimethylsiloxanes; this is a key property for formulating effective antifoam [3]. Foam control is critical here as in many other industries, as excessive foaming slows processes and can reduce volume efficiency.

Organosiloxane materials can be found throughout the processing of pulp and paper, from the digestion of wood chips to the finishing and recycling of papers. Some examples are: As digester additives, silicones improve the impregnation of active alkali in the wood chips and improve the cooking. As antifoams, silicones help de-airing or drainage in the pulp washing and papermaking processes. As additives, silicones contribute in the finishing process of paper and tissues. In the recycling of papers, silicones act as de-inking aids.

In the textile industries, silicones are used in all stages of the process, on the fiber during production, on the fabric and/or directly on the finished goods. Silicones are applied from different delivery systems to provide various benefits like lubrication, softening, foam control or hydrophobic coatings.

Silicones and household cleaning applications have been associated for more than 50 years, particularly in the laundry area, where the main use is foam control in consumer washing machines and fabric softening. But silicones are also used to provide extra benefits such as fabric dewatering, anti-wrinkle characteristics, ease of ironing or improved water absorbency. Silicone additives have also been developed to reduce fabric mechanical losses over time or to improve perfume release.

Silicones are widely used in the coating industries as materials to protect and preserve but also to bring style to a wide variety of applications in our daily lives. The unique combination of properties of silicones is well suited to coating applications. Two families of products are used: silicone polymers as additives and silicone resins as the main component, or binder. At low levels, silicone polymers are used to ease application of paints. The surface properties of silicones enable a paint to wet a substrate easily and give it a smooth appearance once dry. In contrast to the low-level use of silicone polymers as additives, silicone resins can be major components of the coating. Here they are used as binders or co-binders, imparting important benefits such as durability throughout the life of the coating.

Silicone sealants and adhesives as used in the construction industry were introduced approximately forty years ago, and many of the silicones applied in the early days are still performing today. Products are available in a variety of forms, from paste-like materials to flowable adhesives. Both single- and multi-component versions are available, each with several different cure chemistries. These materials have properties which include excellent weather and thermal stability, ozone and oxidation resistance, extreme low temperature flexibility, high gas permeability, good electrical properties, physiological inertness and curability by a variety of methods at both elevated and ambient temperatures. Because of their low surface energy, they wet most substrates, even under difficult conditions, and when formulated with suitable adhesion promoters, they exhibit very good adhesion.

Before 1943, planes could maintain high altitudes for only a few minutes before ignition losses due to moisture condensing in the engines. Simple thickened PDMS grease was the solution and an early example of the excellent dielectric properties of silicones. This application also illustrates key properties of silicones in the electronic industries like hydrophobicity and high dielectric breakdown (keeping moisture away and avoiding loss of high voltage/low current signals), as well as their resistance to low or high temperatures, which allow use in harsh and critical environments [4]. Today, despite a higher cost to acquire, the number of applications involving silicones continues to increase, in some instances driven by Moore's law (chip complexity doubling approximately every two years), but also by tighter specifications.

Silicones are used in the plastics industry as additives for improving the processing and surface properties of plastics, as well as the rubber phase in a novel family of thermoplastic vulcanizate (TPV) materials. As additives, silicones, and in particular polydimethylsiloxane (PDMS), are used to improve mold filling, surface appearance, mold release, surface lubricity and wear resistance. As the rubber portion of a TPV, the cross-linked silicone rubber imparts novel properties, such as lower hardness, reduced coefficient of friction and improved low and high temperature properties. Low molecular weight PDMS polymers, with viscosities less than 1000 cSt, are used extensively by the plastics industry as external release agents applied on the mold surface prior to injection molding [5]. The applications of silicones in the plastics industry continue to grow as more benefits are identified by combining the unique properties of thermoplastics and silicone.

Silicones used in personal care applications are of diverse types, including cyclic, linear, or organo-functional polydimethylsiloxanes (PDMS), as well as silicone elastomer dispersions and resins. This wide range of molecules provides benefits that impact the performance of almost every type of beauty product, conferring attributes such as good spreading, film forming, wash-off resistance, skin feel, volatility and permeability.

Silicone materials celebrate 60 years of use in medical applications. Quickly after their commercial availability in 1946, methylchlorosilanes were described to treat glassware to prevent blood from clotting [6]. At the same time, Dr. F. Lahey implanted a silicone elastomer tube for duct repair in biliary surgery [7]. Since these pioneers, the interest for silicones in medical applications has remained because of their recognized biocompatibility. Silicones are used today in many life-saving medical devices like pacemakers or hydrocephalic shunts [8]. Silicones are also used in many pharmaceutical applications from process aids like tubing used to manufacture pharmaceuticals, to excipients in topical formulations or adhesives to affix transdermal drug delivery systems [9]. They also have found use as active pharmaceutical ingredients in products such as antacid and antiflatulent formulations [10, 11].

6.4   Silicon-Based Materials as Flame Retardant Materials

Synthetic polymers are a crucial part of today's modern world; they can be found nearly everywhere. Today, synthetic polymers materials are rapidly replacing more traditional materials such as metals, ceramics, and natural polymers such as wood, cotton, natural rubber etc. However, one weak aspect of synthetic polymer materials compared with other materials is that polymers are combustible. Thus, the majority of polymer-containing end-products (e.g., cables, fuses, carpets, furniture cabinets, vehicle interiors, etc.) must have a satisfactory degree of fire resistance to ensure public safety from fire. This usually results in flame retardants being added to the synthetic polymer. An ideal flame retardant should be easy to incorporate into the polymer matrix, and be compatible with plastics (i.e. not bleed out), and not alter its mechanical properties. Furthermore, it should exhibit good light stability, and be resistant to aging and hydrolysis. It should be matched to the decomposition temperature of the polymer, i.e. its effect must begin below the decomposition temperature of the plastic and continue over the whole range of its decomposition. It must not cause corrosion, must be temperature resistant, effective in small amounts, odorless and without harmful physiological effects. It must also emit only low levels of smoke and toxic gases and finally be as cheap as possible. No current flame retardant additive is ideal and fulfills all the criteria of the above specifications. However these criteria have led researchers to countless flame retardant formulations. This leads to why silicon-based flame retardants, and silicon-based materials, have been used in flame retardant formulations.

There is a renewed interest in using silicon-based flame retardants as substitutes for the halogens or phosphorus. Almost all forms of silicon have been explored: silicones, silicas, organosilanes, silsesquioxanes, and silicates. Among the silicon based materials, silicones have excellent flame retardant properties. First, they are thermally stable and resistant to high temperatures, and, when exposed to combustion conditions, they end up forming non-flammable residues. Specifically, silicon-based materials and flame retardants are already in their highest oxidation state, or close to it (SiO_2) such that they cannot combust further, and can form inorganic glassy/ceramic residues which further resist flame and heat. Finally silicone pendant end groups can easily be changed to fit the polymer in which they are incorporated. Silicone materials have been produced commercially since the beginning of the 1940s. Over the past 60 years, silicone materials have grown into a billion dollar industry, and are used in many applications in civil engineering, construction building, electrical, transportation, aerospace, defense, textiles, and cosmetic industries [12]. The dominant polymer in the silicone industry is polydimethylsiloxane (PDMS). Most of the studies based on developing silicone based flame retardants were initiated by working on PDMS and its derivatives to check its usage as flame retardant or developing block copolymers of PDMS [13–19]. PDMS is used because it can form SiO_2 during burning, and therefore sets up combustion resistant residues when exposed to fire [20].

There is a continuous growing interest in halogen free flame retardants due to the active or upcoming regulations in EU and US, which require tighter tests to be passed, but also require use of greener chemistry (see Chapter 1 of this book). For this purpose, researchers from academia and industry are making efforts to use various elements and materials made out of these elements to produce new environmentally friendly and efficient flame retardants from phosphorus (P), nitrogen (N), boron (B). Silicon is also a promising material to be used in future flame retardant formulations, but when compared to other mentioned elements, less work and research has been performed on silicon as a flame retardant additive. Regardless, there is a continuous interest on developing silicon based additives, and use of silicones (such as PDMS) as the main polymer for the fire safety application. Use of silicon-based additives and silicon based polymers have been growing in recent years, and in the following sections, current use and recent developments will be discussed.

6.4.1   Inorganic Silicon-Based Flame Retardants

6.4.1.1   Silicon Dioxide (SiO_2) (Silica)

Silicon dioxide, also known as silica, is an oxide of silicon with the chemical formula SiO_2. It has been known since ancient times. Silica is most commonly found in nature as sand or quartz, as well as in the cell walls of diatoms. Silica gel, fumed and fused silica have been tested and used in different polymers as flame retardants. Depending upon the specific form of silica, and how it interacts with the polymer matrix during burning, the silica can either migrate through the polymer and form oxide rich protective surfaces or it can increase polymer viscosity creating chars. Regardless of how the silica interacts with the polymer, the net effect is a condensed phase flame retardant mechanism wherein the mass loss rate (fuel pyrolysis rate) of the polymer is delayed, which in turn slows the rate of burning. In some cases, this phenomena can also have some other physical effects on material burning, including acting as an anti-drip additive in select polymers.

Silica has been extensively studied to date and is a common co-additive in formulations to help with enhancement of inorganic protective residues. In one example, it was observed that time to ignition (TI) and the fire performance index (FPI) increase with the incorporation of SiO_2 into the ethylene vinyl acetate (EVA)/$Mg(OH)_2$ formulations. Besides, SiO_2 significantly decreased the peak rate of heat release (pHRR) during burning and the composites achieve a UL-94 V-0 rating [21]. SiO_2 exhibits a good synergy and could partially replace $Mg(OH)_2$ in halogen free flame retardant EVA formulations especially for cable applications. Silica has also been combined with phosphorus compounds [22, 23], with good effect.

The effect of silica gel structure on the flammability properties of polypropylene (PP) has been investigated by Gilman *et al.* [24]. Three different types of silica gels with different pore volume, particle size and surface silanol concentration were used. Cone calorimeter tests revealed the dramatic effect of silica gel pore volume on the heat release rate (HRR) of PP containing 10 wt.% silica. It was observed that use of high pore volume silica led to a significant reduction in the HRR. There was no noticeable effect of particle size on the flammability properties but a noticeable effect of the surface silanol concentration was observed. The authors explain the reduction of the flammability at higher silica gel pore volume by the possibility offered by larger pores to accommodate macromolecular PP chains or by the increase in molten viscosity during pyrolysis, which can trap or slow volatilization and the evolution of degradation products during fire.

Performances of various types of silica, silica gel, fumed silica and fused silica as flame retardants in non-char-forming thermoplastics (e.g., polypropylene) and polar char forming thermoplastics (e.g., polyethylene oxide) were investigated by Kashiwagi *et al.* [25]. It was concluded that the incorporation of low density, large surface area silica, such as fumed silica (140 and 255 m^2/g) and silica gel (400 m^2/g) in polypropylene and polyethylene oxide significantly reduced the heat release rate and mass loss rate. In the meantime, the addition of fused silica with lower surface area did not reduce the flammability properties as much as the other silica samples.

A similar study was performed using PMMA as the matrix polymer [26]. In this work, two types of silica (fused silica and silica gel) were incorporated in two different molecular weight PMMA samples. Results concluded that viscosity control is a key factor in the formation of the efficient protective char layer.

Silica showed synergy in terms of flame retardancy when combined with magnesium hydroxide (MDH), zinc borate and multi-wall carbon nano-tubes (MWNT) in polyolefins [27].

6.4.1.2   Wollastonite

Wollastonite, is a naturally occurring mineral and is also known as calcium metasilicate. It has a shape of pure white, non-hydrous needle shaped crystals. The particle lengths are typically larger than the widths by a factor of between one and two, but the aspect ratio (diameter divided by thickness) can be much higher, up to 15. Consequently, wollastonite, due to its reinforcing property can compete with or partially replace other reinforcing fillers and fibers as its cost is competitive [28]. Wollastonite is used in composite materials to increase mechanical properties such as tensile, flexural, and impact strength, as well as to increase dimensional stability and minimize distortion at elevated temperatures. Application of wollastonite as flame retardant in PDMS has also been patented by several authors. Nicholson *et al.* [29] from Dow Corning incorporated 21.1 wt.% of wollastonite into 66.4 wt.% of dimethylvinylsiloxy terminated dimethyl siloxane to obtain a cured silicone foam exhibiting high flame resistance, forming hard ceramized char with few cracks on burning. Moreover, for wire and cable coating applications, Shephard [30] from Dow Corning proposed a curable silicone composition, made by mixing ingredients comprised of: 30-90 wt.% of a heat-curable non halogenated organosiloxane polymer, containing at least 2 alkenyl groups per molecule, 1-65 wt.% of a reinforcing silica filler, 5-70 wt.% of wollastonite having an average particle size of 2-30 μm and aspect ratio of at least 3:1, and curing component sufficient to cure the composition (a peroxide catalyst). These formulations exhibited hard and strong char formation and also reduced heat release rate and reduced time to ignition values. High consistency rubbers were systematically formulated for plenum cable coatings with different sizes of wollastonite. A sample containing wollastonite with an average particle size of 12 μm and a particle size range of 1-393 μm was compared with a sample containing wollastonite with an average particle size of 10 μm, and a particle size range of 1-119 μm. Both samples showed very similar HRR peaks without significant difference in the char structures of the two materials [26]. Compositions with less than about 5 wt.% of wollastonite did not exhibit strong char formation and low heat release rate. George *et al.* [31] used modified wollastonite (with functional alkoxysilanes) to improve flame retardancy of PDMS by adding 3.5 wt.% of wollastonite to 65 wt.% of dimethylvinylsiloxy terminated dimethyl siloxane. This formulation resulted in a ceramified char and reduced peak heat release rate with good cohesivity that is required for cable application. As a commercial example Nyco claimed that their advanced wollastonite products trademarked Nyglos® have successfully replaced milled glass fiber used as one of the primary reinforcements in polyamide 6/66 and other engineering alloys to meet the requirements for UL-94 V0 flame retardant properties. There have been other studies as well where wollastonite was part of a broader formulation to impart flame retardant enhancements with other flame retardant additives, ranging from mineral fillers to other inorganic oxides/carbonates [32–36].

6.4.1.3   Magadiite

Magadiite is a white, hydrous sodium silicate mineral ($NaSi_7O_{13}(OH)_3$· $4(H_2O)$) which precipitates from alkali brines as an evaporite phase. Magadiite silicates usually have excess negative charge, which is balanced by the exchangeable cations in the gallery space. It is similar to montmorillonite (MMT); the major difference is that montmorillonite is an aluminosilicate, while magadiite contains only silicate. Studies showed that organically modified magadiite was reported as effective as organically treated MMT (o-MMT) in EVA wire and cable applications [37–39]. Unlike o-MMT, o-Magadiite gives excellent improvement in mechanical properties but no improvement in HHR in PS [40]. Magadiite was also studied in epoxy and polyurea systems by itself (with no other flame retardants added), with some reduction in heat release noted in polyurea systems, but no effect in epoxy [41]. As with many layered silicate materials, uniform nanoscale and macroscale dispersion of Magadiite is critical to it having flame retardant effects. To date, Magadiite has remained a laboratory studied material, with no commercial products yet known. It is unclear the commercial status of this layered silicate beyond lab use, and publications using his material have mostly ceased since 2015.

6.4.1.4   Sepiolite

Sepiolite has crystalline structure incorporating channels like zeolite and is a clay mineral, a complex magnesium silicate. Typical formula for sepiolite is $Mg_4Si_6O_{15}(OH)_2·6H_2O$. Absorbed water is bonded by hydrogen bonds at the external surface or within the channels, called zeolitic water and crystal water, respectively. It can be present in fibrous, fine-particulate, and solid forms. The size of the fibers varies widely, but in most cases they are 10-5,000 nm long, 10-30 nm wide, and 5-10 nm thick. It was shown that modified sepiolite is as effective as nanoclay in EVA in terms of cone calorimeter tests. Unlike organically modified MMT, the char of a sepiolite containing EVA formulation shows much less cracking [42]. A flame retardant system composed of sepiolite and IFR system (ammonium polyphosphate modified with aminosilane coupling agents and combined with melamine and dipentaerithritol) was processed by melt compounding to flame retard PP resin [43]. Sepiolite has been grafted with flame retardant molecules (such as phosphorus groups) to also impart some additional flame retardant effects, with mixed results [44]. Sepiolite has also been studied with other flame retardants (such as ammonium polyphosphate or aluminum alkylphosphinate) and as mixed layered silicate/mixed nanoparticle systems [45–49]. As with many nanoparticle silicates, sepiolite has yet to be commercialized for actual flame retardant use, but it does show potential to be used as part of an overall flame retardant system.

6.4.1.5   Kaolin

Kaolin is a dioctahedral 1:1 layered clay mineral and its structural formula is $Al_2Si_2O_5(OH)$. It is also called as china clay or porcelain earth and each layer consists of a tetrahedral sheet in which silicon atoms are tetrahedrally coordinated by oxygen atoms; and an octahedral sheet where aluminum atoms are octahedrally coordinated to hydroxyl groups and share apical oxygen from the silica tetrahedral sheet. Such typical structure of clay in kaolin crystal indicates that kaolin could have similar flame retardancy to that of montmorillonite. There are two varieties of kaolin, i.e. the naturally occurring (hydrous form) and calcined kaolin (anhydrous form). Calcined kaolin is obtained by heating the clay above 600°C; therefore this variety is harder than the hydrous material. Addition of calcined kaolin (28 wt.%) into PDMS improves its fire retardancy properties and strength [50]. PDMS containing kaolin exhibits a greater mass loss at lower temperature compared to PDMS alone. Kaolin also lends itself to being aligned in processing to form kaolin "rich" papers when combined with cellulose, which imparts some flame retardant barrier benefits [51]. Otherwise, Kaolin, like many of the other layered silicates discussed in this chapter, can be combined with a wide range of other additives for potential fire protection/flame retardant benefit [52–56]. It has also been used as the layered silicate in layer-by-layer coatings for flame retardant applications (see discussion later in this chapter for an overview of this technology, as well as Chapter 8 of this book) [57, 58].

6.4.1.6   Mica

Mica [59] belongs to a group of aluminosilicate minerals characterized by a layered structure which can be cleaved to give thin, flexible sheets. The most common and commercially available classes of mica are muscovite and phlogopite. Muscovite mica is 2:1 layered aluminosilicate ($KAl_2(Si_3Al)O_{10}(OH)_2$). Each 2:1 layer consists of two tetrahedral silica sheets sandwiching an alumina octahedral sheet of about 1 nm thick. Since on average, one Si atom out of four in the tetrahedral sheets is replaced by Al, the layers are negatively charged. These charges are compensated by interlayer cations, mostly potassium, and the layers are held together in stacks by electrostatic and Van der Waals forces. Phlogopite mica is a trioctahedral alkali aluminum silicate ($KMg_3(Si_3Al)O_{10}(OH)_2$). Phlogopite has a layered structure of magnesium aluminum silicate sheets weakly bonded together by layers of potassium ions. Both mica types are typically present in the form of thin plates or flakes with sharply defined edges. Mica is chemically inert and thermally stable up to 600°C where dehydroxylation takes place. It can be used as flame retardant extender and char promoter. Imerys claims that loading levels of 30-35 wt.% ammonium polyphosphate (APP), a halogen free flame retardant system for polyolefins, is commonly required to meet UL-94 requirements such as V-0 ratings and one third of APP could be replaced with mica having same FR performance and better mechanical properties [60]. It was also concluded that as particle size decreases, surface area increases, providing improved flame retardant properties. Mica is also used in the formulations of flame retardant cables for electric & electronic applications, and has been recently combined with silicone rubber for additional flame retardant enhancement in combination with ammonium polyphosphate and aluminum hydroxide [61].

6.4.1.7   Talc

Talc is a naturally occurring hydrated magnesium sheet silicate, $3MgO \cdot 4SiO_2 \cdot H_2O$ [62]. The elementary sheet is composed of a layer of magnesium-oxygen/hydroxyl octahedral, sandwiched between two layers of silicon–oxygen tetrahedral. The main or basal surfaces of this elementary sheet do not contain hydroxyl groups or active ions, which explains its hydrophobic nature and inertness. Most talcs are lamellar in nature, they are chemically inert, organophilic and water-repellent to a great extent. Talc progressively loses its hydroxyl groups above 900°C, and above 1050°C, it re-crystallizes into different forms of enstatite (anhydrous magnesium silicate) [62]. Talc usually works in synergy with metal hydroxides for improved fire resistance. Talc limits burnable gas emissions and oxygen diffusion by foaming effect, resulting in delayed combustion, creates a physical barrier effect limiting heat and mass transfer. It is a good char promoter, improves ash cohesion and also generates less smoke [63]. Talc is a very common additive for polymers, especially polyolefins, and when used as a bulk filler in polyolefins, will effectively flame retard the highly flammable polyolefin by diluting the total amount of fuel available. Essentially, the non-flammable talc replaces flammable polyolefin. Such bulk-filler use does not impart passing of strict flame retardant tests, but, it does lower overall flammability of the material. Talc has been used with other flame retardant additives, especially metal hydroxides [64–68].

6.4.1.8   Halloysite

Halloysite is a 1:1 aluminosilicate clay mineral with the empirical formula $Al_2Si_2O_5(OH)_4$ mined from natural deposits. In terms of chemistry, halloysite is the same as kaolinite except that the sheets are rolled into tubes. The asymmetrical sheets, when rolled up, give the outside of the tube a silica-like chemistry and the inside an alumina like chemistry.

The wide applicability of halloysite as a fire-retardant is a combination of the following effects:

- Releases water above 400°C to quench the flame
- Endothermic decomposition removes heat from the fire
- Char formation due to high surface area
- Synergy with glass fiber further improves char integrity

Work on halloysite in glass fiber reinforced PET revealed a synergy between halloysite and glass fibers [69]. When either reinforcement was used alone, FR testing produced a char of low integrity. However, the combination of halloysite and glass produced a strong, contiguous char, which is beneficial for fire retardancy. This synergy between glass fiber and halloysite was recently confirmed in polypropylene and is expected to apply to other systems such as polyamide and PBT, which are important commercial thermoplastics [70].

Cone calorimetry studies showed halloysite is effective in reducing the peak heat release rate [69]. In this instance PP is depicted but the same has been seen for EVA, LLDPE, EPDM and other host matrices. Halloysite functions as a synergist or even as a stand-alone fire retardant where V-0 rating can be achieved in some cases. For polymers like polycarbonate that have some degree of intrinsic FR, only a small addition of (< 2 wt.%) halloysite can take the rating from V-2 to V-0 while retaining good mechanical properties [71]. The hollow tubes can be loaded with FR synergist [70]. For example, liquid fire retardants normally plasticize polymers causing strength and modulus to drop. By adding the liquid FR inside the tubes, one can retain the FR effect and reinforce the polymer.

Similar to other layered silicates, especially the layered silicate nanocomposites described below, Halloysite has yet to be used in any commercial application, despite many attempts to commercialize it for use and despite its claims of success with a wide range of flame retardant chemistries [72–77]. It is unclear if the reason for the lack of commercialization is due to cost of the material, or some other issue.

6.4.1.9   *Layered Silicate Nanocomposites*

Nanoscale materials and technologies are of great interest to researchers worldwide, mainly due to their exceptional property enhancement and potential application in many fields. Polymer-layered silicate (PLS) nanocomposites, as a new class of filled polymers with ultrafine phase dimensions, offer the potential to combine the advantages of both organic and inorganic materials, such as light-weight, flexibility, high strength and heat stability, which are difficult to be obtained separately from the individual components. Furthermore, because of the nano-phase distribution, as well as the synergism between polymer and the layered silicate, PLS nanocomposites exhibit enhanced flame retardation, barrier properties and ablation resistance, which are not observed in either of their components as conventional composites. Mica, fluoromica, hectorite, fluorohectorite, saponite, bentonite, etc. are widely used for layered silicate nanocomposites, but the greatest commercial interest is on montmorillonite (MMT), which belongs to the structural family known as the 2:1 phyllosilicates. In order to favor the dispersion of the clay nanolayers within the polymer matrix, a modification of natural clays using organic cations (alkylammonium, alkylphosphonium and alkylimidazol(idin)ium cations) is often carried out.

Hu *et al.* [78] studied a form-stable phase change material based on high density polyethylene (HDPE), paraffin, organophilic montmorillonite (o-MMT), and intumescent flame retardant (IFR) hybrids using the melt mixing technique. Flame retardant composites produced a large amount of char residue at 700°C. In addition, a synergistic effect between o-MMT and IFR led to the decrease in the heat release rate (HRR), contributing to improvement of the flammability performance. Liu and Huang [79, 80] studied the flame retardancy of HIPS in the presence of o-MMT system. It was found that 20 wt.% of o-MMT can effectively reduce the HRR by 60%.

It gives more char yield than does pristine HIPS. During decomposition, heat transfer promotes thermal decomposition of the organomodifier and the creation of strongly protonic catalytic sites onto the clay surface, which can catalyze the formation of a stable char residue. Therefore accumulation of the clay on the surface of the material acts as a protective barrier that limits heat transfer into the material, volatilization of combustible degradation products and diffusion of oxygen into the material. Also the authors suggested that the horizontal o-MMT layer orientation had better flame retardancy than the vertical layer orientation. HIPS/o-MMT composites with horizontal layer orientation gave a lower heat release rate and mass loss rate than the HIPS/o-MMT composites with vertical layer orientation due to the better barrier effect resulting from the horizontal o-MMT layers during combustion. Ma et al. [81] prepared ABS/clay and ABS-g-MAH/clay nanocomposites by melt blending, and found that the flammability of ABS/clay nanocomposites was strongly affected by the morphologies of the clay network. As for HRR and pHRR, a better performance for ABS-g-MAH/clay nanocomposites than ABS/clay nanocomposites was found. From the results of dynamic rheological measurements, it was found that the clay network structure was formed in ABS-g-MAH/clay nanocomposites, which strongly affected the flame retardant properties of the nanocomposites. The clay network improves the melt viscosity and results in restraint on the mobility of the polymer chains during combustion, which leads to the improvement of flame retardancy for the nanocomposites. The addition of o-MMT could improve the flame retardant properties of polyurethane foam. Compared with neat foam, the heat release rates (HRR), the total heat release, the mass loss, and the mass loss rates of the composites had a great decrease, for example, the peak HRR was reduced to 55% of neat foam [82]. Song et al. prepared the PA6/o-MMT nanocomposite using the MH and red phosphorus (RP) as flame retardants and o-MMT as synergist via melt blending. 2 wt.% o-MMT, 6 wt.% MDH, and 5 wt.% RP gave a V-0 rating and the LOI value reached 31 and also an increase in tensile strength was observed.

The main fire retardancy mechanisms in polymer/clay nanocomposites is the formation of a barrier against heat and volatiles formation of a clay-rich layer at the material surface, followed by char formation, together with increased melt viscosity for exfoliated nanocomposites. These mechanisms modify the fire properties of the polymer nanocomposite, sometimes improving them and in some cases worsening them, depending on the type of fire test used. For instance, in cone calorimetry, the incorporation of nanoclays generally retards and decreases the peak heat release rate, but does not reduce the total heat involved and may also decrease the time to ignition. The increased melt viscosity in exfoliated nanocomposites prevents dripping and promotes char formation. However, the char formed at the surface of the burning sample in UL-94 and limiting oxygen index (LOI) tests is not effective enough to stop the flame and the sample continues to burn slowly, ultimately displaying poor flame retardant performances when nanocomposites are used by themselves with no other FR additives.

Wilkie *et al.* correlated the change in the degradation pathway of various polymers by incorporation of nanoclay to the heat release rate peak measured by cone calorimetry (Table 6.1). It can be concluded that with the presence of nanoclay, HRR could be decreased up to 70% depending on

Table 6.1 Effects of the incorporation of nanoclay on the thermal degradation pathway of the polymer and reduction of HRR values [83].

Polymer	Degradation pathway of virgin polymer	Degradation change in presence of clay	HRR reduction (%)
PA6	Intra-aminolysis/ acidolysis, random scission	Intra-aminolysis/ acidolysis, random scission	50–70
PS, HIPS	β-Scission (chain end and middle)	Recombination, random scission	40–70
EVA	Chain striping, disproportionation	Hydrogen abstraction, random scission	50–70
SAN, ABS	β-Scission (chain end and middle)	Recombination, random scission	20–50
PE	Disproportionation	Hydrogen abstraction	20–40
PP	β-Scission, disproportionation	Random scission	20–50
PAN	Cyclization, random scission	No change	<10
PMMA	β-Scission	No change	20–30

the polymer. Due to this high effectiveness and the usually low amount of loading, nanoclays are effective flame retardant additives where reduction of HRR is desired depending to the type of the application where the polymeric composite will be used.

In Figure 6.2, schematic representation of catalysis charring mechanism of PP/clay nanocomposite is given to illustrate the mechanism in more detail which is generally valid for general polyolefin o-MMT composites.

Of final note, layered silicates have been used to create conformal protective coatings for solid polymers and textiles using a process known as "layer-by-layer" (LbL) formation. This process yields a wide range of potential chemistries that forms the clays in aligned layers that provides notable fire protection against open flame and radiant heat sources. Please see Chapter 8 of this book for more details on this approach, which is very extensive and has shown great promise to date.

In summary, layered silicate nanocomposites are a nearly universal synergist for other flame retardants [85]. This class of materials has been extensively studied to date [86], but still has yet to be extensively commercialized. The main reasons for the lack of commercialization appear to be

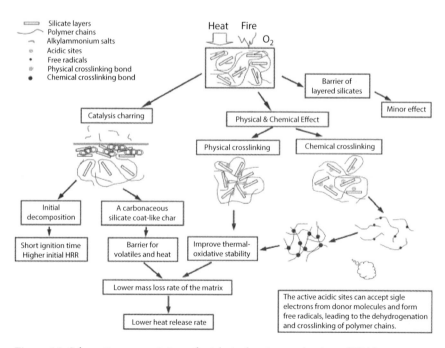

Figure 6.2 Schematic representation of catalysis charring mechanism of PP/clay nanocomposite during combustion [84].

limitations on the thermal stability of the organoclay when compounding above 200 °C (due to the commonly used alkyl ammonium organic treatment decomposing above this temperature) and some concerns about nanoparticle environmental effects.

6.4.1.10   Sodium Silicate

Sodium silicate, commonly known as water glass or sodium metasilicate, has been used in a wide range of flame retardant applications, especially as part of fire protection systems for pipes and firestop gaps in buildings. It does absorb a significant amount of water, which can limit its effectiveness over time. This class of water soluble silica does show notable effectiveness in cellulosic materials, especially wood [87, 88]. Finally, water soluble alkali silicates can be used in geopolymer formulations to manufacture composites that are nearly impervious to fire damage [89–91].

6.4.1.11   Silsesquioxane

Polyhedral oligomeric silsesquioxane (POSS) is an inorganic silica like nano-cage $((RSiO_{1.5})_8)$ surrounded by eight organic groups located at the corners that enhance its compatibility with organic polymers. During combustion of such a polymer composite, POSS acts as a precursor forming thermally stable ceramic materials at high temperature. These inorganic nano-cages are also referred to as pre-ceramic compounds.

While a wide variety of nano structures are described in the literature, essentially two types of POSS have been studied with respect to flame retardancy: bearing either 8 identical R groups (R=methyl, phenyl, isobutyl or isooctyl) or 7 R groups of the same nature and one functional R^I group such as an ester, silane, isocyanate, methacrylate, alcohol, epoxide or amine. This wide range of R and R^I groups enable the selective use of (functionalized) POSS according to the chemical nature of the polymer matrix. The (reactive) functionality of the R^I group can not only improve the compatibility between the dispersed nano-cages and the polymer matrix but also enable either chemical grafting of reactive polymer chains or initiation of polymerization reactions from the POSS surface via the so-called "grafting from" technique.

Much like other inorganic fillers, the introduction of POSS into polymers could enhance both melt viscosity and the mechanical properties of the matrices. In addition to this, as a pre-ceramic compound, POSS could have an effect on the thermal stability and combustion performances by reducing the quantity of heat release during combustion.

Hu and his co-workers have done much work on the flame retardancy of polymer/POSS composites. An octa(tetramethylammonium) POSS (octaTMA-POSS) has been used to prepare PS/POSS composites through melt mixing by Liu *et al.* [92]. Flammability of the composites has been evaluated by the cone calorimeter. Although the value of (peak) heat release rate, concentration, and release rate of carbon monoxide (CO) of the composites could be decreased, the level of reduction was not enough to prepare practical flame retardant materials. Another POSS, trisilanolphenyl POSS (TPOSS) has been incorporated in PC via melt blending and the combustion behavior of PC/TPOSS composites have also been evaluated through cone calorimetry [48]. Results suggest that the addition of TPOSS significantly reduce the peak heat release rate of the composites, and the addition at 2 wt.% decreases the maximum from 492 of the original PC to 267 kW/m^2. Further, a commercial oligomeric bisphenol-A bis(diphenyl phosphate) (BDP) has been incorporated in PC/TPOSS composites by He and co-workers [93]. Results indicated that combination of TPOSS with BDP in an appropriate ratio (2:3 by weight) could enhance both the thermal stability and flame retardancy of the composites. POSS continues to be studied but to date has yet to be commercialized in any flame retardant application, even though there are commercial sources of POSS for a wide range of potential applications [94].

6.4.2   Organic Silicone-Based Flame Retardants

6.4.2.1   *Polyorganosiloxanes*

Polysiloxane polymers have been around since the early 1940's. Of these groups of polymers, the most important is polydimethylsiloxane (PDMS). PDMS has one of the most flexible backbone chain known [95]. This flexibility stems from the relatively large bond distance for the Si-O bond, 1.64 Å [96]. The length of the Si-O bond provides for increased spatial separation of the neighboring organic substituents in polysiloxanes, which, in turn results in significantly reduced steric hindrance and the relief of molecular strain that would otherwise occur [97]. This flexibility of the backbone of PDMS seems to make it ideal for the formation of a micellular structure. The T_g of PDMS is around -125°C, one of the lowest of any recorded polymer [98].

Typically PDMS is made via ring opening polymerization of octamethyltetrasiloxane using diethyl ether as a solvent, and sulfuric acid as a catalyst [99]. The thermal properties of PDMS are some of its most characteristic and technologically important features. Due to the high flexibility of the

Si-O bond, PDMS retains its flexibility even at high temperatures. Typical irreversible decomposition of polysiloxanes can reach over 350°C.

Dow Corning RM 4-7051 and RM 4-7081 are two different polysiloxanes developed by Dow Corning to be used in thermoplastics. These additives, in the form of white powders, have been shown to significantly reduce the rate of heat release and the rate of smoke and carbon monoxide evolution from burning plastics, such as polystyrene, polypropylene, polyethylene, polycarbonate [100]. In Figure 6.3, heat release rate comparison of neat polystyrene (PS), 1 wt.% Dow Corning RM 4-7081 containing PS and 3 wt.% Dow Corning RM 4-7081 containing PS are given. These amounts of polysiloxanes decreased the HRR 35% and 65%, respectively.

These silicone additives also markedly reduce the peak rate of heat release and the peak evolution rate of carbon monoxide and smoke when used with other conventional FR additives, including halogenated compounds, phosphorus compounds, and water-evolving inorganics such as $Mg(OH)_2$.

Abarca *et al.* [101] used HTT 1800 as a flame retardant for polystyrene, using it as a monomer together with styrene to synthesize a hybrid polymer HTT 1800, which is a patented liquid polysilazane-based coating resin produced by Clariant GmbH. It is a material capable of withstanding peak operating temperatures of 1800°C.

Silicones have shown effectiveness in many other polymers as well, including as char enhancement for poly(methylmethacrylate)/polycarbonate blends [102, 103], and commercially in the "CASICO" (Calcium Silicon Carbonate) system sold by Borealis. The CASICO system uses calcium

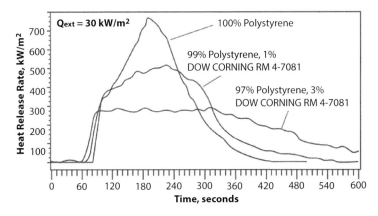

Figure 6.3 Heat release rate of polystyrene modified with Dow Corning RM 4-7081 resin modifier [100].

carbonate and PDMS in combination with a polyethylene-co-acrylic acid system to yield enhancement in heat release reduction as well as improved char yields during burning [104–107]. This system is often used in wire and cable applications [108]. Organosilicones have been combined in a wide range of polymers with other flame retardants as well to enhance flammability properties [20]. Further, the low flammability of PDMS and its propensity to form SiO$_2$ during burning lends itself to use in applications requiring high levels of flame retardancy, where the PDMS is the base polymer that is flame retarded further for wire and cable applications (power control cables in power plants), and for aerospace applications (aircraft cabin interior materials, sealing gaskets, etc.).

6.4.2.2   Silanes

Silicones with methyl and phenyl groups have excellent dispersion in polymers, such as polycarbonate. Aromatic thermoplastics like polycarbonate and its derivatives, ABS and PS are made flame retardant using a silicone derivative [109]. The silicone, a branched chain structure, with a phenyl rich chain and methyl group at the chain end, is found to be effective in retarding the combustion of these thermoplastics, and particularly so for PC. It is believed that rather than methyl containing linear silicon, aromatic containing silicon is a better candidate for aromatic thermoplastics used in electronic products, because silicone derivatives are highly heat resistance and very soluble in these plastics with considerable heat inputs, PC combined with silicone forms a highly flame resistant char according to route (a) in Figure 6.4. High temperatures also promote dehydrogenation of the phenyl groups (see Figure 6.4 (b)). Finally silica particles are formed by thermal degradation of the silicon and remain at the surface within the char, therefore creating a highly oxidation resistant char.

The results show that the silicon finely dispersed in the PC moves to the surface during combustion, and then formed a highly flame retarding char barrier on it. They are the ideal replacement for environmentally hazardous halogen flame retardants.

The effect of the presence of methyl and phenyl groups in the branched silicone additives is given in Figure 6.5. Although the decomposition of the polymer starts at lower temperatures in the presence of both only methyl group and methyl-phenyl groups, the charring is significantly improved. When methyl and phenyl groups are present together, char yield is higher than the case where only methyl groups are present. Although the methyl group can mitigate more easily to the PC surface, due to its low carbon content compared to phenyl group it is not as effective as the phenyl group to

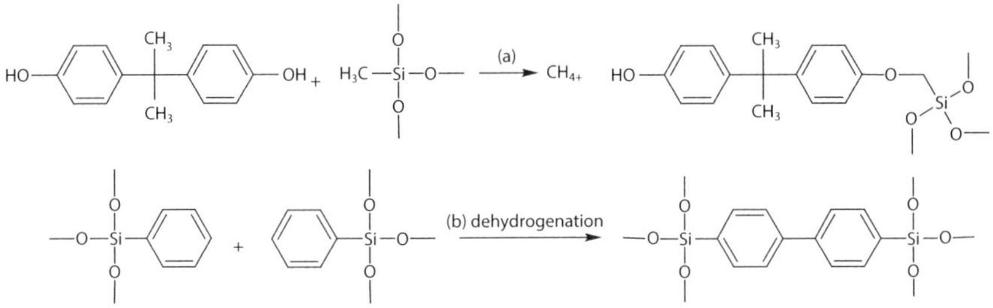

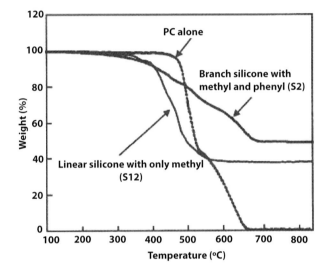

Figure 6.4 Char formation in PC incorporated with silicone.

Figure 6.5 Char formation in PC incorporated with silicone [110].

form a strong char. Addition of phenyl groups to the formulation increases the thermal stability and also supplies the required carbon source to form an effective char.

6.4.3 Other Silicone-Based Flame Retardants

Besides previously mentioned inorganic and organic silicone based materials, several silicone based materials were tried in flame retardant formulations like silicate glasses [111–115], borosilicate [116], borosiloxane [117–119], functional group containing silanes, etc.

Polycarbosilane (PCS), polysilastyrene (PSS), and polysilsesquioxane preceramic polymers are also used to blend with various thermoplastics.

They can reduce the peak heat release rate (HRR) and average HRR, but the total heat released remains unchanged. The primary reason for the lower HRR for the blends is the reduced mass loss rate; i.e., the rate at which fuel is released into the gas phase is slowed by the presence of the ceramic char.

Silicone coatings with dispersed carbon nanotubes have been introduced as Nanocyl's ThermoCyl® to give fire protection to a wide variety of substrates, such as plastics, cables, textiles, foams, metals, and wood. Coatings as thin as 100 nm, have been shown effective and they do not appear to be intumescent coatings [120].

6.4.4   Silicone/Silica Protective Coatings

Along with using silicones and organosilicon compounds as additives for polymer flame retardancy, there have been several studies showing the use of crosslinkable silicones and plasma/vapor phase deposited silicones onto plastics to impart flame retardant benefits [121–130]. In general, these coatings impart a coating that is non-flammable (or has very low flammability) which resists initial flame exposure, resulting in extended times to ignition and some subsequent fire protection after the coating fails/splits open from the flame exposure. The coatings usually are quite thin and have some flexibility to them once deposited on the plastic, but often require some additional flame retardant underneath the coating in the main plastic to be effective in regulatory fire protection tests [20, 121].

6.5   Mode of Actions of Silicone-Based Flame Retardants and Practical Use Considerations

6.5.1   Silicon Dioxide

Silicon dioxide (or silica) has a high melting point (1600-1700°C), but, it can form a low melting silicaceous barrier that functions as heat and mass transfer barrier in the presence of other flame retardants (such as MDH, borates, phosphates, alkali metals). Besides, addition of smaller amounts of silica could increase the melt viscosity of the molten polymer during burning and this helps with the prevention of dripping and, as a result, spread of fire. If a higher loading of silica is incorporated, this can result in the wicking effect which also accelerates dripping during fire and decreases flame retardancy of the matrix polymer. So there is a threshold level of effective loading depending to the type of the matrix polymer. Fumed silicas impart additional hydrogen bonding with increases in local polymer viscosity that yield reductions in mass loss rate/fuel pyrolysis which in turn reduce heat release [131], but, there are limits to how much fumed silica can be added to a polymer before the increases in melt viscosity during compounding prevent the material from being processed in a practical manner. In general, silicas are to be used in combination with other flame retardant additives to obtain satisfactory performance in regulatory fire tests, and never used alone. The exception to this may be the use of silicas in low flammability silicone resins/rubber materials where the base polymer matrix is already of low enough flammability and the silica is enhancing properties. Like all solid additives, compatibilizers may be needed to help the silica disperse in the polymer matrix, and, maintain a good balance of properties in the final system. Of final note, it should be pointed out that some silicas present inhalation hazards in their dry powder forms, and specialized ventilation/loss-in-weight feeder systems will be needed to meter them into extruders for melt compounding operations.

6.5.2   Silicate-Based Minerals

Silicate minerals generally have high melting points. This is beneficial as they do not decompose during processing with the polymer matrix. Their efficiency as flame retardant fillers is closely related to their morphology and crystal structure. Lamellar silicates (with high aspect ratio) will form effective silicaceous barrier. Acicular (needle-like) silicates tend to maintain the integrity of the char. The flame retardancy effect of silicate minerals by formation of a char as a barrier could be enhanced using borates, phosphates, or alkali metals as synergists. For example, borates such as zinc borate, melamine borate or boric acid can function as a flux and form eutectic mixture (lower melting point) at the particle-particle interface between silicate minerals from medium to high temperatures. At high temperatures, borates, such as zinc borate and boric acid, can undergo fusion with silicate minerals to form low melting borosilicate glasses or ceramics.

For layered silicate nanocomposites, the flame retardant mechanism has been extensively studied, and the net effect is a reduction in mass loss rate which in turn lowers heat release. The layered silicates set networks which greatly increase the melt viscosity of the thermally decomposing polymer and as the polymer pyrolyzes away, they form layers silicate rich barriers which in turn further slow fuel from pyrolyzing [132, 133]. Layered silicates must be combined with other flame retardants to be effective in passing regulatory tests, as has been reviewed extensively [85, 86].

All silicate based minerals may require compatibilizers to disperse properly into a polymer matrix, otherwise they may remain as agglomerates which limits their flame retardant effectiveness, and more often causes negative impacts on mechanical properties. Similar to silica systems, specialized feeding and ventilation systems are required for safe handling of silicate based minerals. Finally, organoclays used in making polymer nanocomposites, unless they have been treated with imidazolium treatments, have an upper use temperature of 200 °C, and cannot be compounded above that temperature. Compounding above 200 °C will result in thermal decomposition of the alkyl ammonium treatment on most organoclays which will in turn make the nanocomposite turn into a microcomposite during melt compounding, along with potentially causing other degradation in properties in the final polymer formulation.

6.5.3   Silicones

Silicone acts as a flame retardant through condensed phase by the formation of a protective silicaceous heat and mass barrier. The beneficial effect of flame retardancy is most evident in reduction of pHRR in cone calorimetry. Beside the condensed phase effect, with the attachment of phosphorous or nitrogen containing functional groups to the structure silicone based flame retardants could also gain effect in gas phase and they could become much more efficient. The specific chemical reactions between silicone and polymer are dependent upon the polymer matrix that the silicone is present in.

Silicones come in a variety of forms which generally makes formulation or compounding of these materials into other polymers rather easy. Some care must be taken with partially reactive silicones so that the reaction does not occur during melt compounding (resulting in crosslinking/large increases in melt viscosity). Otherwise, there is enough diversity of organic chemical structure functionality to find a silicone that can be made to be fully compatible with the polymer of choice. Silicones are to be considered as co-additives with other flame retardants, and not to be the sole flame retardant for a polymer, with the exception of use with polymers that already have inherently low heat release (polyimides, polyarylketones, silicones, halogenated polymers, etc.).

6.6   Future Trends in Silicon-Based Flame Retardants

Given the continued desire in the marketplace for non-halogenated flame retardants, it is highly likely that silicon based flame retardant use will grow in the coming decades, driven by increased need for synthetic materials in many markets (electrical and electronic, aerospace, building and construction, transportation). Further, it is expected that silicone material use will increase as well, further enhancing the use of this material due to its thermal durability and ability to serve as a flame retardant material.

The general trend for new flame retardants in general is to go toward polymeric or reactive flame retardants, vs. small-molecule non-reactive additives. Small molecule additives (flame retardants, anti-oxidants, UV stabilizers) can migrate out/be separated from the original polymer it was added to, and this migration causes two problems. The first problem is that the protective benefit brought by the additive is lost to the original plastic. The second problem is that this chemical can end up in the environment where is may have negative persistence, bioaccumulation, and toxicity (PBT) profiles. Reactive flame retardants get rid of this problem completely as the flame retardant is covalently bound to the polymer it is added to. Polymeric flame retardants are more likely to remain with the polymer they were added to, and if they do migrate into the environment, they have a low potential for bioaccumulation. These trends are likely to carry over into silicon-based flame retardants as well. This means that polymeric organosilicon compounds will find increased use, as will silicone compounds that can react with a polymer during polymerization, or can covalently graft onto the polymer. Inorganic additives are likely to continue being used if they have favorable PBT profiles, but some additives may be disfavored from use depending upon how they do/do not migrate out of the polymer over time. Considerations for inorganic additive release upon drilling, grinding, or sanding of a polymer may also need to be factored into the use of silicates and silicas. However, coatings that contain these inorganic additives are likely to continue to grow in use because many of these coatings (such as the before-mentioned LbL coatings) lock the inorganic silicas/silicates into the coating structure, making them difficult to access for accidental environmental exposure. Further, sol-gel/plasma deposited silicon-based flame retardant coatings should expect to be used, provided they show durability against scratch/mar damage from consumer use, and no debonding from surfaces when said surfaces are flexed/moved during consumer use.

Commercial use of silicon-based flame retardants is expected to continue in wire and cable and aerospace, with strong growth in both areas. Mass transportation may also see some use of these materials, if costs remain reasonable vs. other commodity materials used today. Electrical and Electronic applications will only see growth if silicon based flame retardants bring an enhanced fire protection benefit when used in combination with existing commercial flame retardants. For building and construction, silicones are likely to be parts of formulations for building materials requiring fire protection. It will be interesting to see if sodium silicate use increases as a flame retardant for wood when wood use in building materials increases, as it is expected to do so in the next decade. For specific polymers, silicon-based flame retardant use in polycarbonate, polyolefins, epoxies, and polyurethanes is expected to continue growth, or at least maintain market share. Commercial use however will be tempered by chemical use regulations in the European Union and in the United States, including new data that does/does not give a particular chemical a low impact PBT profile. See chapter 1 of this book for more insight into the current and future trends on flame retardant and chemical use regulation.

6.7 Summary and Conclusions

This chapter aims at an overview of silicon based flame retardants. Silicones are greatly acknowledged for their thermal and thermo oxidative stabilities, which compared to most carbon based polymers, are superior. They also do not emit corrosive gases during fire, and are low heat release materials, which is a key benefit in fire protection. Based upon research to date, the flame retardant mechanism for silicon-based flame retardants is condensed phase, with an emphasis on enhanced char formation, or formation of silica-rich layers which slow down mass loss rate and/or act as thermal insulators against further thermal damage. There continues to be a wide range of research on new uses for inorganic silicon-based flame retardants, as well as polymeric and reactive organo-silicon compounds, with the latter having the ability to gain additional benefits via reaction with phosphorus, nitrogen, or boron atoms during combustion. Commercial development is more focused on polymeric organosilicon compounds rather than reactive compounds with Si-P, Si-N, or Si-B bonds due to the costs associated with forming these new chemistries. Still, combinations of organosilicon compounds with organophophorus, nitrogen, and boron compounds (and also with inorganic phosphorus and boron compounds) are being researched, and will continue to be researched.

The main conclusion to make about silicon-based flame retardants is that they are to be used with other flame retardant chemicals/materials to yield enhancements in fire protection as they rarely get used by themselves. With the exception of using polymeric silicones as a low-heat release material, silicon-based flame retardants do not typically have enough efficacy to flame retard polymeric additives by themselves, or at least not at loading levels to be practical when considering a balance of properties. Instead, silicon-based flame retardants are an excellent source of co-additives to consider to obtain better fire protection performance while maintaining a good balance of other properties. The other added benefit is that silicon compounds tend to not interfere with refractive indices of most polymers, meaning they can be a solution toward maintaining transparency while gaining fire protection. For the above reasons, silicon-based flame retardants will continue to be a very useful tool for fire safety scientists when combined with other flame retardant chemistries.

References

1. Rochow, E.G., *Silicon and Silicones*, Springer-Verlag, Berlin, 1987.
2. Roth, C.A., Silyation of Organic Chemicals, *Ind. Eng. Chem. Prod. Res. Dev.*, 11, 134, 1972.
3. Smith, A.L., *The Analytical Chemistry of Silicones*, p. 210, John Wiley & Sons Inc., New York, 1991.
4. Mollie, J.P., Silicone Materials for Electronic Components and Circuit Protection, in: *Plastics for Electronics*, 2nd Ed., p. 25, Kluwer Academic Publishers, Dordrecht, Netherlands 1999.
5. Clark, P.J., Modification of Polymer Surfaces by Silicone Technology, in: *Polymer Surfaces*, p. 235, John Wiley & Sons Inc., New York, 1978.
6. Jaques, L.B., Fidlar, E., Feldsted, E.T., MacDonald, A.G., Silicones and Blood Coagulation, *Can. Med. Assoc. J.*, 55, 26, 1946.
7. Lahey, F.H., Comments made following the speech "Results from using vitallium tubes in biliary surgery," read by Pearse HE before the American Surgical Association, Hot Springs, VA. *Ann. Surg.*, 124, 1027, 1946.
8. Curtis, J.M. and Colas, A., Dow Corning® Silicone Biomaterials, History, Chemistry & Medical Applications of Silicones, in: *Biomaterials Science*, 2nd Edition, p. 80, Elsevier, London, UK, 2004.
9. Leeper, H.M. and Wright, R.M., Elastomers in Medicine, *Rubber Chem. Technol.*, 56, 3, 523, 1983.
10. https://www.medicineindia.org/pharmacology-for-generic/1376/dimethicone (accessed 06/17/21).
11. Rider, J. and Moeller, H., Use of silicone in the treatment of intestinal gas and bloating, *JAMA*, 174, 2052, 1960.
12. Biron, M., *Silicones or siloxanes applications. Techniques de l'Ingenieur*, p. N2882, 2007. https://www.techniques-ingenieur.fr/base-documentaire/materiaux-th11/materiaux-a-proprietes-mecaniques-42535210/silicones-ou-siloxanes-n2882/.
13. Mansouri, J., Burford, R.P., Cheng, Y.B., Hanu, L., Formation of strong ceramified ash from silicone-based compositions. *J. Mater. Sci.*, 40, 5741, 2005.

14. Camino, G., Lomakin, S., Lazzari, M., PDMS Thermal Degradation Part 1. Kinetic Aspects, *Polym.*, 42, 6, 2395, 2001.

15. Radhakrishnan, T.S., New method for evaluation of kinetic parameters and mechanism of degradation from pyrolysis–GC studies: Thermal degradation of polydimethylsiloxanes, *J. Appl. Polym. Sci.*, 73, 441, 1999.

16. Grassie, N. and Macfarlane, I.G., The thermal degradation of polysiloxanes—I. Poly(dimethylsiloxane), *Eur. Polym. J.*, 14, 875, 1978.

17. Camino, G., Lomakin, S., Lageard, M., Thermal polydimethylsiloxane degradation. Part 2. The degradation mechanisms, *Polym.*, 43, 7, 2011, 2002.

18. Grassie, N. and Francey, K.F., The thermal degradation of polysiloxanes—Part 3: Poly(dimethyl/methyl phenyl siloxane), *Polym. Degrad. Stab.*, 2, 1, 53, 1980.

19. Grassie, N., Francey, K.F., Macfarlane, I.G., The thermal degradation of polysiloxanes—Part 4: Poly(dimethyl/diphenyl siloxane), *Polym. Degrad. Stab*, 2, 1, 67, 1980.

20. Hamdani, S., Longuet, C., Perrin, D., Lopez-cuesta, J.-M., Ganachaud, F., Flame retardancy of silicone-based materials. *Polym. Degrad. Stab.*, 94, 465–495, 2009.

21. Huang, H., Tian, M., Liu, L., He, Z., Chen, L.Z., Effects of silicon additive as synergists of Mg(OH)2 on the flammability of ethylene vinyl acetate copolymer, *J. Appl. Polym. Sci.*, 99, 3203, 2006.

22. Liu, Y.-L., Chou, C.-I., The effect of silicon sources on the mechanism of phosphorus-silicon synergism of flame retardation of epoxy resins. *Polym. Degrad. Stab.*, 90, 515–522, 2005.

23. Zhao, D., Shen, Y., Wang, T., Ceramifiable EVA/APP/SGF composites for improved ceramifiable properties. *Polym. Degrad. Stab.*, 150, 140–147, 2018.

24. Gilman, J.W., Kashiwagi, T., Harris, Jr., R.H., Lomakin, S., Lichetenhan, J.D., Jones, P., Bolf, A., *Chemistry and Technology of Polymer Additives*, S. Al-Malaika, C. Wilkie, C.A. Golovoy (Eds.), p. 135, Blackwell Science, London, 1999.

25. Kashiwagi, T., Gilman, J.W., Butler, K.M., Harris, R.H., Shields, J.R., *Asano, A.*, Flame retardant mechanism of silica gel/silica, *Fire. Mater*, 24, 277, 2000.

26. Kashiwagi, T., Shields, J.R., Harris, Jr., R.H., Davis, R.D., Flame-retardant mechanism of silica: Effects of resin molecular weight, *J. Appl. Polym. Sci.*, 87, 205, 2003.

27. Beyer, G., Carbon Nanotubes as Flame Retardants for Polymers, *Fire. Mater*, 26, 291, 2002.

28. Pritchard, G., Fillers, in: *Plastics additives: an A–Z reference*, G. Pritchard (Ed.), p. 241, Chaall, United Kingdom, 1998.

29. W.R. Nicholson, L. Rapson, K. Shephard, Flame retardant silicone foams, US Patent 6,084,002, 2000.

30. K.L. Shephard, Flame resistant silicone rubber wire and cable coating composition, US Patent 6,239,378, 2001.

31. C. George, A. Pouchelon, R. Thiria, Composition polyorganosiloxanes vulcanisables a chaud utilisable notamment pour la fabrication de *fils* ou cables electriques, French Patent 2,899,905, 2006.

32. Hamdani, S., Longuet, C., Lopez-Cuesta, J.-M., Ganachaud, F., Calcium and aluminium-based fillers as flame-retardant additives in silicone matrices. I. Blend preparation and thermal properties. *Polym. Degrad. Stab.*, *95*, 1911–1919, 2010.

33. Hamdani-Devarennes, S., Longuet, C., Sonnier, R., Ganachaud, F., Lopez-Cuesta, J.-M., Calcium and aluminum-based fillers as flame-retardant additives in silicone matrices. III. Investigations on fire reaction. *Polym. Degrad. Stab.*, *98*, 2021–2032, 2013.

34. Zhengzhou, W., Xin, G., Wenfeng, L., Epoxy resin/cyanate ester composites containing DOPO and wollastonite with simultaneously improved flame retardancy and thermal resistance. *High Perform. Polym.*, *32*, 710–718, 2020.

35. Liu, Y., Li, J., Wang, Q., The Investigation of Melamine Polyphosphate Flame Retardant Polyamide-6/Inorganic Silicferous Filler with Different Geometrical Form. *J. App. Polym. Sci.*, *113*, 2046–2051, 2009.

36. Zhang, Y., Liu, Y., Wang, Q., Synergistic effect of melamine polyphosphate with macromolecular charring agent novolac in wollastonite filled PA66. *J. App. Polym. Sci.*, *116*, 45–49, 2010.

37. J.M. Cogen, A.B. Morgan, S.T. Lin, U.S. Patent Application 20060142460.

38. Morgan, A.B., Whaley, P.D., Lin, T.S., Cogen, J.M., The Effects of Inorganic-Organic Cations on EVA-Magadiite Nanocomposite Flammability, in: *Fire and Polymers IV: Materials and Concepts for Hazard Prevention ACS Symposium Series #922*, C.A. Wilkie and G.L. Nelson (Eds.), pp. 48–60, American Chemical Society, Washington DC, 2005.

39. Morgan, A.B., Cogen, J.M., Opperman, R.S., Harris, J.D., The effectiveness of magnesium carbonate-based flame retardants for poly(ethylene-co-vinyl acetate) and poly(ethylene-co-ethyl acrylate). *Fire Mater.*, *31*, 387–410, 2007.

40. Wang, D., Jiang, D.D., Pabst, J., Han, Z., Wang, J., Wilkie, C.A., Polystyrene magadiite nanocomposites, *Polym. Eng. Sci.*, *44*, 1122, 2004.

41. Mariappan, T., Yi, D., Chakraborty, A., Singha, N.K., Wilkie, C.A., Thermal stability and fire retardancy of polyurea and epoxy nanocomposites using organically modified magadiite. *J. Fire Sci.*, *32*, 346–361, 2014.

42. Beyer, G., *Fire Retardancy and Protection Materials (FRPM05)*, BAM, (Federal Institute for Materials Research and Testing), Berlin, Germany, September 7-9, 2005.

43. Huang, N.H., Chen, Z.J., Wang, J.Q., Wei, P., *Express Polym. Lett.*, *4*, 12, 743, 2010.

44. Jiang, P., Zhang, S., Bourbigot, S., Chen, Z., Duquesne, S., Casetta, M., Surface grafting of sepiolite with a phosphaphenanthrene derivative and its flame-retardant mechanism on PLA nanocomposites. *Polym. Degrad. Stab.*, *165*, 68–79, 2019.

45. Vahabi, H., Lin, Q., Vagner, C. *et al.*, Investigation of thermal stability and flammability of poly(methyl methacrylate) composites by combination of APP with ZrO$_2$, sepiolite or MMT. *Polym. Degrad. Stab.*, *124*, 60–67, 2016.

46. Zhan, Z., Xu, M., Li, B., Synergistic effects of sepiolite on the flame retardant properties and thermal degradation behaviors of polyamide 66/aluminum diethylphosphinate composites. *Polym. Degrad. Stab.*, *117*, 66–74, 2015.

47. Xu, Z., Zhou, H., Yan, L., Jia, H., Comparative study of the fire protection performance and thermal stability of intumescent fire-retardant coatings filled with three types of clay nano-fillers. *Fire Mater.*, *44*, 112–120, 2020.

48. de Juan, S., Zhang, J., Acuña, P., Nie, S., Liu, Z., Zhang, W., Puertas, M.L., Esteban-Cubillo, A., Santarén, J., Wang, D.-Y., An efficient approach to improving fire retardancy and smoke suppression for intumescent flame-retardant polypropylene composites via incorporating organo-modified sepiolite. *Fire Mater.*, *43*, 961–970, 2019.

49. Hapuarachchi, T.D., Bilotti, E., Reynolds, C.T., Peijs, T., The synergistic performance of multiwalled carbon nanotubes and sepiolite nanoclays as flame retardants for unsaturated polyester. *Fire Mater.*, *35*, 157–169, 2011.

50. Genovese, A. and Shanks, R.A., Fire performance of poly(dimethyl siloxane) composites evaluated by cone calorimetry, *Compos. Part A: Appl. Sci. Manuf*, *39*, 2, 398, 2008.

51. Castro, D.O., Karim, Z., Medina, L., Haggstrom, J.-O., Carosio, F., Svedberg, A., Wagberg, L., Soderberg, D., Berglund, L.A., The use of a pilot-scale continuous paper process for fire retardant cellulose-kaolinite nanocomposites. *Composites Sci. Technol.*, *162*, 215–224, 2018.

52. Ullah, S., Ahmad, F., Shariff, A.M., Bustam, M.A., Synergistic effects of kaolin clay on intumescent fire retardant coating composition for fire protection of structural steel substrate. *Polym. Degrad. Stab.*, *110*, 91–103, 2014.

53. Huang, H., Tian, M., Liu, L., He, Z., Chen, Z., Zhang, L., Effects of Silicon Additive as Synergists of Mg(OH)$_2$ on the Flammability of Ethylene Vinyl Acetate Copolymer. *J. App. Polym. Sci.*, *99*, 3203–3209, 2005.

54. Shehata, A.B., Hassan, M.A., Darwish, N.A., Kaolin Modified with New Resin-Iron Chelate as Flame Retardant System for Polypropylene". *J. App. Polym. Sci.*, *92*, 3119–3125, 2004.

55. Batistella, M., Otazaghine, B., Sonnier, R. *et al.*, Fire retardancy of ethylene vinyl acetate/ultrafine kaolinite composites. *Polym. Degrad. Stab.*, *100*, 54–62, 2014.

56. Shang, K., Liao, W., Wang, J., Wang, Y.-T., Wang, Y.-Z., Schiraldi, D.A., Nonflammable Alginate Nanocomposite Aerogels Prepared by a Simple Freeze-Drying and Post-Cross-Linking Method. *ACS Appl. Mater. Interfaces*, *8*, 642–650, 2016.

57. Chang, S., Slopek, R.P., Condon, B., Grunlan, J.C., Surface Coating for Flame-Retardant Behavior of Cotton Fabric Using a Continuious Layer-by-Layer Process. *Ind. Eng. Chem. Res.*, *53*, 3805–3812, 2014.

58. Liu, X., Qin, S., Li, H., Sun, J., Gu, X., Zhang, S., Grunlan, J.C., "Combination Intumescent and Kaolin-Filled Multilayer Nanocoatings that Reduce Polyurethane Flammability". *Macromol. Mater. Eng.*, *304*, 1800531, 2019.

59. Briggs, C.C., *Mica:plastics additives: an A–Z reference*, G. Pritchard (Ed.), p. 459, Chapman & Hall, London, 1998.

60. https://www.imerys-performance-minerals.com/your-market/plastics/fire-resistance-0 (accessed 06/17/21).

61. Hu, S., Tan, Z.-W., Chen, F., Li, J.-G., Shen, Q., Huang, Z.-X., Zhang, L.-M., Flame-retardant properties and synergistic effect of ammonium polyphosphate/aluminum hydroxide/mica/silicone rubber composites. *Fire Mater.*, *44*, 673–682, 2020.

62. Pritchard, G., Fillers, in: *Plastics additives: an A–Z reference*, G. Pritchard (Ed.), p. 241, Chaall, United Kingdom, 1998.

63. Rothon, R.N., *Particulate Filled Polymer Composites*, p. 70, Rapra Technology Limited, Shawbury, United Kingdom, 2003.

64. Clerc, L., Ferry, L., Leroy, E., Lopez-Cuesta, J.-M., Influence of talc physical properties on the fire retarding behaviour of (ethylene-vinyl acetate copolymer/magnesium hydroxide/talc) composites. *Polym. Degrad. Stab.*, *88*, 504–511, 2005.

65. Genovese, A. and Shanks, R.A., Structural and thermal interpretation of the synergy and interactions between the fire retardants magnesium hydroxide and zinc borate. *Polym. Degrad. Stab.*, *92*, 2–13, 2007.

66. Wawrzyn, E., Schartel, B., Karrasch, A., Jager, C., Flame-retarded bisphenol A polycarbonate/silicon rubber/bisphenol A bis(diphenyl phosphate): Adding inorganic additives. *Polym. Degrad. Stab.*, *106*, 74–87, 2014.

67. Weil, E.D. and Levchik, S.V., Flame Retardants in Commercial Use or Development for Polyolefins. *J. Fire Sci.*, *26*, 5–42, 2008.

68. Kim S., Flame Retardancy and Smoke Suppression of Magnesium Hydroxide Filled Polyethylene. *J. Polym. Sci., Part B.*, *41*, 936–944, 2003.

69. Johnson, S.L., Guggenheim, S., Koster Van Groos, A.F., Thermal Stability of Halloysite by High-Pressure Differential Thermal Analysis, *Clay Clay Miner*, vol. 38, p. 477, 1990.

70. Zeiotun, A. and De Armitt, C., Halloysite: Natural, Reinforcing and Fire Resistant, in: *23rd Annual Conference on Recent Advances in Flame Retardancy of Polymeric Materials*, Stamford, US, May 21–23, 2012.

71. Stoch, L. and Waclawska, I. Dehydroxylation of kaolinite group minerals I. Kinetics of dehydroxylation of kaolinite and halloysite, *J. Therm. Anal. Calorim.*, 20, 291, 1981.

72. Goda, E.S., Yoon, K.R., El-sayed, S.H., Hong, S.E., Halloysite nanotubes as smart flame retardant and economic reinforcing materials: A review. *Thermochimica Acta*, *669*, 173–184, 2018.

73. Lecouvet, B., Sclavons, M., Bailly, C., Bourbigot, S., A comprehensive study of the synergistic flame retardant mechanisms of halloystie in intumescent polypropylene. *Polym. Degrad. Stab.*, *98*, 2268–2281, 2013.

74. Marney, D.C.O., Russell, L.J., Wu, D.Y., Nguyen, T., Cramm, D., Rigopoulos, N., Wright, N., Greaves, M., The suitability of halloysite nanotubes as a fire retardant for nylon 6. *Polym. Degrad. Stab.*, *93*, 1971–1978, 2008.

75. Marney, D.C.O., Yang, W., Russell, L.J., Shen, S.Z., Nguyen, T., Yuan, Q., Varley, R., Li, S., Phosphorus intercalation of halloysite nanotubes for enhanced fire properties of polyamide 6. *Polym. Adv. Technol.*, *23*, 1564–1571, 2012.

76. Nakamura, R., Netravali, A.N., Morgan, A.B., Nyden, M.R., Gilman, J.W., Effect of halloysite nanotubes on mechanical properties and flammability of soy protein based green composites. *Fire Mater.*, *37*, 75–90, 2013.

77. Lecouvet, B., Sclavons, M., Bourbigot, S., Bailly, C., Thermal and flammability properties of polyethersulfone/halloysite nanocomposites prepared by melt compounding. *Polym. Degrad. Stab.*, *98*, 1993–2004, 2013.

78. Cai, Y., Hu, Y., Song, L., Kong, Q., Yang, R., Zhang, Y., Chen, Z., Fan, W., Preparation and flammability of high density polyethylene/paraffin/organophilic montmorillonite hybrids as a form stable phase change material, *Energy Conv. Manag*, 48, 462, 2007.

79. Liu, X.F., Zhang, J., Zhang, H.P., Studies on flame retardancy of HIPS/montmorillonite composites, *Acta Polym. Sin.*, 5, 650, 2004.

80. Huang, L., Wang, C., Zhang, J.J., *Qingdao Univ. Sci. Tech.*, 27, 219, 2006 [Title unavailable, article in Chinese].

81. Ma, H.Y., Tong, L.F., Xu, Z.B., Fang, Z.P., Clay network in ABS-graft-MAH nanocomposites: Rheology and flammability, *Polym. Degrad. Stab*, 92, 1439, 2007.

82. Song, L., Hu, Y., Lin, Z.H., Xuan, S.Y., Wang, S.F., Chen, Z.Y., Fan, W.C., Preparation and properties of halogen-free flame-retarded polyamide 6/organoclay nanocomposite, *Polym. Degrad. Stab*, 86, 535, 2004.

83. Jang, B.N., Costache, M., Wilkie, C.A., The relationship between thermal degradation behavior of polymer and the fire retardancy of polymer/clay nanocomposites, *Polym*, 46, 10678, 2005.

84. Qin, H., Zhang, S., Zhao, C., Hu, G., Yang, M., Flame retardant mechanism of polymer/clay nanocomposites based on polypropylene, *Polym*, 46, 19, 8386, 2005.

85. Morgan, A.B., Flame retarded polymer layered silicate nanocomposites: a review of commercial and open literature systems. *Polym. Adv. Technol.*, *17*, 206–217, 2006.

86. A.B. Morgan and C.A. Wilkie (Eds.), Flame Retardant Polymer Nanocomposites, John Wiley & Sons, Hoboken, NJ, 2007.

87. Pereyra, A.M., Giudice, C.A., Flame-retardant impregnants for woods based on alkaline silicates. *Fire Saf. J.*, *44*, 497–503, 2009.

88. Samal, R. and Sahoo, P.K., Development of a Biodegradable Rice Straw-g-poly(methylmethacrylate)/Sodium Silicate Composite Flame Retardant. *J. App. Polym. Sci.*, *113*, 3710–3715, 2009.

89. Lyon, R.E., Balaguru, P.N., Foden, A., Sorathia, U., Davidovits, J., Davidovics, M., Fire-resistant Aluminosilicate Composites. *Fire Mater.*, *21*, 67–73, 1997.

90. Giancaspro, J., Papakonstantinou, C., Balaguru, P., Fire resistance of inorganic sawdust biocomposite. *Compos. Sci. Technol.*, 68, 1895–1902, 2008.

91. Sakkas, K., Nomikos, P., Sofianos, A., D., Panias, Sodium-based fire resistant geopolymer for passive fire protection. *Fire Mater.*, 39, 259–270, 2015.

92. Liu, L., Hu, Y., Song, L., Nazare, S., He, S.Q., Hull, R. Combustion and thermal properties of OctaTMA-POSS/PS composites, *J., Mater. Sci.*, 42, 4325, 2007.

93. He, Q., Song, L., Hu, Y., Zhou, S.Combustion and thermal properties of OctaTMA-POSS/PS composites, *J. Mater. Sci.*, 44, 1308, 2009.

94. https://hybridplastics.com/ (accessed 08/15/20)

95. Zeigler, J.M. and Gordon Fearon, F.W.G., *Silicon-Based Polymer Science: A comprehensive resource*, p. 32, ACS, Washington DC, 1990.

96. Voronkov, M.G., Mileshkevicli, V.P., Yuzhelevskii, Y.A., *The Siloxane Bond: physical properties and chemical transformations*, p. 19, Consultants Bureau, New York, 1978.

97. Thomas, T.H. and Kendrick, T.C. Thermal analysis of polydimethylsiloxanes. I. Thermal degradation in controlled atmospheres, *J. Polym. Sci. Part A-2.*, 8, 10, 1823, 1970.

98. Deanin, R., *Polymer Structure, Properties and Applications*, p. 150, Cahners, Boston, 1972.

99. Jones, R.G., Ando, W., Chojnowski, J., *Silicon-Containing Polymers - The Science and Technology of Their Synthesis and Applications*, Verlag, Berlin, Germany, p. 476, 2000.

100. Buch, R., Page, W., Romenesko, D., *Silicone-Based Additives for Thermoplastic Resins Providing Improved Mechanical, Processing and Fire Properties*, Dow Corning Corporation, Midland, MI, USA, 2001. https://www.yumpu.com/en/document/view/10924413/silicone-based-additives-for-thermoplastic-resins-providing- [accessed 06/17/21].

101. Silvia, A.C., Abarca, O.F.A., André, L.G., Gilvan, P., Barroso, S., Coan, T., Ricardo, G.M., Machado, A.F., Synthesis and thermal characterization of silicon-based hybrid polymer, *Chem. Eng. Transac.*, 32, 1621, 2013.

102. Perret, B. and Schartel, B., The effect of different impact modifiers in halogen-free flame retarded polymer blends – I. Pyrolysis. *Polym. Degrad. Stab.*, 94, 2194–2203, 2009.

103. Perret, B. and Schartel, B., The effect of different impact modifiers in halogen-free flame retarded polymer blends – II. Fire behaviour. *Polym. Degrad. Stab.*, 94, 2204–2212, 2009.

104. Kramer, K.H., Blomqvist, P., Hees, P.V., Gedde, U.W., On the intumescence of ethylene-acrylate copolymers blended with chalk and silicone. *Polym. Degrad. Stab.*, 92, 1899–1910, 2007.

105. Hermansson, A., Hjertberg, T., Sultan, B.-A., Linking the flame-retardant mechanisms of an ethylene-acrylate copolymer, chalk and silicone elastomer system with its intumescent behavior. *Fire Mater.*, 29, 407–423, 2005.

106. Hermansson, A.L., Hjertberg, T., Sultan, B.-A., Influence of the Structure of Acyrlate Groups on the Flame Retardant Behavior of Ethylene Acyrlate Copolymers Modified with Chalk and Silicone Elastomer. *J. Fire Sci.*, *25*, 287–319, 2007.

107. Realinho, V., Antunes, M., Velasco, J.I., Enhanced fire behavior of Casico-based foams. *Polym. Degrad. Stab.*, *128*, 260–268, 2016.

108. Emmanuelsson, V., Simonson, M., Gevert, T., The effect of accelerated ageing of building wires. *Fire Mater.*, *31*, 311–326, 2007.

109. Shinomiya, T., Sato, I., Iji, M., Serizawa, S., *Flame-retardant polycarbonate resin composition*, Sumitomo Dow Ltd and NEC Corp., Japan, 2002.

110. Hamdani, S., Longuet, C., Perrin, D., Lopez-Cuesta, J.M., Ganachaud, F., Flame retardancy of silicone-based materials, *Polym. Degrad. Stab.*, 94, 4, 465, 2009.

111. Weil, E.D., Fire-Protective and Flame-Retardant Coatings – A State-of-the-Art Review. *J. Fire Sci.*, *29*, 259–296, 2011.

112. Alongi, J. and Malucelli, G., Thermal stability, flame retardancy and abrasion resistance of cotton and cotton-linen blends by sol-gel silica coatings containing alumina micro- or nano-particles. *Polym. Degrad. Stab.*, *98*, 1428–1438, 2013.

113. Kandola, B.K. and Pornwannachai, W., Enhancement of Passive Fire Protection Ability of Inorganic Fire Retardants in Vinyl Ester Resin Using Glass Frit Synergists. *J. Fire Sci.*, *28*, 357–381, 2010.

114. Yu, D., Kleemeier, M., Wu, G.M., Schartel, B., Liu, W.Q., Hartwig, A., A low melting organic-inorganic glass and its effect on flame retardancy of clay/epoxy composites. *Polymer*, *52*, 2120–2131, 2011.

115. Wu, G.M., Schartel, B., Yu, D., Kleemeier, M., Hartwig, A., Synergistic fire retardancy in layered-silicate nanocomposites combined with low-melting phenylsiloxane glass. *J. Fire Sci.*, *30*, 69–87, 2012.

116. Dogan, M. and Bayramli, E., Synergistic effect of boron containing substances on flame retardancy and thermal stability of clay containing intumescent polypropylene nanoclay composites. *Polym. Adv. Technol.*, *22*, 1628–1632, 2011.

117. Marosi, Gy., Marton, A., Szep, A., Csontos, I., Keszei, S., Zimonyi, E., Toth, A., Almeras, X., Le Bras, M., Fire retardancy effect of migration in polypropylene nanocomposites induced by modified interlayer. *Polym. Degrad. Stab.*, *82*, 379–385, 2003.

118. Marosi, Gy., Anna, P., Marton, A., Bertalan, Gy., Bota, A., Toth, A., Mohai, M., Racz, I., Flame-retarded Polyolefin Systems of Controlled Interphase. *Polym. Adv. Technol.*, *13*, 1–9, 2002.

119. Anna, P., Marosi, Gy., Bertalan, Gy., Marton, A., Szep, Structure-Property Relationship in Flame Retardant Polymers. *A. J. Macromol. Sci. Part B*, *41*, 1321–1330, 2002.

120. Mahy M., Silicones and Carbon Nanotubes: from antistatic to fire barrier and fouling release Coatings, in: *Silicone Elastomers 2009*, Hamburg, Germany, October 7–8, 2009.

121. Jimenez, M., Gallou, H., Duquesne, S., Jama, C., Bourbigot, S., Couillens, X., Speroni, F., New routes to flame retard polyamide 6,6 for electrical applications. *J. Fire Sci.*, *30*, 535–551, 2012.

122. Horrocks, A.R., Flame retardant challenges for textiles and fibres: New Chemistry versus innovatory solutions. *Polym. Degrad. Stab.*, *96*, 377–392, 2011.

123. Alongi, J., Carosio, F., Malucelli, G., Current emerging techniques to impart flame retardancy to fabrics: An overview. *Polym. Degrad. Stab.*, *106*, 138–149, 2014.

124. Lazar, S.T., Kolibaba, T.J., Grunlan, J.C., Flame-retardant surface treatments. *Nat. Rev. Mater.*, 5, pp. 259-275, 2020.

125. Schartel, B., Kuhn, G., Mix, R., Friedrich, J., Surface Controlled Fire Retardancy of Polymers Using Plasma Polymerisation. *Macromol. Mater. Eng.*, *287*, 579–582, 2002.

126. Bourbigot, S., Jama, C., Le Bras, M., Delobel, R., Dessaux, O., Goudmand, P., New approach to flame retardancy using plasma assisted surface polymerisation techniques. *Polym. Degrad. Stab.*, *66*, 153–155, 1999.

127. Quede, A., Mutel, B., Supiot, P., Jama, C., Dessaux, O., Delobel, R., Characterization of organosilicon films synthesized by N_2-PACVD. Application to fire retardant properties of coated polymers. *Surf. Coat. Technol.*, *180-181*, 265–270, 2004.

128. Horrocks, A.R., Nazaré, S., Masood, R. *et al.*, Surface modification of fabrics for improved flash-fire resistance using atmospheric pressure plasma in the presence of a functionalized clay and polysiloxane. *Polym. Adv. Technol.*, *22*, 22–29, 2011.

129. Totolin, V., Sarmadi, M., Manolache, S.O., Denes, F.S., Environmentally Friendly Flame-Retardant Materials Produced by Atmosphereic Pressure Plasma Modifications. *J. App. Polym. Sci.*, *124*, 116–122, 2012.

130. Hilt, F., Gherardi, N., Duday, D., Berne, A., Choquet, P., Efficient Flame Retardant Thin Films Synthesized by Atmospheric Pressure PECVD through the High Co-deposition Rate of Hexamethyldisiloxane and Triethylphosphate on Polycarbonate and Polyamide-6 Substrates. *ACS Appl. Mater. Interfaces*, 8, 12422–12433, 2016.

131. Kashiwagi, T., Gilman, J.W., Butler, K.M., Harris, R.H., Shields, J.R., Asano, A., Flame Retardant Mechanism of Silica Gel/Silica. *Fire Mater.*, 24, 277–289, 2000.

132. Gilman, J.W., Jackson, C.L., Morgan, A.B., Harris, R., Manias, E., Giannelis, E.P., Wuthenow, M., Hilton, D., Phillips, S.H., Flammability Properties of Polymer-Layered Silicate Nanocomposites. Polypropylene and Polystyrene Nanocomposites. *Chem. Mater.*, *12*, 1866–1873, 2000.

133. Gilman, J.W., Harris, R.H., Shields, J.R., Kashiwagi, T., Morgan, A.B., A study of the flammability reduction mechanism of polystyrene-layered silicate nanocomposite: layered silicate reinforced carbonaceous char. *Polym. Adv. Technol.*, *17*, 263–271, 2006.

7

Boron-Based Flame Retardants in Non-Halogen Based Polymers

Kelvin K. Shen

U.S. Borax, Huntington Beach, California, USA

Abstract

Boron compounds such as boric acid and borax are well known fire retardants for wood/cellulosic products and coatings. However, the use of zinc borates, ammonium pentaborate, melamine borate, boron phosphate, and other boron derivatives in polymers has become prominent only since early1990's. In recent years, boron nitride and boronic acid derivatives have attracted considerable attention in fire retardancy. Boron-based flame retardants have been reviewed extensively during the past eight years. This chapter will review recent development of boron based flame retardant during the last 3-4 years. Boron-based flame retardants and new applications that have commercial importance, as well as newly developed compounds with potential of becoming commercial will be reviewed.

Keywords: Borate, boric acid, zinc borate, boron nitride, ammonium pentaborate, boronic acid, boron phosphate, boron-doped graphene, flame retardant, smoke suppressant, afterglow suppressant, arcing/tracking suppressant

7.1 Introduction

Boron-based compounds such as borax are well-known flame retardant in cellulosic textiles, in use as far back as 1735. However, the use of boron compounds such as zinc borate, ammonium pentaborate, melamine borate, boric oxide, boron phosphate in plastics and rubbers has become prominent only since early 1990s. In recent years, research and publications

Email: KelvinKShen@gmail.com

Alexander B. Morgan (ed.) Non-Halogenated Flame Retardant Handbook 2nd edition, (309–336)

related to the use of boron-based flame retardants continue to attract considerable attention. For example, SciFinder ACS search displays at least about 300 publications (not including conference presentations) related to boron flame retardants during 2018-2020 alone.

The use of boron-based compounds in both halogen-containing and non-halogen based polymers were extensively reviewed recently, including this version of the Chapter in the 1st edition of this book [1–4]. This chapter only covers the major use of boron-based flame retardants in non-halogen based polymer systems that are developed or published around 2018-2020. The chemical/physical properties, end-use applications, fire test performance, as well as modes of action of these boron-based flame retardants will be reviewed. In particular, synergistic/beneficial interaction between "boron and metal hydroxide", "boron and nitrogen", "boron and phosphorus", "boron and carbon", "boron and silicon", and "boron and carbon" in fire test performances will be illustrated.

The market demand for halogen-free fire retardant polymers has been increasing steadily in applications such as electrical/electronics (E/E), transportation, and construction products. Since boron-based flame retardants have a wide spectrum of applications, only representative examples are presented in this chapter.

7.2   Major Functions of Borates in Flame Retardancy

In non-halogenated flame retardant polymers, boron-based compounds act as multifunctional flame retardants, providing many useful properties, including the following:

- Promote char/residue formation
- Stabilize the char and prevent dripping
- Inhibit oxidation of the char (afterglow suppression, etc.)
- Stabilize the char by forming glassy barrier
- Most borates release significant amount of water to provide flame retardancy
- Provide anti-tracking (i.e. Comparative Tracking Index–CTI) and anti-arcing properties of base polymer
- Effective flux and glass former for ceramification
- Display synergy with nitrogen, phosphorus, silicon compounds in fire test performances

- Suppress smoke and carbon monoxide formation
- Provide cross-linking between OH groups and thus char formation

While acting as flame retardants, borates can also provide additional functions depending on the polymer used:

- As a nucleating agent for polymer foam formation
- Provide corrosion inhibition function
- Provide UV protection
- IR screening
- Nuclear shielding for thermal neutron
- Photo cure catalyst

7.3 Major Commercial Boron-Based Flame Retardants and Their Applications

Table 7.1 illustrates the major boron-based commercial flame retardants, their properties, and applications. Based on its water solubility, temperature at which it releases water, and chemical property of the cationic moiety, formulators can select a particular borate for their specific flame retardancy requirements. For example, one normally selects a water insoluble borate for engineering polymer applications and water soluble borate for cellulosic applications [3].

7.4 Properties and Applications of Boron-Base Flame Retardants

In this section, boron-based flame retardants are discussed by specific boron chemical, their most recent uses, and some insight into how these flame retardants are used today.

7.4.1 Boric Acid [$B_2O_3 \cdot 3H_2O/B(OH)_3$], Boric Oxide ($B_2O_3$)

Boric acid (commercially available as *Optibor®* from U.S. Borax) is a white triclinic crystal that is soluble in water (5.46 wt. %), alcohols, and glycerin. It is a weak acid and has a pH of 4 (saturated solution at room temperature). Upon heating from 75 °C to around 125 °C, it loses part of its water

Table 7.1 Major commercial boron-based flame retardants.

Chemical name	Formula (Typical B_2O_3 wt.%)	Starting dehydration temp. (°C)	Water solubility (wt.%, ~25 °C)	Applications
Borax Pentahydrate	$Na_2O \cdot 2B_2O_3 \cdot 5H_2O$ (49.0 %)	65	4.4	Wood/cellulose/cotton, urethane, coating
Borax Decahydrate	$Na_2O \cdot 2B_2O_3 \cdot 10H_2O$ (37.5 %)	~45	5.8	Wood/cellulose
Boric Acid	$B_2O_3 \cdot 3H_2O$ (56.6 %)	70	5.5	Wood/cellulose/cotton, urethane, coating, PS foam, phenolic foam
Boric Oxide	B_2O_3 (98.5 %)	-	3.1	Engineering plastics
Anhydrous Borax	$Na_2O \cdot 2B_2O_3$ (68.8 %)	-	3.1	Urethane, wire & cable
Disodium Octaborate Tetrahdrate (*Polybor®*)	$Na_2O \cdot 4B_2O_3 \cdot 4H_2O$ (67.3 %)	40	9.7 (20 °C)	Wood products, cotton
Calcium Borate (Colemanite)	$2CaO \cdot 3B_2O_3 \cdot 5H_2O$ (44 - 48%)	290	0.2	Rubber modified asphalt roofing membrane

(Continued)

Table 7.1 Major commercial boron-based flame retardants. (*Continued*)

Chemical name	Formula (Typical B_2O_3 wt.%)	Starting dehydration temp. (°C)	Water solubility (wt.%, ~25 °C)	Applications
Zinc Borates				Various polymers, elastomers, coatings, sealants/caulkings
Firebrake®ZB	$2ZnO \cdot 3B_2O_3 \cdot 3.5H_2O$ (48%)	290	<0.28	
Firebrake®500	$2ZnO \cdot 3B_2O_3$ (54%)	Anhydrous	-	
Firebrake®415	$4ZnO \cdot B_2O_3 \cdot H_2O$ (17%)	~415	(discontinued)	
ZB-223	$2ZnO \cdot 2B_2O_3 \cdot 3H_2O$ (39%)	200	-	
Ammonium Pentaborate	$(NH_4)_2O \cdot 5B_2O_3 \cdot 8H_2O$ (64.6 %)	120	10.9	Epoxy, urethane, coating
Melamine diborate	$(C_3H_8N_6)O \cdot B_2O_3 \cdot 2H_2O$ (22.0 %)	130	0.7	Epoxy intumescent coating, cotton textile
Boron Phosphate	BPO_4 (18.7 % as B)	NA	low	PPE/polyamide, PPE/ HIPS, PO
Hexagonal Boron Nitride (h-BN)	BN (43.1% as B)	NA	Insoluble	Thermosets, thermoplastics, elastomers in E/E

of hydration to form metaboric acid (HBO_2). The metaboric acid can be further dehydrated to boric oxide (B_2O_3) at around 260-270 °C.

Boric oxide, also known as anhydrous boric acid, is a hard glassy material, which softens at about 325 °C, and melts at about 450-465°C. It can, however, absorb water and revert back to boric acid; however, this normally does not affect its fire retardancy performance. Thus when used in polymers that are processed at above 75 °C, boric acid will start to dehydrate to form metaboric acid and/or boric oxide. Boric oxide and boric acid are not recommended for use in non-polar hydrocarbon polymers, because boric acid (or boric acid formed via boric oxide hydration by moisture) may lead to migration to the polymer surface.

7.4.2   Alkaline Metal Borate

7.4.2.1   Borax Pentahydrate ($Na_2O \cdot 2B_2O_3 \cdot 5H_2O$), Borax Decahydrate ($Na_2O \cdot 2B_2O_3 \cdot 10H_2O$)

Borax Decahydrate (also called borax) is slightly soluble in cold water (4.71% by wt. at 20°C) and very soluble in hot water (30% at 60°C). It has a pH of 9.24 (1% solution at ambient temperature) and exhibits excellent buffering property. As a crystalline material, borax decahydrate is stable under normal storage condition. It will slowly lose water of crystallization if exposed to a warm and dry atmosphere. Conversely, exposure to a humid atmosphere can cause recrystallization at particle contact point, thus resulting in caking.

Borax Pentahydrate (commercially available as *Neobor®* from U.S. Borax) is the most common form of sodium borate used in a variety of industries. Its advantages vs. borax lie in the lower transportation, handling, and storage cost of a more concentrated product. Borax pentahydrate readily effloresces upon heating. It starts to dehydrate at about 65°C, loses all water of hydration when heated above 320°C, and fuses when heated above 740°C. In water, it hydrolyzes to give a mildly alkaline solution with excellent buffering properties.

Due to their low dehydration temperature and water solubility, sodium borates (except the anhydrous sodium borate) are normally only used as flame retardants in cellulose insulation (such as ground-up newspaper), wood timber, textiles, urethane foam, and coatings.

Borax pentahydrate and decahydrate are effective flame retardants for wood/cellulosic materials in terms of surface flammability. However, due to the Na_2O moiety, they promote smoldering combustion in cellulose. Thus, in cellulosic material and wood products, it is commonly used in combination with boric acid, which is an effective smoldering inhibitor.

In addition, a combination of boric acid and sodium borate can also result in significantly higher water solubility.

7.4.2.2 Disodium Octaborate Tetrahydrate ($Na_2O \cdot 4B_2O_3 \cdot 4H_2O$)

This unique form of sodium borate (known as *Polybor®* from U.S. Borax) is an amorphous material and thus can be dissolved into water rapidly (solubility 9.7 wt. % at room temperature and 21.9% at 30 °C). It is particularly effective in reducing the flammability of wood/cellulose/paper products.

Recent Application Examples of Boric Acid and Sodium Borate – Boric acid and the sodium borates are commonly used as flame retardants in epoxy, phenolics, PU foam, various textiles, and cellulosic/wood products [1–4]. In addition, in some unique application, boric acid has been used as an anti-arcing agent in electrical fuse application. Newer examples of use are discussed below.

Phenolic Foam – Boric acid is commonly used as a flame retardant in phenolics. Improving the toughness of phenolic foam has been a subject of considerable interest. Xu *et al.* reported modified foam was prepared by using polyurethane prepolymer (8%) and boric acid (10%). The formulation yielded excellent mechanical properties and flame test performance as evidenced by significant increase in limiting oxygen index (LOI) and drastic reduction in peak heat release rate (pHRR), total heat release (THR), and total smoke production (TSP) in the Cone Calorimeter test [5].

Cellulose/Particle Boards – Recently the use of graphene or its hybrid as a flame retardant has attracted considerable attention. Losic *et al.* reported using graphene-borate composite comprising hydrated sodium metaborate intercalating rGO (reduced graphene oxide, Figure 7.1; see also 7.4.8.1 Graphene section) layers for flame retarding cellulosic products. The authors claimed it has potential for use in commercial particle boards and medium density fiberboard [6].

Rigid Polyurethane Foam (RPUF) – Recently, Gao et.al. reported the use of a combination of graphene oxide, boric acid, and 3-aminopropyltriethoxysilane to prepare a functionalized graphene oxide (fGO). The silanol reacted with OH group of graphene and acting as a bridge between boric acid and GO. This fGO was incorporated flame retardant RPUF as a synergist for expandable graphite (EG) and dimethyl methyltriphosphonate (DMMP). The LOI reached 28.1% with 0.25 phr of fGO and 7.5 phr EG/2.5 phr DMMP, and the UL 94 test reached V-0. Remarkably, in the Cone Calorimeter test, the pHHR of the fGO foam relative to RPUF foam reduced by 33% and the THR reduced by 25.1%. In addition, fGO enhanced the mechanical properties of RPUF [7].

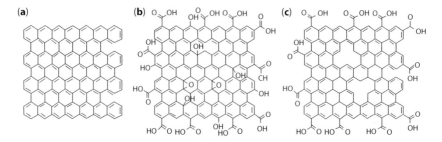

Figure 7.1 Schematic illustration of a) graphene, b) graphene oxide (GO), c) reduced graphene oxide (rGO). Source: J. Kim, F. Kim, J. Huang, Materials Today, 13, pp 28-38 (2010).

Polyamide 66 Fabrics – Flame retardant PA 66 fabrics with laundry resistance has long been a challenge for the textile industry. Hu's group reported the use of borate cross-linked thin coatings containing chitosan (CS, Figure 7.2) and phytic acid (PA, Figure 7.3) via a layer-by-layer assembly to develop a durable flame retardant PA 66 fabric [8]. The amino group of CS is believed to react with PA 66 backbone. Borate is believed

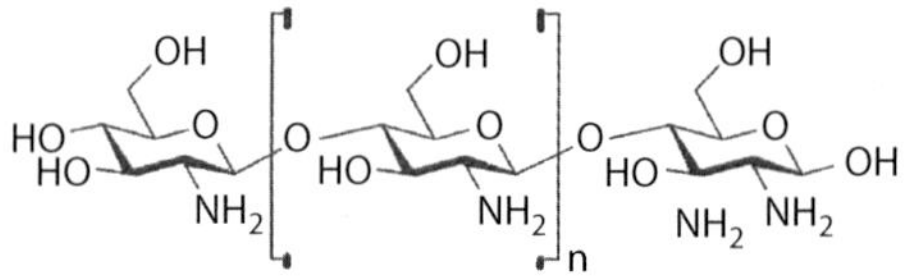

Figure 7.2 Chitosan.

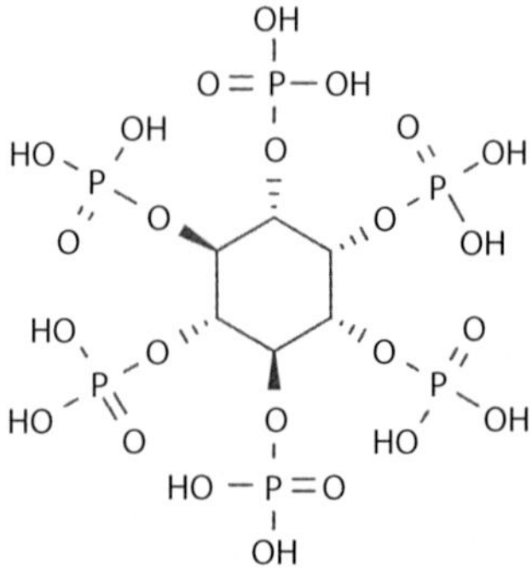

Figure 7.3 Phytic acid.

to cross-link CS and PA that exhibited better thermal resistance and, more importantly, it can also enhance the stability of the coating in sodium dodecyl sulfate solution and results in improvement laundry resistance.

7.4.3   Alkaline-Earth Metal Borate

7.4.3.1   Calcium Borates ($xCaO \cdot yB_2O_3 \cdot zH_2O$)

Among all of the known calcium borates, the natural mineral colemanite ($2CaO \cdot 3B_2O_3 \cdot 5H_2O$) is the most well known in the field of flame retardants (offered by American Borate Company, Sibelco, and Etimine). All calcium borates have low dehydration temperatures (around 115-200 °C) except colemanite, which has a dehydration temperature of 290-300 °C. It is used extensively in rubber modified asphalt roofing–membrane at loadings in the range of 12-14%. However it does not work well in high sloped roofing. This borate was also promoted for use in polyolefin wire and cable application without much success.

7.4.3.2   Magnesium Borate ($xMgO \cdot yB_2O_3 \cdot zH_2O$)

Due to its high charge to size ratio, the Mg^{2+} cation has a strong tendency to include water in its coordination sphere. Thus, most synthetic magnesium borates contain non-hydroxyl water which can cause them to have low dehydration temperature. For use in plastics, these magnesium borates have to be fully or partially pre-dehydrated. Kyocera reported the use of an unspecified magnesium borate in silica-filled epoxy/phenolic for electronic packaging. The addition of 25% magnesium borate resulted in a V-0 (1.6 mm) formulation with good moldability and high temperature reliability [3]. Preparation of magnesium borate ($2MgO \cdot B_2O_3 \cdot 1.5H_2O$) nano-rod and nano-wire were reported by Zhang *et al.* Compressed wood dust containing 20% nano-wire gave LOI of 29.3% vs. 23.8 % for non-nano magnesium borate. This type of nano-magnesium borate is still at academic research stage [9].

7.4.4   Transition Metal Borates

7.4.4.1   Zinc Borates ($xZnO \cdot yB_2O_3 \cdot zH_2O$)

Among all of the boron-containing fire retardants used in polymers, zinc borate has the most commercial importance. Major commercial zinc borates are shown in Table 7.1. The global suppliers include U.S. Borax,

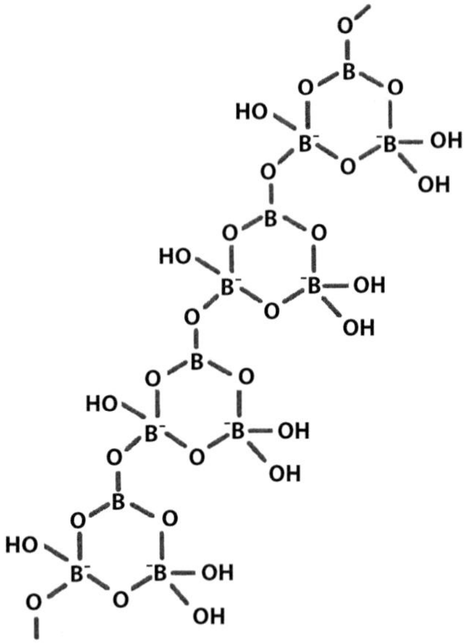

Figure 7.4 Molecular structure of *Firebrake*®ZB (zinc borate) (zinc ions complexing to dihydroxyborate anion is not displayed).

Laderello, Etimine, and Chinese suppliers. Lanxess (former Great Lakes Chemical) discontinued the supply of zinc borate.

Firebrake®ZB ($2ZnO \cdot 3B_2O_3 \cdot 3.5H_2O$) and *Firebrake*®500 ($2ZnO \cdot 3B_2O_3$) (from U.S. Borax) are the two predominant zinc borates in the flame retardant market. *Firebrake* ZB has no interstitial water and is stable to 290-300°C (Figure 7.4). It has a typical median particle size of 8-9 microns. Due to the demand of high production throughput and thin-walled electrical parts, engineering plastics are processed at increasingly higher temperatures. To meet this market demand, U.S. Borax also developed an anhydrous zinc borate that is stable to more than 500°C. *Firebrake*®415 ($4ZnO \cdot B_2O_3 \cdot H_2O$) stable to >415 °C is now discontinued.

7.4.4.1.1 *Firebrake* ZB ($2ZnO \cdot 3B_2O_3 \cdot 3.5H_2O$) and *Firebrake* 500 ($2ZnO \cdot 3B_2O_3$)

These zinc borates have been used in non-halogenated polyamide, polyolefin, epoxy, phenolics, and various elastomers. This zinc borate functions as a flame retardant, smoke suppressant, afterglow suppressant, and anti-arcing agent [1–4, 10]. *Firebrake* 500 can also be used in engineering thermoplastic polymers that are processed at 300 °C or above (such as polyphthalicamide, polyether-sulfone, etc.).

Polyolefins – In non-halogen containing polyolefins, depending application, high loadings of alumina trihydrate/aluminum hydroxide (ATH) or magnesium hydroxide (MDH) are required (30-70 wt.%) in polyolefin. Recent developmental efforts have focused on co-additives of ATH or MDH with the aim of reducing total loading and developing stronger char/residue formation. Additional details on ATH and MDH can be found in Chapter 3 of this book. The use of a combination of metal hydroxide and zinc borate in wire and cable is quite common and previously reviewed [3, 10].

   a. Polypropylene (PP) Copolymer – Recently it was claimed that the use of a combination of PP copolymer (~26%)/Polybutene (~6 %)/MDH (60%, Huber)/Zinc Borate (7%, *Firebrake* ZB-XF, extra fine grade)/Stabilizers can meet UL 2043 (a fire and smoke test for materials in air handling areas, like plenum spaces) [11, 12]. At this high filler loading, the use of polybutene is critical for the improved melt index and processability. The formulation is intended for manufacturing of discrete articles such as plastic connector in heating, ventilation, and air conditioning (HVAC) duct assembly or a plastic conduit in plenum where the articles have to meet pHRR 100 KW or less, peak normalized OD (Optical Density) of 0.50 or less, and an average normalized OD of 0.15 or less (at 60 KW/m^2 heat flux).
   b. Ethylene-propylene diene monomer (EPDM) – Elastomer articles used in European rail vehicles are required to meet the flammability standard EN 45545-2 that covers flame propagation, smoke density, smoke toxicity, and heat release rate. Mestan claimed that a combination of EPDM 80 parts/EVA 20/carbon black 50/paraffinic oil 10/MDH 90/Zinc Borate 10/silicic acid and kaolinite 15 can meet that EN 45545-2 HL3 (Hazard Level 3, for subway, couchette car) as well as other critical physical properties. It is remarkable the formulation requires very low loading of flame retardants and achieves low smoke requirement. The use of paraffinic oil rather than mineral oil (containing aromatic moiety) is also critical for the low smoke [13].

Polyamides – Zinc borates have been used extensively in polyamides and polyphthaliamide (PPA or High Temperature Nylon). Their major applications are electrical connectors, micro-switches, bobbings, micro-sensors, and many other E/E parts. *Firebrake* ZB can be processed at temperature up to 300 °C and *Firebrake* 500 can be processed at above 300 °C [10]. The latter can be used in PA 66 as well as in PPA.

    a.   Polyamide 66 and Polyamide 6 – For glass fiber reinforced PA 66, Clairant reported the use of a combination of aluminum diethylphosphinate (DEPAL), melamine polyphosphate (MPP), and *Firebrake* 500 to achieve V-0 (down to 0.4 mm). Recently, Hoerold *et al.* reported the use of zinc diethylphophinate (DEPZN) in place of DEPAL in the combination resulted in not only V-0 at 0.4 mm but also better CTI (Comparative Tracking Index), less exudation, less corrosion, better melt flow, and better impact and notched impact resistance [14]. Similar benefits were observed in glass fiber reinforced PA 6 with the same combination.

    b.   Polyamide 12 – PA12 is a semi-crystalline material with longer aliphatic chain than that of polyamide 66 (PA 66). They have low moisture absorption/good dimensional stability, good flexibility, good abrasion, UV resistance, and good retention of performance over time. But they do have poorer impact resistance, lower dielectric strength, and higher cost relative to PA 66. Table 7.2 illustrates the use of aluminum diethylphosphinate (18%, Exolit 1230 from Clariant) and zinc borate (2%, *Firebrake* ZB) in unreinforced PA12 to achieve UL 94 V-0 (1.5 mm). Table 7.3 illustrates the use of Exolit 1312 (22%, a blend of DEPAL/MPP/Zinc Borate) in glass reinforced PA12 to achieve V-0 (0.8 mm) [15].

    c.   Polyamide 12 (3D Printing) – Among the different additive manufacturing techniques (i.e. 3D printing) available for commercial production, Selective Laser Sintering (SLS) allows the production of complex 3D shapes for aerospace and defense industries. The SLS is based on selective fusion and recrystallization of polymer powders layer by layer (generally with a CO_2 laser). Lopez-Cuesta and co-workers reported a systematic evaluation of flame retardant additives in semi-crystalline PA12 for powder flowability, thermal degradation when subjected to laser beam, and final porosity of the finished articles [16]. Ammonium polyphosphate, zinc diehtylphosphinate (Exolit OP 950), zinc borate (*Firebrake* ZB), and pentaerythritol ester of methane

Table 7.2 Flame retardant unreinforced polyamide [15].

Examples (wt.%)				
Components	**1**	**2**	**3**	**4**
PA 12	80	80	80	80
Al Diethylphosphinate (Exolit OP 1230 from Clariant)	20	10	10	18
Melamine Cyanurate	-	10	-	-
Melamine Pyrophosphate	-	-	10	-
Zinc Borate (*Firebrake* ZB from U.S. Borax)	-	-	-	2
Properties				
UL 94 (1.5 mm)	V-2	V-2	V-2	V-0
Burning Time (s)	5/2	4/3	7/3	2/2
Burning Drops	Yes	Yes	Yes	No
Dripping but no burning drops	Yes	Yes	Yes	Yes
LOI (%)	28	28.5	28	32

Source: Claraint

phosphonic acid (Aflammit PCO900) were evaluated individually, as well as compositions consisting of 2-3 FR components. It was found that, for compositions in which only one flame retardant was used, the zinc borate gave the best fire test performance. Moreover, for binary and ternary compositions, the zinc borate was able to form zinc phosphate and borophosphate to retain nearly all of the phosphorus in the composition, and resulted in good fire test performances.

Epoxy (carbon fiber reinforced composite) – Carbon fiber reinforced epoxy is used extensively in the transportation industry (aircraft, high performance automotive). It is mostly processed via a RTM (resin transfer molding technique). It is also well known that carbon fiber can be oxidized and form respirable fiber fragment after fire (Figure 7.5). Eibl evaluated the effects of different fire conditions on the potential formation of

Table 7.3 Flame retardant of GF reinforced polyamide[15].

Examples (wt.%)			
Components	**1**	**2**	**3**
PA 12	47.7	42.7	37.7
Glass Fiber HP3610	30	35	40
Exolit OP1213 (a blend of DEPAL/MPP/Zinc Borate)	20	10	10
Licowax E	0.3	0.3	0.3
Properties			
LOI (%)	31.6	32.6	34.4
UL 94 (0.8mm) Burn Time (s)	V-2 97	V-1 77	V-0 35
UL 94 (1.6 mm) Burn Time (s)	V-1 70	V-1 68	V-1 78
UL 94 (3.2 mm) Burn Time (s)	V-0 20	V-0 22	V-0 15
Impact Strength	60	50	44
E-Modulus	8,917	10,414	11,714
Tensile Strength	91.7	95.4	97.6
Elongation at Break (%)	4.6	3.6	2.7

Source: Clariant.

fiber fragments that can cause safety concern [17]. It was discovered that effective fiber protection was achieved by flame retardants acting beyond 600 °C, forming thermally resistant layers such as zinc borate (Figure 7.6). It should be noted that, in RTM, the epoxy viscosity should be kept low, the loading level of solid inorganic additive should be kept low, and particle size should be small, as solid particles can be "filtered out" by the carbon fibers if the particles are too large in size. While it is not always true, in general, particles smaller than 1 micron will be able to migrate through carbon fibers in a RTM process.

Silicone Elastomer – Commercially zinc borate has been used in silicone wire and cables, and insulating material applications. It is known

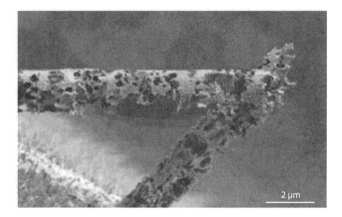

Figure 7.5 SEM of an IM7 carbon fiber after 20 min thermal treatment at 600 °C (Source: S. Eibl, Fire and Materials 2017).

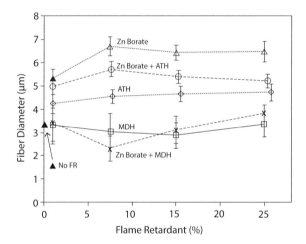

Figure 7.6 Fiber diameter at the surface of the irradiated side of 2-mm thick RTM6/ G0939 epoxy panels modified with various flame retardants after combustion in a cone calorimeter irradiated at 60 kW/m² for 20 minutes, measured by SEM; initial diameter -7µ; Epoxy RTM laminate (2.0 mm) with 8 piles of carbon fabrics (Source: S. Eibl, Fire and Materials, 2017, Vol. 42, Issue 7, p. 808-807).

that borate and silica (derived from silicone) or other mineral fillers (such as ATH) can result in ceramification during polymer combustion [4, 18]. Among all of the borates evaluated, zinc borate provides the best electrical properties [19]. The effects of zinc borates content and the combustion temperature on the properties of the ceramifiable silicone rubber were recently reported [20].

Coatings (water-based) – Flame retardant transparent coating has always been a challenge for the coating industry. Armstrong World Industries claimed the use of a transparent flame retardant coating for interior wood building panels. The inorganic coating consists of a top coat of phosphoric acid/*Firebrake* zinc borate/$CaCO_3$/$MgCO_3$/triethanolamine (pH 2.5) and an intermediate coat of phosphoric acid/boric oxide/$CaCO_3$/$Al_2(CO_3)_3$/$MgCO_3$/fumed silica/triethanolamine. This coating on wood substrate is believed to be able to meet the Flame Spread Rating of Class A in the ASTM E-84 tunnel test. It is intended for maintaining superior aesthetic appearance [21].

7.4.4.1.2   Miscellaneous Metal Borates

A variety of transition and other metal borates are reported in the literature. These include aluminum borate, silver borate, iron borate, copper borate, nickel borate, strontium borate, lead borate, zirconium borate, manganese borate, etc. [3]. Acicular aluminum borate was claimed for use in PA 66 [22].

7.4.5   Nitrogen-Containing Borates

7.4.5.1   Melamine Diborate [(C₃H₈N₆)O·B₂O₃·2H₂O)]/ (C₃H₆N₆·2H₃BO₃)

7.4.5.1   Melamine Diborate [($C_3H_8N_6$)O·B_2O_3·$2H_2O$)]/ ($C_3H_6N_6$·$2H_3BO_3$)

Melamine diborate (MB), known in the fire retardant trade as melamine borate, is a white powder, which can be prepared readily from melamine and boric acid. It is partly soluble in water and acts as an afterglow suppressant and char promoter in epoxy and cellulosic materials (Figure 7.7). The dehydration temperature starts at around 130 °C.

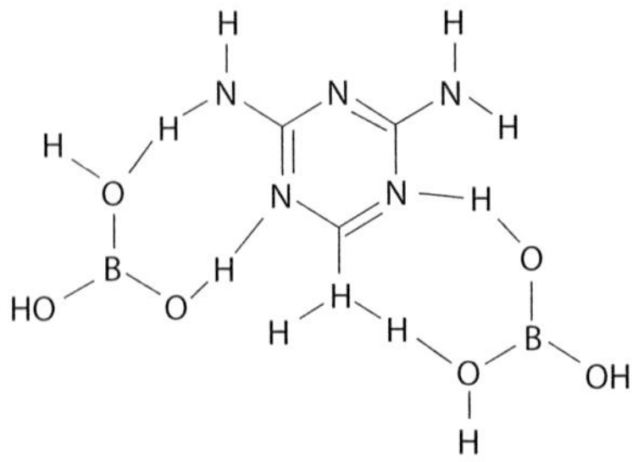

Figure 7.7 Melamine diborate.

7.4.5.2   Ammonium Pentaborate [(NH$_4$)$_2$O·5B$_2$O$_3$·8H$_2$O)]

Ammonium pentaborate (APB) has a water solubility of 10.9 wt.% at 25°C. Upon thermal decomposition, APB first gives off a large amount of water starting at about 120 °C. At about 200 °C, it starts to give off ammonia and, at about 450 °C, it is all converted to boric oxide which is a glass former. APB functions as both inorganic blowing agent and a glass forming fire retardant.

$$(NH_4)_2O·5B_2O_3·8H_2O \rightarrow 2NH_3\uparrow + 9H_2O\uparrow + 5B_2O_3$$

APB has been used in paper products, PU foam for anechoic chamber, and epoxy intumescent coatings. APB's usage in polymers however is limited by its high water solubility (10.9%) and low dehydration temperature.

Epoxy Intumescent Coating – Previously, PPG claimed the use of APB in conjunction with ammonium polyphosphate (APP), zinc borate, silica, talc, mineral fibber, etc. in epoxy intumescent coating for protection of steel against hydrocarbon fires. Recently, Wu *et al.* reported the use of high loading of APB with no zinc borate in the coating. The formulation consisted of epoxy (25)/Polyamide (18)/APP (6)/melamine (9)/APB (25)/tris(1-choloro-2-propyl)phosphate (TCPP, 5)/others [23]. This formulation has an excellent low temperature resistance properties and good UL 1709 fire test performance for coated structural steel application [24].

7.4.5.3   Boron Nitride (h-BN)

Hexagonal boron nitride (h-BN) can normally be prepared from reaction of boric acid and urea or melamine. It is commonly referred to as "white graphite", because of its platy hexagonal structure similar to graphite with boron nitride atoms replacing carbon atoms. Under high pressure and at 1600 °C, the h-BN is converted to cubic BN, which has a diamond-like structure.

The h-BN platelets can be exfoliated to generate nanosheet (Figure 7.8). It has been found to be a good flame retardant for many polymer systems. It is stable in inert or reducing atmosphere to about 2700 °C and in oxidizing atmosphere stable to 850 °C.

There is a tremendous growing demand for thermally conductive and electrically insulating polymer materials. This includes many electrical/elctronic devices such mobile phone, light emitting diodes (LED) for display/lighting, and many electrical vehicle parts. The h-BN is an electrically insulating and highly heat conductive additive having high anisotropic thermal conductivity properties that are desired for many applications.

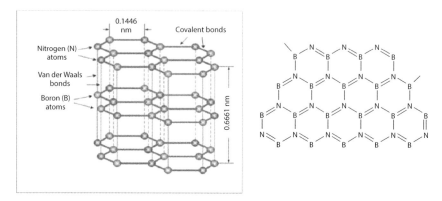

Figure 7.8 Hexagonal boron nitride lattice and nanosheet (on the right). (left source: American Chemical Society).

Acrylics – 3M reported the use of h-BN platelets in acrylics (~73% loading) with 3D printing by fused filaments fabrication (FDM). Carefully controlled thickness of the fused filament allows the fabrication of electrical parts with either in-plane or through-plane orientation with maximum thermal conductivity [25].

Polyurethane Foam – It was reported that h-BN platelet can be partially hydrolyzed to hydroxylated nanosheet via a sonication-assisted process. This functionalized h-BN is water compatible. Grunlan, *et al.* reported that the hydroxylated h-BN nanoparticles were conformally deposited on the cell wall of the flexible foam with polyethylenimine in a single 82 nm bilayer via a layer-by-layer assembly (the foam was pretreated with polyacrylic acid). The treated foam resulted in reduction of pHRR (54%), THR (20%), and gas emission. The PEI/h-BN coating also protects PU foam from weathering and only modestly alters mechanical property of the foam [26].

Polyurethane (WPU- water-borne polyurethane) – Red phosphorus (RP, amorphous) is commonly used as an effective flame retardant in polyamide 66, unsaturated polyester, polyolefin foam, etc. Black phosphorus (BP) has a 2D layered structure similar to that of graphite. Under normal conditions, RP is less stable than the thermodynamically stable black phosphorus (BP). The use of black phosphorus (BP) and boron nitride (h-BN) nanosheets as a flame retardant in WPU was recently reported [27]. The LOI of the BP (0.2 wt.%)/BN(0.2 wt.%)/WPU composite increased from 21.7% to 33.8% compared with pure WPU. It was demonstrated that, in the Cone Calorimeter test, the residue of BP/BN/WPU was denser - about 10 times more that of pure WPU. Efficient flame retardancy at low loading is attributed to the synergistic effects of BP and h-BN.

Natural Rubber (NR) – Rapid development of high-performance electronic devices has focusing on miniaturization, integration, and high power consumption has led to requirement of effective thermal management. Lu *et al.* reported the use of a high loading of h-BN and graphene oxide (~200:10) in NR latex to achieve in-plane thermal conductivity of 16 W/(m·K) vs. less than 10 W/(m·K) for general thermal interface materials [28]. More importantly, the rGO-BN-NR film can still maintain impressive mechanical properties (rGO stands for reduced graphene oxide, see Figure 7.1 and 7.4.8.1 Graphene section).

Unsaturated Polyester (UP) – Song and Hu *et al.* reported the synthesis of h-BN nanosheets containing phosphorus, nitrogen, and silicone as high performance nano-fillers in UP. At 3% loading in UP, there were obvious reduction of 28% and 38% in pHRR and THR, respectively [29].

7.4.5.4   Ammonium Borophosphate

It was reported that boiling a mixture of boric acid and diammonium phosphate resulted in a solid that has flame retardancy effects on unsaturated polyester. The solid was not analyzed and could be a ammonium borophosphate (ABP) [30]. Forg and Hoppe reported the synthesis of ammonium borophosphate $(NH_4)_2[B_2P_3O_{11}(OH)]$ via a hydrothermal method. Unlike ammonium polyphosphate, the material is stable to about 330 °C. This remarkable thermal stability could render it to be a very useful flame retardant for polymers processed at high temperatures [31].

7.4.6   Phosphorus-Containing Borates

7.4.6.1   Boron Phosphate (BPO$_4$)

Boron phosphate, an inorganic polymer of empirical formula BPO_4 (BPO) can be prepared by heating phosphoric and boric acids to calcining temperature. It is a white infusible solid that vaporizes at 1450-1462°C without decomposition. As opposed to the tri-valency in boric oxide, most of the boron atoms in boron phosphate are tetra-coordinated (i.e. as BO_4^- group, Figure 7.9). This low cost material have been claimed for use as a flame retardant in polymers such as epoxy,

TPU (thermoplastic polyurethane), polyimide, etc.–Commercially, BPO has been used in PPO-HIPS (polyphenylene oxide-high impact polystyrene alloy) and PPO/PA (polyphenylene oxide-polyamide alloy). In these alloys, it is normally used in conjunction with other flame retardant such as phosphate esters. Hao *et al.* reported that boron phosphate can act

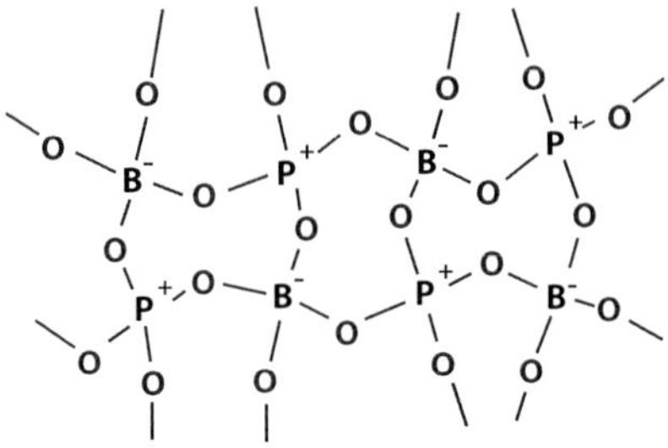

Figure 7.9 Boron phosphate.

as an acid catalyst to promote char formation and the boron moiety can stabilize the char [32].

7.4.6.2   Metal Borophosphate

BASF reported the preparation of zinc and magnesium borophosphates. These compounds were found to exhibit high thermal stability and are useful as flame retardants for plastics. Partial replacement of Exolit 1312 (a mixture of phosphinate, MPP, and zinc borate from Clariant) with zinc borophosphate (15:5 ratio) can still maintain the UL 94 V-0 at 0.8 mm in glass reinforced PA 66 [3]. The advantage of using metal borophosphate is that it contains both boron and phosphorus. For ease of processing, it could partially replace both MPP and zinc borate [33]. It is believed that metal borophosphate is predominately a condensed phase flame retardant. In the presence of aluminum diethylphosphinate, boron phosphate and aluminum phosphate may be formed to render a strong char residue.

7.4.7   Silicon-Containing Borates

Borate and silica are known to function synergistically as a flame retardant by forming borosilicate ceramic/glass residue as the results below indicate.

7.4.7.1   Borosilicate Glass and Frits

Borosilicate glass and frits are a range of glasses/ceramics based on boric oxide, silica, and metal oxides. They have excellent thermal shock resistance and chemical resistance. The glasses with low melting points are normally used as a co-additive in a flame retardant formulation [3]. A recent report presented a systematic evaluation of a combination of IFR (piperazine pyrophosphate/ZnO/hollow glass microsphere (a sodium borosilicate glass)/proprietary polyphosphate/compatibilizer in PP. It was concluded that partial replacement of IFR with hollow glass (~2-4%) resulted in clear improvement of LOI, UL-94, and Cone Calorimeter tests. The Scanning Electron Microscope-Energy Dispersive Spectroscopy (SEM-EDS) and Cone Calorimeter test results suggested that silicate can migrate to the outer surface and promote the formation of a compact char [34].

7.4.8   Carbon-Containing Boron or Borates

7.4.8.1   Graphene (Boron-Doped)

Graphene has excellent thermal/electrical conductivities and mechanical properties in polymer composites. Incorporation of graphene at <5% in polymers can result in reduction of heat release and smoke production. Graphene can be oxidized at around 300 °C in air. To improve its compatibility with polymers, it normally has to be functionalized such as being converted to graphene oxide (GO) and reduced graphene (rGO) [35].

Epoxy – Boron is known to inhibit oxidation of carbon species including carbon fiber, carbon nanotubes (CNT), diamond, etc. Substitutional boron doping of carbon material is well known.

Wang and Hu *et al.* reported the preparation of boron-doped graphene oxide (GO) by annealing boric acid and GO at 900 °C in argon. The boron content of BG (boron-doped graphene) is about 3%, and it is thermally stable to about 500 °C. In their study, in order to overcome the brittleness and flammability of epoxy resins (non-reinforced), a cardanol-benzoxazine monomer (CBz) was prepared. The incorporation of CBz can improve both the flame retardancy and impact strength of epoxy resin simultaneously. For example, with the addition of 10% CBz, the epoxy/CBz sample displays 31% in LOI as well as a V-2 rating in the UL-94 test. In contrast, when combining 8 wt% CBz and 2% of BG, the sample can meet V-0 rating (3 mm). In the Cone Calorimeter test, neat epoxy displayed a pHRR of 1262 kw/m^2 and the composite with 15% CBz exhibited pHRR of 920 kw/m^2 whereas with 13% of CBz and 2% of BG resulted in 650 kw/m^2. In the presence of BG, the composite also displayed better total heat release rate, less total smoke production, and much better char integrity. The superior flame retardancy of Epoxy/CBz/BG was ascribed to BG enhancing the char formation/char stability [36].

7.4.8.2   Boric Acid Esters [B(OR)$_3$]

In general, boric acid esters (also called borester) derived from alcohols and boric acid are hydrolytically unstable for general use in plastics/elastomers, wood, paper, or cotton. The reaction between polyvinyl alcohol and boric acid via x-linking between polymer chains by borate can lead to significant reduction in flammability as evidenced by increase in LOI and char yields [1].

7.4.8.3   Boronic Acid [ArB(OH)$_2$]

Boronic acids are known to release water on thermolysis, thereby leading to the formation of boroxines or boronic anhydrides. Unlike boric acid ester, B-C bond in boronic acid is not readily hydrolysable. Morgan and Benin's group did a lot of pioneering work on the use of boronic acid derivatives as flame retardants for polyurethane materials [1].

Boronic acid derivatives are not yet used as flame retardants in commercial application. However, there have been many recent research reports and patents in this field.

Epoxy – Liu et al. reported the preparation of hexakis(4-boronic acid-phenoxy)-cyclophosphazene (i.e. boronic acid group at the para position of the phenoxy moiety) (CP-6B). At 3.0% loading, this material in epoxy achieved LOI of 30.8% and UL-94 V-0 at 3.0 mm. Interestingly enough, when CP-6B is used in conjunction with 0.5% magnesium hydroxide, a much better char yield (increased from 14.6% to 17.2%), as well as better pHRR and THR in the Cone Calorimeter test [37].

Poly(lactic acid) (PLA) – Fei et al. reported the preparation of a macromolecule with p-formylphenylboronic acid, pentaerythritol, and tris(2-aminoethyl)amine (Figure 7.10). With a very low loading (~2%) in PLA, this microporous macromolecule can enhance remarkably the flammability performance such as LOI 31.5%, UL-94 V-0 and it showed reductions

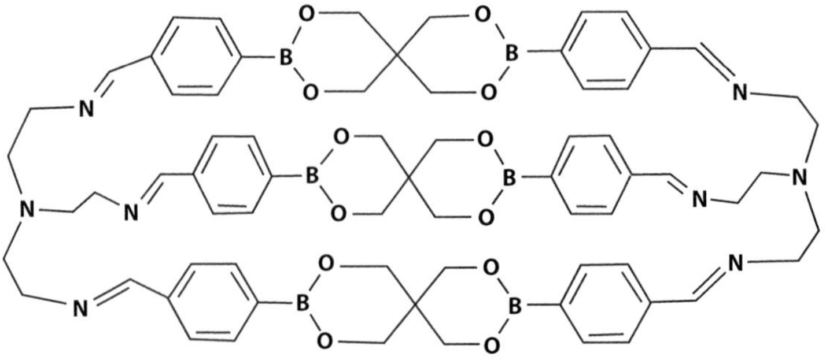

Figure 7.10 Macromolecule prepared from p-formylphenylboronic acid, pentaerythrirol, and tris(2-aminoethyl)amine (Source: B. Fei et al., Chemical Engineering Journal, 402 (2020)126209).

in pHRR, THR, total smoke release (TSR), and CO emissions based on the Cone Calorimeter test. In addition, it can improve UV protection [38].

In a separate study, the same research group reported grafting azophenylboronic acid to GO followed by *in-situ* reduction/intercalation with sodium metaborate. The resulting hybrid displayed substantial improvement in FR performance of PLA [39].

7.4.8.4   Boron Carbide (B_4C)

Boron carbide is produced industrially by the carbo-thermal reduction of B_2O_3 in an electric arc furnace. It is a black powder and has a melting point of 2445°C. In dry air, at around 400-500 °C, B_4C can be oxidized to B_2O_3. In water-air system, it can be oxidized by water to produce boric oxide, C/CO_2, and hydrogen.

$$B_4C \text{ (s)} + 4O_2 \text{ (g)} \rightarrow 2B_2O_3(l) + CO_2 \text{ (g)}$$

$$B_4C \text{ (s)} + 6H_2O \text{ (g)} \rightarrow 2B_2O_3 + C(s) + 6H_2 \text{ (g)}$$

It was reported that the addition of boron carbide (10-15 wt %) in a variety of intumescent coatings containing ammonium polyphosphate and blowing agent resulted in the following benefits [3]:

- forming a uniform and hard foamed layer.
- preventing excessive expansion on heating.
- retarding oxidative weight loss of the foamed layer at 1000°C or even higher.
- improving weight retention, compression strength, and peel strength during fire.

7.5   Mode of Actions of Boron-Based Flame Retardants

- Most borates can release significant amount of water to suppress flaming combustion
- In wood/cellulosic products, boric acid can increase the thermal stability of wood and suppress the mass loss by inhibiting cellulose fragmentation (i.e. alleviate levoglucosan formation).
- Borates can function as a flux/sintering aid for fillers/additives to form a porous insulative barrier

- Borates can stabilize the char by forming a glassy layer and prevent dripping
- In PU, it is believed boric oxide/boric acid released may react with the diol and/or isocyanate. The decomposition fragments from PU produce a highly cross-linked borate ester and possibly boron-nitrogen polymer that can reduce the rate of formation of flammable volatiles and result in intumescent char.

$$B_2O_3 + 6ROH \rightarrow 2B(OR)_3 + 3H_2O$$

$$H_3BO_3 + 3RNCO \rightarrow B(NHR)_3 + 3CO_2$$

- Boron-Nitrogen- When borate is used in a nitrogen-containing polymer or in conjunction with nitrogen-containing co-additive, it is postulated that boron nitride is formed during polymer combustion.
- Boron-Phosphorus- In the presence of phosphorus compounds such as APP and boric acid or zinc borate, one can generate boron phosphate or metal borophosphate that can help to maintain the integrity of the char. Schartel *et al.* reported that addition of MPP/zinc borate to aluminum diethylphosphinate in polyamide causes the change of mode of action from gas phase to a mixed gas/condensed phase. They detected the formation of boron phosphate and aluminum phosphate in the condensed phase.
- Boron-Silicon – Borates such as *Firebrake* zinc borate can function as a flux and form eutectic mixture (lower melting point) at the particle-particle interface between silicas (or silicate minerals) at medium high temperatures. At high temperatures, borates can undergo fusion with silica or silicate minerals. It is believed that borosilicate glass or ceramics is formed.
- Boron-Carbon – Boron can inhibit the oxidation of carbon species via glassy borate formation on the char/carbon surface, or via electron withdrawing capabilities of the boron sp^2 orbital.

7.6   Conclusions

- Boron-based flame retardants have very broad applications. They are multifunctional fire retardants and can function as flame retardants, smoke suppressants, afterglow suppressants, anti-tracking/anti-arcing agents, etc.

- Borates function as flame retardants predominately in condensed phase.
- Borates have been commercially successful in both plastics and elastomers. The fire test performance of borates can be enhanced with the use of co-additives such as metal hydroxides, N, P, Si-containing, and carbon species.
- Borates alone can also function as flame retardant in char forming polymers such as polyethersulfone, polyimide, etc.
- Hexagonal boron nitride will play an increasingly important role in thermal management and flame retardancy in E/E applications.
- Boronic acid derivatives will continue to attract considerable attention in future flame retardant research, especially as non-halogenated flame retardant use increases due to decreasing demand for halogenated flame retardants.

References

1. Shen, K.K. *et al.*, Chapter 9. Boron-based flame retardants and fire retardancy, in: *Fire Retardancy of Polymeric Materials*, 2nd Edition, C.A. Wilkie and A.B. Morgan (Eds.), pp. 207–237, CRC/Taylor & Francis, Boca Raton, FL, 2010.

2. Shen, K.K., Chapter 11- Review of recent advances on the use of boron-based flame retardants, in: *Polymer Green Flame Retardants*, C.D. Papaspyrides and P. Kiliaris (Eds.), pp. 367–388, Elsevier, Amsterdam, Netherlands, 2014.

3. Shen, K.K., Chapter 6. Boron-based flame retardants in non-halogen based polymers, in: *Non-Halogenated Flame Retardant Handbook*, A.B. Morgan and C.A. Wilkie (Eds.), pp. 201–235, Scrivener Publishing, Salem, Massachusetts/Wiley, Hoboken, New Jersey, 2014.

4. Shen, K.K., Chapter 6. Recent Advances in Boron-Based Flame Retardants, in: *Flame Retardant Polymeric Materials: A Handbook*, Y. Hu and X. Wang (Eds.), pp. 97–117, Taylor & Francis, Boca Raton, FL, 2020.

5. Xu, W., Chen, R., Xu, J., Wang, G., Chen, C., Yan, H., Preparation and mechanism of polyurethane prepolymer and boric acid co-modified phenolic foam composite: mechanical properties, thermal stability, and flame retardant properties. *Polym. Adv. Technol.*, Vol. 30, p. 1738–1750, 2019.

6. Nine, M.J., Tran, D.N.H., Tung, T.T., S.Kabiri, D., Losic, Graphene-Borate as an efficient fire retardant for cellulosic material with multiple and synergistic modes of action, *ACS Appl. Mater. Interfaces*, Vol. 9, p. 10160–10168, 2017.

7. Ming, G., Li, J., Zhou, X., A flame retardant rigid polyurethane foam system including functionalized graphene oxide. *Polym. Compos.*, 40, S2, 1–9, 2018.

8. Kundu, C.K., Wang, X., Song, L., Hu, Y., Borate-cross-linked, layer by layer assembly of green, photoelectrolytes on polyamide 66 fabrics for flame-retardant treatment. *Prog. Org. Coat.*, 121, 173–181, 2018.

9. Zhang, L. and Liu, Z., Preparation of $2MgO \cdot B_2O_3 \cdot 1.5H_2O$ nano materials and evaluation of their flame retardant properties by a thermal decomposition kinetic method. *J. Therm. Anal. Calorim.*, 129, 715–719, 2017.

10. Shen, K.K., Kochesfahani, S., Jouffret, F., Zinc borates as multifunctional polymer additives. *Polym. Adv. Technol.*, Vol. 19, p. 469–474, 2008.

11. Surraco, S., and Davidson, D., Flame retardant compositions, U.S. Patent Application 2020/0032042 A1, 2020.

12. Standard for fire test for heat and visible smoke release for discrete products and their accessories installed in air-handling spaces, 2013, UL 2043 (Underwriter Laboratory, Northbrook, Illinois).

13. Mestan G., Flame retardant polymer composition, U.S. Patent Application 2020/0032030 A1 2020.

14. Hoerold, S., and Schlosser, E., Flame retardant mixtures for thermoplastic polymers, PCT Patent Application, WO 2020/165017 A1, 2020.

15. Source: Clariant

16. Batistella, M., Regazzi, A., Pucci, M.F., Lopez-Cuesta, J.-M., Kadri, Q., Bordeaux, D., Ayme, F., Selective laser sintering of polyamide 12/flame retardant compositions. *Polym. Degrad. Stab.*, 181, 109318, 2020.

17. Eibl, S., Potential for the formation of respirable fibers in carbon fiber reinforced plastic materials after combustion. *Fire and Mater.*, Vol. 41, Issue 7, p. 808-816, 2017.

18. Li, Y., Hu, S., Wang, D.-Y., Polymer-based ceramifiable composites for flame retardant applications: A review. *Composite Commun.*, 21, 100405, 2020.

19. Guo, H., Chen, X., Zhang, Y., Improving the mechanical and electrical properties of ceramifiable silicone rubber/halloysite composites and their ceramic residue by incorporating different borates. *Polymers*, 2018, 10, 388.

20. Song, J., Huang, Z., Qin, Y., Li, X., Thermal decomposition and ceramification of ceramifiable silicone rubber composite with hydrated zinc borate. *Materials*, 2019, 12, 1591.

21. Hughes, J., Wang, M., Flame retardant clear coating for building panels, U.S. Patent Application 2018/0347185 A1 (2018), assigned to Armstrong World Industries.

22. Takeda, T., Kirikoshi, H., Flame retardant polyamide resin composition, U.S. Patent Publication 2002/0002228 (2002).

23. Wu, H., Zhu, L., Yang, L., Intumescent coating composition, U.S. Patent Application US 2020/0181429 A1 (2020).

24. Standard for rapid rise fire tests of protection materials for structural steel, 2017, UL 1709 (Underwriter Laboratory, Northbrook, Illinois).

25. Schadel, R. M., Unibel, K. B., Wildhack, S., 3D printed component comprising a thermoplastic composite material and boron nitride, method for making a 3D printed component and use of 3D printed component, WO 2020/004236 A1 (2020), assigned to 3M.

26. Davesne, A., Lazar, S., Bellayer, S., Qin, S., Grunlan, J.C., Bourbigot, S., Jimenez, M., et.al., Hexagonal boron nitride platelet-based nanocoating for fire protection. *ACS Appl. Nano Mater.*, 2, 5450–5459, 2019.

27. Yin, S., Ren, X., Lian, P., Zhu, Y., Mei, Y., "Synergistic effects of black phosphorus/boron nitride nanosheets on enhancing the flame retardant properties of water-borne polyurethane and its flame retardant mechanism,". *Polymer*, 12, 1487, 2020.

28. Li, J., Zhao, X., Wu, W., Zhang, Z., Xian, Y., Lin, Y., Lu, Y., Zhang, Li, Advanced flexible r-GO-BN natural rubber films with high thermal conductivity for Improved thermal management capabilities. *Carbon*, 162, 46–55, 2020.

29. Wang, D., Mu, X., Cai, W., Song, L., Ma, C., Hu, Y., Constructing phosphorus, nitrogen, silico-co-contained boron nitride nanosheets to reinforce flame retardant properties of unsaturated polyester resin. *Composites Part A*, 109, p. 546–554, 2018.

30. D.S. Sperber, Flame retardant formulations and methods relating thereto, U.S. Patent Appl. 2017/0066970 (2017).

31. Forg, K. and Hoppe, H.A., Synthesis, crystal structure of a new structure type, and thermal analysis of ammonium borophosphate $(NH_4)_2[B_2P_3O_{11}(OH)]$. *Z. Anorg. Allg. Chem.*, 643, p. 766–771, 2017.

32. Li, Y., Wang, Y., Yang, X., Liu, X., Yang, Y., Hao, J., Acidity regulation of boron phosphate flame retardant and its catalyzing carbonization mechanism in epoxy resin. *J. Therm Anal. Calorim*, 129, 1481–1494, 2017; X., Liu, Y., Zhou, H., Peng, J., Hao, Catalyzing charring effect of solid acid boron phosphate on dipentaerythritol during the thermal degradation and combustion. *Polym. Degrad. Stab.*, Vol. 119, p. 242–250, 2015.

33. B. Ewald and H. Hibst, borophosphate, borate phosphate, and metal borophosphate as novel flame proofing additives for plastics, U.S. Patent 8,841373 B2 (2014). assigned to BASF, 2014.

34. Kang, B., Yang, X., Lu, X., Effect of hollow glass microsphere on the flame retardancy and combustion behavior of intumescent flame retardant polypropylene composite. *Polym. Bull.*, Vol. 77, p. 4307–4324, 2020.

35. Wang, X., Guo, W., Chapter 9. Graphene/polymer composites: A new class of fire retardant, in: *Flame Retardant Polymeric Materials: A Handbook*, Y. Hu and X. Wang (Eds.), pp. 151–180, 2020, Taylor & Francis, (Boca Raton, Fl).

36. Guo, W., Wang, X., Gangireddy, C.S.R., Wan, J., Pang, Y., Xin, W., Song, L., Hu, Y., Cardanol derived benzoxazine in combination with boron-doped graphene toward simultaneously improved toughening and flame retardant epoxy. *Compos. Part A*, 116, 13–23, 2019.

37. Ai, L., Chen, S., Zeng, J., Yang, L., Liu, P., Synergistic flame retardant effect of an intumescent flame retardant containing boron and magnesium hydroxide. *ACS Omega*, 4, 3314–3321, 2019.

38. Tawiah, B., Zhou, Y., Yuen, R.K.K., Sun, J., Fei, B., Microporous boron based intumescent marcrocycle flame retardant for poly(lactic acid) with excellent UV protection. *Chem. Eng. J.*, 402, 126209, 2020.

39. Tawiah, B., Yu, B., Yuen, R.k.k., Hu, Y., Wei, R., Xin, J.H., Fei, B., Highly efficient flame retardant and smoke suppression mechanism of boron modified graphene oxide/poly(lactic acid) nanocomposites. *Carbon*, 50, 8–20, 2019.

8

Non-Halogenated Conformal Flame Retardant Coatings

Federico Carosio

Dipartimento di Scienza Applicata e Tecnologia, Politecnico di Torino, Alessandria Campus, Italy

Abstract

In recent years the development of water-based conformal coatings has been demonstrated as a viable strategy capable of producing substantial improvements of the flame retardant (FR) characteristics of many different substrates such as textiles, flexible foams and thin films. This chapter will focus on the description of the most recent techniques based on the general principle of adsorption of the functional FR constituents (i.e. polyelectrolytes and nanoparticles) from aqueous media. The aim is to provide the reader with a selection of the most efficient FR treatments developed so far for textiles (natural and synthetic), porous substrates (flexible/rigid foams and wood) and thin films. The coatings are selected based on the main FR properties achieved, the adopted FR mechanism and the possibility to confer additional properties beside flame retardancy. Important technological advances that could be employed in order to step from lab- to pre-industrial-/industrial-scale are also highlighted and discussed.

Keywords: Layer by Layer, nanocoatings, textiles, foams, nanoparticles, polyelectrolytes, polyelectrolyte complexes

List of Acronyms

APP	Ammonium polyphoshpate
BL	Bi-layer
BOH	Bohemite
BPEI	Branched branched polyethylenimine

Email: federico.carosio@polito.it

Alexander B. Morgan (ed.) Non-Halogenated Flame Retardant Handbook 2nd edition, (337–412)
© 2022 Scrivener Publishing LLC

BTCA	1,2,3,4–butanetetracarboxylic acid
CNF	Cellulose nano fibrils
CNT	Carbon nanotubes
CS	Chitosan
CVD	Chemical vapor deposition
DNA	Deoxiribonucleic acid
GO	Graphene oxide
HCCP	Hexachlorocyclotriphosphazene
HFST	Horizontal flame spread test
HN	Halloysite nanotubes
HRR	Heat realease rate
IR	Infrared
LbL	Layer by Layer
LOI	Limit oxigen index
Mel	Melamine
MMT	Montmorillonite
MWCNT	Multi wall carbon nanotubes
NP	Nanoparticle
OSA	Oxidized sodium alginate
PA	Polyamide
PAA	Poly(acrylic acid)
PAAm	Poly(acrylamide)
PAAs	Poly(amido amines)
PAH	Poly(allyl amine)
PAN	Poly acrylonitrile
PC	Polycarbonate
P-CNF	Phosphorilated nanocellulose
PDAC	Poly(diallyldimethylammonium chloride)
PE	Polyelectrolyte
PECs	Polyelectrolyte complexes
PECVD	Plasma enhanced chemical vapor deposition
PET	Polyethylene terephthalate
PhA	Phytic acid
pkHRR	peak of heat release rate
PLA	Polylactic acid
PPA	Poly(phosphoric acid)
PS	Polystirene
PSP	Poly(sodium phosphate)
PU	Poly(urethane)
PVD	Physical vapor deposition
PVPA	Poly(vinyl phosphoric acid)

PVS   Poly(vinyl sulfonic acid sodium salt)
QL   Quad-layer
SA   Sodium alginate
SEA   specific extintion area
SEP   Sepiolite
SNP   Silica nanoparticles
ST   Starch
TA   Tannic acid
THR   Total heat realease
TL   Tri-layer
TSR   total smoke release
TTI   Time to ignition
VFST   Vertical flame spread test
VMT   Vermiculite
XPS   X-ray photoelectron spectroscopy

8.1   Introduction to Conformal Coatings: The Role of Surface During Combustion

Polymer modification by melt compounding with specific additives or fillers is a widely employed and easy strategy to produced materials capable of meeting the physical, mechanical and chemical requirements linked to a specific application. This approach is also employed to confer flame retardant (FR) properties to the selected polymer. While this has been the approach of choice for many decades, in recent years, the concept of surface engineered flame retardants has arisen and has been proposed as valuable alternative to confer FR properties [1]. The simple principle is that by confining the FR chemicals on the surface by means of nanoscale or nanostructured coatings it possible to achieve the desired FR characteristics while addressing some drawbacks associated to melt compounding such as detrimental effects on the polymer mechanical properties. This stems from the fact that polymer flammability is a property where the surface can play a key role. Indeed, as schematized in Figure 8.1, during combustion the heat is transmitted through the surface to the condensed phase where it triggers thermal decomposition processes that lead to the release of volatile products that diffuse towards the surface and the gas phase, feeding the flame.

The key role of the surface becomes also apparent by looking at some of the most efficient FR systems produced by bulk inclusion. Polymer-clay nanocomposites can achieve good flame retardant properties by building up, during combustion, a protective barrier made of nanoparticles embedded in a charred

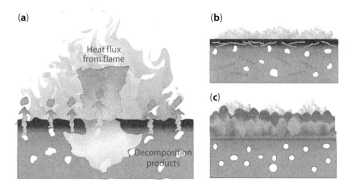

Figure 8.1 Schematic representation of: (a) heat and mass exchange at the surface of a burning polymer, (b) polymer clay nanocomposite FR action and (c) intumescent systems FR action.

residue at the polymer/flame interface [2–6]. Similarly, intumescent formulations, can produce an expanded and thermally stable structure that efficiently shields the polymer from the heat produced by the flame [7]. A downside to both solutions is represented by the need to produce the protective layer during the early stages of combustion as a consequence of polymer surface layer decomposition. This aspect from one side might limit the efficiency of the FR action (e.g. time is required to achieve an optimal protection) and on the other side makes it impossible to achieve a flame retardant effect in the pre-ignition phase. This latter point is directly addressed by the surface approach. Indeed, by confining the FR on the surface there is no need to build a protective layer as everything that is needed is already in the right place and a FR effect could be achieved at the beginning of the exposure to a small flame or a heat flux. To this extent, it should be noted that the approach described in this chapter does not represent a sort of panacea to the flame retardancy of polymers but rather an effective solution for certain classes of products such as textiles, foams and thin films. Indeed, these substrates are characterized by high surface to bulk ratios meaning that there is plenty of surface available for the deposition of a functional coating that should protect a limited amount of bulk polymer. By reducing the surface to bulk ratio to less favorable values the efficiency of the deposited coatings might be reduced to the point of being negligible. With that being said, the deposition of conformal coatings is capable of producing substantial improvements of the flame retardant characteristics of the above-mentioned classes of polymers. While there are several approaches, developed in the late 90' or early 2000, capable of yielding FR coatings, this chapter will focus on the description of the most recent techniques based on the general principle of adsorption of the functional FR constituents (i.e. polyelectrolytes and nanoparticles) from aqueous media.

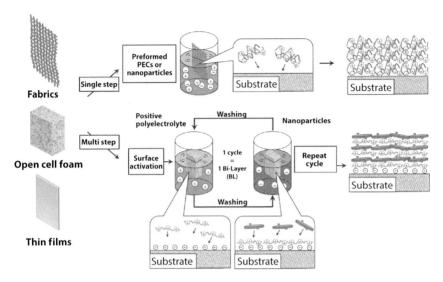

Figure 8.2 Schematic representation of a single step and a multi step deposition of FR coatings encompassing polyelectrolytes, nanoparticles and complexes.

They can be roughly divided, from a technological point of view, into two general categories (Figure 8.2):

> *Single step*: the substrate of choice is immerged into the aqueous polyelectrolyte (PE) or nanoparticle (NP) solution/ suspension in order to let the adsorption of the selected chemicals [8]. This approach started as a very simple nanoparticle impregnation and exhaustion process but has recently developed in a more refined deposition process where different PEs or NPs bearing opposite charge in water are premixed together under precise conditions in order to achieve water-stable polyelectrolyte-complexes (PECs) that are subsequently deposited on the substrate [9]. This latter represents an evolution that aims at overcoming the lengthy nature of the multi-step approach described below while maintaining the undeniable positive characteristics conferred by the formation of complexes.

> *Multi step*: the deposition of the functional FR coating is achieved by subsequent adsorption steps. Although the basic principle is the same as in the single step approach, the multi-step approach allows one to exploit specific interactions occurring among chemicals employed in subsequent

solution/suspension baths. The Layer-by-Layer (LbL) deposition represent the most employed incarnation of the multi step approach [10]. Indeed, the LbL allows for the construction of multi-layered coatings on the basis of the electrostatic attraction occurring between charged PEs or NPs in water. Although it is possible to exploit other interactions such as hydrogen bonding and hydrophobic interactions, the formation of polyanion-polycation complexes at the substrate/water interface is the most widely employed interaction for the construction of LbL coatings [11].

Besides the above mentioned approaches there are other techniques capable of depositing flame retardant coatings: physical and chemical vapor depositions, plasma-assisted deposition, sol-gel approach and conventional intumescent coatings. These approaches are here briefly revised in order to provide a technological context for this chapter (see Chapter 10 for additional information). Physical vapor deposition (PVD) and chemical vapor deposition (CVD) are vacuum based technologies where a thin coating is deposited on the substrate from a vapor phase [12]. In PVD the vapor is produced by either sublimation or evaporation of a target and it is subsequently deposited by condensation on the substrate [13]. Similarly, in a general CVD, the vapor contains one or more volatile precursor that deposit the desired coating by single- or multi-step chemical reactions at the substrate surface [14]. Both techniques have been extensively employed for the deposition of functional coatings on a plethora of substrates. FR properties were also addressed either by depositing functional FR coatings on textiles or by synthetizing hybrid FR additives via the surface functionalization of commercially available particles [15]. Their ability to uniformly coat complex substrates offers great potentials in the FR field; however, the batch nature of the process and the need for a vacuum environment might limit the competitiveness of PVD and CVD with respect to other processes. Plasma technology exploits the 4th state of matter to provide the proper reactive environment for the deposition and growth of functional coatings. The chemical nature of polymers naturally limits the choice of processing to cold plasma (low pressure or atmospheric). This is normally employed to promote the polymerization/grafting of functional chemicals that are either previously adsorbed on the substrate or directly vaporized in the deposition chamber (this latter being an hybrid with CVD, known as plasma enhanced chemical vapor deposition, PECVD). By this approach FR coatings have been deposited on fabrics, foams and thin films [16–19]. Thick PA 66 plates have also been investigated with scarce results due to the low surface to bulk ratio [20]. Recently, atmospheric plasma has been presented as valuable alternative capable of overcoming the drawbacks associated to a vacuum based process [21]. Coupled with a continuous process for fabrics, atmospheric plasma offers the possibility for a "dry" textile FR finishing capable of minimizing water requirements, effluent production and drying procedures [22]. In addition, plasma could also be employed as surface activation pre-treatment or as post crosslinking step. In this latter case the role of the technology is fundamental for the final properties of the coating (i.e. proper adhesion or improved durability) but does not directly confer flame retardant properties.

The sol-gel is a versatile wet-chemical approach for the synthesis of completely inorganic or hybrid organic–inorganic chemical structures. The process normally involves a two steps hydrolysis-condensation reaction of a (semi)metal alkoxides (e.g. tetraethoxysilane, tetramethoxysilane) producing a colloidal solution ("*sol*") that can further develop into an integrated network ("*gel*") of either discrete particles or network polymers as a function of the employed conditions (e.g. water/alkoxide ratio, pH, time, temperature) [23]. Initially employed for the synthesis of metal oxides particles characterized by high levels of chemical homogeneity and novel compositions [24], it was then quickly extended to the deposition of functional coatings including flame retardant solutions for textiles [25, 26]. In a typical process the fabric is impregnated with the "*sol*" and then dried in order to consolidate the inorganic network into a dense coating. The simple deposition of an inorganic coating results, most of the times, in unsatisfying FR properties. However, thanks to the versatility of the process, phosphorus and/or nitrogen-containing compounds can be easily incorporated within the structure granting a synergistic or joint FR effect [27, 28]. Intumescent coatings have been widely developed since the '70s as fire protective solutions for wood boards and steel structures. Typically deposited as paints with thicknesses ranging from hundreds of microns to a few mm, intumescent formulations containing three main components (i.e. an acid, a carbon source and a blowing agent) can easily produce an expanded insulating charred foam when exposed to a direct flame or high temperatures. The produced structure exerts the FR effect by limiting heat transfer and, subsequently, by extending the time needed to reach the substrate critical temperature (e.g. the temperature corresponding to drastic drop in steel mechanical properties. Intumescent systems have been continuously developed and optimized from both the formulation and physical properties of the expanded structure points of view. These advances are discussed in a dedicated chapter of this book (chapter 4). From an overall point of view, the single- and multi-step processes described in this chapter represent a possible evolution of the above discussed technologies. Indeed, both approaches exploit the main FR mechanisms of the above described techniques at reduced length scale while adding versatility. For instance, an intumescent formulation can be replicated by carefully selecting the components employed in single- or multi-step approaches thus achieving a micro intumescent behavior. This stems from the broad selection of FR reagents available and the simplicity of both approaches. It is indeed possible to deposit synthetic or natural polyelectrolytes (e.g. ammonium polyphosphate (APP), chitosan (CS), etc...), nanoparticles of different shapes (e.g. silica nanospheres (SNP), halloysite nanotubes (HN), montmorillonite nanoplatelets (MMT), etc...), carbon-based materials (e.g. nanotubes, nanographites, graphenes, etc...), biomacromolecules (caseins, deoxyribonucleic acids, etc.) and M-xenes ($Ti_3C_2T_x$) [29]. In addition to that, both approaches are also attractive from a "green" point of view as they rely on diluted aqueous solutions/suspensions (generally $\leq 1\%$ wt, with the possibility of recycling the solvent/deposition baths after use) and operate at room temperature and atmospheric pressure. The LbL further adds to this by providing additional benefits:

1. *Complex phase formation*: during the LbL assembly the selected polyanion and polycation "react" at the substrate surface in order to form non soluble complexes where an intimate and stoichiometric mixture of the two components is achieved [30]. The so produced complex phase allows for synergistic interactions during heating or upon exposure to a direct flame thus producing a FR effect superior to any theoretical mixture of non-interacting components (Figure 8.3) [31–33]. This is one of the main reasons for the high FR efficiency demonstrated by LbL assembled coatings.

2. *Tunable process*: the deposition parameters (pH, ionic strength, temperature, time, etc...) can deeply affect the outcome of the LbL deposition in terms of coating thickness (ranging from few nm up to hundreds of nm per deposited layer) and composition (it is possible to change the relative amounts of coating components) thus further increasing the versatility of the process [29].

3. *Multi-functionality*: for any of the coating constituents listed above it is possible to find conditions that allow for a LbL assembly. This opens up to a nearly infinite set of possibilities. By carefully selecting the reactants it is thus possible to

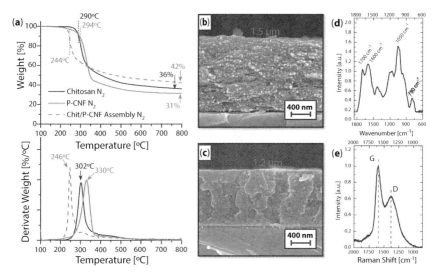

Figure 8.3 Characterization from ref [33] performed on CS/P-CNF LbL assembly: (a) TGA of neat components and their LbL assembly (b) 100 BL coating, (c) 100 BL coating after exposure to the cone heat flux (35 kW/m2), (d) IR and (e) Raman spectra of the charred residue. (Reprinted from Carbohydrate polymers, 202, Carosio, F., Ghanadpour, M., Alongi, J., Wågberg, L., Layer-by-layer-assembled chitosan/phosphorylated cellulose nanofibrils as a bio-based and flame protecting nano-exoskeleton on PU foams, 479–487, Copyright (2018), with permission from Elsevier.)

incorporate, in addition to flame retardancy, extra functionalities such as: super-hydrophobicity, electrical conductivity, anti-bacterial properties and uv-shielding [34–37].

The single step deposition of water stable preformed polyelectrolytes or polyelectrolyte/nanoparticles complexes provides similar benefits, mostly linked to the above discussed complex phase formation. There are, however, some constraints with respect to the processing parameters and the possibility of achieving a multifunctional coating. Indeed, complex formation requires precise mixing conditions (concentration, pH, ionic strength) so the possible compositions of a water stable complex phase are rather limited with respect to LbL [38]. There are, up to now, several reviews that would serve as complete collection of most if not all the published papers on these single- and multi-steps approaches [8, 39–46]. In order to provide a complementary overview to the current scientific literature, the aim of this chapter is to provide the reader with a selection of the most efficient treatments as a function of the selected substrate (i.e. fabrics, foams or thin films). Selection is based on the main

FR properties achieved by each system and the adopted FR mechanism. Important technological advances that could be employed in order to step from lab- to pre-industrial-/industrial-scale are also highlighted in this chapter, something that is not covered in the scientific reviews to date.

8.2   Fabrics

Fabrics represent the first substrate that was coated by means of single- or multi-step approaches aiming at improving their FR properties. This was a natural start for these new technologies since the use of a surface approach to fabrics FR was an already employed and well-established practice at both lab- and industrial-scale. Indeed, for many natural fibers (e.g. cotton, animal fibers) this is the main viable approach to confer extra properties to the fabric [47]. Differently, synthetic fibers produced by melt- or wet-processing might involve the use of additional approaches such as the bulk inclusion of FR additives or copolymerization strategies where the FR chemical is incorporated within the polymer chain [48]. While both strategies can result in efficient FR performances there are some possible drawbacks (such as limited processability, difficulty in fiber production, and decreased mechanical properties) that might render the surface approach a more viable option. Indeed, the latter has the potential advantage of not influencing the substrate properties related, for example, to fabric mechanical properties/comfort and optical properties/color. In order to achieve this, it is important that the treatment is designed in such a way to deposit a homogeneous and thin coating on each fiber avoiding a detrimental fiber/fiber bridging phenomenon that might increase the fabric stiffness. This also directly affects the achieved FR properties while also contributing to a limited influence of the coating on the fabric optical properties. It is worthy mentioning that while nanoscale coatings comprising polyelectrolytes and nanoparticles have the potentiality of not altering the color of the substrate, this ability is strictly dependent on the choice of the coating constituents. Reagents such as tannins, graphenes, etc. that bear chromophore groups will inevitably result in colored coatings. With that being said, the wide selection of reagents available for the deposition allows for the design of functional FR coatings while preserving the desired physical properties as a function of the chosen application and fabric. Although some papers report of universal treatments capable of simultaneously conferring FR properties to different kinds of fabrics (e.g. Cotton, Cot/PET, PET) [49, 50], the most efficient widely employed approach is to design the coating composition in order to match the substrate chemistry and burning characteristics. For this reason, and to better present the most performing solutions in the field of textiles, this paragraph is further divided in order address three main classes of fabrics: natural, synthetic and blends.

8.2.1 Natural Fabrics

Natural fabrics such as cotton, ramie, silk and wool have been used extensively as substrate for the water-based deposition of polyelectrolytes and nanoparticles. This high interest is strongly related to their widespread application in the marked due to their valued comfort, luxurious look and renewability [51]. Being produced from fibers occurring in nature (such as animals' coats, plant seeds, leaves and stems) these kinds of fabrics shows peculiar characteristics influencing both the deposition process and the burning behavior, which in turns affects the coating FR action and performances. From the processing point of view the hydrophilicity and surface roughness of natural fibers represent a clear advantage for the water-based approaches described in this chapter. Indeed, the presence of polar functionalities such as hydroxyl groups (cotton, ramie, silk and wool), amide bonds (silk and wool) and amines/carboxylic acids groups (wool) allows for not only a good wettability and homogeneous coating deposition but also the possibility of covalent crosslinking with the fiber surface. In addition, the irregular surface morphology that is a function of fiber type and production processes, also increases the coating adhesion by means of providing physical anchoring points. From the thermal stability and combustion point of view, both cellulose- and protein-based fibers show an intrinsic char-forming ability [52]. This can produce a compact charred residue in the case of cotton and ramie and an expanded/intumescent like structure in the case of silk and wool. This intrinsic char forming behavior is, of course, insufficient to confer satisfactory FR characteristics on its own but certainly provides a good starting feature to be exploited by a functional FR treatment [53]. It is worthy highlighting that, in the particular case of cotton and ramie, the formation of a compact charred residue during combustion might provide a solid support for the coating, preserving its structural integrity and thus allowing for optimal FR effects. Conversely, a conspicuous change in fiber morphology after the exposure to a flame or heat flux either by small intumescence or by melting (as will be discussed for synthetic fibers) can certainly have detrimental effects on the coating structure and the ability to provide the desired FR protection. For these reason the coating design in terms of composition and deposition condition should be carefully planned in order to obtain the best FR properties. With all the positive characteristics mentioned above, it is thus with no surprise that the first studies dealing with the surface deposition of water-based FR coatings were mainly focused on cotton. These studies focused on the deposition of inorganic structures comprising nanoparticles in order to produce a shielding barrier to heat and volatiles during combustion mimicking the FR mechanism of bulk polymer-clay nanocomposites [8]. Similarly Li *et al.* deposited a LbL assembly comprising branched polyethylenimine (BPEI) and either laponite or MMT clays [54]. Both assemblies deposited a brick and mortar like structure characterized by clay nanoplatelets organized parallel to the fiber surface and embedded in a polymer matrix. Although the treated fabrics were not able to stop flame spread, vertical flammability test (ASTM D6413) results pointed out the formation of a coherent charred residue resembling the original texture of the fabrics. Similar results were also achieved by Laufer *et al.* by means of a LbL structure based on silica nanoparticles [55]. These first papers clearly pointed out that the presence of a conformal coating can improve the char forming ability of cotton while also exerting a somewhat limited FR effect likely linked to the formation of a clay rich structure limiting heat and mass transfer at the condensed phase/gas phase interface. Subsequent studies fully disclosed the potentialities of water based techniques by developing coating designs capable of greatly enhancing the fiber char forming abilities while also providing additional FR effects (i.e. the built up of a protective barrier by either micro-intumescence or by extensive charring of the coating). This has been achieved, most of the time, by introducing different phosphorous containing compounds within the coating structure. The first proof of concepts were developed on cotton fabrics and then the coating composition was adapted and extended to ramie, silk and wool. Table 8.1 collects a selection of the main FR treatments developed for each kind of natural textile, the main results and scientific developments associated to each work are discussed in the following. In 2011, Li *et al.* reported the first LbL intumescent system for cotton fabrics. The coating composition was designed to mimic a classical intumescent formulation. Indeed, the employed positively charged poly(allylamine) PAH acted as carbon rich compound and a blowing agent while the negatively charged poly (sodium phosphate) SPS behaved as acid source [56]. The system showed a two regime LbL growth with the first 10 BL depositing a thin coating in which thickness is then greatly expanded with the deposition of additional 10 BL. At 20 BL the treated cotton showed a self-extinguishing behavior during VFST (ASTM D6413) while also greatly improving the cotton char forming abilities (Figure 8.4).

Post combustion residue analysis evidenced the formation, only at 20 BL, of an intumescent like structure characterized by micro-scale expanded bubbles. This work also indirectly shows that a minimum coating thickness

Table 8.1 Main FR conformal coating solutions for natural fabrics.

Substrate	Coating composition - Approach	Main FR results	Minimum add on required	Durability	Ref
Cotton	PAAm/PSP (LbL)	Self-extinguishment VFST (ASTM D6413)	17.5% at 20 BL	N.A.	[56]
	ST/PPA (LbL)	Self-extinguishment HFST, cone calorimeter pkHRR (up to -35%) THR (up to -40%)	5% at 2 BL	N.A.	[60]
	CS/APP+UV-curable latex (LbL)	Self-extinguishment HFST, cone calorimeter THR (-28%)	5% at 3 BL	1 cycle (water 65°C for 1h)	[62]
	BPEI/PSP-complex (single step)	Self-extinguishment VFST (ASTM D6413)	16.5%	N.A.	[63]
	BPEI/HCCP (LbL)	Self-extinguishment VFST (ASTM D6413)	23 % at 1 BL	1 washing cycle (standard nonphosphate detergent 49°C for 45 min)	[64]
	DNA (single step)	Self-extinguishment HFST, no ignition during cone calorimetry	10%	N.A.	[66]

(Continued)

Table 8.1 Main FR conformal coating solutions for natural fabrics. (*Continued*)

Substrate	Coating composition - Approach	Main FR results	Minimum add on required	Durability	Ref
	BPEI/PhA (LbL) + Ag nanowires (single step)	Self-extinguishment VFST, LOI 37%, cone calorimeter pkHRR (-41%) THR (-27%)	31.7% at 8BL + Ag nanowires ads.	EMI shielding properties durable to 20 washing cycles (water commercial detergent solution, 40 °C for 30 min)	[86]
	GO/Casein (LbL) + reduction NaBH$_4$ + APP (single step)	LOI 23.6%, cone calorimeter pkHRR (-65%)	26.3% at 10 BL	Partially durable to 5 washing cycles (commercial detergent solution, 45 min total washing time)	[89]
Ramie	BPEI/APP (LbL)	Self-extinguishment VFST (ASTM D6413)	16% at 20 BL	N.A.	[58]
	BPEI/APP (spray LbL)	Self-extinguishment VFST (ASTM D6413)	19% at 15BL	N.A.	[59]
Silk	Casein-Metal ions (single step)	Self-extinguishment VFST (ASTM D6413), LOI 28%, smoke density (-35-48%)	N.A.	20 washing cycles (commercial detergent solution, 40 °C for 30 min)	[70]

(*Continued*)

Table 8.1 Main FR conformal coating solutions for natural fabrics. (*Continued*)

Substrate	Coating composition - Approach	Main FR results	Minimum add on required	Durability	Ref
	CS/Vitamin B2 (LbL)	Self-extinguishment VFST (GB/T 5455), LOI 32.8%	N.A. 5 or 10 BL	N.A.	[71]
	TA (single step)	Self-extinguishment VFST (ASTM D6413), LOI 27%	c.a. 20%	20 washing cycles (commercial detergent solution, 40 °C for 30 min)	[74]
	GO (single step) + reduction L-ascrobic acid	self-extinguishment VFTS (ASTM D6413), LOI 27.5%, smoke density (-43%)	N.A.	5 washing cycles (commercial detergent solution, 40 °C for 50 min total washing time)	[88]
Wool	PhA (single step)	Self-extinguishment VFST (ASTM D6413), LOI≈30%	c.a. 9-10%	N.A.	[77]
	CS/PhA-complex (single step)	Self-extinguishment VFST (ASTM D6413), LOI≈33% (reduced to 28 after 10 washing cycles)	c.a. 20%	10 washing cycles (water commercial detergent solution, 40 °C for 30 min)	[79]

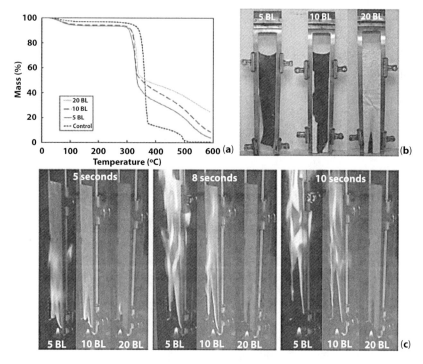

Figure 8.4 Results from ref [56] describing cotton treated by PAH/SPS LbL assembly: (a) TGA in air (b) residues collected after VFST and (c) images of treated fabrics during VFST. Reproduced with permission. [56] Copyright 2011, Wiley-VCH.

is required in order to achieve a performing micro intumescent behavior. A more recent study performed on model cellulose substrates further confirm this by demonstrating that the coating thickness also affects the size and number of the intumescent structures produced [57]. Similarly, branched polyethylene imine was combined with ammonium polyphosphate on ramie fabrics achieving, at 20 BL deposited, a self-extinguishing behavior during vertical flame tests (ASTM D6413) [58]. The same coating composition was also deposited by spray LbL achieving a reduction of BL needed to the desired FR effect (10-15 BL) and also a potential reduction in processing time [59]. This work also highlights the importance of the amount of substances deposited per layer and the total coating add-on. Indeed, spraying consistently achieved increased add-ons with respect to dipping at the same BL number suggesting that a threshold in FR-properties exists as a function of the selected LbL formulation. The main limitations of the systems described above are certainly linked to the high number of BL required (20) and the absence of a durability evaluation. Subsequent works based on similar coating compositions addressed these two latter points. A LbL coating based on starch and polyphosphoric acid was found capable of delivering FR properties with only 2 or 4 deposited BL [60]. This study, based on cotton, also addresses the effects of different fabrics grammages (namely 100, 200 and 400 g/m^2) on the FR performances achieved by the coatings. It is indeed known that the density of the fabric can influence both its char forming ability and burning behavior [61]. 2BL (5 % add-on) were found capable of self-extinguishing the flame during horizontal flame spread tests regardless of the cotton substrate adopted. The coating FR mechanism was related to an improved cotton char forming ability coupled with an extensive charring of the coating towards a dense and compact layer rather than expanded structures. This work proved that it is possible to reduce the number of deposited BL while retaining FR properties. The durability to laundering of the coating was not investigated; however, the same research group proposed a durable FR assembly based on the inclusion of a UV-curable latex within an intumescent system comprising CS and APP [62]. A photo-cured 3BL sample was found capable of withstanding washing (65°C for 1h) while maintaining a self-extinguishing behavior by horizontal flame spread tests. These works achieved a FR effect also thanks to the intimate/molecular-scale mixture of the components employed granted by the formation of a complex phases during LbL assembly. In order to exploit this peculiar feature while limiting the number of treatment steps the deposition of poly-electrolyte complexes was developed. To this aim, BPEI and PSP have been extensively studied for cotton FR. These two components can be pre-mixed in controlled pH conditions in order to achieve stable PECs solutions/suspensions [63]. A one step deposition followed by curing in acidic buffer (pH 2-5) can grant self-extinguishing behavior during vertical flame test (ASTM D6413). Morphological investigations showed a dependence of the coating FR behavior as a function of the pH adopted for curing, and pH 2 was found as the best condition to achieve a micro intumescence behavior and a more efficient FR coating (Figure 8.5).

A subsequent study proved that the substitution of poly sodium phosphate with hexachlorocyclotriphosphazene (HCCP) can impart self extinguishing behavior (ASTM D6413) while also granting durability to one washing cycle [64]. The increase use of water soluble polysaccharides such as chitosan and starch as layer constituents showed the potential of biomacromolecules as bio-based FR chemicals [65]. This was quickly extended to proteins, nucleic acids, vitamins and phenols. A landmark paper reported the use of deoxyribonucleic acid (DNA) as all in one intumescent system for cotton [66]. Indeed, the DNA phosphate-deoxyribose backbone can act as both carbon and acid source while the nitrogen containing bases can behave as blowing agents [67]. 10%-wt was reported as the minimum

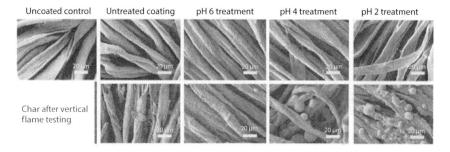

Figure 8.5 Results from ref [63] describing cotton treated by BPEI/PSP-complexes. SEM micrographs showing the surface morphology after the PECs deposition and VFST as a function of different post-curing treatments. The formation of intumescent-like structure for samples post treated at pH 2 is apparent. (Reprinted from Polymer Degradation and Stability, 114, Haile, M., Fincher, C., Fomete, S., Grunlan, J.C., Water-soluble polyelectrolyte complexes that extinguish fire on cotton fabric when deposited as pH-cured nanocoating, 60-64, Copyright (2015), with permission from Elsevier.).

add-on in order to achieve a self-extinguishing behavior while also reducing the release of volatiles to the limits of ignitability by cone calorimetry. The use of simple adsorption of caseins or a LbL assembly comprising chitosan/DNA was also proposed as a FR solution for cotton although the FR results achieved were not as satisfactory as the one of DNA alone [68, 69]. Conversely, metal ions complexed caseins have been found to impart substantial FR properties to silk. Among the ions studied in this work, Ti^{4+} allowed to self-extinguish the flame by vertical flame spread test (ASTM D6413) and reach an LOI of 28% with minimum impact on the fabric color [70]. The washing fastness to 20 washing cycles was also evaluated, pointing out a good durability of the coating to maintain FR performances. Another work performed on silk employed the LbL assembly of chitosan and a phosphorous containing vitamin (B12) [71]. The assembled coatings achieved self-extinguishment by VFST (GB/T 5455) after 5 BL. The same coating at 10 BL also showed a good antibacterial activity thus demonstrating the potentialities of LbL in delivering multifunctional coatings. Among the most recently studied and employed biomacromolecules it is important to mention tannic acid and phytic acid [72, 73]. The former has been deposited on silk as char former/promoter agent by means of single step adsorption simultaneously granting self-extinguishing behavior, antibacterial and antioxidant activity as well as the durability of the achieved properties to 20 washing cycles [74]. The latter has been extensively studied on natural fabrics (mainly cotton and wool) as a stand alone FR treatment, as LbL/PECS constituents or in combination with sol-gel derived coatings [75, 76]. When deposited on wool, phytic acid (PhA) can grant a

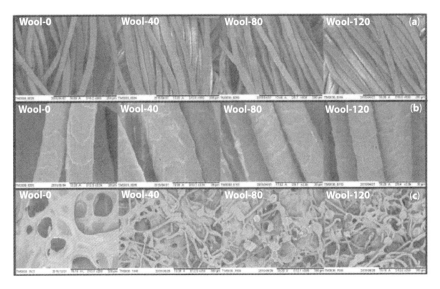

Figure 8.6 Results from ref [77] describing wool treated by PhA. SEM micrographs showing the surface morphology after the treatment and VFST at different coating add-ons. The presence of PhA resulted in a micro intumescent effect. Reproduced from [77] under the terms of the CC-BY Creative Commons Attribution 4.0 (https:// creativecommons.org/licenses/by/4.0). Copyright 2016, The Authors, published by MDPI.

self-extinguishing behavior (ASTM D6413) at 9-10% wt add-on by a micro intumescent effect as reported in Figure 8.6 [77].

The treatment based on PhA alone shows scarce durability to washing. This issue was addressed by a post crosslinking strategy employing 1,2,3,4–butanetetracarboxylic acid (BTCA) that resulted in FR fabrics capable of withstanding 30 washing cycles [78]. Similarly, the use of pH cured CS/PhA complexes endowed the treatment with durable (10 washing cycles) FR properties [79]. On the basis of the many proof of concepts available in the literature a synthetic route was developed towards the production of polyelectrolytes mimicking the structures and FR action of bio-based compounds. To this aim, linear polyamidoamines (PAAs) have been presented as a family of synthetic water-born polymers which structural versatility allows to build in a single macromolecule all the components normally present in bio-based FR formulation (e.g. carboxyl-, amine- and guanidine groups) [80]. Similarly to natural bio macromolecules, PAAs can easily endow cotton with excellent FR properties by granting self-extinguishing behavior when flammability is tested in horizontal and vertical configurations [81–84]. The durability of this kind of treatments has not yet been evaluated but it is likely that a cross-linking strategy would be required to ensure FR properties are maintained through multiple washing cycles.

From an overall point of view, and also by comparing the treatment formulations reported in Table 8.1, it is apparent that the general FR mechanism employed for natural fabrics rely on the combination of improved fiber char forming and micro intumescence features. The use of nanoparticles alone has been demonstrated to deliver less efficient FR results. Attempts at combining intumescent-like coatings with nanoparticle rich structures require a careful design of the treatment in order to avoid detrimental effects or inefficient FR with respect to what could have been obtained by an intumescent coating alone [85]. On the other hand, a successful implementation of nanoparticles can open up to additional features as demonstrated by the use of Ag nanowires in combination with a BPEI/PhA LbL assembly on cotton [86]. The presence of the nanowires did not alter the intumescent FR action of the coating and provided additional EMI shielding characteristics to the fabric (Figure 8.7).

Graphene related materials have been employed as well. GO certainly represents the most suitable component for FR water based coatings due to the ease of preparation and hydrophilicity [87]. To this aim silk was coated by single step adsorption of GO [88]. A post treatment employing ascorbic acid was then employed to reduced GO. The treated fabrics showed self-extinguishment during VFTS (ASTM D6413) and LOI up to 27.5%. A considerable reduction in smoke density was also achieved (-43.4%). This treatment was found

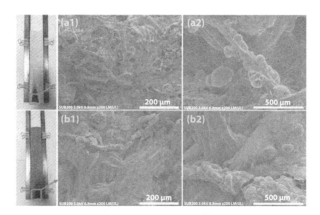

Figure 8.7 Results from ref [86] describing cotton treated by BPEI/PhA LbL followed by the adsorption of Ag nanowires. SEM micrographs showed that the presence of Ag nanowires (b1 and b2) did not influence the intumescent features of the BPEI/PhA assembly (a1 and a2). (Reprinted from Chemical Engineering Journal, 273, Zhang, Y., Tian, W., Liu, L., Cheng, W., Wang, W., Liew, K. M., Wang, B., Hu, Y., Eco-friendly flame retardant and electromagnetic interference shielding cotton fabrics with multi-layered coatings, 1077-1090, Copyright (2019), with permission from Elsevier.)

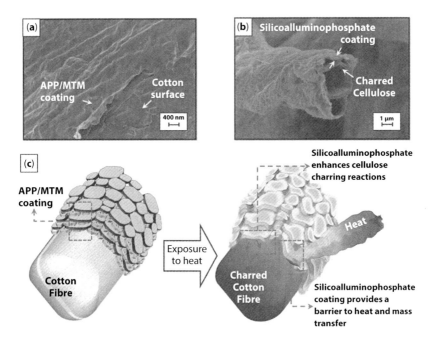

Figure 8.8 Results from ref [90] describing cotton treated by APP/MMT nanocoating. SEM micrographs show the coating morphology before (a) and after (b) exposure to the flame while the schematization (c) describes the coating reaction to flame exposure. Reproduced from [90] under the terms of the CC-BY Creative Commons Attribution 4.0 (https://creativecommons.org/licenses/by/4.0). Copyright 2016, The Authors, published by MDPI.

durable to 10 washing cycles. Similarly, An *et al.* employed a LbL approach to build a casein/GO coating on cotton followed by reduction with NaBH$_4$ and final adsorption of APP [89]. The presence of the assembly increased LOI to 23.6% and reduced pkHRR values (-65%) during cone calorimetry tests. Durability was evaluated showing a partial reduction in the achieved FR properties. The authors also proved the antistatic properties (enabled by the electrical properties of the chemically reduced GO) of the fabrics thus demonstrating the possibility to achieve a multi functional coating. Differently, an intumescent-like system could be assembled from inorganic components [90]. To this aim, APP/MMT nanocoatings were assembled on cotton by multi-step adsorption capable of self-extingushing the flame and granting no ignition during cone calorimetry test (35kW/m^2). Post combustion residue evaluation evidenced the formation of a silicoalluminophosphate expanded structure capable of improving cotton char forming ability and also exerting a barrier function towards volatile and heat (Figure 8.8).

8.2.2 Synthetic Fabrics and Blends

Synthetic fibers are normally conceived as a more performing and less expensive alternative to natural fibers such as cotton or silk. Their synthetic characteristics allow them to be strong and durable, dry faster than other textiles while also showing better dye-ability. For these reasons a good portion of the marked is shared by synthetic commodity textiles with polyethylene terephthalate (PET), poly amides (nylon 6 and 6,6) and acrylics (PAN) being the most common ones. As far as the coating deposition processability and burning characteristics are considered, this class of fabrics shows remarkable and worth mentioning differences from their natural counterparts. Indeed, these polymers normally show a poor affinity towards water as demonstrated by water contact angle values falling in the 70-80° range, which is close to the limits of wettability (contact angle <90°) [91, 92]. The shape of the fibers and the structure/texture of the fabric can influence how water is absorbed; however, the coating adhesion and growth is certainly reduced with respect to natural fibers. For this reason the use of a surface activation step (for example: plasma technology, chemical etching, adsorption of a primer layer, etc..) is normally employed in order to improve the coating homogeneity and efficiency [92, 93]. From the burning behavior point of view, the synthetic characteristic of the fibers would normally imply that melting will occur upon exposure to an heat flux, small flame and during fabric combustion giving rise to the melt-dripping known for synthetic commodity textiles. This phenomenon is affected by many parameters such as the polymer considered, the properties of the fabric (i.e. grammage and structure) and the testing conditions (e.g. horizontal vs vertical positioning). While during a flammability test the occurrence of melt dripping can self-extinguish the flame before the complete combustion of the fabric, (i.e., the dripping material pulls the flame away from the rest of the fabric) it is normally considered a dangerous threat as the droplets can easily ignite other flammable materials and increase fire spread. Further, the dripping material can land on people and adhere/cling to skin, leading to 2nd and 3rd degree burns. Another important behavior to consider is the so defined "runaway effect" where a low grammage fabric would retract and literally run away from the flame due to melting and recovering of processing tensions. This behavior, most likely to happen in horizontal configuration, might lead to a difficult evaluation of the flammability rating (according to specific standards). Specifically, even though the heat heavily damages the fabric, no ignition or melt dripping occurs. From the coating point of view, the melting and possible retracting of the substrate poses a dangerous threat to its structural integrity and limits the efficiency of the FR effect. For example, an intumescent formulation might have no time to build an expanded protective structure as the coating is dismantled and partially embedded by the melting/retracting fibers. This might lead to a non-linear, and sometimes, counterintuitive relationship between the coating add-on and the achieved FR performances. Indeed, at low add-ons the coating might improve the FR behavior of the fabric by, for instance, reducing or preventing melt dripping while allowing for a "controlled" run away effect. Conversely, at high add-ons the protection of the coating might prevent the fiber structural collapsing and fully exert its FR potential. Intermediate add-ons could lead to a situation where the fibers cannot run away from the flame and the coating FR action is not enough to prevent flame spread [94]. This peculiar behavior is described in one of the first papers presenting the LbL of SNP nanoparticles as FR solution for PET fabrics (Figure 8.9) [95]. Table 8.2 collects the main FR solutions for synthetic fabrics based on conformal coatings.

Indeed, this work clearly shows how an assembly comprising positively and negatively charged nanoparticles can effectively grant self-extinguishing behavior during vertical flame tests (ASTM D6413) and prolonged ignition times (up to +45 %, + 99 s) during forced combustion tests when deposited

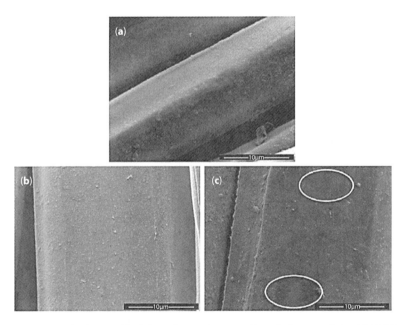

Figure 8.9 Results from ref [95] describing PET treated by Al2O3 coated SNP/SNP LbL. The SEM micrograph shows the production of a conformal coating on the PET fibers. The presence of uncoated spots points out a limited stability of the coating at high BL number. (Reprinted from Polymer Degradation and Stability, 96 (5), Carosio, F., Laufer, G., Alongi, J., Camino, G., Grunlan, J. C., Layer-by-layer assembly of silica-based flame retardant thin film on PET fabric, 745-750, Copyright (2011), with permission from Elsevier.)

Table 8.2 Main FR conformal coating solutions for synthetic fabrics.

Substrate	Coating composition - Approach	Main FR results	Minimum add on required	Durability	Ref
PET	Al_2O_3 coated SNP/SNP (LbL)	No melt-dripping Self-extinguishment VFST(ASTM D6413), TTI (+45 %, + 99 s)	N.A., 5 BL	N.A.	[95]
	Al_2O_3 coated SNP/SNP (LbL-spray)	No melt-dripping Self-extinguishment VFST, TTI (+20%, 18s), pkHRR(-30%), TSR (-30%)	N.A., 5 BL	N.A.	[96]
	MMT-Plasma activation (Single step)	TTI (+103%, 164s), THR (-28%)	N.A.	N.A.	[97]
	CS/SA/Cu^{2+}/CS (LbL)	Self-extinguishment VFST	N.A, 2QL	N.A.	[98]
	BPEI/OSA + post crosslinking (LbL)	No melt-dripping self-extinguishment HFST, pkHRR (-45%) THR (-22%)	9.4 % at 15 BL	12 washing cycle (water, standard nonphosphate detergent 40°C for 12 min)	[99]

(Continued)

Table 8.2 Main FR conformal coating solutions for synthetic fabrics. (*Continued*)

Substrate	Coating composition - Approach	Main FR results	Minimum add on required	Durability	Ref
	PDAC/PAA/PDAC/APP (LbL)	No melt-dripping Self-extinguishment HFST, No melt-dripping VFST	13% at 5QL	N.A.	[50]
	CS/APP (LbL)	No melt-dripping VFST (ASTM D6413), LOI 26.6%	13.6% at 10BL	N.A.	[101]
	guanidine sulfamate+CS/APP	No melt-dripping Self-extinguishment VFST (ASTM D6413), LOI 26%	19.6% at 10BL	N.A.	[102]
PA 6, PA 6.6	APP+MEL+PE-OH (single step)	Reduced afterflame time, no melt-dripping VFST, LOI 27.9%	40%	N.A.	[100]
	CS/PhA/CS/OSA (LbL)	UL-94V1, LOI 26%	9.3% 5 QL	N.A	[103]
	CS/PhA + post crosslinking (LbL)	UL-94V1, LOI 26%	11.3%, 5BL	5 cycles (non-ionic detergent 38°C)	[104]

(Continued)

Table 8.2 Main FR conformal coating solutions for synthetic fabrics. (*Continued*)

Substrate	Coating composition - Approach	Main FR results	Minimum add on required	Durability	Ref
PAN	POSS/APP (LbL)	No melt-dripping HFST,	8.3% at 4 BL	N.A.	[109]
	CS/MMT-complexes (single step)	No melt-dripping Self-extinguishment HFST, TTI (+30%, +15 s), pkHRR(-55%), THR (-48%), TSR (-37%)	10%	N.A.	[112]

at relatively low BL number (i.e. 5-10). Coatings achieved after 20 BL did not reach the same performances despite displaying thicker coatings. This was ascribed to a partial coating instability during combustion leading to a reduced protection effect. A similar coating composition has been also deposited on dense PET mats (490 g/m²) either by dipping or by horizontal spraying. Results from vertical flame tests pointed out the latter approach as the most efficient one in delivering FR properties [96]. These works clearly show the potential of nanoparticles as FR solutions for synthetic fiber and it is clearly in antithesis with what previously described for natural fibers thus further highlighting the differences among the substrates and the FR strategies to be adopted. Similar results were achieved by combining a cold oxygen plasma activation step and the single step adsorption of MMT nanoparticles [97]. The use of plasma prevented the formation of aggregates while allowing for the deposition of a more homogeneous coating. Of the many plasma conditions investigated, 80W and 180s were demonstrated as the optimal pretreatment capable of delivering MMT treated PET fabrics with greatly enhanced ignition times (+ 164 s) during cone calorimetry test. These results were lately demonstrated to be among the best achievable by the single step adsorption of nanoparticles. A possible limitation of both studies is represented by the lack of a durability evaluation. Indeed, without the use of crosslinking strategies (i.e. by using for instance commonly available coupling agents) it is likely that the deposited coatings cannot withstand washing cycles [8]. Following the developments of conformal coatings already described for natural textiles, polyelectrolyte containing assemblies were also developed and applied on synthetic fibers. Two different FR strategies can be easily devised: i) the deposition of char forming coatings based on polysaccharides, and ii) the use of intumescent like formulations. As far as the former is concerned, Liu *et al.* evaluated the deposition of a LbL assembly comprising chitosan and sodium alginate in combination with Cu^{2+} ions in a quad-layer fashion [98]. A system assembled with just sodium alginate and copper ions was also prepared for comparison. Vertical flammability results pointed out the role of alginate and copper as the main constituents delivering a FR effects; indeed, 8 deposition steps are enough to impart self-extinguishing behavior and prevent melt dripping. In a subsequent work Pan *et al.* combined BPEI with oxide sodium alginate (OSA) followed by cross-linking with hypophosphorous acid [99]. The authors also evaluated the effects of a 2-propenamide UV-grafting pre-treatment in order to improve durability. At 15 deposited BL the BPEI/OSA system was found able to grant self-extinguishing behavior during HFST for UV activated samples. These performances were maintained after 12 laundering cycles (Figure 8.10). The same formulation also decreased the pkHRR (-45%) and THR (-22%) during cone calorimetry tests.

Laundering cycle	2	4	6	8	10	12
Melt-dripping	No	No	No	No	No	No
Sample	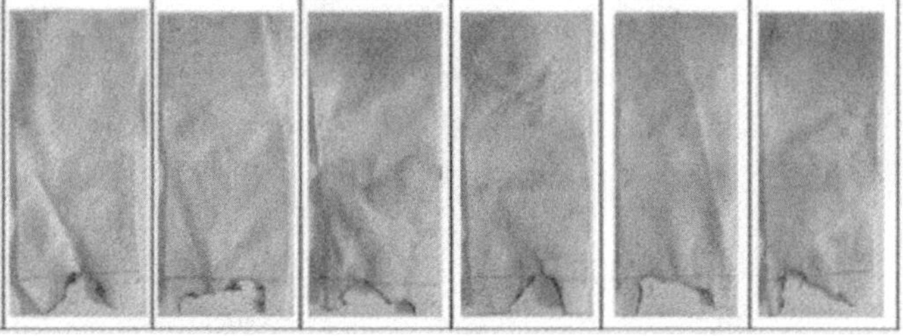					

Figure 8.10 Results from ref [99] describing PET treated by BPEI/OSA LbL. HFST results after different washing cycles performed on coated fabrics post cross-linked with hypophosphorous acid. (Reprinted from Polymer Degradation and Stability, 165, Durable flame retardant treatment of polyethylene terephthalate (PET) fabric with cross-linked layer-by-layer assembled coating, Pan, Y., Liu, L., Song, L., Hu, Y., Wang, W., Zhao, H., 145-152, Copyright (2019), with permission from Elsevier.)

Phosphorous containing intumescent formulations were also developed. An early example based on conventional adsorption of an intumescent like formulation (APP, Melamine and pentaerythritol) was proposed by Li *et al.* on PA 6,6 [100]. The treatment suppressed the melt-dripping phenomenon and strongly reduced the after flame time during vertical flame test; a LOI of 27.9% was also achieved. The formation of intumescent-like structures was confirmed by post combustion residue SEM evaluation. The relatively high add-on (30-40% wt) needed to obtain the above mentioned results points out the need for substantial optimization of the formulation/deposition strategies while also addressing the always needed washing resistance evaluation. As far as LbL is concerned, Alongi *et al.* presented the first example of an intumescent-like formulation on PET in 2012 [49, 50]. The coating composition comprised APP in combination with two fast growing polyelectrolytes (poly(diallyldimethylammonium chloride, PDAC and poly(acrylic acid) PAA) in a QL architecture. Fabrics treated by 1 QL self-extinguished the flame in horizontal flammability tests. However, the same architecture did not prove efficient in vertical configuration. Indeed, even with an increased number of deposited layers (i.e. 5 and 10 QL) the samples were not able to self-extinguish the flame only achieving a suppression of the melt-dripping phenomenon (Figure 8.11).

These results further demonstrate the non-linear relationship between add-on and FR performances. Indeed, although the presence of the coating prevented fiber collapsing/retracting, it was not able to stop flame spread thus

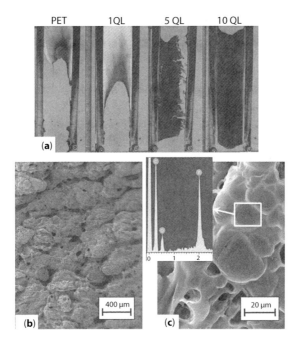

Figure 8.11 Results from ref [45] describing PET treated by PDAC/PAA/PDAC/APP LbL. VFST results showing a detrimental effect due to the addition of an increased number of QL. SEM morphology of a 10 QL samples after VFST showing the formation of intumescent like structures. (Reprinted from Polymer degradation and stability, 98(9), Carosio, F., Alongi, J., Malucelli, G., Flammability and combustion properties of ammonium polyphosphate-/ poly (acrylic acid)-based layer by layer architectures deposited on cotton, polyester and their blends, 1626-1637, Copyright (2014), with permission from Elsevier.)

resulting in increased burning times and reduced final residues. Subsequent works combined APP with CS in a two component architecture [101]. This strategy allows the incorporation of more APP within the coating with respect to the previously discussed QL. By this way, Fang *et al.* demonstrated that it is possible to reduce the char length and suppress the melt-dripping with a 10 BL assembly on PET. The use of a surface activation step by NaOH further allowed for a reduction of the required BL to 5 by increasing the coating deposition efficiency and thus the add-on per layer. Similarly, Jordanov *et al.* assembled a CS/APP coating on PET fibers [102]. The CH solution also included nitrogen-rich and low molecular weight molecules. Guanidine sulfamate was found as the additive capable of conferring the best FR performances and reducing the number of BL needed (from 30 to 10 for the unmodified and guanidine sulfamate modified CS solution, respectively) in order to achieve a self-extinguishing behavior with no melt-dripping during vertical flammability tests (ASTM D6413). Both studies demonstrated that the versatility of

LbL has the potential to tune the coating composition in order to achieve improved and more efficient FR properties. Unfortunately, the durability of the coatings was not evaluated. PhA was also evaluated as another phosphorous source substituting APP [103]. To this aim, Kundu et al. combined PhA, CS and SA in a QL coating on PA 66 fabrics. 5 QL were found able to grant a UL-94 V1 classification to the fabrics, the deposition of additional QL (10 and 15) did not produce further improvements in the flammability rating. As previously commented for APP-based QLs, this coating architecture might suffer from a reduced phosphorus content with respect to the other components. The same authors improved the FR efficiency of the coating by removing the SA component and depositing a CS/PhA assembly [104]. This new system granted a UL-94 V1 rating with only 5 BL; a post crosslinking step performed with sodium tetraborate decahydrate further conferred FR characteristics and durability to 5 washing cycles to a 10BL treated fabric. It should be quickly noted that while the authors used UL-94 V testing on a fabric, this test method is not appropriate for fabrics, although it does give some guidance on vertical flame spread behavior. PhA has been also employed as phosphorous source either chemically bonded to GO or combined with a silane based polyelectolyte for PA and PET fabrics, respectively [105, 106]. Although the results did not perform as well as the previously mentioned PhA containing systems, these two works show that hybrid inorganic/intumescent FR solutions could reach satisfactory properties if properly developed. Other, less successful, intumescent formulations based on PAH/PSP have been deposited on PA 66 [107]. The coating achieved a limited add-on (7.4 %wt) even when 40BL were deposited and produced limited effects, with respect to other LbL systems, on PA char forming ability. The inclusion of TiO_2 nanoparticles did not provide additional benefits [108]. It is worth mentioning that coating composition encompassing PAH/PSP proved to be an extremely efficient FR solution for cotton fabrics thus further highlighting the key role of the substrate and deposition conditions when designing a FR coating formulation [56]. From the literature survey it is apparent that the majority of the FR solutions have been developed for either PET or PA fabrics, which also share a good portion of the global textile market. PAN fibers were rarely evaluated mostly due to the versatility of the wet-spun production process employed for this kind of fabrics that allow additional strategies to implement FR chemicals. Notwithstanding this, there are few examples of conformal coatings applied to PAN worth to be discussed. Octa-ammonium PolyhedralOoligomeric Silsesquioxane (OAmPOSS) was coupled with APP achieving an all inorganic LbL assembly [109]. POSS molecules are synthetic products having a Si-O cage nanostructure (Si_8O_{12}) with specific functional groups linked to Si atoms at the corners of the cage [110]. They are considered as the smallest

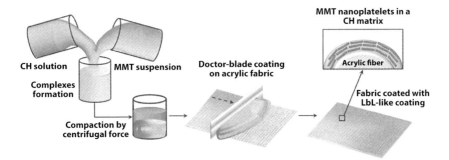

Figure 8.12 Schematization of the procedure adopted in ref [112]. CH and MMT are premixed in order to produce complexes that are compacted by centrifugal forces and doctor-bladed on PAN fabrics. Reproduced from [112] under the terms of the CC-BY Creative Commons Attribution 4.0 (https://creativecommons.org/licenses/by/4.0). Copyright 2018, The Authors, published by MDPI.

particles of silica possible and have been extensively used as bulk FR additives for polymers [111]. The OAmPOSS/APP system, deposited as 4 or 6 BL coating, managed to suppress the melt dripping phenomenon during HFST but without achieving self-extinguishing behavior. Post combustion analyses pointed out an improved fiber char forming ability while also highlighting the absence of intumescent-like structures. This lack of forming intumescent structures during burning negatively impacted the achievable thermal shielding properties and the overall coating FR performance. The same authors later proposed the use of CS/MMT complexes deposited by a one step doctor blading approach (Figure 8.12) [112].

The deposition yielded morphologies typical of a LbL assembled coating demonstrating that PECs have the great potential of reducing the number of deposition steps needed. A 10% wt add-on allowed this PECs to suppress melt dripping and extinguish the flame during HFST. Treated fabrics also showed improved TTI (up to +23 s, +46%) and reduced pkHRR (-62%) when tested by cone calorimetry. Differently from natural fibers it seems that the use of nanoparticles-based formulations is capable of delivering efficient FR characteristics and should thus be considered as a valuable strategy for synthetic fibers. Concerns associated to this option are related to a possible release of nanoparticles during use and washing so the proper evaluation of coating stability and durability to wear and washing cycles represents a mandatory condition to be addressed.

Strictly related to synthetic fabrics is their use in combination with natural ones in order to achieve blends. Blending is normally performed to impart desirable characteristics (e.g strength or durability, reduced cost, achieve special color or texture effects) to the final product by combining the strengths of the two types of fibers considered. For example, when combining polyester with cotton, the result is a durable and affordable material less prone to pilling and static. From the deposition and burning characteristics points of view, a blend will display the pros and cons of each type of fiber employed (e.g. good Vs poor coating affinity, charring Vs melting, etc...). In addition, during combustion, a dangerous detrimental joint effect might occur during combustion. A Cot/PET blend, for example, can catch fire with extreme ease with cotton representing the initial source of fuel, since it decomposes earlier than polyester. This latter will then melt with molten droplets remaining trapped within the charred structure produced by cotton. Although this behavior might limit the melt dripping phenomenon, it provides additional fuel to the overall combustion process. Such a peculiar burning behavior requires a careful design of the FR strategy. For example, to address the sharp difference between the fibers, Lewin proposed the use of a mixture of additives specifically targeting each fiber type [113]. As far as conformal coatings are concerned, a natural practice has been to transfer to the blends the coating formulations developed for single type fabrics. Table 8.3 reports the main FR coatings developed for blends. An early attempt was performed in 2012 with the deposition of CH/APP or Silica/APP coatings on Cot/PET blends [114]. Results clearly pointed out the char forming ability of the CH/APP pair as demonstrated by increased residues collected after vertical flame test and cone calorimetry. Similar results were achieved by hybrid structures encompassing CH/APP and Silica in a single coating formulation [115]. Although these different systems did not achieve a self-extinguishing behavior, these two preliminary works suggested intumescent systems and silicon containing intumescent formulations as promising routes to achieve the desired FR properties. Thus, aiming at an intumescent system Fang *et al.* employed CS/PhA on CoT/PET [116]. 15 and 20 BL were found capable of self-extinguishing the flame during VFST (ASTM D6413). The durability of the 20 BL coating was also evaluated pointing out a limited resistance of the assembly to washing cycles. Similarly, Liu *et al.* employed hypophosphorous acid in order to modify CH [117]. This latter was then assembled with BPEI on Cot/PET fabrics achieving, at 20 deposited BL, a self-extinguishing behavior during horizontal flame spread test. The intumescent behavior of this formulation was confirmed by post combustion SEM investigations. These two systems show the main drawback of requiring an excessive number of deposition steps. Aiming at improving the efficiency of the system and its durability, hypophosphorous acid was then employed as post treatment crosslinking strategy BPEI/OSA LbL assembly build on Cot/PET blends [118]. This approach successfully incorporated phosphorous within the coating structure as demonstrated by IR and XPS. Fabrics treated by 5 or 10

Table 8.3 Main FR conformal coating solutions for blend fabrics.

Substrate	Coating composition - Approach	Main FR results	Minimum add on required	Durability	Ref
Cot/PET	CS/PhA (LbL)	No melt-dripping Self-extinguishment VFST ((ASTM D6413)), LOI 26%	29.9 % at 15BL	Partially durable to 5 washing cycles (synthetic detergent, 49°C for 45 min)	[116]
	BPEI/CS-H₃PO₂ modified (LbL)	No melt-dripping Self-extinguishment HFST,	20.1 % at 20 BL	N.A.	[117]
	BPEI/OSA-H₃PO₂- crosslinked (LbL)	No melt-dripping Self-extinguishment HFST, pkHRR (-77%) and THR (-75%)	4.6% at 10BL	12 cycles (non-phosphate detergent 40°C for 12 min)	[118]
	Modified silane/APP (LbL)	No melt-dripping Self-extinguishment VFST (GB/T 5455), pkHRR (- 38%)	15.5 at 15BL	N.A.	[121]
Cot/Ny	BPEI/APP-complex Mel/ Buffer post treatment (single step)	No melt-dripping Self-extinguishment VFST (ASTM D6413)	19.1%	N.A.	[119]
	siloxane-based polyelectrolyte/PSP (spray-LbL)	No melt-dripping Self-extinguishment VFST (ASTM D6413)	56-89	N.A.	[120]

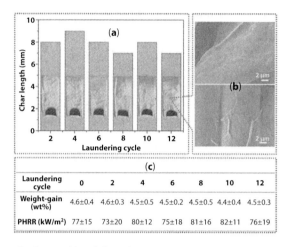

Figure 8.13 Results from ref [118] describing PET treated by BPEI/OSA LbL post cross-linked with hypophosphorous acid. HFST, coating add-on and pkHRR evaluated after different washing cycles. (Reprinted from Carbohydrate polymers, 201, Pan, Y., Liu, L., Wang, X., Song, L., & Hu, Y., Hypophosphorous acid cross-linked layer-by-layer assembly of green polyelectrolytes on polyester-cotton blend fabrics for durable flame-retardant treatment, 1-8. Copyright (2018), with permission from Elsevier.)

BL achieved self-extinguishing behavior during horizontal flame test and a strong reduction in pkHRR (up to 77%) with performances durable up to 12 washing cycles (Figure 8.13). Although these results were achieved with relatively low add-ons (2-4 wt%) it should be noted that the post treatment changed the color of the fabrics from white to yellowish.

As a natural evolution of LbL, PECs have been employed as well. To this aim, Leistner *et al.* presented a procedure to deposit BPEI/APP complexes on Cot/Ny blends [119]. A post treatment with melamine and a buffer allowed the system to obtain a self-extinguishing behavior during vertical flame tests (ASTM D6413). The *in-situ* formation of melamine polyphosphate was reported as the main reason for the observed FR effect. The above commented works demonstrate that the deposition of an intumescent system can prove to be an efficient FR solution for natural/synthetic blends. As an alternative approach, the deposition of hybrid silicon/phosphorous systems was also attempted. To this aim, Narkhede *et al.* employed a spray approach to deposit siloxane-based polyelectrolytes in combination with PSP on Ny/Cot blends [120]. A 20 BL coating achieved self-extinguishing behavior during vertical flame test (ASTM D6413) while showing a somewhat limited intumescent behavior. Similarly, Wang *et al.* employed γ-piperazinylproplym-ethyldimethoxy silane in combination with APP in a LbL assembly on Cot/PET achieving a self-extinguishing behavior (GB/T 5455) after 15 deposited

BL [121]. The same system also produced a consistent decrease in pkHRR (- 38%) as evidenced by cone calorimetry test. From an overall point of view it is apparent that a classical intumescent formulation (i.e. intumescent PECs) might result in the deposition of a more efficient coating. On the other hand, the use of PECs is relatively new in the field of conformal coatings and new hybrid silicon/phosphorous systems might be developed in the near future.

8.2.3 Process Equipment and Related Patents

The transition of a technology from lab- to preindustrial- and full industrial-scale is always a challenging task. Many parameters should be considered,

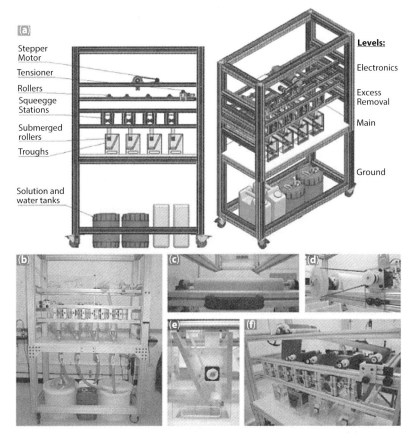

Figure 8.14 Schematization and images of the lab-scale pilot plant employed in refs [126, 127] for the LbL deposition of conformal coating on cotton. Reprinted with permission from Mateos, A. J., Cain, A. A., Grunlan, J. C. (2014). Large-scale continuous immersion system for layer-by-layer deposition of flame retardant and conductive nanocoatings on fabric. Industrial & Engineering Chemistry Research, 53(15), 6409-6416. Copyright (2014) American Chemical Society.

planned and, eventually reworked in order to fit into a totally new dimension/process requirement. The design of a new production process might also be required. The application of LbL and PECs technologies to flame retardancy is described in several patents held by the Texas A&M university [122–124]. These patents mainly describe the technology and the general procedure to achieve a FR nanocoating while providing limited indications concerning possible larger scale solutions. Luckily, in the particular field of textile fabrics there are already fabric treating technologies available at industrial scale. It is indeed extremely common for a fabric to undergo subsequent impregnating processes (e.g. exhaust techniques, padding methods) in order to dye the fabric or to confer FR properties by traditional approaches [125]. Thus, a direct application of some of the FR-solutions employing few depositions steps such as nanoparticle adsorption or PECs described in the previous chapters is certainly feasible and pose limited risks. On the other hand, there exist few scientific papers and a patent reporting of a continuous immersion system capable of treating rolls of fabrics with a given LbL formulation (Figure 8.14) [126, 127].

This equipment has been employed to successfully depositing a flame extinguishing coating on cotton by alternating layers of BPEI with urea and diammonium phosphate coupled with kaolin. A subsequent paper evaluated the differences between a lab scale- and a machine-assembled coating comprising CS and PSP [128]. While results from flammability tests were similar, a morphological evaluation pointed out a substantial difference between the two approaches with "traditional" LbL coatings properly depositing around cotton fibers, whereas the machine assembled a skin-like continuous film on top of the fabric weave. This suggests that the approach clearly needs adjustments to control coating growth and morphology, likely by applying a more aggressive rinsing procedure. Another possibility is to employ a spray assisted deposition. This solution is extremely versatile and has been presented by several scientific papers often in comparison with the conventional dipping procedure. There are also few patents describing the spray procedure to achieve multi-layered nanocoatings [129]. The results demonstrate the advantages of this approach with respect to dipping: i) a faster and more homogeneous deposition procedure, ii) the possibility to increase the coating add-on per layer, iii) the absence of cross-contamination of the solutions and iv) a potentially reduced volume of solution needed. On the other hand, in has to be considered that the spraying of polyelectrolytes/nanoparticle containing solutions should be carefully controlled with ventilation systems to capture nanoparticles that are converted into an aerosol rather than applied to the fabric. An additional problem might be represented by the need to build ad hoc production lines employing spraying in order to replace a previously existing exhaust/padding lines, but this can be addressed via traditional chemical and mechanical engineering design work.

8.3   Porous Materials

Low-density porous materials also fall in the category of substrate with a favorable surface to bulk ratio and thus possible to be treated by conformal coatings. In this field, open cell foams certainly represent the most studied substrate. These foams are widely employed in various applications such as civil, residential and industrial buildings as well as seats cushioning employed in houses (furniture) or transportations (train, bus, aircraft seating). The low density and organic nature of open cell foams, combined with the chemistry of many common foam materials, makes them an extremely easy to ignite material. Conventional solutions to this problem involve the addition of FR chemicals during the foam production process or the use of FR monomers that are thus included in the polymer chain during polymerization [41]. These procedures need to be finely tuned in order to avoid undesired effects on the foaming process and foam final properties. On the other hand, the deposition of conformal coatings appears to be a viable solution to confer the desired FR properties in an efficient way while not

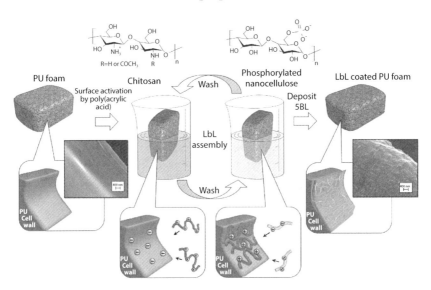

Figure 8.15 Schematization of the LbL deposition process of CS/P-CNF employed in ref [33]. (Reprinted from Carbohydrate polymers, 202, Carosio, F., Ghanadpour, M., Alongi, J., & Wågberg, L. (2018). Layer-by-layer-assembled chitosan/phosphorylated cellulose nanofibrils as a bio-based and flame protecting nano-exoskeleton on PU foams, 479-487, Copyright (2018), with permission from Elsevier.)

reducing foam properties, or, negatively affecting foam production processes. Indeed, the foam open 3D structure and inherent low density allow for the deposition of a FR conformal coating on every surface that can be reached by solution/suspension employed (Figure 8.15).

If successful the deposition is capable of creating a protective exoskeleton that extends through the entire thickness of the foam thus maximizing fire protection. Polyurethane foams represent the first and most widely substrate that was investigated. Through years other foamed substrates (with open or close porosity) have been investigated as well. As already commented for fabrics, the conformal coating should be designed in order to have minimum impact on the foam properties. Indeed, in the particular case of open cell foams, the deposition of an exoskeleton will most likely result in a change in density. This is related to the overall coating add-on and has the potential consequence of affecting the foam acoustic/thermal insulation and mechanical properties. In the following chapters the main FR solution for open or close cell foams are reported. A full sub-chapter (3.1) is dedicated to PU foams due to the high number of FR solutions available for this kind of substrate as demonstrated by a recent review paper focused on PU [39]. When reported, the durability of the coatings to multiple compression cycles is also discussed.

8.3.1   Open Cell PU Foams

Open cell PU foams have been the first foamed substrate treated by FR conformal coatings. They normally find applications in several fields including cushioning for a variety of consumer and commercial products (e.g. bedding, furniture, automotive interiors, carpet underlay and packaging). From the processing point of view the low density and open cell nature results in a rather favorable surface to bulk ratio. Thus, as reported in Figure 8.15, provided that the foam is completely exposed to the solution/suspension, the coating procedure is capable of delivering a continuous protective exoskeleton on each available surface throughout the foam. While polyurethanes are normally considered a group of generally polar polymers [130], as commented for synthetic fibers, the coating deposition can certainly benefit from a surface activation step. Indeed, such a preliminary step is often employed in the literature (e.g by nitric acid or PAA) in order to improve and promote subsequent depositions. This procedure, where the foam is immersed in the activating solution/suspension and squeezed several times, has also the practical functions of completely opening the partially closed pores while removing residual processing products trapped within the structure of the foam. During combustion, the

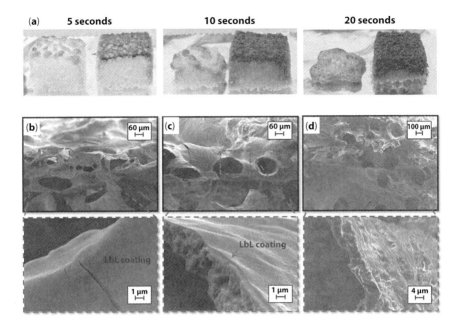

Figure 8.16 Results from ref [33] describing PU foams treated by CS/P-CNF LbL assembly: (a) images of uncoated (left) and coated (right) foams after being exposted for different times to the cone calorimetry heat flux (35 kW/m2). FE-SEM micrographs of LbL-coated PU foam exposed to 35 kW/m2 for 5 (b), 10 (c) and 20 (d) seconds. (Reprinted from Carbohydrate polymers, 202, Carosio, F., Ghanadpour, M., Alongi, J., & Wågberg, L. (2018). Layer-by-layer-assembled chitosan/phosphorylated cellulose nanofibrils as a bio-based and flame protecting nano-exoskeleton on PU foams, 479-487, Copyright (2018), with permission from Elsevier.)

behavior of PU foams pose a severe threat to the coating stability and FR performance. Indeed, upon exposure to a small flame or a heat flux, the PU starts decomposing and goes through a structural collapse that destroys the original 3D structure. This produces a pool of low-viscosity low-molecular weight compounds (mostly the original components employed during the synthesis) that burn with a vigorous flame [131]. During a flammability test this behavior leads to melt dripping that, although pronounced, does not result in a self-extinguishing behavior and the PU foams are normally consumed completely. A conformal coating has thus to act immediately after the exposure to the heat flux begins. Failing in this task will result in a completely inefficient FR action. However, if the coating is designed in a way to maintain structural integrity during the early stages of PU decomposition, substantial fire protection can be easily achieved [33]. Thus, in most, if not all, the FR-coatings formulations the mandatory action is the inhibition of PU structural collapsing that is easily observed from the post

Table 8.4 Main FR conformal coating solutions for PU foams.

Coating type	Coating composition	Main FR results	Add on required	Ref
Spherical-shape	BPEI/PAA stabilized Al(OH)$_3$	pkHRR (-64%)	32% (6BL)	[133]
	CS/Mg(OH)$_2$ nanoparticles	pkHRR (-52.6%), reduced smoke production and density	10.3% (1BL)	[134]
Needle-like shape	Pyrene-BPEI or PAA stabilized MWCNT / PAA or Pyrene-BPEI	self extinguishment HFST (ASTM D5132) (6BL), self extinguishment VFST (ASTM D6413) (9BL), pkHRR (-67%), TSR (-80%)	23-32% (6BL), 44% (9BL)	[135]
	CS/Titanate nanotube/SA	pkHRR (-70%), TSR (-40%)	6% (8TL)	[137]
	BPEI stabilized halloysite/ PAA stabilized halloysite	self extinguishment HFST, pkHRR (-67%), TSR (-70%),	34.2% (5BL)	[138]
	BPEI/ SA stabilized sepiolite	pkHRR (-76%), TSR (-25%)	30% (6BL)	[139]
Platelets	CS/MMT	pkHRR (-52%)	4% (10BL)	[140]
	BPEI/MMT+PAA	53% for pkHRR (-53%) aHRR (-63%) and no melt dripping during real scale chair mockup	N.A. (2.5BL)	[143]
	BPEI/VMT	pkHRR (-54%), TSR (-31%)	3% (1BL)	[144]

(Continued)

Table 8.4 Main FR conformal coating solutions for PU foams. (*Continued*)

Coating type	Coating composition	Main FR results	Add on required	Ref
	CS/VMT	pkHRR (-53%), TSR (-63%), resistance to flame penetration (112kW/m^2 for 900 s)	18% (8BL)	[145]
	BPEI/GO+SA	pkHRR (-73%), TSR (-56%),	25% (12BL)	[146]
	CS/GO	Cone calorimetry behavior close to flammability limits	13.4% (6BL)	[147]
	PDAC/GO	No ignition during cone calorimetry (3,6 BL), self extinguishment HFST (3, 6BL), resistance to flame penetration (6BL)	22% (3BL) 56% (6BL)	[149]
	α-CO(OH)$_2$/SA	pkHRR (-70%), TSR (-33)	4.2% (1BL)	[150]
	CS/MoS$_2$	pkHRR (-73%), TSR (-56%),	8.5% (8BL)	[151]
	CS/Ti$_3$C$_2$	pkHRR (-57.2%), THR (-65.5%),TSR (-71%)	6.9% (8BL)	[152]
Char-forming	CS/SA	pkHRR (-66%)	5.7 % (10BL)	[153]
	CS/PVS	self extinguishment HFST, pKHRR (-52%)	5.5% (10BL)	[155]

(*Continued*)

Table 8.4 Main FR conformal coating solutions for PU foams. (*Continued*)

Coating type	Coating composition	Main FR results	Add on required	Ref
	CS/PAA/CS/PPA	self extinguishment HFST, pkHRR (-55%) at 35 kW/m², pkHRR (-55%) at 50 kW/m², pkHRR (-60%) at 75 kW/m², resistance to the penetration of a butane flame torch	48 % (5 QL)	[156]
Intumescent coatings	PSP/PAH/MMT	pkHRR (-54%)	3% (4TL)	[158]
	CS/VMT + CS/APP	self extinguishment HFST, pkHRR (-66%)	19.9 % (4BL + 20 BL)	[159]
Hybrid coatings	PAA+MMT/PDAC+BOH/ APP+MMT	self self extinguishment HFST, pkHRR (-50%), SEA (-50%) TSR (-34%), durability to 100 compression cycles	37 % (3 layers)	[160]

pkHRR, THR, SEA, TSR refer to cone calorimetry performed at 35 kW/m² unless otherwise specified.

combustion residue that maintains the original shape and 3D structure of the foam. This behavior is depicted in Figure 8.16 where uncoated and coated foams are exposed to the cone calorimeter heat flux for different times.

For this reason, one of the most successful and employed FR-coating formulation relies on the use of micro- or nano-particles in combination with a polyelectrolyte matrix. The first paper reporting a surface approach to PU flammability was published in 2011 by Kim *et al.* [132]. The authors combined carbon nanofibers stabilized by BPEI with PAA. A sample coated by 4BL exhibited a 40% reduction in pkHRR and prevented foam collapsing. This work clearly showed the potential of the surface approach to PU flame retardancy and opened up to a plethora of particle containing coatings. These formulations can be roughly classified as function of the shape of the inorganic component into: spherical, needle-like and platelets. Table 8.4 reports the main results of conformal coatings for PU foams. In the category of spherical particles, hydroxides have been employed. To this aim, PAA stabilized $Al(OH)_3$ has been combined with BPEI in a LbL assembly

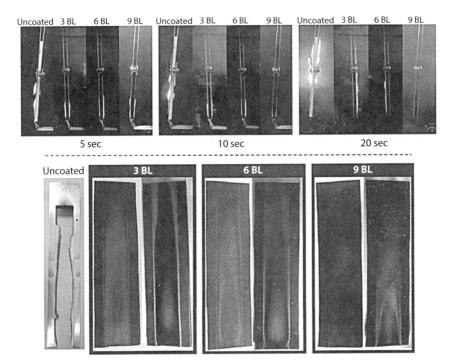

Figure 8.17 VFST results from ref [135] describing PU foams treated by pyrene-BPEI/PAA stabilized MWCNT LbL assembly. Reproduced with permission. [135] Copyright 2016, Wiley-VCH.

[133]. Foams treated by 6BL exhibited reduced pkHRR (-64%) during cone calorimeter tests and prevented foam collapsing during combustion. Similarly, nanoscale sized $Mg(OH)_2$ was employed in combination with CS [134]. This system proved to be an extremely efficient coating at a relatively low deposited layers. Indeed, a single BL reduced the pkHRR by 52.6% while reducing smoke production and density in cone calorimetry testing. As far as needle-like particles are concerned, MWCNT has been stabilized by either pyrene modified BPEI or PAA [135]. The same polyelectrolytes have been used as counterparts in three different LbL assemblies. A 6 BL coating pyrene-BPEI/PAA stabilized MWCNT can drastically decrease the pkHRR and TSR. In addition the same formulation can stop flame spread in horizontal configuration (ASTM D5132). The authors also evaluated vertical flammability (ASTM D6413) pointing out the need for at least 9BL in order to pass the test (Figure 8.17).

A similar coating composition deposited as a TL structure was also attempted [136]. This system achieved less performing FR effects with respect to the one reported in ref [135] likely due to the reduced MWCNT content within the coating. Inorganic titanate and halloysite nanotubes have been employed as well. The former was assembled in a TL structure comprising CS and SA, following a typical formulation for nanoparticle filled LbL coatings [137]. The latter was stabilized by either BPEI or PAA and LbL assembled [138]. The presence of inorganic tubes embedded within a continuous organic coating can easily decrease pkHRR to various extents (-60–70%) during cone calorimeter testing. Interestingly, both components show a propensity to reduce TSR values during cone calorimeter testing indicating that the shape of the particles can play a key role on volatiles release adsorption. In addition, 5BL containing HN nanotubes can stop flame spread when ignited in horizontal configuration. Pan *et al.* further evaluated the effects of SA stabilized sepiolite nanorods in combination with BPEI [139]. With respect to previously commented needle-like containing coatings, this formulation proved to be as efficient in reducing pkHRR values (-76%) but with reduced effects on smoke production (TSR – 25%) again in cone calorimeter testing. This latter result might suggest an effect of the tube inner diameter on the smoke adsorbing effect particle employed. The vast majority of publications employed nanoplatelets. This "success" stems from the shape of the particles and their tendency to assemble flat and parallel to the substrate thus maximizing the surface coverage and protection that can be achieved per deposited layer. Laufer *et al.* assembled a CS/MMT LbL coating characterized by a green composition [140]. The effects of chitosan pH on the coating growth were also evaluated. 10 BL assembled at pH 6 can halve the pkHRR with a relatively low add-on (4%wt). This system clearly shows the potentialities of the LbL in controlling the thickness and composition of the deposited coating by finely tuning the deposition parameters (such as pH, ionic strenght, etc.). On the other hand, it should be pointed out that, the need for 10 BL (i.e. 20 deposition steps) suggest a limited efficiency of the deposited assembly. Subsequent works focused on the construction of formulations where the amount of MMT deposited per layer was maximized. To this aim, Kim *et al.* presented a TL strategy to build fast growing clay filled coatings by employing a TL architecture BPEI/MMT/PAA in comparison with a conventional BPEI/MMT system [141]. The results showed that a TL approach is capable of achieving thicker coatings by exploiting the exponential growth of the BPEI/PAA system and by preventing the release of MMT during the rinsing steps. The FR performances were limited with respect to the previously commented CS/MMT assembly likely due to the excessive presence of PE within the assembly. The effects of solution/suspension parameters were further evaluated highlighting a beneficial role of an increased MMT concentration [142]. The authors also reported how the coating can increase the stiffness of the foam while maintaining its ability of being cyclically deformed. Coated foams subjected to 4 compression cycles kept their original FR properties thus demonstrating a certain degree of durability although after an extremely limited number of compression cycles. This latter aspect is of extreme importance as one of the main application of PU foams (i.e. cushioning) directly implies a multitude of compression cycles (i.e. hundreds/year) over their lifespan. Unfortunately, the durability of the coating is rarely evaluated thus resulting

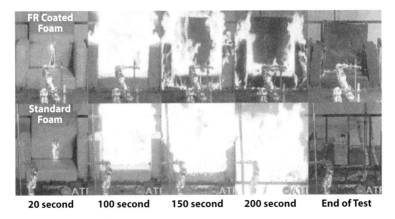

Figure 8.18 Real scale during real scale chair mockup test results from ref [143]. Reprinted with permission from Kim, Y. S., Li, Y. C., Pitts, W. M., Werrel, M., Davis, R. D. (2014). Rapid growing clay coatings to reduce the fire threat of furniture. ACS applied materials & interfaces, 6(3), 2146-2152. Copyright (2014) American Chemical Society.

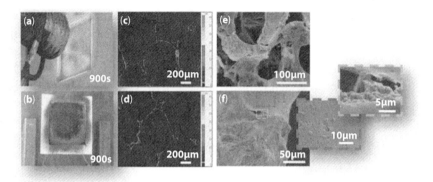

Figure 8.19 Burn-through fire test results of 8 BL CS/VMT coating from ref [145]. Digital images of coated PUF during the burn-through test: (a) front view and (b) back view. Aluminum cross-section EPMA X-ray mapping of coated foam residues after the burn-through fire test: (c) front side and (d) backside of PUF. SEM images of residue from the burn through fire test: (e) front side and (f) back side of PUF. Insets reveal the formation of cracks and bubbles. Reprinted with permission from Lazar, S., Carosio, F., Davesne, A. L., Jimenez, M., Bourbigot, S., Grunlan, J. (2018). Extreme heat shielding of clay/chitosan nanobrick wall on flexible foam. ACS applied materials & interfaces, 10(37), 31686-31696. Copyright (2018) American Chemical Society.

in a missing piece of information for many efficient FR assemblies that have been reported to date. The same research group further developed this strategy by combining MMT and PAA in the same solution [143]. This system proved to be more efficient as demonstrated by real scale test where 2.5 BL considerably deceased the HRR of a treated furniture while also suppressing the melt dripping (Figure 8.18).

In order to improve the efficiency of the added nanoplateletet, MMT was replaced with high aspect ratio VMT [144]. This substitution proved to be beneficial when a LbL system comprising either BPEI or CS is assembled. Indeed, a single BL BPEI/VMT was found able to substantially reduce the pkHRR (-54%) in cone calorimeter testing with a drastic improvement over a BPEI/MMT system. A subsequent work investigated the fire resistance properties of PU foams coated by 8 BL CS/VMT [145]. Beside an expected reduction in HRR values during forced combustion tests, the treated foams were able to withstand flame penetration during burn-through tests performed with an applied heat flux of 112 kW/m² (Figure 8.19). The presence of the coating maintained the structural integrity of the foam successfully insulating the unexposed side with a temperature drop ≈ 160°C/cm.

These results further highlight the improved performances of LbL coatings containing high aspect ratio clays and the possibly of extending the application field of coated PU foams. Graphene related materials have

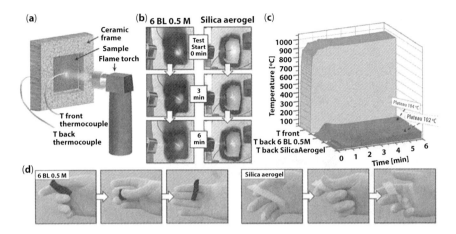

Figure 8.20 Flame penetration test results of 6 BL PDAC/GO 0.5 M coating from ref [149]. (a) schematic representation of the layout adopted for the test, (b) digital pictures of the 6 BL 0.5M and silica aerogel front surface during the test and (c) front and back side temperatures as a function of time for 6 BL 0.5M and silica aerogel. (d) simple bending test performed on 6 BL 0.5M foam and silica aerogel. Reproduced with permission. [149] Copyright 2018, Wiley-VCH.

also been employed as high aspect ratio nanoplates. Pan *et al.* build a LbL system by combining BPEI with a solution of SA containing GO [146]. The author also evaluated the effect of a post reduction treatment employing a thermal treatment. Results showed that the BPEI/GO+SA assembly is capable of delivering FR performances similar to VMT in terms of foam stability, HRR reduction and smoke suppression during cone calorimeter testing. The GO reduction post treatment did not provide any benefit on the coating FR behavior. Similarly, high aspect ratio GO was coupled with CS [147]. Surprisingly, a 6BL coating was found able to push the PU foam to the limits of flammability during cone calorimetry (35 kW/m²) where 50% of the tested samples did not ignite. This was ascribed to the ability of the deposited coating to hinder volatile release upon exposure to a specific heat flux. The same coating prevented melt dripping but was not able to stop flame propagation during HFST tests. The effects of GO particle size were investigated in a subsequent work highlighting how an improved aspect ratio can increase the efficiency of the coating by reducing the amount of coating needed to achieve a FR effect [148]. The authors further improved on the FR performances by replacing CS with PDAC and changing the ionic strength of GO with di ammonium phosphates [149]. A 3BL coating built at modified ionic strength can self-extinguish the flame in HFST and grant a consistent non-igniting behavior by cone calorimetry (35 kW/m²). In addition, when deposited at 6 BL, the treated foams have been found able to withstand the penetration of a flame torch for more than 6 minutes with a temperature gradient greater than 500 °C/cm, similar to inorganic aerogels (Figure 8.20).

Other 2D materials such as mxenes (which are platey-type materials such as Ti_3C_2), MoS_2 and α-Co(OH)$_2$ have been employed as well [150–152]. Results are aligned with what previously discussed and further support the ability of high aspect ratio nanoplates containing coatings in preventing foam collapsing, decreasing HRR and reducing smoke production. Beside the use of nanoparticles, coatings comprising only polyelectrolytes have been developed. These kinds of coatings fall in the categories, already described for fabrics, of charring and intumescent-like assemblies. Polysaccharide-based char forming coatings were achieved by combining SA with CS [153]. This composition can slow down flame spread and suppress melt dripping by HFST as well as considerably reduce pkHRR. Lignin sulfonate has been combined with CS as well but with less satisfactory results [154]. Laufer *et al.* deposited a coating based on CS and poly(vinyl sulfonic acid sodium salt) (PVS) [155]. 10 BL can stop flame spread when ignited in horizontal configuration and produced a reduction in pkHRR (-52%) during cone calorimetry tests. CS was also coupled with PAA and PPA in a QL architecture [156]. The treated foams achieved a self-extinguishing behavior in HFST and proved to be quite efficient in reducing the pkHRR (-50-60%) when tested by cone calorimetry at different heat fluxes (i.e. 35, 50 and 75 kW/m²). In addition, the treated foams were found able to withstand the penetration of a butane–propane flame torch (T exposed side≈870°C). A key point that correlates these last papers is the higher number of layers required in order to achieve an appreciable FR effect with respect to nanoparticle containing systems. This can be explained by considering the foam behavior during combustion. Indeed, the structural collapsing that occurs almost immediately after exposure to an heat source inevitably compromises the integrity of the coating that has limited time to exert its FR effect. On the other hand, nanoparticle-rich coatings normally show better structural integrity, achieved with reduced deposited layers, due to the presence of an inorganic reinforcing agent (compare required BL values in Table 8.4). As a result the vast majority of papers published on the topic mostly refers to this latter coating formulation. The same concept is transferred to intumescent formulations where collapsing occurs before intumescence can commence. For this reason, intumescent-like coatings have been demonstrated to provide substantial FR results only when combined with nanoparticles [157]. Cain *et al.* combined PSP with PAH and MMT in a TL structure [158]. The resulting assembly produced an intumescent behavior with the formation of MMT reinforced expanded structures that prevented

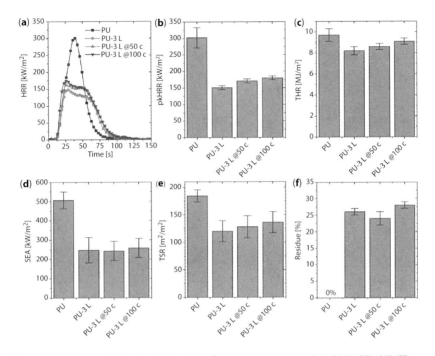

Figure 8.21 Cone calorimetry test results of a PAA+MMT/PDAC+BOH/APP+MMT coating from ref [160]. Cone calorimetry parameters neat PU, treated PU and treated PU foams after 50 or 100 compression cycles: (a) HRR Vs time plot, (b) pkHRR, (c) THR, (d) SEA, (e) TSR and (f) final residue. Reproduced from [160] under the terms of the CC-BY Creative Commons Attribution 4.0 (https://creativecommons.org/licenses/by/4.0). Copyright 2019, The Authors, published by Frontiers.

foam collapsing with reduced pKHRR values. Holder *et al.* proposed a different approach by employing a stacked configuration where an intumescent formulation based on APP and CS is deposited on top of a CS/VMT coating [159]. This configuration prevented the coating structural collapse and allowed for the proper development of a protective expanded structure capable of self-extinguishing the flame in horizontal configuration and considerably reduce pkHRR values. This sort of synergistic FR effect was further demonstrated by comparison with coatings assembled with either only CS/VMT or only CS/APP. Similarly, hybrid coatings combining different FR actions have been achieved by depositing only three layers each one containing a polyelectrolyte and a nanoparticle (i.e. PAA+MMT/PDAC+BOH/APP+MMT) [160]. This structure was found capable of self-extinguishing the flame during flammability tests and considerably reduce pkHRR and smoke optical density during cone calorimetry. The authors also evaluated the durability of these results to 100 compression cycles (Figure 8.21).

SEM investigations pointed out the formation of cracks on the coating upon the first compression but a limited increase in crack dimensions as the number of cycles increased. The compressed foams maintained the original FR properties thus demonstrating a good durability of the coating. This paper, is one of the very few to address durability that is at the basis of many practical applications of PU foams. In addition, although it is possible to somehow extend the presented results to other FR-formulations, it is important to notice that a durability evaluation must be eventually performed in order to access a final application. From an overall point of view, and from the parameters reported in Table 8.4, it is possible to devise a clear trend in the development of conformal coatings and the resulting FR properties. Indeed, the use of high aspect ratio nanoplatelets seems to confer the best FR properties with a reduced number of deposition step. In addition, it is worth highlighting that although it is quite common to achieve substantial reduction in HRR values during cone calorimetry (i.e. pkHRR reduced by 40-60%) there are few formulations that can stop flame propagation during flammability test. This is a sort of limitation to many conformal coatings since this latter aspect is of great importance as PU foams are normally considered to be the one of the first item ignited at the start of a fire [161, 162].

8.3.2 Other Porous Substrates

While it is apparent that most of the research efforts in conformal coatings have been directed towards PU foams, there are other substrates worth mentioning in the category of porous substrates. Yang *et al.* proposed a LbL coating based on CS and APP for melamine foams [163]. 2 BL were found able to coat the 3D structure of the foam and improve its already good FR properties. Indeed, LOI values increased from 34.5 to 47% while cone calorimetry test (heat flux 50 kW/m²) pointed out a considerable decrease in pkHRR (-87%) and THR (-77%). The presence of the coating prevented foam shrinkage during combustion as evaluated by SEM evaluation. This paper demonstrates how the deposition of conformal coatings is capable of enhancing the FR properties of an already performing substrate such as a melamine foam thus potentially extending the application fields of the substrate. The same system was also employed on open cell silicone foams by Deng *et al.* and compared with a CS/MMT assembly [164]. This latter produced a more homogeneous coating yielding the best FR performances and achieving, at 21 BL, an increase in LOI from 20.2 to 25.7% as well as a decrease in pkHRR (-24%) and TSR (-41%) in cone calorimeter testing. Interestingly, the FR properties of the CS/APP system showed an inversely related proportion with the number of deposited BL. This was explained by considering a possible change in CS/APP ratio, and thus a reduction in intumescent features, upon increasing the number of absorbed layers. This latter observation along with the high BL number required by the CS/MMT assembly clearly suggests a limited FR efficiency of these two coatings in reducing the flammability of silicone foams. Recently, a surface approach encompassing GO has been developed where the conformal coating is achieved during the silicone foaming process by employing a GO water suspension as foaming agent [165]. The prepared foams can achieve a LOI close to 32, a self-extinguishing behavior during VFST and considerable reductions in pkHRR (-56%) and TSR (-87%) during cone calorimeter testing. While the approach to achieve a conformal coating does not fit in the definition provided in the introduction section, the results described in the above commented work certainly support the fact that a properly designed conformal coating is capable of delivering high performing FR properties. As far as close cell foams are concerned, there is a limited amount of papers reporting a surface approach to limit their flammability. Most of these papers do not employ conformal coatings but rather a more conventional approach based on paints [166]. Notwithstanding this, a proof of concept was presented on close cell PET foams. To this aim, an

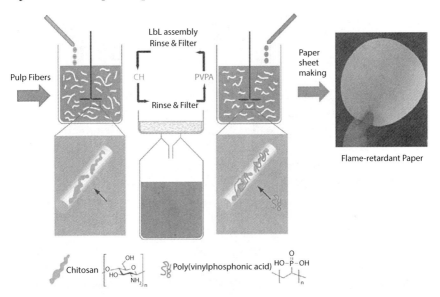

Figure 8.22 Schematic representation of the procedure adopted to coat wood fibers with a CS/PVPA LbL assembly from ref [168]. Reprinted with permission from Koklukaya, O., Carosio, F., Grunlan, J. C., & Wagberg, L. (2015). Flame-retardant paper from wood fibers functionalized via layer-by-layer assembly. ACS applied materials & interfaces, 7(42), 23750-23759. Copyright (2015) American Chemical Society.

intumescent QL system based on PAA/PDAC/APP/PDAC was developed on PET foams [167]. The authors further evaluated the efficiency of DNA as a possible replacement for APP. This latter was found to deliver the best FR properties likely due to a more homogeneous and thick deposition. Indeed, 4 QL containing APP suppressed melt dripping and self-extin-guished the flame in HFST while achieving a reduction (-25%) in pkHRR during cone calorimeter testing. Conversely, a DNA-based QL only par-tially reduced melt-dripping and did not modify the burning behavior during cone calorimetry tests. Post combustion SEM evaluation further confirmed the observed differences in FR by clearly showing an intumes-cent behavior of APP-based coatings and a limited protection for DNA containing ones. Beside synthetic foams, natural porous substrates should be considered as well. This latter, small, category includes paper, cellulose based foams and wood. The concept of FR waterborne conformal coatings can be applied to paper by performing the deposition on the fibers prior to their assembly in the final product (Figure 8.22). To this aim, Koklukaya *et al.* assembled a LbL coating comprising CS and PVPA [168].

The presence of a 20 BL coating improved the char forming ability of the cellulose fibers and yielded a self-extinguishing behavior during HFST. The same samples also achieved a 50% reduction in pkHRR during cone calo-rimetry tests. This paper certainly reports an innovative proof of concept to paper flame retardancy although the high number of BL required represents the main limitation of the proposed approach. In order to address this prob-lem, the same authors further developed a LbL formulation based on PEI at different MW and PSP [169]. 3.5 BL comprising high MW PEI achieved self-extinguishing behavior during HFST while also improving the LOI from 20 to 24%. The same coating assembled on a wet strengthen paper did no achieve the same results thus demonstrating the benefits of performing the LbL assembly on the single fibers prior to their assembly. Similarly, cellulose fiber networks were LbL coated in order to obtain a green alterna-tive to synthetic foams. The LbL deposition was performed on wet-stable cross-linked networks [170]. A QL structure comprising CS/PVSA/CS/NP, where the NP was either SNP, SEP or MMT, was employed as hybrid char forming/inorganic barrier. The use of different NP of different shapes fur-ther allowed to evaluate the effect of shape and aspect ratio on the achieved FR properties. Only SEP and MMT based coatings were found able to self-extinguish the flame in HFST and produce a 47% reduction in pkHRR during cone calorimetry tests. Both formulations were not able to stop flame spread during VFST; however, the presence of MMT preserved the internal portion of the samples whereas fiber network coated with SEP-based QL were completely damaged by the flames. Cellulose based foams

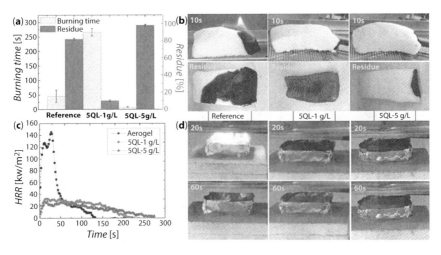

Figure 8.23 Flammability and cone calorimetry results of nanocellulose aerogels treated by CS/PVPA/CS/MMT assembly from ref [174]. Burning time measured during HFST (a), (b) Photographs taken during HFST of the different aerogels, 10 s after removal of methane flame and at the end of the test, (c) Heat release rate curves for the untreated and LbL-treated aerogels, and (d) Photographs of the reference and of the LbL-treated CNF aerogel 20 and 60 s after exposure to the cone heat flux (35 kW/m²). Reprinted with permission from Köklükaya, O., Carosio, F., & Wågberg, L. (2017). Superior flame-resistant cellulose nanofibril aerogels modified with hybrid layer-by-layer coatings. ACS applied materials & interfaces, 9(34), 29082-29092. Copyright (2017) American Chemical Society.

can also be produced by the freeze-drying of nanocellulose containing colloids. These foams normally show intrinsic FR properties due to the inclusion of nanoparticles or other water-soluble FR chemicals in the suspension [171–173]. Alternatively, cross-linked and water-stable nanocellulose foams can be coated by conformal coatings in order to achieve FR properties. To this aim, Koklukaya *et al.* assembled a QL structure comprising CS/PVPA/CS/MMT evaluating the effects of different concentrations on the achieved FR properties [174]. Coatings build up at higher concentration achieved a self-extinguishing behavior during HFST and prevented ignition during forced combustion (cone calorimeter) tests (Figure 8.23).

In addition, modified aerogels were found capable of withstanding the penetration of a propane-butane flame torch (T exposed surface 800–700°C) and achieving impressive temperature drops for a cellulose based material. The results presented in the above mentioned works can certainly compete with what already discussed for open cell PU foams, however it should be mentioned that the processes employed for the production of such cellulose-based foams still need substantial development in order to be competitive with conventional, petroleum-based foams. Another porous substrate worth mentioning in the field of conformal coating is wood. This substrate is widely employed as construction material and some of the strategies employed to confer FR properties already rely on impregnation processes [175]. Zhou et al. developed a LbL assembly based on sodium phytate evaluating the effects of adding TiO_2/ZnO nanoparticles either alone or together [176]. Samples treated by 10 BL were capable of increasing the LOI value to 33% (reference wood = 24.6%) while granting self-extinguishing behavior in a candle like test. A possible drawback of the presented system is certainly represented by the rather long deposition steps (90 min per layer) required to achieve satisfactory FR properties. Rehman et al. considerably reduced the number of deposition steps and the time per step by employing a preliminary surface activation step based on a combined acid/UV treatment [177]. Thanks to this pre-treatment, 3BL of a CNF/VMT assembly yielded a self-extinguishing behavior during UL-94 tests performed in horizontal orientation. This work clearly demonstrates the potentialities of conformal coatings and the beneficial effects of surface pre-treatments in ensuring efficient FR properties. Similarly to

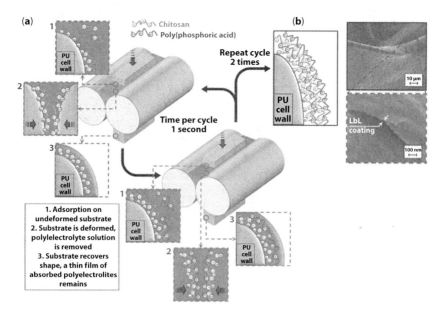

Figure 8.24 Schematic representation of the procedure adopted to LbL coat PU foams employing a CS/PPA coating (a) and SEM images of the resulting coating (b) from ref [180]. Adapted with permission from Carosio, F., Alongi, J. (2016). Ultra-fast layer-by-layer approach for depositing flame retardant coatings on flexible PU foams within seconds. ACS Applied Materials & Interfaces, 8(10), 6315-6319. Copyright (2016) American Chemical Society.

other fields already discussed, PECs were also employed. Kolibaba *et al.* deposited a BPEI/PSP complex followed by a "curing step" in acidic solution [178]. The treated wood successfully self-extinguished the flame after 45 s application of a flame torch. The same coating increased the TTI during cone calorimetry tests. HRR parameters were almost not affected whereas THR was considerably decreased due to an improved wood charring effect.

8.3.3 Process Equipment and Related Patents

Differently from textiles, the production processes of foamed/porous substrates do not normally imply an impregnation step. Similarly to textiles the patents available on the multi-step deposition of conformal coatings mostly focus on the lab-scale process [179]. Given the different kinds of substrates discussed in this chapter it is of interest to present the current processing options and limitations for each subcategory. As far as open cell PU foams are concerned, it is apparent that a multi-step impregnation may be difficult to implement. Although research efforts have been directed

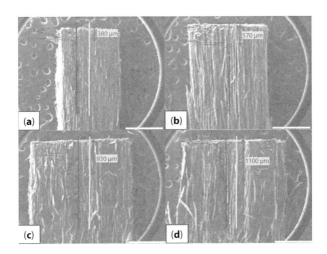

Figure 8.25 Characterization from ref [178] investigating the penetration depth of PECs by impregnation and post curing. EDS traces signifying intensity of the phosphorus Kα peak as a function of distance through cross sections of (a) PEC1min deposition, 1 min cured, (b) PEC1min deposition, 10 min cured, (c) PEC60min deposition, 1 min cured and (d) PEC60min deposition, 10 min cured. White scale bars are all 1 mm. Blue lines are to guide the eye as to the approximate level of background noise where there is no phosphorus signal. Green lines indicate the areas determined to be of relevant phosphorus content, along with the measured length. Reproduced with permission. [178] Copyright 2019, Wiley-VCH.

towards a reduction in the number of layers needed the time needed for the conventional deposition of 2-4 layers might still limit the applicability of conformal coatings in this particular fields. In the case of flexible open cell foams, it has been demonstrated that a fast deposition process (i.e. few seconds per layer) where the foams are impregnated and compressed at the same time is possible and can efficiently produce FR foams (Figure 8.24) [180].

This procedure is, of course, not applicable to open cell rigid foams. For these latter it has been proposed a procedure where the foam is used a sort of filter through which the different solutions/suspensions flow in order to deposit subsequent layers of a conformal coating [174]. A recent patent reports of a batch procedure, for open cell foams, that deposits high performing FR conformal coatings by employing a single deposition step [181]. As far as paper is concerned, given the waterborne characteristic of the production process, the possibility of incorporating adsorption steps of polyelectrolytes and nanoparticles in pilot-scale processes has been investigated and successfully developed [182]. Indeed, there exist a methodology where the chemicals needed to build a conformal coating directly on the fibers are added consecutively in a specially constructed tube reactor consisting of in-line mixers in a pipe-line. This allows for a fast and secure pre-treatment of the fibers prior to their assembly in the final product [183]. Additionally, water based processes using rollers or spray-based systems are known for paper processing. Similarly, coating or impregnation processes are employed in order to confer FR properties to wood [175]. Coatings normally consist in FR paintings that should be reapplied systematically in order to maintain sufficient FR protection. In the impregnation approach, the flame retardant solution is forced inside the porous wood structure. A vacuum step is used to remove air from inside the structure, and the solution of fire retardant chemicals are injected under high pressure (pressure treatment of lumber). The process can be repeated in order to achieve the desired level of protection. This procedure does not operate at room temperature and pressure as it occurs in the case of conformal coating. Such difference is linked to the difficulty in impregnating the structure of wood that normally show a poorly interconnected, and thus difficult to access, porosity. Indeed, as also demonstrated by the literature without the use of pressure and temperature the simple impregnation with polyelectrolytes and nanoparticles often results in the conformal coating being deposited within the outermost layers of wood that are in direct contact with the solution/suspension, and the inner core of the wood is left unprotected (Figure 8.25) [178].

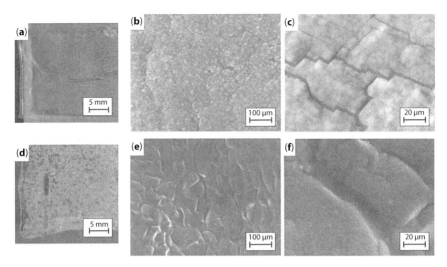

Figure 8.26 Characterization from ref [184] investigating the effects of cone calorimetry exposure for LbL coated PC films. Digital photos and SEM micrographs of 0.2 (a, b, c) and 1 (d, e, f) mm PC treated by 20 BL after 20 second exposition to the cone calorimeter heat flux (50 kW/m2). (Reprinted from European polymer journal, 49(2), Carosio, F., Di Blasio, A., Alongi, J., Malucelli, G., Layer by layer nanoarchitectures for the surface protection of polycarbonate, 397-404, Copyright (2013), with permission from Elsevier.)

8.4   Other Substrates

Conformal coatings have been widely developed in order to improve the FR properties of textiles and foam. However there are some papers reporting on the use of this FR approach on thin plastic films and composites. As far as films are concerned, a silica based LbL coating has been developed on PC evaluating the effects of different film thicknesses (namely 0.2 and 1 mm) on the resulting FR properties [184]. A 20 BL coating was found able to suppress melt dripping and self-extinguish the flame during VFST (UL94-VTM). By cone calorimetry (heat flux 50 kW/m^2) the same coating produced an increase in TTI (+ 29%, + 17 s) with limited reductions in pkHHR values. These results were achieve on 0.2 mm films while thicker substrates did not show substantial FR improvements. This was ascribed to the different surface to bulk ratio as well as different coating stability due to the melting process of the substrate. Indeed, SEM observations performed on PC films exposed to the cone heat flux in the pre ignition time clearly showed how the melting of thinner films resulted in limited damage to the coating whereas thicker films considerably damaged the silica assembly leaving big portion of the surface unprotected (Figure 8.26).

This paper proves, once again, the limits for conformal coatings in protecting substrates were the surface to bulk ratio is low. The same authors improved the durability of the coating by introducing a UV curable waterborne resin within the coating in a QL architecture (BPEI/SiO$_2$/SiO$_2$/UV-resin) [185]. 5 QL UV cured self-extinguished the flame in HFST while also suppressing the melt-dripping. The same performances were maintained after 1 h washing in water at 50°C. The stability of the coating was also tested against a 1 M ammonia solution further proving the improved coating stability after UV-curing. Laachachi *et al.* assembled a PAH/MMT LbL coating on 2 mm PLA sheets [186]. A post diffusion of SPS was employed in order to produce a hybrid inorganic-organic intumescent FR action. Samples treated by 60 BL were found able to considerably delay ignition during cone calorimetry

Figure 8.27 Characterization from ref [188] investigating the resistance to flame torch exposure for LbL coated PS films. Pictures of flame-through torch test 5 s after ignition of 3.2 mm thick PS plates: (a) control, (b) 8-BL CH+tris/MMT film added, or (c) 8-BL CH+tris/VMT film added. Pictures of the PS plates after 10 s flame-through torch test of the (d) control, (e) with a 3-BL CH+tris/MMT film added, or (f) with a 2-BL CH+tris/VMT film added. Reproduced with permission. [188] Copyright 2019, Wiley-VCH.

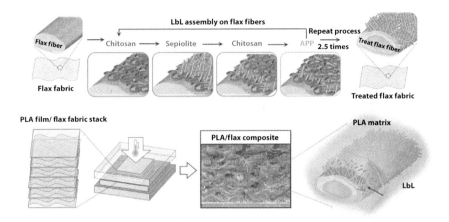

Figure 8.28 Schematic representation of the procedure adopted in ref [191] to coat flax fabrics and assemble them in the final PLA-based composite. (Reprinted from Composites Part B: Engineering, 200, Battegazzore, D., Frache, A., Carosio, F., Layer-by-Layer nanostructured interphase produces mechanically strong and flame retardant bio-composites, 108310, Copyright (2020), with permission from Elsevier.)

tests (TTI + 110%, + 62s) and reduce the pkHRR (-37%). A 30 BL assembly achieved similar results in terms of TTI increase but did not affect pkHRR in a remarkable way (reduction within 10%). Similarly PAH/MMT coatings were deposited on PA sheets (0.5 mm). 20BL achieved a considerable reduction in pkHRR (-60%) by cone calorimetry (25 kW/m^2) [187]. From these preliminary papers, it is apparent that the successful development of high performing conformal coatings on thin films is strictly related to the ability of the selected assembly to deposit thick coatings within a relatively low deposition steps. To this aim, Guin *et al.* evaluated the effects of different thickening agents (salts, amines, alcohols and buffers) on the coating growth of CS/MMT and CS/VMT assemblies [188]. The use of a tris buffer increase the thickness of the deposited coating by one order of magnitude allowing to reach at 8 BL thicknesses of 2170 and 3900 nm for MMT- and VMT-based assemblies, respectively. The FR properties were demonstrated on a 3.2 mm PS plate by depositing either 3BL CS/MMT or 2BL CS/VMT. The treated PS was found able to withstand the application of a butane micro torch (10s) and self-extinguish the flame right after the removal of the impinging flame (Figure 8.27).

These results prove that it is possible to dramatically change the assembly of the selected constituents, and thus the resulting FR properties, by controlling the deposition parameters. Recently the LbL has been employed to finely tune the fiber/matrix interfacial properties in composites. In this approach the LbL deposition on the fibers prior to their incorporation in the final material. Initial works focused on the build up of nanoengineered interphases capable of improving the mechanical properties of carbon fiber thermosets and bio-composites [189, 190]. A subsequent work employed this approach in order to build a nanoengineered interphases capable of simultaneously impart mechanical strength and flame retardant properties to PLA-flax fibers composites [191]. The authors deposited a QL coatings comprising CS/SEP/CS/APP (Figure 8.28).

The deposition of 2.5QL significantly improved the FR properties of the prepared composites yielding an increase in LOI (from 21.5 to 25.4%), a considerable reduction in flame spread rates during HFST (UL94 HB), and substantial reduction in peak of heat release rate (-33%) and maximum average rate of heat emission (-30%) by cone calorimetry. The evaluation of mechanical properties further showed how the presence of a LbL inter-phase can improve modulus while showing limited reductions in flexural strength. This work present a new strategy towards the preparation of composites with simultaneously improved mechanical and FR properties. The achieved results are certainly promising but further developments of this approach are needed in order to meet the fire safety requirements and performances for practical applications.

8.5   Future Trends and Needs

This chapter presented the use of conformal coating as viable FR solution for textiles, porous substrates and films. The results and the multitude of papers available on the topic certainly suggest that conformal coatings are a developing field in the design of new and advanced FR materials. Although the pros and cons of each main FR formulation were discussed in the related paragraph, it worthy to present a short conclusion summarizing currently unmatched needs and future trends. These can be summarized as follows:

1. Durability: there is a clear lack of durability evaluation on the performed treatments. This parameter is evaluated mainly for fabrics where the resistance to washing cycles is normally considered a must in order to approach a practical application. Nevertheless, this parameter is of paramount importance for foams and other substrates as well.

2. Toxicological assessment of the coating: current FR chemicals are under scrutiny because of their known toxic and environmental problems [192]. As a result, the scientific and industrial communities have become aware of the problem. Thus a careful evaluation of the toxicological and release aspects of the deposited conformal coatings must be performed before planning an industrialization step.

3. Address multifunctionality: recent papers often employ the design freedom of LbL and PECs in order to build multifunctional coatings. This certainly adds value to the treatment and the final product and should be considered a viable strategy in order to promote the use of conformal coatings at industrial level.

4. Scale-up and process science is needed for some coating chemistries for some substrates. While applying these coatings to fabrics is well known, application to foams and solid substrates would benefit from additional process research.

Future research and development should focus on the above topics to help conformal coatings become another useful tool for non-halogenated flame retardancy of materials.

References

1. Malucelli, G., Carosio, F., Alongi, J., Fina, A., Frache, A., Camino, G., Materials engineering for surface-confined flame retardancy. *Mater. Sci. Eng. R: Rep.*, 84, 1–20, 2014.

2. Ray, S.S. and Okamoto, M., Polymer/layered silicate nanocomposites: a review from preparation to processing. *Prog. Polym. Sci.*, 28, 11, 1539–1641, 2003.

3. Gilman, J.W., Harris Jr., R.H., Shields, J.R., Kashiwagi, T., Morgan, A.B., A study of the flammability reduction mechanism of polystyrene-layered silicate nanocomposite: layered silicate reinforced carbonaceous char. *Polym. Adv. Technol.*, 17, 4, 263–271, 2006.

4. Gilman, J.W., Jackson, C.L., Morgan, A.B., Harris, R., Manias, E., Giannelis, E.P., Wuthenow, M., Hilton, D., Phillips, S.H., Flammability properties of polymer–layered-silicate nanocomposites. Polypropylene and polystyrene nanocomposites. *Chem. Mater.*, 12, 7, 1866–1873, 2000.

5. Bartholmai, M. and Schartel, B., Layered silicate polymer nanocomposites: new approach or illusion for fire retardancy? Investigations of the potentials

and the tasks using a model system. *Polym. Adv. Technol.*, 15, 7, 355–364, 2004.

6. Schartel, B., Bartholmai, M., Knoll, U., Some comments on the main fire retardancy mechanisms in polymer nanocomposites. *Polym. Adv. Technol.*, 17, 9–10772–777, 2006.

7. Alongi, J., Han, Z., Bourbigot, S., Intumescence: tradition versus novelty. A comprehensive review. *Prog. Polym. Sci.*, 51, 28–73, 2015.

8. Alongi, J., Tata, J., Carosio, F., Rosace, G., Frache, A., Camino, G., A comparative analysis of nanoparticle adsorption as fire-protection approach for fabrics. *Polymers*, 7, 1, 47–68, 2015.

9. Meka, V.S., Sing, M.K., Pichika, M.R., Nali, S.R., Kolapalli, V.R., Kesharwani, P., A comprehensive review on polyelectrolyte complexes. *Drug Discovery Today*, 22, 11, 1697–1706, 2017.

10. Zhang, X., Xu, Y., Zhang, X., Wu, H., Shen, J., Chen, R., Xiong, Y., Li, J., Guo, S., Progress on the layer-by-layer assembly of multilayered polymer composites: Strategy, structural control and applications. *Prog. Polym. Sci.*, 89, 76–107, 2019.

11. Ariga, K., Ahn, E., Park, M., Kim, B.S., Layer-by-layer assembly: recent progress from layered assemblies to layered nanoarchitectonics. *Chem.– Asian J.*, 14, 15, 2553–2566, 2019.

12. Lukaszkowicz, K., *Review of nanocomposite thin films and coatings deposited by PVD and CVD technology*, pp. 145–162, Intech, Rijeka, 2011.

13. Deng, Y., Chen, W., Li, B., Wang, C., Kuang, T., Li, Y., Physical vapor deposition technology for coated cutting tools: A review. *Ceram. Int.*, 46, 11, 18373–18390, 2020.

14. Manawi, Y.M., Samara, A., Al-Ansari, T., Atieh, M.A., A review of carbon nanomaterials' synthesis via the chemical vapor deposition (CVD) method. *Materials*, 11, 5, 822, 2018.

15. Cai, Y., Wu, N., Wei, Q., Zhang, K., Xu, Q., Gao, W., Song, L., Hu, Y., Structure, surface morphology, thermal and flammability characterizations of polyamide6/organic-modified Fe-montmorillonite nanocomposite fibers functionalized by sputter coating of silicon. *Surf. Coat. Technol.*, 203, 3-4, 264–270, 2008.

16. Martin, P.M., *Handbook of deposition technologies for films and coatings: science, applications and technology*, William Andrew, Norwich, NY, 2009.

17. Tsafack, M.J. and Levalois-Grützmacher, J., Flame retardancy of cotton textiles by plasma-induced graft-polymerization (PIGP). *Surf. Coat. Technol.*, 201, 6, 2599–2610, 2006.

18. Jimenez, M., Lesaffre, N., Bellayer, S., Dupretz, R., Vandenbossche, M., Duquesne, S., Bourbigot, S., Novel flame retardant flexible polyurethane foam: plasma induced graft-polymerization of phosphonates. *RSC Adv.*, 5, 78, 63853–63865, 2015.

19. Bourbigot, S., Jama, C., Le Bras, M., Delobel, R., Dessaux, O., Goudmand, P., New approach to flame retardancy using plasma assisted surface polymerisation techniques. *Polym. Degrad. Stab.*, 66, 1, 153–155, 1999.

20. Schartel, B., Kühn, G., Mix, R., Friedrich, J., Surface controlled fire retardancy of polymers using plasma polymerisation. *Macromol. Mater. Eng.*, 287, 9, 579–582, 2002.

21. Shishoo, R., *Plasma technologies for textiles*, pp. 97–122, Woodhead Publ, Ltd., Cambridge, 2007.

22. Horrocks, A.R., Eivazi, S., Ayesh, M., Kandola, B., Environmentally sustainable flame retardant surface treatments for textiles: The potential of a novel atmospheric plasma/UV laser technology. *Fibers*, 6, 2, 31, 2018.

23. Brinker, C., Hurd, A., Schunk, P., Frye, G., Ashley, C., Review of sol-gel thin film formation. *J. Non-Cryst. Solids*, 147, 424–436, 1992.

24. Hench, L.L. and West, J.K., The sol-gel process. *Chem. Rev.*, 90, 1, 33–72, 1990.

25. Ismail, W.N.W., Sol–gel technology for innovative fabric finishing—a review. *J. Sol-Gel Sci. Technol.*, 78, 3, 698–707, 2016.

26. Alongi, J. and Malucelli, G., State of the art and perspectives on sol–gel derived hybrid architectures for flame retardancy of textiles. *J. Mater. Chem.*, 22, 41, 21805–21809, 2012.

27. Alongi, J., Ciobanu, M., Malucelli, G., Novel flame retardant finishing systems for cotton fabrics based on phosphorus-containing compounds and silica derived from sol–gel processes. *Carbohydr. Polym.*, 85, 3, 599–608, 2011.

28. Castellano, A., Colleoni, C., Iacono, G., Mezzi, A., Plutino, M.R., Malucelli, G., Rosace, G., Synthesis and characterization of a phosphorous/nitrogen based sol-gel coating as a novel halogen-and formaldehyde-free flame retardant finishing for cotton fabric. *Polym. Degrad. Stab.*, 162, 148–159, 2019.

29. Decher, G., *Polyelectrolyte multilayers, an overview, Multilayer thin films*, pp. 1–46, Wiley-VCH Verlag, Weinheim, Germany, 2003.

30. Van der Gucht, J., Spruijt, E., Lemmers, M., Stuart, M.A.C., Polyelectrolyte complexes: Bulk phases and colloidal systems. *J. Colloid Interface Sci.*, 361, 2, 407–422, 2011.

31. Negrell-Guirao, C., Carosio, F., Boutevin, B., Cottet, H., Loubat, C., Phosphonated oligoallylamine: Synthesis, characterization in water, and development of layer by layer assembly. *J. Polym. Sci. Part B: Polym. Phys.*, 51, 16, 1244–1251, 2013.

32. Ghanadpour, M., Carosio, F., Wågberg, L., Ultrastrong and flame-resistant freestanding films from nanocelluloses, self-assembled using a layer-by-layer approach. *Appl. Mater. Today*, 9, 229–239, 2017.

33. Carosio, F., Ghanadpour, M., Alongi, J., Wågberg, L., Layer-by-layer-assembled chitosan/phosphorylated cellulose nanofibrils as a bio-based and flame protecting nano-exoskeleton on PU foams. *Carbohydr. Polym.*, 202, 479–487, 2018.

34. Zhang, L., Li, Y., Sun, J., Shen, J., Mechanically stable antireflection and antifogging coatings fabricated by the layer-by-layer deposition process and postcalcination. *Langmuir*, 24, 19, 10851–10857, 2008.

35. Li, Y., Liu, F., Sun, J., A facile layer-by-layer deposition process for the fabrication of highly transparent superhydrophobic coatings. *Chem. Commun.*, 19, 2730–2732, 2009.

36. Li, S., Lin, X., Liu, Y., Li, R., Ren, X., Huang, T.-S., Phosphorus-nitrogen-silicon-based assembly multilayer coating for the preparation of flame retardant and antimicrobial cotton fabric. *Cellulose*, 26, 6, 4213–4223, 2019.

37. Chen, X., Fang, F., Zhang, X., Ding, X., Wang, Y., Chen, L., Tian, X., Flame-retardant, electrically conductive and antimicrobial multifunctional coating on cotton fabric via layer-by-layer assembly technique. *RSC Adv.*, 6, 33, 27669–27676, 2016.

38. Yang, M., Shi, J., Schlenoff, J.B., Control of dynamics in polyelectrolyte complexes by temperature and salt. *Macromolecules*, 52, 5, 1930–1941, 2019.

39. Yang, H., Yu, B., Song, P., Maluk, C., Wang, H., Surface-coating engineering for flame retardant flexible polyurethane foams: A critical review. *Compos. Part B: Eng.*, 176, 107185, 2019.

40. Qiu, X., Li, Z., Li, X., Zhang, Z., Flame retardant coatings prepared using layer by layer assembly: a review. *Chem. Eng. J.*, 334, 108–122, 2018.

41. Alongi, J. and Carosio, F., Flame retardancy of flexible polyurethane foams: traditional approaches versus layer-by-layer assemblies, in: *Novel fire retardant polymers and composite materials* Elsevier, Woodhead Publishing, Sawston, UK, pp. 171–200, 2017.

42. Lazar, S.T., Kolibaba, T.J., Grunlan, J.C., Flame-retardant surface treatments. *Nat. Rev. Mater.*, 5, 4, 259–275, 2020.

43. Alongi, J., Carosio, F., Malucelli, G., Current emerging techniques to impart flame retardancy to fabrics: an overview. *Polym. Degrad. Stab.*, 106, 138–149, 2014.

44. Alongi, J., Carosio, F., Kiekens, P., Recent advances in the design of water based-flame retardant coatings for polyester and polyester-cotton blends. *Polymers*, 8, 10, 357, 2016.

45. Liang, S., Neisius, N.M., Gaan, S., Recent developments in flame retardant polymeric coatings. *Prog. Org. Coat.*, 76, 11, 1642–1665, 2013.

46. Kundu, C.K., Li, Z., Song, L., Hu, Y., An overview of fire retardant treatments for synthetic textiles: special focus on anti-dripping, flame retardant and thermal properties. *Eur. Polym. J.*, 137, 109911, 15 August, 2020.

47. Alongi, J. and Malucelli, G., Cotton flame retardancy: state of the art and future perspectives. *RSC Adv.*, 5, 31, 24239–24263, 2015.

48. Zhao, H.B. and Wang, Y.Z., Design and synthesis of PET-based copolyesters with flame-retardant and antidripping performance. *Macromol. Rapid Commun.*, 38, 23, 1700451, 2017.

49. Alongi, J., Carosio, F., Malucelli, G., Influence of ammonium polyphosphate-/poly (acrylic acid)-based layer by layer architectures on the char formation in cotton, polyester and their blends. *Polym. Degrad. Stab.*, 97, 9, 1644–1653, 2012.

50. Carosio, F., Alongi, J., Malucelli, G., Flammability and combustion properties of ammonium polyphosphate-/poly (acrylic acid)-based layer by layer

architectures deposited on cotton, polyester and their blends. *Polym. Degrad. Stab.*, 98, 9, 1626–1637, 2013.

51. Byrne, C., Technical textiles market–an overview, in: *Handbook of technical textiles*, pp. 1–23, 2000.

52. Horrocks, A.R., Developments in flame retardants for heat and fire resistant textiles—the role of char formation and intumescence. *Polym. Degrad. Stab.*, 54, 2-3, 143–154, 1996.

53. Davies, P.J., Horrocks, A.R., Alderson, A., The sensitisation of thermal decomposition of ammonium polyphosphate by selected metal ions and their potential for improved cotton fabric flame retardancy. *Polym. Degrad. Stab.*, 88, 1, 114–122, 2005.

54. Li, Y.-C., Schulz, J., Mannen, S., Delhom, C., Condon, B., Chang, S., Zammarano, M., Grunlan, J.C., Flame retardant behavior of polyelectrolyte–clay thin film assemblies on cotton fabric. *ACS Nano*, 4, 6, 3325–3337, 2010.

55. Laufer, G., Carosio, F., Martinez, R., Camino, G., Grunlan, J.C., Growth and fire resistance of colloidal silica-polyelectrolyte thin film assemblies. *J. Colloid Interface Sci.*, 356, 1, 69–77, 2011.

56. Li, Y.C., Mannen, S., Morgan, A.B., Chang, S., Yang, Y.H., Condon, B., Grunlan, J.C., Intumescent all-polymer multilayer nanocoating capable of extinguishing flame on fabric. *Adv. Mater.*, 23, 34, 3926–3931, 2011.

57. Köklükaya, O., Karlsson, R. M. P., Carosio, F., & Wågberg, L. The use of model cellulose gel beads to clarify flame-retardant characteristics of layer-by-layer nanocoatings. Carbohydrate Polymers, 255, 117468, 2021.

58. Zhang, T., Yan, H., Wang, L., Fang, Z., Controlled formation of self-extinguishing intumescent coating on ramie fabric via layer-by-layer assembly. *Ind. Eng. Chem. Res.*, 52, 18, 6138–6146, 2013.

59. Zhao, L., Yan, H., Fang, Z., Wang, J., Wang, H., On the flameproof treatment of ramie fabrics using a spray-assisted layer-by-layer technique. *Polym. Degrad. Stab.*, 121, 11–17, 2015.

60. Carosio, F., Fontaine, G., Alongi, J., Bourbigot, S., Starch-based layer by layer assembly: efficient and sustainable approach to cotton fire protection. *ACS Appl. Mater. Interfaces*, 7, 22, 12158–12167, 2015.

61. Alongi, J., Cuttica, F., Carosio, F., Bourbigot, S., How much the fabric grammage may affect cotton combustion? *Cellulose*, 22, 5, 3477–3489, 2015.

62. Carosio, F. and Alongi, J., Few durable layers suppress cotton combustion due to the joint combination of layer by layer assembly and UV-curing. *RSC Adv.*, 5, 87, 71482–71490, 2015.

63. Haile, M., Fincher, C., Fomete, S., Grunlan, J.C., Water-soluble polyelectrolyte complexes that extinguish fire on cotton fabric when deposited as pH-cured nanocoating. *Polym. Degrad. Stab.*, 114, 60–64, 2015.

64. Zhao, B., Kolibaba, T.J., Lazar, S., Grunlan, J.C., Facile two-step phosphazine-based network coating for flame retardant cotton. *Cellulose*, 64, 27, 4123–4132, 2020.

65. Malucelli, G., Bosco, F., Alongi, J., Carosio, F., Di Blasio, A., Mollea, C., Cuttica, F., Casale, A., Biomacromolecules as novel green flame retardant systems for textiles: an overview. *RSC Adv.*, 4, 86, 46024–46039, 2014.

66. Alongi, J., Carletto, R.A., Di Blasio, A., Carosio, F., Bosco, F., Malucelli, G., DNA: a novel, green, natural flame retardant and suppressant for cotton. *J. Mater. Chem. A*, 1, 15, 4779–4785, 2013.

67. Alongi, J., Carletto, R.A., Di Blasio, A., Cuttica, F., Carosio, F., Bosco, F., Malucelli, G., Intrinsic intumescent-like flame retardant properties of DNA-treated cotton fabrics. *Carbohydr. Polym.*, 96, 1, 296–304, 2013.

68. Alongi, J., Carletto, R.A., Bosco, F., Carosio, F., Di Blasio, A., Cuttica, F., Antonucci, V., Giordano, M., Malucelli, G., Caseins and hydrophobins as novel green flame retardants for cotton fabrics. *Polym. Degrad. Stab.*, 99, 111–117, 2014.

69. Carosio, F., Di Blasio, A., Alongi, J., Malucelli, G., Green DNA-based flame retardant coatings assembled through layer by layer. *Polymer*, 54, 19, 5148–5153, 2013.

70. Zhang, W., Wang, M., Guan, J.-P., Tang, R.-C., Qiao, Y.-F., Casein phospho-peptide-metal salts combination: A novel route for imparting the durable flame retardancy to silk. *J. Taiwan Institute Chem. Engineers*, 101, 1–7, 2019.

71. Lv, Z., Hu, Y.-T., Guan, J.-P., Tang, R.-C., Chen, G.-Q., Preparation of a flame retardant, antibacterial, and colored silk fabric with chitosan and vitamin B2 sodium phosphate by electrostatic layer by layer assembly. *Mater. Lett.*, 241, 136–139, 2019.

72. Laufer, G., Kirkland, C., Morgan, A.B., Grunlan, J.C., Intumescent multilayer nanocoating, made with renewable polyelectrolytes, for flame-retardant cotton. *Biomacromolecules*, 13, 9, 2843–2848, 2012.

73. Nam, S., Condon, B.D., Xia, Z., Nagarajan, R., Hinchliffe, D.J., Madison, C.A., Intumescent flame-retardant cotton produced by tannic acid and sodium hydroxide. *J. Anal. Appl. Pyrolysis*, 126, 239–246, 2017.

74. Yang, T.-T., Guan, J.-P., Tang, R.-C., Chen, G., Condensed tannin from Dioscorea cirrhosa tuber as an eco-friendly and durable flame retardant for silk textile. *Ind. Crops Prod.*, 115, 16–25, 2018.

75. Zilke, O., Plohl, D., Opwis, K., Mayer-Gall, T., Gutmann, J.S., A Flame-Retardant Phytic-Acid-Based LbL-Coating for Cotton Using Polyvinylamine. *Polymers*, 12, 5, 1202, 2020.

76. Cheng, X.-W., Liang, C.-X., Guan, J.-P., Yang, X.-H., Tang, R.-C., Flame retardant and hydrophobic properties of novel sol-gel derived phytic acid/silica hybrid organic-inorganic coatings for silk fabric. *Appl. Surf. Sci.*, 427, 69–80, 2018.

77. Cheng, X.-W., Guan, J.-P., Chen, G., Yang, X.-H., Tang, R.-C., Adsorption and flame retardant properties of bio-based phytic acid on wool fabric. *Polymers*, 8, 4, 122, 2016.

78. Cheng, X.-W., Guan, J.-P., Yang, X.-H., Tang, R.-C., Improvement of flame retardancy of silk fabric by bio-based phytic acid, nano-TiO2, and polycarboxylic acid. *Prog. Org. Coat.*, 112, 18–26, 2017.

79. Cheng, X.-W., Guan, J.-P., Yang, X.-H., Tang, R.-C., Yao, F., A bio-resourced phytic acid/chitosan polyelectrolyte complex for the flame retardant treatment of wool fabric. *J. Clean. Prod.*, 223, 342–349, 2019.

80. Ferruti, P., Poly (amidoamine) s: past, present, and perspectives. *J. Polym. Sci. Part A: Polym. Chem.*, 51, 11, 2319–2353, 2013.

81. Manfredi, A., Carosio, F., Ferruti, P., Ranucci, E., Alongi, J., Linear polyamidoamines as novel biocompatible phosphorus-free surface-confined intumescent flame retardants for cotton fabrics. *Polym. Degrad. Stab.*, 151, 52–64, 2018.

82. Manfredi, A., Carosio, F., Ferruti, P., Alongi, J., Ranucci, E., Disulfide-containing polyamidoamines with remarkable flame retardant activity for cotton fabrics. *Polym. Degrad. Stab.*, 156, 1–13, 2018.

83. Beduini, A., Carosio, F., Ferruti, P., Ranucci, E., Alongi, J., Sulfur-based copolymeric polyamidoamines as efficient flame-retardants for cotton. *Polymers*, 11, 11, 1904, 2019.

84. Alongi, J., Ferruti, P., Manfredi, A., Carosio, F., Feng, Z., Hakkarainen, M., Ranucci, E., Superior flame retardancy of cotton by synergetic effect of cellulose-derived nano-graphene oxide carbon dots and disulphide-containing polyamidoamines. *Polym. Degrad. Stab.*, 169, 108993, 2019.

85. Yan, H., Zhao, L., Fang, Z., Wang, H., Construction of multilayer coatings for flame retardancy of ramie fabric using layer-by-layer assembly. *J. Appl. Polym. Sci.*, 134, 48, 45556, 2017.

86. Zhang, Y., Tian, W., Liu, L., Cheng, W., Wang, W., Liew, K.M., Wang, B., Hu, Y., Eco-friendly flame retardant and electromagnetic interference shielding cotton fabrics with multi-layered coatings. *Chem. Eng. J.*, 372, 1077–1090, 2019.

87. Zhang, X., Shen, Q., Zhang, X., Pan, H., Lu, Y., Graphene oxide-filled multilayer coating to improve flame-retardant and smoke suppression properties of flexible polyurethane foam. *J. Mater. Sci.*, 51, 23, 10361–10374, 2016.

88. Ji, Y., Chen, G., Xing, T., Rational design and preparation of flame retardant silk fabrics coated with reduced graphene oxide. *Appl. Surf. Sci.*, 474, 203–210, 2019.

89. An, W., Ma, J., Xu, Q., Fan, Q., Flame retardant, antistatic cotton fabrics crafted by layer-by-layer assembly. *Cellulose*, 27, 14, 8457–8469, 2020.

90. Alongi, J. and Carosio, F., All-inorganic intumescent nanocoating containing montmorillonite nanoplatelets in ammonium polyphosphate matrix capable of preventing cotton ignition. *Polymers*, 8, 12, 430, 2016.

91. Panda, S.R. and De, S., Preparation, characterization and antifouling properties of polyacrylonitrile/polyurethane blend membranes for water purification. *RSC Adv.*, 5, 30, 23599–23612, 2015.

92. Leroux, F., Perwuelz, A., Campagne, C., Behary, N., Atmospheric air-plasma treatments of polyester textile structures. *J. Adhes. Sci. Technol.*, 20, 9, 939–957, 2006.

93. Shahidi, S., Wiener, J., Ghoranneviss, M., Surface modification methods for improving the dyeability of textile fabrics, in: *Eco-friendly textile dyeing and finishing*, vol. 10, p. 53911, 2013.

94. Carosio, F., Di Pierro, A., Alongi, J., Fina, A., Saracco, G., Controlling the melt dripping of polyester fabrics by tuning the ionic strength of polyhedral oligomeric silsesquioxane and sodium montmorillonite coatings assembled through layer by layer. *J. Colloid Interface Sci.*, 510, 142–151, 2018.

95. Carosio, F., Laufer, G., Alongi, J., Camino, G., Grunlan, J.C., Layer-by-layer assembly of silica-based flame retardant thin film on PET fabric. *Polym. Degrad. Stab.*, 96, 5, 745–750, 2011.

96. Carosio, F., Di Blasio, A., Cuttica, F., Alongi, J., Frache, A., Malucelli, G., Flame retardancy of polyester fabrics treated by spray-assisted layer-by-layer silica architectures. *Ind. Eng. Chem. Res.*, 52, 28, 9544–9550, 2013.

97. Carosio, F., Alongi, J., Frache, A., Influence of surface activation by plasma and nanoparticle adsorption on the morphology, thermal stability and combustion behavior of PET fabrics. *Eur. Polym. J.*, 47, 5, 893–902, 2011.

98. Liu, J. and Xiao, C., Fire-retardant multilayer assembled on polyester fabric from water-soluble chitosan, sodium alginate and divalent metal ion. *Int. J. Biol. Macromol.*, 119, 1083–1089, 2018.

99. Pan, Y., Liu, L., Song, L., Hu, Y., Wang, W., Zhao, H., Durable flame retardant treatment of polyethylene terephthalate (PET) fabric with cross-linked layer-by-layer assembled coating. *Polym. Degrad. Stab.*, 165, 145–152, 2019.

100. Li, L., Chen, G., Liu, W., Li, J., Zhang, S., The anti-dripping intumescent flame retardant finishing for nylon-6, 6 fabric. *Polym. Degrad. Stab.*, 94, 6, 996–1000, 2009.

101. Fang, Y., Liu, X., Tao, X., Intumescent flame retardant and anti-dripping of PET fabrics through layer-by-layer assembly of chitosan and ammonium polyphosphate. *Prog. Org. Coat.*, 134, 162–168, 2019.

102. Jordanov, I., Magovac, E., Fahami, A., Lazar, S., Kolibaba, T., Smith, R.J., Bischof, S., Grunlan, J.C., Flame retardant polyester fabric from nitrogen-rich low molecular weight additives within intumescent nanocoating. *Polym. Degrad. Stab.*, 170, 108998, 2019.

103. Kundu, C.K., Wang, W., Zhou, S., Wang, X., Sheng, H., Pan, Y., Song, L., Hu, Y., A green approach to constructing multilayered nanocoating for flame retardant treatment of polyamide 66 fabric from chitosan and sodium alginate. *Carbohydr. Polym.*, 166, 131–138, 2017.

104. Kundu, C.K., Wang, X., Song, L., Hu, Y., Borate cross-linked layer-by-layer assembly of green polyelectrolytes on polyamide 66 fabrics for flame-retardant treatment. *Prog. Org. Coat.*, 121, 173–181, 2018.

105. Kundu, C.K., Li, Z., Li, X., Zhang, Z., Hu, Y., Graphene oxide functionalized biomolecules for improved flame retardancy of Polyamide 66 fabrics with intact physical properties. *Int. J. Biol. Macromol.*, 156, 362–371, 2020.

106. Tao, Y., Liu, C., Li, P., Wang, B., Xu, Y.-J., Jiang, Z.-M., Liu, Y., Zhu, P., A flame-retardant PET fabric coating: Flammability, anti-dripping properties, and flame-retardant mechanism. *Prog. Org. Coat.*, 150, 105971, 2021.

107. Apaydin, K., Laachachi, A., Ball, V., Jimenez, M., Bourbigot, S., Toniazzo, V., Ruch, D., Intumescent coating of (polyallylamine-polyphosphates) deposited on polyamide fabrics via layer-by-layer technique. *Polym. Degrad. Stab.*, 106, 158–164, 2014.

108. Apaydin, K., Laachachi, A., Ball, V., Jimenez, M., Bourbigot, S., Ruch, D., Layer-by-layer deposition of a TiO2-filled intumescent coating and its effect on the flame retardancy of polyamide and polyester fabrics. *Colloids Surf. A: Physicochem. Eng. Asp.*, 469, 1–10, 2015.

109. Carosio, F. and Alongi, J., Influence of layer by layer coatings containing octapropylammonium polyhedral oligomeric silsesquioxane and ammonium polyphosphate on the thermal stability and flammability of acrylic fabrics. *J. Anal. Appl. Pyrolysis*, 119, 114–123, 2016.

110. Li, G., Wang, L., Ni, H., Pittman, C.U., Polyhedral oligomeric silsesquioxane (POSS) polymers and copolymers: a review. *J. Inorg. Organomet. Polym.*, 11, 3, 123–154, 2001.

111. Zhang, W., Camino, G., Yang, R., Polymer/polyhedral oligomeric silsesquioxane (POSS) nanocomposites: An overview of fire retardance. *Prog. Polym. Sci.*, 67, 77–125, 2017.

112. Carosio, F. and Alongi, J., Flame retardant multilayered coatings on acrylic fabrics prepared by one-step deposition of chitosan/montmorillonite complexes. *Fibers*, 6, 2, 36, 2018.

113. Lewin, M. and Sello, S.B., *Chemical Processing of Fibers and Fabrics: Functional Finishes, Part B*, Marcel Dekker, New York, NY, 1984.

114. Carosio, F., Alongi, J., Malucelli, G., Layer by layer ammonium polyphosphate-based coatings for flame retardancy of polyester–cotton blends. *Carbohydr. Polym.*, 88, 4, 1460–1469, 2012.

115. Alongi, J., Carosio, F., Malucelli, G., Layer by layer complex architectures based on ammonium polyphosphate, chitosan and silica on polyester-cotton blends: flammability and combustion behaviour. *Cellulose*, 19, 3, 1041–1050, 2012.

116. Fang, Y., Sun, W., Li, J., Liu, H., Liu, X., Eco-friendly flame retardant and dripping-resistant of polyester/cotton blend fabrics through layer-by-layer assembly fully bio-based chitosan/phytic acid coating. *Int. J. Biol. Macromol.*, 175, 140–146, 2021.

117. Liu, L., Pan, Y., Wang, Z., Hou, Y., Gui, Z., Hu, Y., Layer-by-layer assembly of hypophosphorous acid-modified chitosan based coating for flame-retardant polyester–cotton blends. *Ind. Eng. Chem. Res.*, 56, 34, 9429–9436, 2017.

118. Pan, Y., Liu, L., Wang, X., Song, L., Hu, Y., Hypophosphorous acid cross-linked layer-by-layer assembly of green polyelectrolytes on polyester-cotton

blend fabrics for durable flame-retardant treatment. *Carbohydr. Polym.*, 201, 1–8, 2018.

119. Leistner, M., Haile, M., Rohmer, S., Abu-Odeh, A., Grunlan, J.C., Water-soluble polyelectrolyte complex nanocoating for flame retardant nylon-cotton fabric. *Polym. Degrad. Stab.*, 122, 1–7, 2015.

120. Narkhede, M., Thota, S., Mosurkal, R., Muller, W.S., Kumar, J., Layer-by-layer assembly of halogen-free polymeric materials on nylon/cotton blend for flame retardant applications. *Fire Mater.*, 40, 2, 206–218, 2016.

121. Wang, B., Xu, Y.-J., Li, P., Zhang, F.-Q., Liu, Y., Zhu, P., Flame-retardant polyester/cotton blend with phosphorus/nitrogen/silicon-containing nano-coating by layer-by-layer assembly. *Appl. Surf. Sci.*, 509, 145323, 2020.

122. Grunlan, J. C., Leistner, M., & Haile, M. M. Coating method for forming flame retardant substrate. *U.S. Patent No. 10,150,142.* Washington, DC: U.S. Patent and Trademark Office, 2018.

123. Grunlan, J. C., & Guin, T. C. Flame retardant nanocoated substrate. *U.S. Patent No. 10,343,185.* Washington, DC: U.S. Patent and Trademark Office, 2019.

124. Grunlan, J. C. Multilayer coating for flame retardant substrates. *U.S. Patent No. 9,540,764.* Washington, DC: U.S. Patent and Trademark Office, 2017.

125. Haji, A., Malek, R., Mazaheri, F., Comparative study of exhaustion and pad-steam methods for improvement of handle, dye uptake and water absorption of polyester/cotton fabric. *Chem. Ind. Chem. Eng. Q.*, 17, 3, 359–365, 2011.

126. Chang, S., Slopek, R.P., Condon, B., Grunlan, J.C., Surface coating for flame-retardant behavior of cotton fabric using a continuous layer-by-layer process. *Ind. Eng. Chem. Res.*, 53, 10, 3805–3812, 2014.

127. Mehrabi, A., Akhave, J., Licon, M., & Koch, C. Continuous process for manufacturing electrostatically self-assembled coatings. *U.S. Patent Application No. 10/439,657,* 2004.

128. Mateos, A.J., Cain, A.A., Grunlan, J.C., Large-scale continuous immersion system for layer-by-layer deposition of flame retardant and conductive nano-coatings on fabric. *Ind. Eng. Chem. Res.*, 53, 15, 6409–6416, 2014.

129. Krogman, K. C., Hammond, P. T., & Zacharia, N. S. Automated layer by layer spray technology. *U.S. Patent No. 8,234,998.* Washington, DC: U.S. Patent and Trademark Office, 2012.

130. Król, P. and Król, B., Surface free energy of polyurethane coatings with improved hydrophobicity. *Colloid Polym. Sci.*, 290, 10, 879–893, 2012.

131. Kraemer, R.H., Zammarano, M., Linteris, G.T., Gedde, U.W., Gilman, J.W., Heat release and structural collapse of flexible polyurethane foam. *Polym. Degrad. Stab.*, 95, 6, 1115–1122, 2010.

132. Kim, Y.S., Davis, R., Cain, A.A., Grunlan, J.C., Development of layer-by-layer assembled carbon nanofiber-filled coatings to reduce polyurethane foam flammability. *Polymer*, 52, 13, 2847–2855, 2011.

133. Haile, M., Fomete, S., Lopez, I.D., Grunlan, J.C., Aluminum hydroxide multilayer assembly capable of extinguishing flame on polyurethane foam. *J. Mater. Sci.*, 51, 1, 375–381, 2016.

134. Pan, Y., Zhan, J., Pan, H., Wang, W., Ge, H., Song, L., Hu, Y., A novel and effective method to fabricate flame retardant and smoke suppressed flexible polyurethane foam. *RSC Adv.*, 5, 83, 67878–67885, 2015.

135. Holder, K.M., Cain, A.A., Plummer, M.G., Stevens, B.E., Odenborg, P.K., Morgan, A.B., Grunlan, J.C., Carbon nanotube multilayer nanocoatings prevent flame spread on flexible polyurethane foam. *Macromol. Mater. Eng.*, 301, 6, 665–673, 2016.

136. Kim, Y.S. and Davis, R., Multi-walled carbon nanotube layer-by-layer coatings with a trilayer structure to reduce foam flammability. *Thin Solid Films*, 550, 184–189, 2014.

137. Pan, H., Wang, W., Pan, Y., Song, L., Hu, Y., Liew, K.M., Formation of layer-by-layer assembled titanate nanotubes filled coating on flexible polyurethane foam with improved flame retardant and smoke suppression properties. *ACS Appl. Mater. Interfaces*, 7, 1, 101–111, 2015.

138. Smith, R.J., Holder, K.M., Ruiz, S., Hahn, W., Song, Y., Lvov, Y.M., Grunlan, J.C., Environmentally benign halloysite nanotube multilayer assembly significantly reduces polyurethane flammability. *Adv. Funct. Mater.*, 28, 27, 1703289, 2018.

139. Pan, Y., Liu, L., Cai, W., Hu, Y., Jiang, S., Zhao, H., Effect of layer-by-layer self-assembled sepiolite-based nanocoating on flame retardant and smoke suppressant properties of flexible polyurethane foam. *Appl. Clay Sci.*, 168, 230–236, 2019.

140. Laufer, G., Kirkland, C., Cain, A.A., Grunlan, J.C., Clay–chitosan nanobrick walls: completely renewable gas barrier and flame-retardant nanocoatings. *ACS Appl. Mater. Interfaces*, 4, 3, 1643–1649, 2012.

141. Kim, Y.S., Harris, R., Davis, R., Innovative approach to rapid growth of highly clay-filled coatings on porous polyurethane foam. *ACS Macro Lett.*, 1, 7, 820–824, 2012.

142. Li, Y.-C., Kim, Y.S., Shields, J., Davis, R., Controlling polyurethane foam flammability and mechanical behaviour by tailoring the composition of clay-based multilayer nanocoatings. *J. Mater. Chem. A*, 1, 41, 12987–12997, 2013.

143. Kim, Y.S., Li, Y.-C., Pitts, W.M., Werrel, M., Davis, R.D., Rapid growing clay coatings to reduce the fire threat of furniture. *ACS Appl. Mater. Interfaces*, 6, 3, 2146–2152, 2014.

144. Cain, A., Plummer, M., Murray, S., Bolling, L., Regev, O., Grunlan, J.C., Iron-containing, high aspect ratio clay as nanoarmor that imparts substantial thermal/flame protection to polyurethane with a single electrostatically-deposited bilayer. *J. Mater. Chem. A*, 2, 41, 17609–17617, 2014.

145. Lazar, S., Carosio, F., Davesne, A.-L., Jimenez, M., Bourbigot, S., Grunlan, J., Extreme heat shielding of clay/chitosan nanobrick wall on flexible foam. *ACS Appl. Mater. Interfaces*, 10, 37, 31686–31696, 2018.

146. Pan, H., Yu, B., Wang, W., Pan, Y., Song, L., Hu, Y., Comparative study of layer by layer assembled multilayer films based on graphene oxide and reduced

graphene oxide on flexible polyurethane foam: flame retardant and smoke suppression properties. *RSC Adv.*, 6, 115, 114304–114312, 2016.

147. Maddalena, L., Carosio, F., Gomez, J., Saracco, G., Fina, A., Layer-by-layer assembly of efficient flame retardant coatings based on high aspect ratio graphene oxide and chitosan capable of preventing ignition of PU foam. *Polym. Degrad. Stab.*, 152, 1–9, 2018.

148. Maddalena, L., Gomez, J., Fina, A., Carosio, F., Effects of Graphite Oxide Nanoparticle Size on the Functional Properties of Layer-by-Layer Coated Flexible Foams. *Nanomaterials*, 11, 2, 266, 2021.

149. Carosio, F., Maddalena, L., Gomez, J., Saracco, G., Fina, A., Graphene oxide exoskeleton to produce self-extinguishing, nonignitable, and flame resistant flexible foams: a mechanically tough alternative to inorganic aerogels. *Adv. Mater. Interfaces*, 5, 23, 1801288, 2018.

150. Mu, X., Yuan, B., Pan, Y., Feng, X., Duan, L., Zong, R., Hu, Y., A single α-cobalt hydroxide/sodium alginate bilayer layer-by-layer assembly for conferring flame retardancy to flexible polyurethane foams. *Mater. Chem. Phys.*, 191, 52–61, 2017.

151. Pan, H., Shen, Q., Zhang, Z., Yu, B., Lu, Y., MoS 2-filled coating on flexible polyurethane foam via layer-by-layer assembly technique: flame-retardant and smoke suppression properties. *J. Mater. Sci.*, 53, 12, 9340–9349, 2018.

152. Lin, B., Yuen, A.C.Y., Li, A., Zhang, Y., Chen, T.B.Y., Yu, B., Lee, E.W.M., Peng, S., Yang, W., Lu, H.-D., MXene/chitosan nanocoating for flexible polyurethane foam towards remarkable fire hazards reductions. *J. Hazard. Mater.*, 381, 120952, 2020.

153. Wang, X., Pan, Y.-T., Wan, J.-T., Wang, D.-Y., An eco-friendly way to fire retardant flexible polyurethane foam: layer-by-layer assembly of fully bio-based substances. *RSC Adv.*, 4, 86, 46164–46169, 2014.

154. Pan, Y., Zhan, J., Pan, H., Wang, W., Tang, G., Song, L., Hu, Y., Effect of fully biobased coatings constructed via layer-by-layer assembly of chitosan and lignosulfonate on the thermal, flame retardant, and mechanical properties of flexible polyurethane foam. *ACS Sustain. Chem. Eng.*, 4, 3, 1431–1438, 2016.

155. Laufer, G., Kirkland, C., Morgan, A.B., Grunlan, J.C., Exceptionally flame retardant sulfur-based multilayer nanocoating for polyurethane prepared from aqueous polyelectrolyte solutions. *ACS Macro Lett.*, 2, 5, 361–365, 2013.

156. Carosio, F., Di Blasio, A., Cuttica, F., Alongi, J., Malucelli, G., Self-assembled hybrid nanoarchitectures deposited on poly (urethane) foams capable of chemically adapting to extreme heat. *RSC Adv.*, 4, 32, 16674–16680, 2014.

157. Lin, B., Yuen, A.C.Y., Chen, T.B.Y., Yu, B., Yang, W., Zhang, J., Yao, Y., Wu, S., Wang, C.H., Yeoh, G.H., Experimental and numerical perspective on the fire performance of MXene/Chitosan/Phytic acid coated flexible polyurethane foam. *Sci. Rep.*, 11, 1, 1–13, 2021.

158. Cain, A.A., Nolen, C.R., Li, Y.-C., Davis, R., Grunlan, J.C., Phosphorous-filled nanobrick wall multilayer thin film eliminates polyurethane melt

dripping and reduces heat release associated with fire. *Polym. Degrad. Stab.*, 98, 12, 2645–2652, 2013.

159. Holder, K., Huff, M., Cosio, M., Grunlan, J., Intumescing multilayer thin film deposited on clay-based nanobrick wall to produce self-extinguishing flame retardant polyurethane. *J. Mater. Sci.*, 50, 6, 2451–2458, 2015.

160. Carosio, F. and Fina, A., Three organic/inorganic nanolayers on flexible foam allow retaining superior flame retardancy performance upon mechanical compression cycles. *Front. Mater.*, 6, 20, 2019.

161. Singh, H. and Jain, A., Ignition, combustion, toxicity, and fire retardancy of polyurethane foams: a comprehensive review. *J. Appl. Polym. Sci.*, 111, 2, 1115–1143, 2009.

162. Morgan, A.B., Revisiting flexible polyurethane foam flammability in furniture and bedding in the United States. *Fire Mater.*, 45, 1, 68–80, 2021.

163. Yang, J.-C., Cao, Z.-J., Wang, Y.-Z., Schiraldi, D.A., Ammonium polyphosphate-based nanocoating for melamine foam towards high flame retardancy and anti-shrinkage in fire. *Polymer*, 66, 86–93, 2015.

164. Deng, S.-B., Liao, W., Yang, J.-C., Cao, Z.-J., Wang, Y.-Z., Flame-retardant and smoke-suppressed silicone foams with chitosan-based nanocoatings. *Ind. Eng. Chem. Res.*, 55, 27, 7239–7248, 2016.

165. Li, Y., Cao, C.-F., Li, S.-N., Huang, N.-J., Mao, M., Zhang, J.-W., Wang, P.-H., Guo, K.-Y., Gong, L.-X., Zhang, G.-D., *In situ* reactive self-assembly of a graphene oxide nano-coating in polymer foam materials with synergistic fire shielding properties. *J. Mater. Chem. A*, 7, 47, 27032–27040, 2019.

166. Wang, S., Wang, X., Wang, X., Li, H., Sun, J., Sun, W., Yao, Y., Gu, X., Zhang, S., Surface coated rigid polyurethane foam with durable flame retardancy and improved mechanical property. *Chem. Eng. J.*, 385, 123755, 2020.

167. Carosio, F., Cuttica, F., Di Blasio, A., Alongi, J., Malucelli, G., , Layer by layer assembly of flame retardant thin films on closed cell PET foams: Efficiency of ammonium polyphosphate. *Polym. Degrad. Stab.*, 113, 189–196, 2015.

168. Koklukaya, O., Carosio, F., Grunlan, J.C., Wagberg, L., Flame-retardant paper from wood fibers functionalized via layer-by-layer assembly. *ACS Appl. Mater. Interfaces*, 7, 42, 23750–23759, 2015.

169. Köklükaya, O., Carosio, F., Wågberg, L., Tailoring flame-retardancy and strength of papers via layer-by-layer treatment of cellulose fibers. *Cellulose*, 25, 4, 2691–2709, 2018.

170. Köklükaya, O., Carosio, F., Durán, V.L., Wågberg, L., Layer-by-layer modified low density cellulose fiber networks: A sustainable and fireproof alternative to petroleum based foams. *Carbohydr. Polym.*, 230, 115616, 2020.

171. Chen, H.-B. and Schiraldi, D.A., Flammability of polymer/clay aerogel composites: an overview. *Polym. Rev.*, 59, 1, 1–24, 2019.

172. Medina, L., Carosio, F., Berglund, L.A., Recyclable nanocomposite foams of Poly (vinyl alcohol), clay and cellulose nanofibrils–Mechanical properties and flame retardancy. *Compos. Sci. Technol.*, 182, 107762, 2019.

173. Ghanadpour, M., Wicklein, B., Carosio, F., Wågberg, L., All-natural and highly flame-resistant freeze-cast foams based on phosphorylated cellulose nanofibrils. *Nanoscale*, 10, 8, 4085–4095, 2018.

174. Koklukaya, O., Carosio, F., Wågberg, L., Superior flame-resistant cellulose nanofibril aerogels modified with hybrid layer-by-layer coatings. *ACS Appl. Mater. Interfaces*, 9, 34, 29082–29092, 2017.

175. Lowden, L.A. and Hull, T.R., Flammability behaviour of wood and a review of the methods for its reduction. *Fire Sci. Rev.*, 2, 1, 1–19, 2013.

176. Zhou, L. and Fu, Y., Flame-retardant wood composites based on immobilizing with chitosan/sodium phytate/nano-TiO2-ZnO coatings via layer-by-layer self-assembly. *Coatings*, 10, 3, 296, 2020.

177. Rehman, Z.U., Niaz, A.K., Song, J.-I., Koo, B.H., Excellent Fire Retardant Properties of CNF/VMT Based LBL Coatings Deposited on Polypropylene and Wood-Ply. *Polymers*, 13, 2, 303, 2021.

178. Kolibaba, T.J. and Grunlan, J.C., Environmentally benign polyelectrolyte complex that renders wood flame retardant and mechanically strengthened. *Macromol. Mater. Eng.*, 304, 8, 1900179, 2019.

179. Grunlan, J. C. Multilayer coating for flame retardant foam or fabric. *U.S. Patent No. 9,540,763*. Washington, DC: U.S. Patent and Trademark Office, 2017.

180. Carosio, F. and Alongi, J., Ultra-fast layer-by-layer approach for depositing flame retardant coatings on flexible PU foams within seconds. *ACS Appl. Mater. Interfaces*, 8, 10, 6315–6319, 2016.

181. Fina, A., Carosio, F., Saracco, G., A method for superficially coating polymeric foams in order to improve their flame reaction and the related superficially coated flame resistant polymeric foams, WO2019058186, PCT/IB2018/054571, 2017.

182. Pettersson, G., Höglund, H., Norgren, S., Sjöberg, J., Peng, F., Hallgren, H., Moberg, A., Ljungqvist, C.-H., Bergström, J., Solberg, D., Strong and bulky paperboard sheets from surface modified CTMP, manufactured at low energy. *Nord. Pulp Paper Res. J.*, 30, 2, 318–324, 2015.

183. Marais, A., Wågberg, L., Enarsson, L.-E., Pettersson, G., Lindström, T., Pilot-scale papermaking using Layer-by-Layer treated fibres; comparison between the effects of beating and of sequential addition of polymeric additives. *Nord. Pulp Paper Res. J.*, 31, 2, 308–314, 2016.

184. Carosio, F., Di Blasio, A., Alongi, J., Malucelli, G., Layer by layer nanoarchitectures for the surface protection of polycarbonate. *Eur. Polym. J.*, 49, 2, 397–404, 2013.

185. Alongi, J., Di Blasio, A., Carosio, F., Malucelli, G., UV-cured hybrid organic–inorganic layer by layer assemblies: effect on the flame retardancy of polycarbonate films. *Polym. Degrad. Stab.*, 107, 74–81, 2014.

186. Laachachi, A., Ball, V., Apaydin, K., Toniazzo, V., Ruch, D., Diffusion of polyphosphates into (poly (allylamine)-montmorillonite) multilayer films: flame retardant-intumescent films with improved oxygen barrier. *Langmuir*, 27, 22, 13879–13887, 2011.

187. Apaydin, K., Laachachi, A., Ball, V., Jimenez, M., Bourbigot, S., Toniazzo, V., Ruch, D., Polyallylamine–montmorillonite as super flame retardant coating assemblies by layer-by layer deposition on polyamide. *Polym. Degrad. Stab.*, 98, 2, 627–634, 2013.

188. Guin, T., Krecker, M., Milhorn, A., Hagen, D.A., Stevens, B., Grunlan, J.C., Exceptional flame resistance and gas barrier with thick multilayer nanobrick wall thin films. *Adv. Mater. Interfaces*, 2, 11, 1500214, 2015.

189. Zhao, M., Meng, L., Ma, L., Ma, L., Yang, X., Huang, Y., Ryu, J.E., Shankar, A., Li, T., Yan, C., Layer-by-layer grafting CNTs onto carbon fibers surface for enhancing the interfacial properties of epoxy resin composites. *Compos. Sci. Technol.*, 154, 28–36, 2018.

190. Battegazzore, D., Frache, A., Carosio, F., Sustainable and high performing biocomposites with chitosan/sepiolite layer-by-layer nanoengineered interphases. *ACS Sustain. Chem. Eng.*, 6, 8, 9601–9605, 2018.

191. Battegazzore, D., Frache, A., Carosio, F., Layer-by-Layer nanostructured interphase produces mechanically strong and flame retardant bio-composites. *Compos. Part B: Eng.*, 200, 108310, 2020.

192. Stieger, G., Scheringer, M., Ng, C.A., Hungerbühler, K., Assessing the persistence, bioaccumulation potential and toxicity of brominated flame retardants: Data availability and quality for 36 alternative brominated flame retardants. *Chemosphere*, 116, 118–123, 2014.

9

Multicomponent Flame Retardants

Bernhard Schartel

Bundesanstalt für Materialforschung und -prüfung (BAM), Unter den Eichen 87, Berlin, Germany

Abstract

The important take home message of this chapter: When multicomponent flame retardant systems are applied to polymeric materials, it becomes possible to address multiple fire properties, increase efficiency, and minimize flame retardant use to maximize polymer property balance. Flame retardants are combined or used together with adjuvants or synergists; fibers and fillers make a crucial contribution to their fire properties. Multicomponent systems are discussed in their capacity as an overall powerful strategy for achieving and optimizing non-halogenated flame-retardant polymeric materials.

Keywords: Flame retardants, synergy, adjuvants, fillers, fibres

9.1 The Need for Multicomponent Flame Retardants

Modern flame retardant polymeric materials are often based on halogen-free multicomponent flame retardants. Two flame retardants are combined, or flame retardants are used in combination with adjuvants or synergists to enhance efficiency, reduce the amount of flame retardant required, or just to reduce costs. Since adding flame retardants often tends to worsen the material properties of polymeric materials, such as mechanical strength, glass transition temperatures or processability, multicomponent systems are used as an approach to optimize property profiles. Fibers, fillers, and several other additives make a further crucial contribution to the fire properties of a polymeric material. This chapter will illuminate

Email: bernhard.schartel@bam.de

Alexander B. Morgan (ed.) Non-Halogenated Flame Retardant Handbook 2nd edition, (413–474)
© 2022 Scrivener Publishing LLC

the concept of multicomponent systems in flame retardancy by explaining the need for a multicomponent approach and the main phenomena it achieves. Multicomponent systems are discussed in their capacity as an overall powerful strategy for achieving and optimizing actual and future flame-retardant polymeric materials.

Flame retardant polymeric materials are used as technical or high-performance materials in the transportation sector, for instance for shipping, railway vehicles, and aviation, which demand flame retarded lightweight polymeric materials. They are widely used in electrical engineering and electronics for housings, cables, switches and circuit boards, and as often cost-efficient flame retardant building products. The market increasingly demands environmentally friendly and halogen-free solutions, and in the future will also require an increase in sustainability. What is more, the markets in question are highly competitive and innovative. There is constant pressure to optimize and improve the flame-retardant materials on offer, which should ideally be systems that simultaneously reduce flammability, smoke production, and the release of toxic fire effluent (particulates and gases), while also reducing costs, improving mechanical properties, prolonging durability (mainly stability against hydrolysis and photo-oxidation), adjusting color, and improving processability. On top of all of this, they will be in even greater demand if they are also compostable or recyclable. In the electrical engineering sector, for instance, product cycles may require an optimized generation once every 3 to 5 years to maintain market share. The importance of applying multicomponent systems and developing better multicomponent systems to achieve optimized property profile cannot be overemphasized when it comes to commercial flame-retardant polymeric materials.

Clearly, the possible combinations of established commercially available flame retardants, synergists, adjuvants, fillers - as well as their concentration, particle size distribution, and so forth - present seemingly infinite possibilities to proceed and advance in thousands of different directions. Yet frankly, it is generally not the possibilities that drive innovation in the application of multicomponent systems, but the needs of the market. Several of the most important flame retardants determine the need for multicomponent systems, as the efficiency of their flame retardant effects shows a nonlinear dependence on increasing concentrations. In other words, it is not always profitable or even possible to achieve the desired flame retardancy by merely increasing the amount of a given flame retardant. Figure 9.1 illustrates the various common tasks for phosphorus flame retardants functioning mainly in the gas phase, for intumescent systems and for metal hydrates. Phosphorus flame retardants usually show a flame

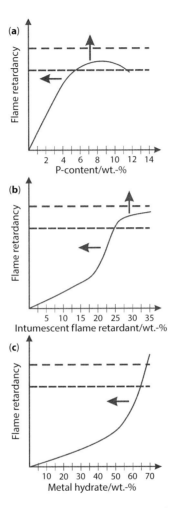

Figure 9.1 Nonlinear dependence of the flame-retardancy effect on the concentration (a) phosphorus flame retardant working in the gas phase, (b) intumescent flame retardant, and (c) metal hydrate. Dashed lines represent two different required flame retardancy levels, arrows the goals for applying multicomponent systems.

retardancy that levels off with increasing concentrations, particularly when flame inhibition is the main flame retardant mode of action. Flame retardancy may even pass through a maximum before fire accelerating effects set in when combustible phosphorus in the flame overcompensates for any flame retardancy. As red phosphorus is reported to show such LOI behavior in different polymers several times [1, 2], it makes little sense to add more than 7-10 wt.-% red phosphorus. Similar behavior was reported for aluminum diethylphosphinate in polyester, indicating that around 12-14 wt.-%, the improvement to LOI levels off, an effect that is also seen in epoxy resin and in polyolefins [2–4]. The gap to the flame retardancy required from flame-retardant polymeric materials is much larger for polyolefins than for technical or high-performance polymers (Figure 9.1a); phosphorus flame retardant must be combined with other flame retardants to reach the demanded flame retardancy effect [4, 5]. Charring, intumescent, or protective layer systems typically show a S-shape or a step function over concentration (Figure 9.1b). In low amounts the systems function as simple charring systems. Total fuel release and fuel release rate are reduced since carbon remains in the carbonaceous char. At a certain threshold concentration the protective layer function of the fire residue becomes effective enough to prevent the specimen from complete pyrolysis [2, 6, 7]. In some cases, thermal insulation by the fire residue becomes strong enough to prevent the pyrolysis front from passing through the entire specimen [8]. Depending on the flame retardancy demanded, it may be necessary to optimize the multicomponent intumescent system to reduce the concentration needed to reach the threshold or to increase the level of flame retardancy after the threshold (Figure 9.1b). Indeed, in flame retardant polyolefin materials such as polypropylene (PP)/charring agent/ammonium polyphosphate (APP) one of the most important research goals is to decrease the amount of flame retarded mixture needed by applying synergists. Thus, plenty of synergists are proposed for intumescent systems in polymeric materials, including several of the flame retardants discussed throughout this book, such as boron compounds, phosphorus compounds, silicone, nanoparticles, metal oxides, and fibers and fillers [9–19]. The third important, well-known task is determined by the nonlinear dependency of the flame retardancy achieved with metal hydrates [20, 21]. Quite a large amount of metal hydrate (see Chapter 3 of this book) in polyolefins often more than 60 wt.-% of aluminum trihydrate, must be applied to achieve UL 94 V-0 classifications (Figure 9.1c). The increase in flame retardancy until these very large amounts are added is moderate, so that the entire curve resembles an exponential behavior. Several dozen different approaches are proposed to shift the strong increase to lower concentrations [20–22], many of which have been commercialized. Metal hydrates, such as aluminum trihydrate, are inexpensive compared to other flame retardants so that their use in polyolefins are their main application; the combination with synergist in multicomponent systems is often crucial to reduce the filling degree of the polymeric material.

The second main motivation for multicomponent systems is to optimize the property profiles of the flame retarded polymeric material. Obviously, for flame retardant technical polymers or flame retardant high-performance polymeric materials a dozen properties must fulfill multiple demanding technical specifications at any given time. Approaches using only one flame retardant often are often detrimental to other key properties such as mechanical strength, glass transition temperature, color, processability, or durability. Tailored and synergistic multicomponent systems provide routes to increase efficiency and thus enable a reduction in the amount of a distinct flame retardant needed, or in the overall filler content. The reduction of the content contributes to a better property profile. Further multicomponent systems are used as alternatives to replacing certain flame retardants. Several halogen-free multicomponent systems were ultimately applied mainly to replace former halogenated solutions. For instance, the remarkable story of success of metal phosphinate salts originates from applying multicomponent flame retardant mixtures in halogen-free flame retardant polyamides and polyesters for electrical engineering [23, 34].

The third main force driving the development and application of multicomponent flame retardant polymers is cost reduction. Material costs are the most important factor in the area of commodities and technical polymers. As R. Lyon demonstrated well, when plotting fire hazard over the bulk price of polymers [35], an important component of the flame retardancy of polymers is how well they compete with more expensive polymers that have higher intrinsic flame retardancy. Flame retarded polypropylene (FR-PP) was developed to compete with polyamides (PAs), and flame retarded polycarbonate/poly(acrylonitrile-co-butadiene-co-styrene) blends (FR-PC/ABS) to compete with high-performance polymers. Consequently, flame retardants must be very economical for polyolefins and low-priced for technical polymers; only for high-performance polymers does the role of the bulk price of the flame retardant lose some of its dominance. Multicomponent systems exploiting the combination with less expensive flame retardants or inexpensive fillers are a successful way to reduce costs.

Clearly, several of these reasons overlap and coincide with each other, such that cost is usually the most important property within any given property profile.

When it comes to developing or optimizing a flame-retardant polymer material, as mentioned previously in this chapter, the huge number of possible variations and combinations of flame retardants, synergists, adjuvants, and fillers, along with their concentration, particle size distribution, and so forth, also determine typical research projects. Even following one distinct idea still requires that a wide range of different variations be compared, and combinations and concentrations optimized. Research and development projects approaching application usually entail much screening of several alternative approaches, combinations, and additives. For instance, during projects performed in close contact with industrial partners, our working group investigated various inorganic additives belonging to different groups, such as layered materials, metal hydroxides, metal oxides/carbonate and metal borates in bisphenol A polycarbonate/ silicon rubber/bisphenol A bis(diphenyl phosphate) (PC/SiR/BDP) [36], multicomponent flame retardant systems containing aluminum diethyl-phosphinate and another flame retardant and synergist in thermoplastic styrene–ethylene–butylene–styrene elastomers [4], different combinations of flame retardant and adjuvants in wood plastic composites [37], and different double and triple combinations of flame retardants in thermo-plastic polyurethane [38]. Even if an ingenious empirical concept reduces the number of systems to be investigated, a comprehensive screening of a variety of systems generally remains. Thus, in a typical example for industrial research and development, the patent DE 42 31 774 A1 [39]. protects a thermoplastic polymer flame retarded with a combination of 0.1 – 30 parts per weight phosphorous flame retardant and 0.1 – 15 parts per weight aluminosilicate. The special idea proposed by this patent covers an overwhelming matrix of components. Thermoplastic polymers include all kinds of homo- or (graft-)copolmers, polymer blends and compos-ites (based on fibers but also with poly(tetrafluoroethylene) PTFE fibers), which are based on alkene containing monomers such as propylene, vinyl acetate, styrene, 2-methyl styrene, acrylnitriles, methaacrylnitriles, methyl methacrylate, methacrylate, maleic acid anhydride, maleinimide, chloro-prene, butadiene, isoprene and C1-C8 alkyl acrylates; or on bifunctional compounds such as bisphenol A polycarbonate (PC), polyesters (PET, PBT, etc), polysulfones (PSU), polyethersulfones (PES), and polyetherketones (PEEK, PEK, etc.). All kinds of phosphorous compounds are proposed as flame retardants, including various phosphine oxides, phosphinates, phosphonates and phosphates with or without halogen. Aluminosilicates include all kinds of zeolites according to $M_{2/n}O \cdot Al_2O_3 \cdot xSiO_2 \cdot yH_2O$ (M = hydrogen or metal ion of the 1st to 5th main group or 1st to 8th subgroup), all kinds of layered silicates (kaolin types, spinel types, serpentine types, and montmorillonite types), and types containing Cu^{2+}-, Cn^{2+}-, and Ni^{2+}, including all mixtures and modified compounds [39]. When it comes to commercial systems, complex, synergistic, multifunctional multicompo-nent systems are the standard, not the exception.

What is more, the almost infinite feasible variations and combinations of flame retardants, synergists, adjuvants, fillers, as well as their concentration, particle size distribution, and so forth demand preselection, guidelines, rapid screening and accelerated assessment, so that "high-throughput" approaches are proposed [40–49]. Considering that fire behavior is not a material property, but the response of a defined specimen in a certain fire scenario, some intrinsic problems become obvious, such as limited possibilities to reduce the specimen size or to perform parallel testing. Nevertheless, developing multicomponent systems generates strong demands to accelerate tools in one way or the other.

9.2   Concepts

In practice, combinations showing superposition or slight antagonistic combinations often do the job well. They are successfully applied to optimize property profiles, including by reducing costs. However, the dominant concept for multicomponent flame retardant is synergy. The flame-retardant effect achieved by the combination of additives or measures is greater than the simple sum of the effects (superposition) caused by the individual contributions. Thus, using synergistic combinations is the most effective route to reduce the content of flame retardant needed. The term "synergy" is often misused as synonym for any kind of "improvement," but the concept of synergy in flame retardancy has been well defined, stated, and reviewed several times at length over recent decades [50–54]. Synergism is defined and thus observed in the measure of fire performance. Synergism in a fire property can be due to quite different physical and/or chemical phenomena. There are specific molecular effects that cause synergy, for instance, the occurrence of chemical reactions which are only possible in the multicomponent system, such as the formation of metal phosphates remaining in the residue rather than the release of phosphorus into the vapor phase. In other systems, material properties that control the response in distinct fire tests are optimized. Efficiency is increased due to chemical interactions or physical phenomena that adjust key parameters such the decomposition temperature range, dispersion, or viscosity of the melt. Enhancing flame retardant modes of action play a role by mechanically reinforcing fire residues or by closing the surface of the fire residue through glass formation. Indeed, many synergies are exploited in today's flame retardant polymeric products: metal hydroxides + synergists, phosphorus–nitrogen, flame retardant + anti-dripping agent, flame inhibition + free radical generators causing enhanced decomposition/melt flow. Consequently, several groups of additives are used regularly as adjuvants or synergists to achieve synergy in performance, such as zinc additives (zinc borates, zinc sulfide, zinc oxide, zinc stannate, zinc hydroxy stannate), boron sub-stances, metal/metal oxides, silica/silicone/siloxane/silicate fillers, sulfur compounds, and poly(tetrafluoroethylene) fibers [55–58].

Further, combining condensed phase modes of action and gas phase modes of action is concluded to be a general route to synergy, well explained as an intrinsic feature by distinguishing the combustion in the flame and the pyrolysis in the condensed phase as two coupled chemical reactions [54, 59, 60]. What is more, the use of systems exploiting "pseudo-synergy" is prevalent [53]. Although no additional synergistic mechanism occurs, the efficiency of flame retardancy is crucially increased for a combination of flame retardants by optimizing the concentration and mixing ratio when the single flame retardants show non-linear behavior as their concentra-tions are increased. This effect becomes obvious when systems are com-pared in which the total amount of flame retardant is fixed at a certain level (Figure 9.2). An adjusted mixture can outperform one with the same amount of flame retardants. Synergism is believed to be accompanied by surprising performance, and thus a bargain to get a patent exploiting the great potential for commercial systems.

A second general concept is the tailored system. The main flame retardant modes of action such as charring, protection layer formation/intumescence, and flame inhibition originate from specific chemical reac-tions. Their efficiency can be enhanced by increasing the reactivity and yield through adjustments to the systems. Corresponding decomposition temperature ranges of the flame retardant and the polymer are a key crite-rion for selecting promising flame retardants, not only for efficient flame inhibition, but also for charring [61, 62]. Multicomponent flame retardants can be used to shift the decomposition temperature range or to broaden it [63]. Synergists like metal oxides are reported to change the reactivity of chemical reactions in the condensed phase due to catalysis, salt formation, or stabilization, or to change the conditions for chemical reactions, such as the pH value [56, 64].

Multicomponent flame retardant systems are used to realize multifunc-tional performance in the polymeric material of interest. Flame retardants are combined with smoke suppressant to reduce flammability and smoke production [56, 58, 65]. Carbon fillers are proposed to achieved flame retardancy and electric/thermal conductivity. Several fibers, fillers such as carbon black or talc, and nanoparticles are applied or proposed to enhance both flame retardancy and mechanical properties. At the end of the day, optimizing the performance profile is usually achieved by applying multi-functional multicomponent additives.

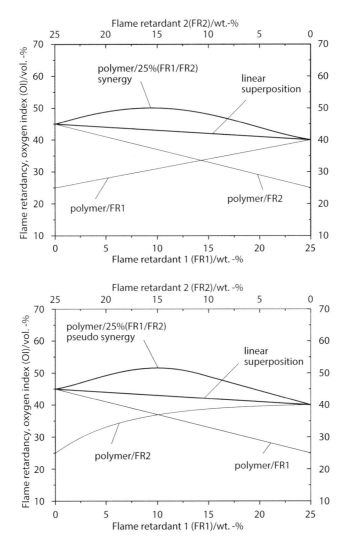

Figure 9.2 Synergy, an additional mechanism improves the efficiency so that the performance of the mixture outperforms the linear superposition of the individual contributions, and pseudo synergy, adding the nonlinear behaviors outperforms the linear mixing rule, for flame retardants FR1 and FR2.

In a lot of flame-retardant polymers, it turns out that synergistic and tailored systems are based on something known as "residue design." Charring itself, or in other words reducing the amount of carbon-based fuel released by forming a carbonaceous residue, is often only half the story. Adjuvant or synergistic combinations are used to improve the protective properties of the fire residue. These protective properties are controlled much more by the properties of the fire residue than by its amount. Most fire residue properties are strongly linked to the morphology of the residue, such as closed surface, multicellular or other thermal insulating structures, and the thermal and mechanical stability of the residue. Charring aims to reduce the total mass loss and thus the heat evolved (THE) in a fire test (fire load) as well as the mass loss rate and heat release rate via the factor $(1-\mu)$, where μ is the char yield between 0 and 1 [66–68]. The

$$THE \sim x h_c^0 (1-\mu) m_0$$

and the

$$HRR(t) \sim \theta(t) x h_c^0 (1-\mu)$$

with the effective heat of combustion of the volatiles, h_c^0, and the combustion efficiency of the flame χ. At first glance, protective layer properties may reduce only the rates, such as mass loss rate, heat release rate (HRR) or the peak of heat release rate (PHRR) by a factor $\theta(t)$ without changing the total heat evolved (total heat release, THR, at flameout or end of test = total heat evolved, THE). Nanocomposites such as polyolefins/layered silicate are good examples illustrating this description [69]. 5 – 10 wt.-% nanoparticles, well dispersed in polypropylene, polyethylene, and polystyrene, switch the systems from materials burning without any relevant residue to materials that form a protective residual layer. The PHRR can be reduced by 50–70%, whereas the reduction in THE is restricted to 5–10% [70]. In many systems, residue design in order to improve the residual protective layer, combined with other flame retardant modes of action such as fuel dilution, cooling, charring, or flame inhibition, is a most powerful approach to optimize efficient flame retardant systems. General residue design approaches for creating an enhanced protective layer entail inorganic-organic residues or phosphorous or nitrogenous carbonaceous char with better thermal stability. Mechanically reinforced chars, crusts and glassy coatings provide for closed surfaces. Moreover, the better the protective layer, the greater is the chance that the mass loss rate or HRR of a burning specimen drops below the critical value for self-extinguishing. The burning stops before all of the material is pyrolyzed. All kinds of intumescent systems are based on multicomponent systems realizing the residue design of thermally insulating morphologies. Efficient intumescent systems of the kind realized in intumescent coatings exploit endothermic decomposition and chemical interactions to absorb heat and to produce bubbles and viscous residue, causing the material to swell into a foam fire residue (see Chapter 4 of this book for more on intumescent phenomena). Usually non-flammable gases are released, some of which are used to form a multicellular carbonaceous or silicate layer structure. Several systems first show the formation of a grayish black foam-like carbonaceous char, which shrinks when high temperatures are applied for longer periods, converting into an inorganic sponge (Figure 9.3, left and middle) [71]. Intumescent systems are complex multicomponent systems consisting of several ingredients such as an acid source, a charring agent, a blowing agent, and several adjuvants. Up to seven different decomposition steps may be observed [72]. Commercially available systems consist of mixtures of distinct components such as binding agents, carbonizing substances, non-flammable gas-producing agents, dehydrating compounds and esterification catalysts, dispersion agents, degassing additives used for coatings, reinforcing filler, and fibers and synergists for mechanical and thermal stabilization of the residue. These intumescent systems are used in a polymer matrix such as polyolefins or with a binder in intumescent coatings [9, 73, 74]. The basic foundations of intumescence such as the chemical components, thermal and rheological mechanisms are well known (Chapter 4 of this book). However, several chemical and physical phenomena act together, with most of them activated thermally. There is plenty of room to optimize the reactivity and yield of the chemical reactions as well as to adjust the viscosity and optimize the residue properties and morphology to produce clearly improved thermal insulation [9, 75–79]. It should be noted that efficient thermal insulation is delivered not only by foam-like multicellular

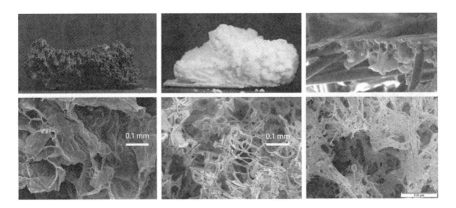

Figure 9.3 Fire residue design: Complex fire residue morphologies.

carbonaceous systems, but also by inorganic sponges, namely layered structures such as those yielded by expandable graphite on the microscopic level, but also other layered structures on the scale of delaminating long-fiber–reinforced composites or multicomponent laminates [80–82]. Complex residues of multicomponent flame retardant polymeric materials are described as consisting of a layered crust covering a rather hollow fiber stabilized space like a skin (Figure 9.3, right) [8, 83].

9.3   Combination with Fillers

The fire behavior of fiber reinforced polymer composites clearly differs from the fire behavior of their corresponding polymer matrix [3, 84–88]. Even though inorganic fibers such as glass fibers and carbon fibers are rather inert with respect to the mainly anaerobic pyrolysis taking place during burning, they strongly influence flammability. They change heat absorption and transfer within the condensed phase, so that the changed heating behavior of the materials influences ignition and flame spread [89, 90]. Composites show a dramatically changed melt flow and dripping of pyrolyzing melts during burning; in some setups wicking (transportation of fuel due to capillary forces) may even occur [91, 92]. Performance in several fire tests is changed significantly, with one notable example being how composites can affect the classification in UL 94, a reaction to small flame test. The amount of fire residue in a composite increases dramatically; in other words, the fire load decreases due to fuel replacement (polymer) by fiber in the composite. The morphology and properties of the fire residue are changed. In sum, flame retardancy concepts tailored to composites are needed. Furthermore, tasks that are specific for composites, such as structural integrity in fire, come into focus [93–95]. Achieving both the demanded flame retardancy and superior mechanical properties is a further intrinsic challenge that demands outstanding solutions, sometimes even thinking-out-of-the-box approaches such as phase selective filled multicomponent laminates [96].

Halogen-free flame retarded short-fiber–reinforced technical polymers such as polyesters and polyamides play a major role in electrical engineering. Short glass fibers are typically used in the range of 20–40 wt.-%. According to their higher density of around 2.6 g/cm³, the vol.-% of the filling is less than half of the wt.-%; the fire load of a test specimen is reduced to just 8–15%. The changed viscosity, including potential wicking effects, usually results in poorer performance in fire tests that monitor the reaction to small flame, such as UL 94 and oxygen index or in glow wire testing. Reduced melt flow and reduced dripping worsen the performance in other tests as well, including for PP-GF in comparison to PP in FMVSS 302. In other fire tests, such as in the cone calorimeter, short fibers contribute crucially to the amount of fire residue; in non-charring systems, such as polyolefins, polyesters, and polyamides, nearly the entire polymer matrix is pyrolyzed, with the residue consisting of 80–97 wt.-% glass fiber. The remaining fiber mat changes burning behavior, as a PHHR occurs at the beginning of burning before the residue drastically reduces the HRR. The PHRR is drastically reduced. In systems that tend to charring, the fibers increase the amount of residue and may improve the properties of the residue. Successful flame retardant systems combine flame inhibition with a relevant protective layer, based, for instance, on a mechanically stabilized glass fiber mat functioning as barrier for heat [97]; alternatively, they target residue design in the condensed phase, for instance, using the glass fiber to stabilize thermally insulating morphologies [8].

Long-fiber–reinforced materials are usually characterized by higher filling grades of around 60 vol.-%, anisotropic properties and laminate behavior. The reduction in fire load becomes a major effect as well as the increase in residue. Controlled delamination of single layers can be used as a thermally insulating structure, with woven fiber resulting in a superior residue [98]. In terms of the morphology and properties of the fire residue, the mats of long fibers become dominant. The tight fiber layers are reported to be so dense that the release of gas is disrupted, temporally and spatially. Such a disruptive gas release together with flame inhibition can yield to phenomenon in UL 94 that is called the "blowing out" effect, where the pyrolysis gases are released with enough pressure to "blow out" the flame due to creation of a fuel-rich/oxygen starved event at the flame front [99–101]. Successful flame retardants usually combine flame inhibition with minor charring accompanied with the formation of a protective layer, ideally one that has been optimized [102–104].

Large amounts of fibers or fillers are used to replace fuel. This strategy is used widely in the fire protection of flame retardant polyolefins, such as for some PP-based materials that achieve a V0 rating in UL 94. The combination of flame retardants with fibers/fillers that reduce fire load is crucial to enabling a material to achieve the demanded flame retardancy. Further, there are materials such as sheet molding composites (SMC) and bulk molding compounds (BMC), but also other materials, in which only a small amount of polymeric binder is used to glue incombustible fillers together; such materials may even approach a classification as incombustible when glass fibers and metal hydrates are combined to filling grades most probable > 90 wt.-%. SMC and BMC systems with a limiting oxygen index of 80 vol.-% and even with 100 vol.-% are commercialized [105–108].

There is increasing demand for flame retardant natural-fiber–reinforced polymers [109–111] and biocomposites in the transportation sector [112, 113], but also in building construction [114]. Natural fibers are not inert under the pyrolytic conditions of a burning specimen. Nevertheless, since they usually show charring and their effective heat of combustion is typically between 12–19 MJ/kg, they cause relevant fuel dilution in the condensed phases as well as in the gas phase, since their heat of combustion is crucially lower than most of the common and technical polymers. Tailored flame retardancy attempts to use them as charring agents, and the fiber residue to reinforce the char. Flame retardant coatings and soaking fibers are reported [115–117]. In biocomposites, synergistic combinations that remain compostable are demanded [118], and in the future, increasingly sustainable solutions that utilize industrial waste [119–121].

Inorganic fillers are often used in a wide range of thermoplastics for mechanical reinforcement. Their impact on fire behavior is often significant, and sometimes relevant [122]. They are inert fillers that change heat capacity, replace fuel, and contribute to residue. Some show sintering (fusing into glassy/ceramic residues) or relevant mechanical reinforcement of the char. Some inorganic fillers release water or CO_2 when heated to higher temperatures, and nearly all of them change the viscosity of melts, melt flow and dripping behavior in flammability tests. Talc is one of the most typical substances in this class. Talc is natural magnesium silicate; its properties differ from source to source according to the particle size and aspect ratio. Originally used for mechanical reinforcement, talc has a crucial influence on fire behavior in most systems; combinations between talc and phosphorous flame retardant can show synergistic performance [56, 123]. Melt flow and dripping behavior is strongly hindered, fire residue mechanical reinforced, and in some systems chemical interactions are reported [124, 125]. The lamellar behavior of talc is believed to form barriers [126, 127].

Most polymers in most fire tests burn based on anaerobic pyrolysis feeding the flame with volatile fuel. Combustion, the exothermic reaction with oxygen, is assumed to take place solely in the flame. Thus, not only carbon fibers, but all kinds of carbon fillers function as inert fillers during burning [128–130]. A huge amount of carbon black is used in high amounts in various elastomers. However, the impact on fire properties is reported to be limited [131]. Some years ago carbon nanotubes were introduced in flame retardancy applications [132–135]. Like other anisotropic nanoparticles, well dispersed carbon nanotubes change the melt viscosity, usually preventing dripping in UL 94. The dispersion of nanoparticles in non-charring polymers works as prerequisite for forming a protective fire residue layer [136]. Burning behavior is changed, dramatically reducing the PHRR in the cone calorimeter. In charring polymers, nanoparticles can improve the properties of fire residues. More recently graphene [129, 137], graphene oxides, functionalized graphene, and functionalized graphene oxide are proposed as nanofillers with great potential in flame retardancy applications [138–140]. Hardly any of these graphene materials consist of single graphene sheets; instead, they are used as multilayer graphene, with stacks of 5 to 20 graphene layers, comprising extraordinary nanometer-thick plates with an impressive aspect ratio. Well dispersed, these graphene nanofillers are believed to outperform nanotubes in performance and should become available at a lower price in the future. Carbon nanoparticles usually show a relevant impact on fire behavior in terms of maximum heat release rate, and are reported to exhibit flame retardant effects; their use as synergists is proposed in combination with flame retardants. Their filler character is usually obvious through a limited reduction in fire load [141, 142]. Although carbon nanoparticles alone are not flame retardant, when cleverly functionalized or used in combination with flame retardants they can be impressive multifunctional adjuvants and synergists with respect to mechanical properties, thermal and electrical conductivity and flame retardancy [143–146].

Well dispersed, anisotropic nanoparticles proposed as synergist for flame retardancy, such as layered silicates [147], oligomeric silsesquioxane [148–150], layered double hydroxides [151], sepiolite [152], halloysite[153, 154], boron nitride [155], zirconium phosphate [156], or boehmite nanocrystals [157], change the melt viscosity considerably; for instance, dripping in UL 94 is usually prevented. The dispersion of the nanoparticles in non-charring polymers works as a prerequisite for forming residual protective layers [158–160]. Burning behavior is changed; in the cone calorimeter the PHRR of non-charring polymers is reduced dramatically [69, 161, 162]. In some charring polymers and intumescent systems, nanoparticles may improve the properties of the fire residues [163, 164]. Nanocomposites for flame retardancy would require industrial mass production that exhibits a clear nanoscience character. Nanocomposites outperform the analogous microcomposites by far. Although they are proposed as a revolutionary new approach for flame retardancy [165], here they are discussed as fillers, in keeping with their inert filler character with respect to fire load, time to ignition, and reaction to small flame. Nevertheless, nanoparticles often show a distinct impact on fire behavior; their flame-retardant effects are apparent, and they are successfully implemented as synergists in combination with flame retardants [166–171]. Although nanoparticles alone do not provide sufficient flame retardancy and combinations are not beneficial in every application, in some multicomponent systems nanoparticles are applied as impressive multifunctional synergists with respect to mechanical properties and flame retardancy [172]. Although proposed right at the start of nanocomposite development in the second half of the 1990s, the combination with phosphorus flame retardants has generated all kinds of outcomes: commercial systems, innovative ideas, and deleterious combinations [173–177]. At the beginning of our century, the combination of layered silicate and ATH was commercialized [178, 179]. This combination enabled the reduction of filler content from 65 wt.-% to < 55 wt.-% in EVA.

9.4   Adjuvants

Most of the commercially successful multicomponent systems applying well-established adjuvants are also strongly synergistic. Indeed, several of the groups of additives discussed contain examples for adjuvants and synergists with respect to flame retardancy. Thus, discussing a group of additives under "adjuvants" here or as "synergists" in the following paragraph is somewhat arbitrary. I have tried to summarize additives that mainly add or enable flame retardancy modes of action here under "adjuvants," and additives that mainly optimize a flame retardancy function under "synergists" in the following paragraph.

Melt flow and dripping are crucial properties that influence the response of test specimen in several of the most important fire tests for polymeric materials. When a small flame or a glowing wire is applied, removing fuel and heat by melt flow and dripping can result in efficient extinguishing [180]. There are not only thin foils and fibers, but several successful products on the market with V-2 or V-0 non-flaming dripping behavior. Also, in other fire tests using a small burner, such as for building products or in the automotive industry, what is sometimes called the "retreat" effect is used to pass the mandatory test. Dripping agents are often used as adjuvants, with flame retardants causing flame inhibition or strong fuel dilution [181–184]. The combination of adjuvants enhancing dripping and flame inhibition is a very old idea that was applied to halogenated systems [185], the popularity of which is increasing in phosphorus-based multicomponent systems. The most successful dripping agents are radical generators in the condensed phase. They enhance early decomposition reactions in polyolefins and polystyrene changing the intermediate product balance, increasing the liquid fraction with low viscosity. The mass loss rate in thermogravimetry often does not show any major change, but the viscosity of the pyrolyzing melt does. Further, the burning in fire tests is influenced less by melt flow and dripping, such that the cone calorimeter may show hardly any change to some increase in heat release or to some shifting in time to ignition, whereas the UL 94 classification or LOI changes dramatically. Additives based on N-alkoxy amine, with a chemical structure closely related to antioxidants [186, 187], particularly Flamestab ®NOR 116 a monomeric N-alkoxy hindered amine stabilizer, is widely introduced as dripping agent not only in halogenated but also in several halogen-free flame retardant combinations. Several new systems are proposed, such as symmetrical and asymmetrical azoalkanes, along with related azoxy, hydrazone and azine derivatives [188, 189], sulfenamides [190], and oxyimide [191] systems. Recently, more and more combinations have been proposed, some of which show synergistic effects that demand more complex modes of action, such as additional activity in the gas phase.

In most of the flame retardant technical thermoplastics, such as polyamides, polyesters, or polycarbonate blends, dripping must be avoided to achieve a UL 94 V-0 non-dripping classification. Sometimes enhanced charring and crosslinking prevent dripping in fire tests [192–196], sometimes this effect comes from inorganic fillers [197] such as talc [123]. In a multitude of flame retarded thermoplastics, 0.1–0.8 wt.-% microfibrillated polytetrafluoroethylene (PTFE) is added as an anti-dripping agent [43, 46]. In some systems where the flame retardant also works as plasticizer, combination with an anti-dripping agent is essential to enable the flame retardant to work in the condensed phase [198, 199]. Anti-dripping agents such as microfibrillated polytetrafluoroethylene (PTFE) provide quite clever tailored solutions. The plasticizing effect, or generally low viscosity, is often desired during the processing of thermoplastics, such as during extrusion and injection molding. Thus, anti-dripping agents introduce flow limits at low shear rates to prevent dripping without increasing the low viscosity needed for the high shear rates typical for processing [199].

One of the most important classes of adjuvant is charring agents. Some flame retardants are typically used together with charring agents or adjuvants that combine charring and blowing properties, in particular ammonium polyphosphate (APP) in polyolefins [56]. In some systems that do not provide any interactions with phosphoric acid, without a charring agent, fire residue would consist only of polyphosphate, not phosphate-stabilized carbonaceous char. Of course, the combinations of phosphorus flame retardant and charring agent in these systems are strongly synergistic. One of the standard charring agents used with APP is pentaerythritol (PER) and its derivatives [200]. Another charring agent is polyphenylether (PPE), formerly called polyphenyloxide (PPO), synthesized via self-condensation of 2,6-xylenol, is processed as a blend with polystyrene to yield high-impact polystyrene HIPS/PPE [56]. Several sources propose adding novolac to thermoplastics [201, 202], some adding polyamides or polyimides [203]. Charring hyperbranched structures and dendrimers have been proposed as well as condensation products of melamine, morpholine and piperazine [58, 204–207]. Triazines and piperazine are proposed as charring groups in different molecules [208–211]. Following the trend toward replacing petrochemicals with natural renewable resources, bio-based charring agents are increasingly favored, such as lignin, vanillin, tannic acid, starch, cellulose, or cyclodextrin and their derivatives [212–219].

Blowing agents such as melamine, additives containing melamine and its derivatives are a very successful class of additives, because these substances often provide manifold benefits in multicomponent systems (see Chapter 5 of this book) [220, 221]. The conversion to non-volatile char-enhancing products and the release of NH_3 spread over a certain temperature range generate much more than simple fuel dilution in the condensed and gas phases, but make melamine the most important blowing agent in commercial intumescent systems. Melamine cyanurate and melamine salts are reported to combine fuel dilution with enhanced melt flow [184, 222, 223].

There is a strong need to add smoke suppressants for applications in transportation, particularly because several flame retardants efficiently reduce ignitability, flame spread and heat release via flame inhibition, but intrinsically exhibit increases in yields of CO and smoke. In some systems even the total smoke release is increased under forced flaming conditions. Although poly(vinyl chloride) (PVC) is not a halogen-free flame-retarded polymer, the use of smoke suppressants in PVC is the most comprehensive, intensively studied, and best understood benchmark for applying smoke suppressants [224, 225]; moreover, several smoke suppressants started their carrier in PVC and are now widely used in halogen-free materials. Metal hydrates are applied to reduce smoke production [226, 227], as well as zinc borate, zinc stannate, all kinds of catalytic materials such as metal oxides [228], and nanoparticles are also believed to harbor potential as a smoke suppressant [75, 155, 229–231]. The understanding of smoke suppression is often insufficient, because the production of smoke is complex and controlled by both the pyrolysis of the polymeric material and the combustion of the volatiles, as well how the smoke is measured in different test methods. Both pyrolysis and combustion change under different fire scenarios such as ventilation, heat flux, and temperature. The interactions reducing smoke production may be quite specific and differ from one system to the other. Catalysis, and perhaps specific chemical reactions are also believed to play an important role in reducing aromatic volatiles by charring or reducing the size of volatiles through further decomposition [232–234]. As changing the residence time and temperature in the pyrolysis zone changes pyrolysis products, protective layers also influence smoke production [235, 236].

9.5   Synergists

The performance of nanofillers and most of the examples of adjuvants and synergists discussed is quite specific; in other words, the same additive or combination of groups may work in synergy in one system, while their effectivity might be limited in another system. Further, several of the groups of additives discussed include examples of typical fillers as well as adjuvants and synergists. Thus, discussing carbon and inorganic nanofillers as fillers and groups of additives here under "synergists" or in the previous paragraph under "adjuvants" is somewhat arbitrary as discussed in the previous section of this chapter. Several classes contain crossover materials. Nevertheless, I have decided to discuss each group only once in detail, summarizing additives that mainly optimize a flame retardancy function here under the term "synergists," and additives that mainly add or enable flame retardancy modes of action under "adjuvants" in the previous paragraph.

Several synergistic combinations with phosphorus are reported. The most popular rule of thumb is the phosphorus-nitrogen (P-N) synergy. However, as it is an oversimplification to call P and N flame retardant elements, there is hardly a general P-N synergism. Specific phosphorous and nitrogenous groups are applied to realize specific chemical reactions that cause flame retardant modes of action, such as fuel dilution, charring and flame inhibition. The impression of a general P-N-synergism is most probably due to the fact that so many different combinations of phosphorous and nitrogenous structures are used; several different phenomena contribute to the concept of P-N synergism. Several phosphorous flame retardants enhance the charring of amides quite efficiently [56]. Amides are even proposed as charring agents for phosphorous flame retardants [163]. Several N-containing molecules result in the endothermic formation and release of NH_3 [237, 238]. In flame retardant combinations, nitrogenous groups often support phosphorous flame retardants crucially through fuel dilution in the gas phase. The combination of aluminum diethylphosphinate and melamine cyanurate in polybutylesters may be one example [239], as is the combination of melamine cyanurate with RDP in thermoplastic polyurethane [56]. Nitrogenous molecules such as melamine are widely used as blowing agents in intumescent systems. For instance, combining ammonium polyphosphate with melamine is the synergistic combination of ingredients in most of the intumescent coatings used. P-N bonds or the formation of P-N bonds essentially provides for thermal stabilization of the char in charring systems [240]. Phosphoramidates and phosphazene are interesting classes of flame retardant similar to phosphine oxides, phosphinates, phosphonates, and phosphates [58, 241–245].

Although phosphorus-sulfur (P-S) synergy and phosphorus-silicone (P-Si) synergy are well known, attempts to exploit these phenomena have recently increased [246–248]. P-Si systems cover a wide range of performance. In some systems, whenever molecules containing Si were applied, they behaved like inert fillers, with a negligible influence on fire behavior. Further, in oligomeric or polymeric Si-substances such as silicone oils and polysiloxanes, degrees of branching and side groups control the pyrolysis into volatiles or into inorganic residue. However, in some systems, the potential of inorganic-organic residues with superior protective layer properties are spawning synergistic solutions [249–255]. In several combinations the formation of silicophosphates is documented, and thus such combinations are proposed to be crucial for improving the stability of fire residue [256]. Special silicone synergists are under development and proposed for application in glass-fiber–reinforced polyamides flame retarded with aluminum diethylphosphinate, but they also work in natural-fiber–reinforced biocomposites, in synergism with aluminum diethylphosphinate [257, 258]. Incidentally, the key flame retardant mode of action of the successful nanocomposites is not the inorganic residue itself, but the formation of an inorganic-organic residual protective layer [259, 260]. P-S synergism is obvious and impressive when the flame inhibition of a phosphorous flame retardant is combined with disulfides as a radical generator in the condensed phase, analogous to the combination of halogen and disulfides in polystyrene [261]. Analogously, sulfenamides are proposed as extremely efficient synergists in combination with phosphorous flame retardants in polyolefins [190]. However, sulfurous groups harbor the potential for several flame retardant modes of action and synergistic impacts in combinations with phosphorous flame retardants [262]. Organo-/polysulfones are known to release SO_2, functioning to dilute fuel in the gas phase, while the remaining structure simultaneously contributes to charring [263–266].

All kinds of boron compounds (see Chapter 7 of this book), including boric acid that is also described as main flame retardant in some systems, boric oxide and borates [58, 267–269], particularly zinc borate, are applied as synergists in multicomponent systems [270–272]. Zinc borate is used together with metal hydrates such as ATH in EVA and other polymers to improve the protective layer properties of the fire residue [273], but also together with phosphorous flame retardants. Borates release water, sinter and form glasses. In combination with phosphorous flame retardants, for instance arylphosphates with zinc borate in PC blends, the formation of borate networks and borophosphates are observed; the inorganic-organic fire residue is proposed to improve flame retardancy [274, 275]. Similar chemical reactions resulting in an improved protective residual layer are observed when zinc borate is combined with phosphorous flame retardants [276]. Melamine borates have also been proposed [277].

Metal compounds (see Chapter 10 of this book), mainly metal oxides, are used as synergists in flame retardant polymeric materials [229, 278–281]. Sometimes they are even multifunctional, improving thermal stability and flame retardancy. Zinc oxide of less than 1 wt.% is often used in thermoplastics, such as in combination with red phosphorus in polyamides and polyesters [56, 64]. However, all kinds of metal oxides harbor the potential to catalyze charring, for instance, in polyesters, or to enhance the condensed phase action of phosphorous flame retardant. These distinct effects can be enhanced using nanoparticle with larger specific surfaces [282–287]. All kinds of metal compounds, such as oxides, acetates, acetyl acetonates, borates and sulfates, are proposed as synergists in intumescent systems applied in polypropylene [288]. TiO_2 has become a standard ingredient in intumescent coatings and intumescent systems, due to its beneficial reaction with APP to form metal phosphates, stabilizing the intumescent residue as well as improving its properties [289–292]. More complex structures such as flame retardant – metal oxide hybrids, metal complexes and metal organic frameworks have been proposed more recently, however [293, 294].

All kinds of synergistic combinations of flame retardants and nanoparticles are proposed [295–298]. Carbon and inorganic nanofillers were already discussed under section 9.3 of this chapter. However, in the few successful commercialized systems, the synergistic systems use nanocomposites. The best example is the above mentioned combination of organically modified layered silicate with ATH, which enables a beneficial reduction in the amount of filler needed to achieve the demanded flame retardancy [188, 189]. Layered plate-like nanoparticles, tube-shaped nanoparticles, zeolites, but also talc and fibers are reported to act as synergists in intumescent systems, when the changed viscosity favors intumescence, and the particles stabilize the residue due to physical or chemical interactions [18, 79, 299–309]. Although usually reported as success, the beneficial use of nanoparticle in intumescent systems is somewhat tricky. The strong reinforcing influence on the viscosity of the pyrolyzing melt can also effectively prevent successful intumescent char formation (see Chapter 4 of this book).

Glass-forming substances are proposed as synergists in halogen-free flame retardant multicomponent systems. They are applied to optimize the properties of the fire residue by closing its surface. Borates and phosphates form glasses, too, of course, so that glass formation is a phenomenon achieved by several flame retardants. Several products are commercialized as glass-forming flame retardants, consisting of distinct mixtures of glass frits and fumed glass used as synergists [56, 310–312]. The combination of melamine cyanurate and porous glass powder is proposed for unfilled nylon 6. It has been argued that adjusting the softening temperatures is the key to achieving efficient systems [313–315].

Apart from the P-Si synergy, silicone, siloxane, fumed silica and amorphous SiO_2 (see Chapter 6 of this book) have been proposed as synergists in all kinds of systems [316–320]. Yet additives containing Si exhibit quite a wide range of performance, from decomposing into volatiles, inert filler character, to very efficient synergism. Side groups and branching make major differences. However, in some systems the potential of inorganic-organic residues with superior protective layer properties are spawning synergistic solutions. Sub-micrometer spherical particles of amorphous SiO_2, are used intensively in synergistic combinations with ATH in the cable industry and in rubbers for conveyor belts [321, 322].

Metal hydrates (Chapter 3 of this book) are economical, environmentally friendly, and reduce not only fire risks but also smoke production. Obviously, these are very good reasons to apply metal hydrates in polyolefins and more recently in biopolymers. In polyolefins, replacing fuel by very high content of filler is a crucial part of the flame retardant mode of action. Considering both the high flame retardant efficiency demanded in polyolefins and the limited effectiveness of metal hydrates, filling content of over 60 wt.-% is necessary to achieve the demanded flame retardancy. With a density of 2.42 g/cm^3, over 25% of the fire load is thus reduced by fuel dilution in the condensed phase. As the high content of filler deteriorates the mechanical property profile of the materials, metal hydrates like ATH are frequently used together with synergists to reduce the necessary amount of filler. Mixtures with other fillers or fibers are proposed, including layered silicate, nanotubes, amorphous silicon dioxide, silicone, metal borates, zinc stannate/zinc hydroxstannates, carbonates, diatomite, melamine, and charring agents [20, 21, 323–326]. Modification and various coatings are proposed [327], such as silane, melamine-formaldehyde resin, melamine/novolac resin, poly-methyl- methacrylate, and other sophisticated ideas [328].

Clearly, in practice what is proposed is not the combination of one flame retardant and one adjuvant or, one flame retardant and one synergist, but multicomponent systems consisting of flame retardants, fillers, adjuvants and synergists [329].

9.6   Combinations of Different Flame Retardants

There is proof of phosphorus-phosphorus (P-P) synergism [53]. The synergism is achieved mainly due to the combination of different flame retardancy modes of action in the gas and condensed phase, flame inhibition, and charring/protective layer formation [330]. For instance, aluminum diethylphosphinate is combined with melamine polyphosphate or ammonium polyphosphate [56], and aluminum hypophosphite with resorcinol diphosphate (RDP). Enhancing the charring in the condensed phase is usually controlled by specific chemical reactions between the partly decomposed polymer and partly decomposed flame retardant [61]. Further phosphorous flame retardants are known to be sensible to chemical interactions that shift their pyrolysis products from inorganic phosphates to phosphates, enhancing the amount and thermal stability of char, or to the release of phosphorous volatiles. Thus, in several systems not only simple superpositions of the effects of the different flame retardants occur, but also "new" chemical interactions. In some systems these interactions contribute to synergistic performance or are the main reason for synergy. For instance, the combination of aluminum diethylphosphinate and melamine polyphosphate is reported to show heavy reactions between the flame retardants, improving the flame retardant mode of action in the condensed phase [331–333]. A reaction between two phosphorous additives improving the condensed phase effectivity has been observed in other systems as well [334]. In the case of arylphosphate–arylphosphate synergism in polycarbonate (PC) blends, the intermediate products exhibiting beneficial higher thermal stability are proposed to cause the synergy [335].

As metal hydrates are economical, environmentally friendly, provide water for desired hydrolysis, reduce smoke, provide inorganic residue, metal ions and metal oxides, they harbor plenty of features that make them interesting for multicomponent flame retardant mixtures. In combinations of metal hydrates and phosphorus flame retardants functioning mainly in the gas phase, the synergistic interaction of different flame retardant modes of action is encouraging: release of phosphorus resulting in flame inhibition in the gas phase, release of water resulting in cooling and fuel dilution, formation of a inorganic residue. However, metal hydrates also react with phosphorous flame retardants to form metal phosphates [336], improving the protective residue properties, for instance in red phosphorus/Mg(OH)$_2$ or APP/Mg(OH)$_2$ combinations [5, 276, 337, 338]. Due to its higher decomposition temperature, magnesium hydrate exhibits this phenomenon much more strongly than ATH. The combination of ATH with a small amount of phosphorous flame retardant is also reported, for instance ATH with RDP in thermoplastic polyurethane (TPU) for power & communication cables [56]. The combination of metal hydrate and a phosphorous flame retardant to decrease the filler content can also be proposed for other polymers such as EVA, and for the combination with the mineral huntite/hydromagnesite [339, 340]. Combinations of higher content 30 wt.-% of boehmite or ATH with reasonably small amounts of reactive organophosphorus, DOPO derivatives, APP, aluminum diethylphosphinate, zinc diethylphosphinate, or hypophosphite are proposed for epoxy resins and polyester resins [341–343]. Although the combination of metal hydrates and phosphorus flame retardants may be prevalent, also other combinations such as with expandable graphite were proposed [344, 345]. The combination of a large amount of metal hydrate and other flame retardants and synergists is proposed for ceramifying systems [346, 347].

Expandable graphite is a fantastic flame retardant, often showing extraordinary efficiency in fire tests such as the cone calorimeter [348]. Yet in other fire tests, and sometimes even in the cone calorimeter, the fire residue is too fluffy and detaches from the sample. Expandable graphite is widely used in combination with phosphorous flame retardants; the combination is often proposed for biopolymers and wood plastic composites [349–353]. Massive synergy occurs, mainly due to the fixation of the expanded graphite protective layer, sometimes also due to the combination of the thermal insulating residue and flame inhibition [354, 355]. The combination of expandable graphite and phosphorus flame retardants is a prevalent approach in hard and flexible polyurethane foams [356–361]. Hereby the additive and reactive phosphorus flame retardants were reported.

There are also examples for combining three and more flame retardants, for instance expandable graphite, phosphorous flame retardant and metal hydrate, or expandable graphite and two phosphorous flame retardants [345, 362]. The multicomponent synergistic flame retardant system of melamine polyphosphate (MPP), melamine cyanurate (MC) and aluminum diethylphosphinate (AlPi) is proposed for thermoplastic polyurethane (TPU) [363].

Synergistic combinations are proposed which exploit different particle shapes or sizes, and also broaden the decomposition temperature range. The broad decomposition temperature range of natural mixtures of huntite-hydromagnesite is reported to be a key factor in their success [364]. Mixtures of ATH, Mg(OH), and boehmite are proposed. With respect to protective layer formation, synergistic combinations of different filler sizes and shapes are reported. ATH with a bimodular size distribution is commercialized, the combination of layered silicate with carbon nanotubes is reported [365], as well as synergy in layered silicate and spherical amorphous SiO_2 [366].

Flame retarded materials can be refined with intumescent coating, but also other non-intumescent flame retardant coatings [367–375]. A successful commercial flame retardant composites is reported to consist of fiber mats soaked with flame retardants, thermoset resins modified with flame retardant, and intumescent coatings applied on the surface [376, 377].

9.7   Combinations of Different Flame-Retardant Groups in One Flame Retardant

Although recently proposed as a successful new concept [378], and spawning some of the innovative and promising developments of commercial flame retardants [379–383], the combination of different flame retardant groups in one flame retardant is well established in some of the flame retardants in use for several decades, such as ammonium polyphosphate and melamine (poly)phosphate.

In melamine polyphosphate or ammonium polyphosphate combinations, phosphorus working in the condensed phase is combined with nitrogenous groups causing fuel dilution in the gas phase or working as a blowing agent [56]. The combination is established not only for gas production, which effects fuel dilution, but also for charring groups, for instance with bis(melamine) salt of pentaerythritol bis(acid phosphate) [383, 384]. The idea is to achieve intumescence, usually through a mixture of different additives, with a single flame retardant. Replacing multicomponent intumescent systems with one flame retardant functioning as the acid source, charring agent and blowing agent is indeed very interesting [385–388]. Several mono-component intumescent flame retardants were proposed [387, 389–392]. The idea of slightly upgrading phosphorous flame retardants by adding functional groups such as ethylenediamine to achieve more efficient intumescence is a very successful concept that has been proposed and used over the last four decades and discussed in recent papers [393, 394].

The functionalization of bio-based polymers with phosphorus, or with phosphorus and nitrogen such as phosphorylated lignin, can be described as combining P with a charring agent [395–400]. Phosphorylated biopolymers are proposed not only for biopolymers such as PLA, but also for non-charring polyolefins. With natural phosphorous substances such as DNA or phytic acid, the door is open to completely bio-based intumescent flame retardants [401–407]. Phosphorylated vegetable oils are believed to combine flame inhibition and dripping agent in polyester [408]. 9,10-di hydro-9-oxa-10-phosphaphenanthrene-10-oxide – cellulose – acrylate is proposed for transparent flame retardant coatings [409].

Analogous to combining two different phosphorous flame retardants, two different phosphorus groups can be combined in a single flame retardant. The combination of phosphine oxide or phosphinate with phosphonate or phosphate, respectively, harbors the potential for combining efficient flame inhibition and charring [410, 411]. Several phosphine oxides and phosphinate groups, such as side groups consisting of DOPO or DOPO derivatives, efficiently release phosphorus volatiles that generate radical trapping in the flame. Several phosphonate and phosphate group are precursors for acids that catalyze and stabilize carbonaceous char.

In some works, the combination of phosphorus and sulfur is proposed [412–415]. The combination of P and N is widely used, not only in the form of melamine derivatives and salts, but also molecules containing triazine–phosphorus [416–420], or structures based on P-N bonds such as phosphazene, phosphoramines, or phosphoramidates [380, 421–425]. P-N combinations in flame retardants have always attracted much attention in academic and industrial research. No matter how much effort is needed to synthesize such compounds, they harbor benefits such as high thermal stability of the flame retardant or the fire residue, and potential for flame inhibition, fuel dilution and charring [247]. Incorporating silicone or borate groups together with phosphorus were proposed [426–429]. Due to the higher costs of the matrix, in particular in epoxy resins, there are all kinds of complex structures proposed [301]. Recently, molecules containing all kinds of complex combinations of P with N, S, Si-, and boron have been reported [430–437].

More recently, functionalized carbon particles and carbon nanoparticles to promote them to full-fledged flame retardants have been proposed [143, 438–440]. In particular, functionalized graphene and graphene oxides are proposed to work as flame retardants;[6, 58, 148, 438, 441–445]. however, upgrading synergistic nanofillers to stand-alone flame retardants is also being attempted for other nanoparticles [446].

9.8   Conclusion

Although the main flame retardant modes of action are known, in practice, our detailed scientific understanding usually falls short when it comes to complex and synergistic multicomponent systems, crucial minor optimizations, or quantification in terms of specific fire properties. The description of the flame retardancy yielded by multicomponent systems usually remains vague and fragmentary. This chapter has attempted to provide thought-provoking impulses on how multicomponent systems can be utilized and to direct the development of future flame retardant polymer products. The benefits of multicomponent systems can be quite clear, and will be a definite area of research and commercial non-halogenated flame retardant material development in the near and far future.

References

1. Levchik, G.F., Levchik, S.V., Camino, G., Weil, E.D., Fire retardant action of red phosphorus in Nylon 6, in: *Fire Retardancy of Polymers the Use of Intumescence*, M. Le Bras, G. Camino, S. Bourbigot, R. Delobel (Eds.), pp. 304–315, The Royal Society of Chemistry, London, UK, 1998.
2. Rabe, S., Chuenban, Y., Schartel, B., Exploring the Modes of Action of Phosphorus-Based Flame Retardants in Polymeric Systems. *Materials*, 10, 455, 2017.
3. Brehme, S., Köppl, T., Schartel, B., Altstädt, V., Competition in Aluminium Phosphinate-based Halogen-free Flame Retardancy of Poly(butylene terephthalate) and Its Glass-fibre Composites. *e-Polymers*, 14, 193–208, 2014.
4. Langfeld, K., Wilke, A., Sut, A., Greiser, S., Ulmer, B., Andrievici, V., Limbach, P., Bastian, M., Schartel., B., Halogen-free Fire Retardant styrene-ethylene-butylene-styrene-based Thermoplastic Elastomers Using Synergistic Aluminum Diethylphosphinate-based Combinations. *J. Fire Sci.*, 33, 157–177, 2015.
5. Braun, U. and Schartel, B., Flame Retardant Mechanisms of Red Phosphorus and Magnesium Hydroxide in High Impact Polystyrene. *Macromol. Chem. Phys.*, 205, 2185–2196, 2004.
6. Wang, X., Xing, W., Feng, X., Yu, B., Song, L., Hu, Y., Functionalization of graphene with grafted polyphosphamide for flame retardant epoxy composites: Synthesis, flammability and mechanism. *Polym. Chem.*, 5, 1145–1154, 2014.
7. Eckel, T., The most important flame retardant plastics, in: *Plastics Flammability Handbook*, J. Troitzsch (Ed.), pp. 158–172, Hanser, Munich, Germany, 2004.

8. Braun, U., Schartel, B., Fichera, M.A., Jäger, C., Flame Retardancy Mechanisms of Aluminium Phosphinate in Combination with Melamine Polyphosphate and Zinc Borate in Glass-fibre Reinforced Polyamide 6,6. *Polym. Degrad. Stab.*, 92, 1528–1545, 2007.

9. Alongi, J., Han, Z.D., Bourbigot, S., Intumescence: Tradition versus novelty. A comprehensive review. *Prog. Polym. Sci.*, 51, 28–73, 2015.

10. Bourbigot, S., Le Bras, M., Bréant, P., Trémillon, J.M., Delobel, R., Zeolites:new synergistic agents for intumescent fire retardant thermoplastic formulations – criteria for the choice of the zeolite. *Fire Mater.*, 20, 145–154, 1996.

11. Vannier, A., Duquesne, S., Bourbigot, S., Castrovinci, A., Camino, G., Delobel, R., The use of POSS as synergist in intumescent recycledpoly(ethylene terephthalate). *Polym. Degrad. Stab.*, 93, 818–826, 2008.

12. Anna, P., Marosi, G., Csontos, I., Bourbigot, S., Le Bras, M., Delobel, R., Influence of modified rheology on the efficiency of intumescent flameretardant systems. *Polym. Degrad. Stab.*, 74, 423–426, 2001.

13. Qian, Y., Wei, P., Jiang, P., Zhao, X., Yu, H., Synthesis of a novel hybrid synergistic flame retardant and its application in PP/IFR. *Polym. Degrad. Stab.*, 96, 1134–1140, 2011.

14. Liu, Y., Wang, J.S., Deng, C.L., Wang, D.Y., Song, Y.P., Wang, Y.-Z., The synergistic flame-retardant effect of O-MMT on the intumescent flame-retardant PP/CA/APP systems. *Polym. Adv. Technol.*, 21, 789–796, 2010.

15. Ren, Q., Wan, C., Zhang, Y., Li, J., An investigation into synergistic effects of rare earth oxides on intumescent flame retardancy of poly-propylene/poly (octylene-co-ethylene) blends. *Polym. Adv. Technol.*, 22, 1414–1421, 2011.

16. Lin, M., Li, B., Li, Q., Li, S., Zhang, S., Synergistic effect of metal oxides on the flame retardancy and thermal degradation of novel intumescent flameretardant thermoplastic polyurethanes. *J. Appl. Polym. Sci.*, 121, 1951–6190, 2011.

17. Wang, X., Wu, L., Li, J., A study on the performance of intumescent flameretarded polypropylene with nano-ZrO2. *J. Fire Sci.*, 29, 227–242, 2011.

18. Liu, Y., Zhao, J., DengL, C.L., Chen, L., Wang, D.Y., Y.-Z.Wang Flameretardant effect of sepiolite on an intumescent flame-retardant polypropylene system. *Ind. Eng. Chem. Res.*, 50, 2047–54, 2011.

19. Wang, J.S., Liu, Y., Zhao, H.B., Liu, J., Wang, D.Y., Song, Y.P., Wang, Y.-Z., Metal compound-enhanced flame retardancy of intumescent epoxy resins containing ammonium polyphosphate. *Polym. Degrad. Stab.*, 94, 625–631, 2009.

20. Horn Jr., W.E., Inorganic Hydroxides and Hydroxycarbonates: Their Function and Use as Flame-Retardant Additives, in: *Fire Retardancy of Polymeric Materials*, A.F. Grand and C.A. Wilkie (Eds.), pp. 285–352, Marcel Dekker, New York, 2000, chap 9.

21. Hornsby, P., Fire-Retardant Fillers, in: *Fire Retardancy of Polymeric Materials*, 2nd Edition, C.A. Wilkie and A.B. Morgan (Eds.), pp. 163–185, CRC Press, Boca Raton, 2010, chap 7.

22. Hull, T.R., Witkowski, A., Hollingbery, L., Fire retardant action of mineral fillers. *Polym. Degrad. Stab.*, 96, 1462–1469, 2011.
23. Herwig, W., Kleiner, H.J., Sabel, H.D., European Patent EP0006568A1, 1980.
24. Kleiner, H.J., Budzinsky, W., Kirsch, G., US Patent 5773556A, 1998.
25. Kleiner, H.J., Budzinsky, W., Kirsch, G., Flameproofed polyester molding composition- US Patent 5780534A, 1998.
26. Kleiner, H.J., Budzinsky, W., Kirsch, G., Salts of phosphoric acids and use therof as flame retardants in plastics- US Patent 5891226A, 1999.
27. Kleiner, H.J., US Patent 6194605B1, 2001.
28. Kleiner, H.J., US Patent 6211402 B1, 2001.
29. Schlosser, E., Nass, B., Wanzke, W., US Patent 6,255,371B1, 2001.
30. Hörold, S., US Patent 6420459B1, 2002.
31. Jenewein, E., Kleiner, H.J., Wanzke, W., Budzinsky, W., US Patent 6365071B1, 2002.
32. Kleiner, H.J., US Patent 6414185B2, 2002.
33. Klatt, M., Leutner, B., Nam, M., Fisch, H., US Patent 6503969B1, 2003.
34. Knop, S., Sicken, M., Hoerold, S., EP Patent 1403309A1, 2004.
35. Lyon, R.E. and Janssens, M.L., Flammability, in: *Encyclopedia of Polymer Science and Technology*, H.F. Mark (Ed.), Wiley, Hoboken, 2005.
36. Wawrzyn, E., Schartel, B., Karrasch, A., Jäger, C., Flame-retarded bisphenol A polycarbonate/silicon rubber/bisphenol A bis(diphenyl phosphate): Adding inorganic additives. *Polym. Degrad. Stab.*, 106, 74–87, 2014.
37. Yin, H., Sypaseuth, F.D., Schubert, M., Schoch, R., Bastian, M., Schartel, B., Routes to halogen-free flame-retardant polypropylene wood plastic composites. *Polym. Adv. Technol.*, 30, 187–202, 2019.
38. Sut, A., Metzsch-Zilligen, E., Großhauser, M., Pfaendner, R., Schartel, B., Rapid mass calorimeter as a high-throughput screening method for the development of flame-retarded TPU. *Polym. Degrad. Stab.*, 156, 43–58, 2018.
39. T. Eckel, P. Ooms, D. Wittmann, H.-J. Buysch, German patent. A 1, 1994.
40. Gilman, J.W., Bourbigot, S., Shields, J.R., Nyden, M., Kashiwagi, T., Davis, R.D., Vanderhart, D.L., Demory, W., Wilkie, C.A., Morgan, A.B., Harris, J., Lyon, R.E., High throughput methods for polymer nanocomposites research: extrusion, NMR characterization and flammability property screening. *J. Mater. Sci.*, 38, 4451–4460, 2003.
41. Chigwada, G. and Wilkie, C.A., Synergy between conventional phosphorus fire retardants and organically-modified clays can lead to fire retardancy of styrenics. *Polym. Degrad. Stab.*, 80, 551–557, 2003.
42. Wilkie, C.A., Chigwada, G., Gilman, J.W., Lyon, R.E., High-throughput techniques for the evaluation of fire retardancy. *J. Mater. Chem.*, 16, 2023–2029, 2006.
43. Gilman, J.W., Brassell, L.D., Davis, R.D., Shields, J.R., Harris, R.H., Wentz, D., Reitz, R., Becker, A., Chi, C., Evaluation of iron carbonate as a flame retardant for polyolefins, in: *Advances in the Flame Retardancy of Polymeric Materials:*

Current perspectives presented at FRPM'05, B. Schartel (Ed.), pp. 131–140, Books on Demand, Germany, 2007.

44. Davis, R.D., Lyon, R.E., Takemori, M.T., Eidelman, N., High throughput techniques for fire resistant materials development, in: *Fire Retardancy of Polymeric Materials*, Second edition, C.A. Wilkie and A.B. Morgan, (Eds.), pp. 421–451, CRC press, Boca Raton, FL, USA, 2010.

45. Lyon, R.E. and Walters, R.N., Pyrolysis combustion flow calorimetry. *J. Anal. Appl. Pyrolys.*, 71, 27–46, 2004.

46. Lyon, R.E., Walters, R.N., Stoliarov, S.I., Screening flame retardants for plastics using microscale combustion calorimetry. *Polym. Eng. Sci.*, 47, 1501–1510, 2007.

47. Schartel, B., Pawlowski, K.H., Lyon, R.E., Pyrolysis combustion flow calorimeter: a tool to assess flame retarded PC/ABS materials? *Thermochim. Act.*, 462, 1–14, 2007.

48. Rabe, S. and Schartel, B., The Rapid Mass Calorimeter: Understanding Reduced-scale Fire Test Results. *Polym. Test.*, 57, 165–174, 2017.

49. Rabe, S. and Schartel, B., The Rapid Mass Calorimeter: A route to High Throughput Fire Testing. *Fire Mater.*, 41, 834–847, 2017.

50. Lewin, M., Synergistic and catalytic effects in flame retardancy of polymeric materials—an overview. *J. Fire Sci.*, 17, 3–19, 1999.

51. Weil, E.D., Additivity, synergism and antagonism in flame retardancy, in: *Flame retardancy of polymeric materials*, W.C. Kuryla and A.J. Papa (Eds.), pp. 185–243, Marcel Dekker, New York, 1975.

52. Lewin, M., Synergism and catalysis in flame retardancy of polymers. *Polym. Adv. Technol.*, 12, 215–222, 2001.

53. Weil, E.D., Synergists, Adjuvants, and Antagonists in Flame-Retardant Systems, in: *Fire Retardancy of Polymeric Materials*, A.F. Grand and C.A. Wilkie (Ed.), pp. 115–145, Marcel Dekker, New York, 2000, chap 4.

54. Schartel, B., Wilkie, C.A., Camino, G., Recommendations on the scientific approach to polymer flame retardancy: Part 2 – concepts. *J. Fire Sci.*, 35, 3–20, 2017.

55. Kind, D.J. and Hull, T.R., A review of candidate fire retardants for polyisoprene. *Polym. Degrad. Stab.*, 97, 201–213, 2012.

56. E.D. Weil and S.V. Levchik, Flame Retardants for Plastics and Textiles, 2nd Edition, Hanser, Munich, 2016.

57. Kashiwagi, T., Shields, J.R., Harris, R.H., Davis, R.D., Flame-retardant mechanism of silica: Effects of resin molecular weight. *J. Appl. Polym, Sci.*, 87, 1541–1553, 2003.

58. Fink, J.K., Flame retardants, Scrivener Publishing and Wiley & Sons, Hoboken, 2020.

59. Khanna, Y.P. and Pearce, E.M., Synergism and flame retardancy, in: *Flame—retardant polymeric materials*, vol. 2, M. Lewin, S.M. Atlas, E.M. Pearce (Eds.), pp. 43–61, Plenum Press, New York, 1978.

60. Camino, B. and Camino, G., The chemical kinetics of the polymer combustion allows for inherent fire retardant synergism. *Polym. Degrad. Stab.*, 160, 142–147, 2019.

61. Schartel, B., Perret, B., Dittrich, B., Ciesielski, M., Krämer, J., Müller, P., Altstädt, V., Zang, L., Döring, M., Flame Retardancy of Polymers: The Role of Specific Reactions in the Condensed Phase. *Macromol. Mater. Engin.*, 301, 9–35, 2016.

62. Perret, B., Pawlowski, K.H., Schartel, B., Fire retardancy mechanisms of arylphosphates in polycarbonate (PC) and PC/acrylonitrile-butadiene-styrene. *J. Therm. Anal. Calorim.*, 97, 949–958, 2009.

63. Hollingbery, L.A., The fire retardant behaviour of huntite and hydromagnesite – A review. *Polym. Degrad. Stab.*, 95, 2213–2225, 2010.

64. Braun, U., Knoll, U., Neubert, D., Schartel, B., Fire Retarding Mechanisms of Phosphorus in Polybutylenterephthalat (PBT) in Combination with Melamine Cyanurat or Zinc Oxide, in: *Advances in the Flame Retardancy of Polymeric Materials: Current Perspectives Presented at FRPM'05*, B. Schartel (Ed.), pp. 35–49, Books on Demand, Norderstedt, 2007.

65. Liu, X., Hao, J.W., Gaan, S., Recent studies on the decomposition and strategies of smoke and toxicity suppression for polyurethane based materials. *RSC Adv.*, 6, 74742–74756, 2016.

66. Lyon, R.E., Plastics and rubbers, in: *Handbook of Building Materials for Fire Protection*, C.A. Harper (Ed.), pp. 3.1–3.51, McGraw-Hill, New York, 2004, Chap. 3.

67. van Krevelen, D.W., Some Basic Aspects of Flame Resistance of Polymeric Materials. *Polymer*, 16, 615–620, 1975.

68. Schartel, B. and Kebelmann, K., Fire Testing for the Development of Flame Retardant Polymeric Materials, in: *Flame Retardant Polymeric Materials A Handbook*, Y. Hu and X. Wang (Eds.), pp. 35–55, CRC press, Boca Raton, 2020, chap. 3.

69. Schartel, B., Bartholmai, M., Knoll, U., Some Comments on the Main Fire Retardancy Mechanisms in Polymer Nanocomposites. *Polym. Adv. Technol.*, 17, 772–777, 2006.

70. Bartholmai, M. and Schartel, B., Layered Silicate Polymer Nanocomposites: New Approach or Illusion for Fire Retardancy? Investigations on the Potential and on the Tasks Using a Model System. *Polym. Adv. Technol.*, 15, 355–364, 2004.

71. Morys, M., Illerhaus, B., Sturm, H., Schartel, B., Variation of Intumescent Coatings Revealing Different Modes of Action for Good Protection Performance. *Fire Technol.*, 53, 1569–1587, 2017.

72. Kunze, R., Schartel, B., Bartholmai, M., Neubert, D., Schriever, R., TG-MS and TG-FTIR Applied for an Unambiguous Thermal Analysis of Intumescent Coatings. *J. Therm. Anal. Calorim.*, 70, 897–909, 2002.

73. Vandersall, H.L., Intumescent coating system, their development and chemistry. *J. Fire Flamm.*, 2, 97–140, 1971.

74. Weil, E.D., Fire-Protective and Flame-Retardant Coatings - A State-of-the-Art Review. *J. Fire Sci.*, 29, 259–296, 2011.

75. Morys, M., Illerhaus, B., Sturm, H., Schartel, B., Revealing the inner secrets of intumescence: Advanced standard time temperature oven (STT Mufu+) - μ-computed tomography approach. *Fire Mater.*, 41, 927–939, 2017.

76. Müller, P., Morys, M., Sut, A., Jäger, C., Illerhaus, B., Schartel, B., Melamine poly(zincphosphate) as flame retardant in epoxy resin: decomposition pathways, molecular mechanisms and morphology of fire residues. *Polym. Degrad. Stabil.*, 130, 307–319, 2016.

77. Muller, M., Bourbigot, S., Duquesne, S., Klein, R., Giannini, G., Lindsay, C., Vlassenbroeck, J., Investigation of the synergy in intumescent polyurethane by 3D computed tomography. *Polym. Degrad. Stabil.*, 98, 1638–1647, 2013.

78. Horacek, H. and Pieh, S., The importance of intumescent systems for fire protection of plastic materials. *Polym. Int.*, 49, 1106–1114, 2000.

79. Yuan, B.H., Fan, A., Yang, M., Chen, X.F., Hu, Y., Bao, C.L., Jiang, S.H., Niu, Y., Zhang, Y., He, S., Dai, H.M., The effects of graphene on the flammability and fire behavior of intumescent flame retardant polypropylene composites at different flame scenarios. *Polym. Degrad. Stab.*, 143, 42–56, 2017.

80. Duquesne, S., Le Bras, M., Bourbigot, S., Delobel, R., Vezin, H., Camino, G., Eling, B., Lindsay, C., Roels, T., Expandable graphite: a fire retardant additive for polyurethane coatings. *Fire Mater.*, 27, 103–117, 2003.

81. Parlevliet, P. and Geistbeck, M., Investigations into lightweight solutions for epoxy composite fire property improvement. *Plast. Rubber Compos.*, 44, 104–10, 2015.

82. Vogelesang, L.B. and Vlot, A., Development of fibre metal laminates for advanced aerospace structures. *J. Mater. Process. Technol.*, 103, 1–5, 2000.

83. Sturm, H., Schartel, B., Weiß, A., Braun, U., SEM/EDX: Advanced Investigation of Structured Fire Residues and Residue Formation. *Polym. Test.*, 31, 606–619, 2012.

84. Horrocks, A.R. and Kandola, B.K., 9 - Flammability and fire resistance of composites, in: *Design and Manufacture of Textile Composites. Woodhead Publishing Series in Textiles*, A.C. Long (Ed.), Woodhead Publishing, Cambridge, UK, 2005.

85. Casu, A., Camino, G., De Giorgi, M., Flath, D., Laudi, A., Morone, V., Effect of glass fibres and fire retardant on the combustion behaviour of composites, glass fibres poly(butylene terephthalate). *Fire Mater.*, 22, 7–14, 1998.

86. Casu, A., Camino, G., Luda, M.P., Degiorgi, M., Mechanisms of Fire Retardantion in Glass-fiber Polymer Composites. *Makromol. Chem. Macromol. Symp.*, 74, 307–310, 1993.

87. Köppl, T., Brehme, S., Wolff-Fabris, F., Altstädt, V., Schartel, B., Döring, M., Structure-property Relationships of Halogen-free Flame Retarded Poly(butylene terephthalate) (PBT) and Glass Fibre Reinforced PBT. *J. Appl. Polym, Sci.*, 124, 9–18, 2012.

88. Perret, B., Schartel, B., Stöß, K., Ciesielski, M., Diederichs, J., Döring, M., Krämer, J., Altstädt, V., Novel DOPO-based flame retardants in high-performance carbon fibre epoxy composites for aviation. *Eur. Polym. J.*, 47, 1081–1089, 2011.

89. Eibl, S., Influence of carbon fibre orientation on reaction-to-fire properties of polymer matrix composites. *Fire Mater.*, 36, 309–324, 2012.

90. Eibl, S., and Swanson, D., Influence of out-of-plane fiber orientation on reaction-to-fire properties of carbon fiber reinforced polymer matrix composites. *Fire Mater.*, 42, 234–243, 2018.

91. Ding, Y., Swann, J.D., Sun, Q., Stoliarov, S.I., Kraemer, R.H., Development of a pyrolysis model for glass fiber reinforced polyamide 66 blended with red phosphorus: Relationship between flammability behavior and material composition. *Compos. Part B-Engin.*, 176, 107263, 2019.

92. Ding, Y. Stoliarov, S.I., Kraemer, R.H., Pyrolysis model development for a polymeric material containing multiple flame retardants: Relationship between heat release rate and material composition. *Combust. Flame*, 202, 43–57, 2019.

93. Feih, S., Mathys, Z., Gibson, A.G., Mouritz, A.P., Modelling the tension and compression strengths of polymer laminates in fire. *Compos. Sci. Technol.*, 67, 551–564, 2007.

94. Mouritz, A.P. and Gibson, A.G., *Fire Properties of Polymer Composite Materials. Book Series: Solid Mechanics and its Applications*, vol. 143, pp. 1–401, Springer, Dordrecht, 2006.

95. Gibson, A.G., Wright, P.N.H., Wu, Y.S., Mouritz, A.P., Mathys, Z., Gardiner, C.P., The integrity of polymer composites during and after fire. *J. Compos. Mater.*, 38, 1283–1307, 2004.

96. Gallo, E., Schartel, B., Acierno, D., Cimino, F., Russo, P., Tailoring the flame retardant and mechanical performances of natural fiber-reinforced biopolymer by multi-component laminate. *Compos. Part B: Engin.*, 44, 112–119, 2013.

97. Braun, U., Bahr, H., Sturm, H., Schartel, B., Flame Retardancy Mechanisms of Metal Phosphinates and Metal Phosphinates in Combination with Melamine Cyanurate in Glass-fiber Reinforced Poly(1,4-butylene terephthalate): The Influence of Metal Cation. *Polym. Adv. Technol.*, 19, 680–692, 2008.

98. Eibl, S., Influence of unwoven roving and woven fabric carbon fiber reinforcements on reaction-to-fire properties of polymer matrix composites. *Fire Mater.*, 44, 557–572, 2020.

99. Zhang, W.C., Li, X.M., Yang, R.J., Blowing-out effect in epoxy composites flame retarded by DOPO-POSS and its correlation with amide curing agents. *Polym. Degrad. Stab.*, 97, 1314–1324, 2012,.

100. Zhang, W.C., Li, X.M., Li, L.M., Yang, R.J., Study of the synergistic effect of silicon and phosphorus on the blowing-out effect of epoxy resin composites. *Polym. Degrad. Stab.*, 97, 1041–1048, 2012.

101. Zhang, W.C., Li, X.M., Yang, R.J., Blowing-out effect and temperature profile in condensed phase in flame retarding epoxy resins by phosphorus-containing oligomeric silsesquioxane. *Polym. Adv. Technol.*, 24, 951–961, 2013.

102. Braun, U., Balabanovich, A.I., Schartel, B., Knoll, U., Artner, J., Ciesielski, M., Döring, M., Perez, R., Sandler, J.K.W., Altstädt, V., Hoffmann, T., Pospiech, D., Influence of the Oxidation State of Phosphorus on the Decomposition and Fire Behaviour of Flame-Retarded Epoxy Resin Composites. *Polymer*, 47, 8495–8508, 2006.

103. Schartel, B., Balabanovich, A.I., Braun, U., Knoll, U., Artner, J., Ciesielski, M., Döring, M., Perez, R., Sandler, J.K.W., Altstädt, V., Hoffmann, T., Pospiech, D., Pyrolysis of Epoxy Resins and Fire Behaviour of Epoxy Resin Composites Flame-Retarded   with   9,10-dihydro-9-oxa-10-phosphaphenanthrene-10-oxide Additives. *J. Appl. Polym. Sci.*, 104, 2260–2269, 2007.

104. Perret, B., Schartel, B., Stöß, K., Ciesielski, M., Diederichs, J., Döring, M., Krämer, J., Altstädt, V., A New Halogen-Free Flame Retardant Based on 9,10-Dihydro-9-oxa-10-phosphaphenanthrene-10-oxide for Epoxy Resins and their Carbon Fiber Composites for the Automotive and Aviation Industries. *Macromol. Mater. Engin.*, 296, 14–30, 2011.

105. Ooms, P. and Lorenz, T., World Premiere: Flame Retardant Plastic BMC 0204 Sets New Standards, in: *25th Annual Conference on Recent Advances in Flame Retardancy of Polymeric Materials*, Stamford, US, May 19–21, 2014.

106. Lorenz, T., Lomix BMC 0204 - neues flammenfestes BMC für den weltweiten Einsatz, in: *2. AVK-Fachtagung – Flammschutz bei Composites-Anwendungen*, Frankfurt, Germany, December 10, 2013.

107. Duroplast SMC 0208 für die Pressverarbeitung, Produktmeldung 12.05.2016, article-253228, Kunststoffe.de, Hanser, München, 2016

108. Stoess, N., Neue Generation flammgeschützter SMC-Materilien, in: *2. AVK-Fachtagung – Flammschutz bei Composites-Anwendungen,* Frankfurt, Germany, December 10, 2013.

109. Sonnier, R., Taguet, A., Ferry, L., Lopez-Cuesta, J.M., Flame Retardancy of Natural Fibers Reinforced Composites, in: *Towards Bio-based Flame Retardant Polymers. Book Series: Springer Briefs in Molecular Science*, R. Sonnier, A. Taguet, L. Ferry, J.-M. LopezCuesta, (Eds.), pp. 73–98, 2018.

110. Chapple, S. and Anandjiwala, R., Flammability of Natural Fiber-reinforced Composites and Strategies for Fire Retardancy: A Review. *J. Thermoplast. Compos. Mater.*, 23, 871–893, 2010.

111. Schartel, B., Braun, U., Schwarz, U., Reinemann, S., Fire retardancy of polypropylene/flax blends. *Polymer*, 44, 6241–6250, 2003.

112. Kozłowski, R. and Władyka-Przybylak, M., Flammability and fire resistance of composites reinforced by natural fibers. *Polym. Adv. Technol.*, 19, 446–453, 2008.

113. Shah, A.U.R., Prabhakar, M.N., Song, J.-I., Current advances in the fire retardancy of natural fiber and bio-based composites–A review. *Int. J. Prec. Engin. Manufac.-Green Technol.*, 4, 247–262, 2017.

114. Hapuarachchi, T.D., Ren, G., Fan, M., Hogg, P.J., Peijs, T., Fire Retardancy of Natural Fibre Reinforced Sheet Moulding Compound. *Appl. Compos. Mater.*, 14, 251–264, 2007.

115. Zhang, L., Li, Z., Pan, Y.-T., Pérez Yáñez, A., Hu, S., Zhang, X.-Q., Wang, R., Wang, D.-Y., Polydopamine induced natural fiber surface functionalization: a way towards flame retardancy of flax/poly(lactic acid) biocomposites. *Compos. Part B: Engin.*, 154, 56–63, 2018.

116. Schirp, A. and Su, S., Effectiveness of pre-treated wood particles and halogen-free flame retardants used in wood-plastic composites. *Polym. Degrad. Stab.*, 126, 81–92, 2016.

117. Hamalainen, K. and Karki, T., Effects of wood flour modification on the fire retardancy of wood-plastic composites. *Eur. J. Wood Wood Prod.*, 72, 703–711, 2014.

118. Gallo, E., Sánchez-Olivares, G., Schartel., B., Flame Retarded Starch-based Biocomposites - Aluminum Hydroxide-Coconut Fiber Synergy. *Polimery*, 58, 395–402, 2013.

119. Sain, M., Park, S.H., Suhara, F., Law., S., Flame retardant and mechanical properties of natural fibre–PP composites containing magnesium hydroxide. *Polym. Degrad. Stab.*, 83, 363–367, 2004.

120. Sanchez-Olivares, G., Sanchez-Solis, A., Calderas, F., Alongi, J., Keratin fibres derived from tannery industry wastes for flame retarded PLA composites. *Polym. Degrad. Stab.*, 140, 42–54, 2017.

121. Rabe, S., Sánchez-Olivares, G., Pérez-Chávez, R., Schartel., B., Natural Keratin and Coconut Fibres from Industrial Wastes in Flame Retarded Thermoplastic Starch Biocomposites. *Materials*, 12, 344, 2019.

122. Batistella, M.A., Sonnier, R., Otazaghine, B., Petter, C.O., Lopez-Cuesta., J.-M., Interactions between kaolinite and phosphinate-based flame retardant in Polyamide 6. *Appl. Clay Sci.*, 157, 248256, 2018.

123. Schartel, B., Richter, K.H., Böhning, M., Synergistic Use of Talc in Halogen-free Flame Retarded Polycarbonate/Acrylonitrile-Butadiene-Styrene Blends, in: *Fire and Polymers VI: New Advances in Flame Retardant Chemistry and Science. ACS symposium series*, vol. 1118, A.B. Morgan, C.A. Wilkie, G.L. Nelson (Eds.), pp. 15–36, ACS, Washington, 2012, chap. 2.

124. Levchik, S.V., Levchik, G.F., Camino, G., Costa, L., Mechanism of Action of Phosphorus-based Flame Retardants in Nylon-6. 2. Ammonium Polyphosphate Talc. *J. Fire Sci.*, 13, 43–58, 1995.

125. Laoutid, F., Ferry, L., Lopez-Cuesta, J.-M., Crespy, A., Flame-retardant action of red phosphorus/magnesium oxide and red phosphorus/iron oxide compositions in recycled PET. *Fire Mater.*, 30, 343–358, 2006.

126. Longerey, M., Lopez-Cuesta, J.-M., Gaudon, P., Crespy., A., Talcs and brominated trimethylphenyl indane/Sb2O3 blend in a PP-PE copolymer. *Polym. Degrad. Stab.*, 64, 489–496, 1999.

127. Durin-France, A., Ferry, L., Lopez-Cuesta, J.-M., Crespy, A., Magnesium hydroxide/zinc borate/talc compositions as flame-retardants in EVA copolymer. *Polym. Int.*, 49, 1101–1105, 2000.

128. Wang, X., Kalali, E.N., Wan, J.T., Wang, D.Y., Carbon-family materials for flame retardant polymeric materials. *Prog. Polym. Sci.*, 69, 22–46, 2017.

129. Dittrich, B., Wartig, K.-A., Hofmann, D., Mülhaupt, R., Schartel, B., Flame retardancy through carbon nanomaterials: Carbon black, multiwall nanotubes, expanded graphite, multi-layer graphene and graphene in polypropylene. *Polym. Degrad. Stab.*, 98, 1495–1505, 2013.

130. Liu, S., Chevali, V.S., Xu, Z.G., Hui, D., Wang, H., A review of extending performance of epoxy resins using carbon nanomaterials. *Compos. Part B-Engin.*, 136, 197–214, 2018.

131. Wen, X., Wang, Y.J., Gong, J., Liu, J., Tian, N.N., Wang, Y.H., Jiang, Z.W., Qiu, J., Tang, T., Thermal and flammability properties of polypropylene/carbon black nanocomposites. *Polym. Degrad. Stab.*, 97, 793–801, 2012.

132. Beyer, G., Short communication: Carbon nanotubes as flame retardants for polymers. *Fire Mater.*, 26, 291–293, 2002.

133. Kashiwagi, T., Grulke, E., Hilding, J., Harris, R., Awad, W., Douglas, J., Thermal degradation and flammability properties of poly(propylene)/carbon nanotube composites. *Macromol. Rapid Commun.*, 23, 761–765, 2002.

134. Kashiwagi, T., Grulke, E., Hilding, J., Groth, K., Harris, R., Butler, K., Shields, J., Kharchenko, S., Douglas, J., Thermal and flammability properties of polypropylene/carbon nanotube nanocomposites. *Polymer*, 45, 4227–4239, 2004.

135. Schartel, B., Pötschke, P., Knoll, U., Abdel-Goad, M., Fire behaviour of polyamide 6/multiwall carbon nanotube nanocomposites. *Eur. Polym. J.*, 41, 1061–1070, 2005.

136. Kashiwagi, T., Du, F.M., Douglas, J.F., Winey, K.I., Harris, R.H., Shields, J.R., Nanoparticle networks reduce the flammability of polymer nanocomposites. *Nat. Mater.*, 4, 928–933, 2005.

137. Wang, X., Song, L., Yang, H.Y., Lu, H.D., Hu, Y., Synergistic Effect of Graphene on Antidripping and Fire Resistance of Intumescent Flame Retardant Poly(butylene succinate) Composites. *Ind. Engin. Chem. Res.*, 50, 5376–5383, 2011.

138. Yu, B., Shi, Y.Q., Yuan, B.H., Qiu, S.L., Xing, W.Y., Hu, W.Z., Song, L., Lo, S.M., Hu., Y., Enhanced thermal and flame retardant properties of flame-retardant-wrapped graphene/epoxy resin nanocomposites. *J. Mater. Chem. A*, 3, 8034–8044, 2015.

139. Xu, W.Z., Liu, L., Zhang, B.L., Hu, Y., Xu, B.L., Effect of Molybdenum Trioxide-Loaded Graphene and Cuprous Oxide-Loaded Graphene on Flame Retardancy and Smoke Suppression of Polyurethane Elastomer. *Ind. Engin. Chem. Res.*, 55, 4930–4941, 2016.

140. Yu, B., Wang, X., Qian, X.D., Xing, W.Y., Yang, H.Y., Ma, L.Y., Lin, Y., Jiang, S.H., Song, L., Hu, Y., Lo, S.M., Functionalized graphene oxide/phosphoramide oligomer hybrids flame retardant prepared via in situ polymerization for improving the fire safety of polypropylene. *RSC Adv.*, 4, 31782–31794, 2014.

141. Zirnstein, B., Tabaka, W., Frasca, D., Schulze, D., Schartel, B., Graphene/Hydrogenated Acrylonitrile-butadiene Rubber Nanocomposites: Dispersion, Curing, Mechanical Reinforcement, Multifunctional Filler. *Polym. Test.*, 66, 268–279, 2018.

142. Frasca, D., Schulze, D., Böhning, M., Krafft, B., Schartel, B., Multilayer Graphene Chlorine Isobutyl Isoprene Rubber Nanocomposites: Influence of the Multilayer Graphene Concentration on Physical and Flame-Retardant Properties. *Rubber Chem. Technol.*, 89, 316–334, 2016.

143. Ma, H.Y., Tong, L.F., Xu, Z.B., Fang, Z.P., Functionalizing carbon nanotubes by grafting on intumescent flame retardant: Nanocomposite synthesis, morphology, rheology, and flammability. *Adv. Funct. Mater.*, 18, 414–421, 2008.

144. Dittrich, B., Wartig, K.-A., Mülhaupt, R., Schartel, B., Flame-retardancy Properties of Intumescent Ammonium Poly(phosphate) and Mineral Filler Magnesium Hydroxide in Combination with Graphene. *Polymers*, 6, 2875–2895, 2014.

145. Dittrich, B., Wartig, K.-A., Hofmann, D., Mülhaupt, R., Schartel., B., Carbon black, multiwall carbon nanotubes, expanded graphite and functionalized graphene flame retarded polypropylene nanocomposites. *Polym. Adv. Technol.*, 24, 916–926, 2013.

146. Zhang, T., Yan, H.Q., Peng, M., Wang, L.L., Ding, H.L., Fang, Z.P., Construction of flame retardant nanocoating on ramie fabric via layer-by-layer assembly of carbon nanotube and ammonium polyphosphate. *Nanoscale*, 5, 3013–3021, 2013.

147. Gilman, J.W., Flammability and thermal stability studies of polymer layered-silicate (clay) nanocomposites. *Appl. Clay Sci.*, 15, 31–49, 1999.

148. Zhang, W.J., Camino, G., Yang, R.J., Polymer/polyhedral oligomeric silsesquioxane (POSS) nanocomposites: An overview of fire retardance. *Prog. Polym. Sci.*, 67, 77–125, 2017.

149. Bourbigot, S., Turf, T., Bellayer, S., Duquesne., S., Polyhedral oligomeric silsesquioxane as flame retardant for thermoplastic polyurethane. *Polym. Degrad. Stab.*, 94, 1230–1237, 2009.

150. Fina, A., Tabuani, D., Frache, A., Camino, G., Polypropylene-polyhedral oligomeric silsesquioxanes (POSS) nanocomposites. *Polymer*, 46, 7855–7866, 2005.

151. Zammarano, M., Franceschi, M., Bellayer, S., Gilman, J.W., Meriani, S., Preparation and flame resistance properties of revolutionary self-extinguishing epoxy nanocomposites based on layered double hydroxides. *Polymer*, 46, 9314–9328, 2005.

152. Alongi, J., Investigation on Flame Retardancy of Poly(ethylene terephthalate) for Plastics and Textiles by Combination of an Organo-modified Sepiolite and Zn Phosphinate. *Fibers Polym.*, 12, 166–173, 2011.

153. Du, M.L., Guo, B.C., Jia, D.M., Thermal stability and flame retardant effects of halloysite nanotubes on poly(propylene). *Eur. Polym. J.*, 42, 1362–1369, 2006.

154. Marney, D.C.O., Russell, L.J., Wu, D.Y., Nguyen, T., Cramm, D., Rigopoulos, N., Wright, N., Greaves, M., The suitability of halloysite nanotubes as a fire retardant for nylon 6. *Polym. Degrad. Stab.*, 93, 1971–1978, 2008.

155. Yu, B., Xing, W.Y., Guo, W.W., Qiu, S.L., Wang, X., Lo, S.M., Hu, Y., Thermal exfoliation of hexagonal boron nitride for effective enhancements on thermal stability, flame retardancy and smoke suppression of epoxy resin nanocomposites via sol-gel process. *J. Mater. Chem. A*, 4, 7330–7340, 2016.

156. Yang, D.D., Hu, Y., Song, L., Nie, S.B., He, S.Q., Cai, Y.B., Catalyzing carbonization function of alpha-ZrP based intumescent fire retardant polypropylene nanocomposites. *Polym. Degrad. Stab.*, 93, 2014–2018, 2008.

157. Pawlowski, K.H. and Schartel, B., Flame Retardancy Mechanisms of Aryl Phosphates in Combination with Boehmite in Bisphenol A Polycarbonate/Acrylonitrile-Butadiene-Styrene Blends. *Polym. Degrad. Stab.*, 93, 657–667, 2008.

158. Gilman, J.W., Flame retardant mechanism of polymer-clay nanocomposites, in: *Flame retardant polymer nanocomposites*, A.B. Morgan and C.A. Wilkie (Eds.), pp. 67–87, John Wiley and Sons, Hoboken, 2007.

159. Kashiwagi, T., Harris, R.H., Zhang, X., Briber, R.M., Cipriano, B.H., Raghavan, S.R., Awad, W.H., Shields, J.R., Flame retardant mechanism of polyamide 6-clay nanocomposites. *Polymer*, 45, 881–891, 2004.

160. Schartel, B., Weiß, A., Sturm, H., Kleemeier, M., Hartwig, A., Vogt, C., Fischer, R.X., Layered Silicate Epoxy Nanocomposites: Formation of the Inorganic-carbonaceous Fire Protection Layer. *Polym. Adv. Technol.*, 22, 1581–1592, 2011.

161. Gilman, J.W., Jackson, C.L., Morgan, A.B., Harris, R., Manias, E., Giannelis, E.P., Wuthenow, M., Hilton, D., Phillips, S.H., Flammability properties of polymer - Layered-silicate nanocomposites. Polypropylene and polystyrene nanocomposites. *Chem. Mater.*, 12, 1866–1873, 2000.

162. Schartel, B., Considerations Regarding Specific Impacts of the Principal Fire Retardancy Mechanisms in Nanocomposites, in: *Flame Retardant Polymer Nanocomposites*, A.B. Morgan and C.A. Wilkie (Eds.), pp. 107–129, John Wiley & Sons, Hoboken, 2007, chap 5.

163. Bourbigot, S., Le Bras, M., Dabrowski, F., Gilman, J.W., Kashiwagi, T., PA-6 clay nanocomposite hybrid as char forming agent in intumescent formulations. *Fire Mater.*, 24, 201–208, 2000.

164. Pappalardo, S., Russo, P., Acierno, D., Rabe, S., Schartel, B., The synergistic effect of organically modified sepiolite in intumescent flame retardant polypropylene. *Eur. Polym. J.*, 76, 196–207, 2016.

165. Gilman, J.W., Kashiwagi, T., Lichtenhan, J.D., Nanocomposites: A revolutionary new flame retardant approach. *SAMPE J.*, 33, 40–46, 1997.

166. He, W.T., Song, P.A., Yu, B., Fang, Z.P., Wang, H., Flame retardant polymeric nanocomposites through the combination of nanomaterials and conventional flame retardants. *Prog. Mater. Sci.*, 114, 100687, 2020.

167. Liu, Y., Gao, Y.S., Wang, Q., Lin, W.R., The synergistic effect of layered double hydroxides with other flame retardant additives for polymer nanocomposites: a critical review. *Dalton Transact.*, 47, 14827–14840, 2018.

168. Lu, H.D., Song, L., Hu, Y.A., A review on flame retardant technology in China. Part II: flame retardant polymeric nanocomposites and coatings. *Polym. Adv. Technol.*, 22, 379–394, 2011.

169. Lopez-Cuesta, J.-M. and Laoutid, F., Multicomponent FR Systems: Polymer Nanocomposites Combined with Additional Materials, in: *Fire Retardancy of Polymeric Materials*, 2nd Edition, C.A. Wilkie and A.B. Morgan (Eds.), pp. 301–328, CRC Press, Boca Raton, 2010, chap. 12.

170. Bourbigot, S., Duquesne, S., Fontaine, G., Bellayer, S., Turf, T., Samyn, F., Characterization and reaction to fire of polymer nanocomposites with and without conventional flame retardants. *Mol. Cryst. Liquid Cryst.*, 486, 1367–1381, 2008.

171. Gilman, J.W. and Kashiwagi, T., Polymer-Layered Silicaste Nanocomposites with conventional Flame Retardants, in: *Polymer-Clay Nanocomposites*, T.J. Pinnavaia and G.W. Beall (Eds.), pp. 193–206, John Wiley & Sons, Chichestrer, 2000, chap. 10.

172. Schartel, B., Fire Retardancy Based on Polymer Layered Silicate Nanocomposites, in: *Advances in Polymeric Nanocomposite*, M. Okamoto (Ed.), pp. 242–257, CMC Publishing, Osaka, 2004, chap 2–3.

173. M. Bödiger, T. Eckel, D. Wittmann, H. Alberts, German Patent DE 19530200 A1, 1997.

174. Schartel, B., Knoll, U., Hartwig, A., Pütz., D., Phosphonium-Modified Layered Silicate Epoxy Resins Nanocomposites and Their Combinations with ATH and Organo-Phosphorus Fire Retardants. *Polym. Adv. Technol.*, 17, 281–293, 2006.

175. Schartel, B., Weiß, A., Mohr, F., Kleemeier, M., Hartwig, A., Braun, U., Flame Retarded Epoxy Resins by Adding Layered Silicate in Combination with the Conventional Protection-Layer-Building Flame Retardants Melamine Borate and Ammonium Polyphosphate. *J. Appl. Polym, Sci.*, 118, 1134–1143, 2010.

176. Chigwada, G., Jash, P., Jiang, D.D., Wilkie, C.A., Fire retardancy of vinyl ester nanocomposites: Synergy with phosphorus-based fire retardants. *Polym. Degrad. Stab.*, 89, 85–100, 2005.

177. M. Klatt, S. Grutke, T. Heitz, V. Rauschenberger, T. Plesnivy, P. Wolf, J.R. Wünsch, M. Fischer, German Patent. 19705998 A1, 1998.

178. Beyer, G., Flame retardant properties of EVA-nanocomposites and improvements by combination of nanofillers with aluminium trihydrate. *Fire Mater.*, 25, 193–197, 2001.

179. N. Schall, T. Engelhardt, H. Simmler-Hübenthal, G. Beyer, German Patent DE 19921 472 A1, 2000.

180. Matzen, M., Marti, J., Oñate, E., Idelsohn, S., Schartel, B., Advanced Experiments and Particle Finite Element Modelling on Dripping V-0 Polypropylene, in: *Conference Proceedings, Fire Mater. 2017, 15th International Conference*, pp. 57–62, Interscience Communications, London, 2017.

181. Buczko, A., Stelzig, T., Bommer, L., Rentsch, D., Heneczkowski, M., Gaan, S., Bridged DOPO derivatives as flame retardants for PA6. *Polym. Degrad. Stab.*, 107, 158–165, 2014.

182. Beach, M.W., Rondan, N.G., Froese, R.D., Gerhart, B.B., Green, J.G., Stobby, B.G., Shmakov, A.G., Shvartsberg, V.M., Korobeinichev, O.P., Studies of degradation enhancement of polystyrene by flame retardant additives. *Polym. Degrad. Stab.*, 93, 1664–1673, 2008.

183. Liu, Y. and Wang, Q., The investigation on the flame retardancy mechanism of nitrogen flame retardant melamine cyanurate in polyamide 6. *J. Polym. Res.*, 16, 583–589, 2009.

184. Casu, A., Camino, G., De Giorgi, M., Flath, D., Morone, V., Zenoni, R., Fire-retardant mechanistic aspects of melamine cyanurate in polyamide copolymer. *Polym. Degrad. Stab.*, 58, 297–302, 1997.

185. Eichhorn, J., Synergism of free radical initiators with self-extinguishing additives in vinyl aromatic polymers. *J. Appl. Polym. Sci.*, 8, 2497–2524, 1964.

186. Srinivasan, R., Gupta, A., Horsey, D., A revolutionary UV stable flame retardant system for polyolefins. *Proceedings of the: International Conference on Additives for Polyolefins*, pp. 63–83, Houston, TX, February 23-25, 1998.

187. Ureyen, M.E., Kaynak, E., Yuksel, G., Flame-retardant effects of cyclic phosphonate with HALS and fumed silica in polypropylene. *J. Appl. Polym, Sci.*, 137, 48308, 2020.

188. Aubert, M., Nicolas, R.C., Pawelec, W., Wilén, C.E., Roth, M., Pfaendner., R., Azoalkanes-novel flame retardants and their structure-property relationship. *Polym. Adv. Technol.*, 22, 1529–1538, 2011.

189. Nicolas, R.C., Wilén, C.E., Roth, M., Pfaendner, R., King, R.E., Azoalkanes: A novel class of flame retardants. *Macromol. Rapid Commun.*, 27, 976–981, 2006.

190. Tirri, T., Aubert, M., Aziz, H., Brusentsev, Y., Pawelec, W., Wilén, C.-E., Sulfenamides in synergistic combination with halogen free flame retardants in polypropylene. *Polym. Degrad. Stab.*, 164, 75–89, 2019.

191. Spiess, B., Metzsch-Zilligen, E., Pfaendner, R., A New Class of Oxyimides: Oxyimide Ethers and their Use as Flame Retardants. *Macromol. Mater. Engin.*, 306, 2000650, 2021.

192. Zhang, N., Zhang, J., Yan, H., Guo, X., Sun, Q., Guo, R., A novel organic-inorganic hybrid K-HBPE@APP performing excellent flame retardancy and smoke suppression for polypropylene. *J. Hazard. Mater.*, 373, 856–86, 2019.

193. Ke, C.H., Li, J., Fang, K.Y., Zhu, Q.L., Zhu, J., Yan, Q., Wang, Y.-Z., Synergistic effect between a novel hyperbranched charring agent and ammonium polyphosphate on the flame retardant and anti-dripping properties of polylactide. *Polym. Degrad. Stab.*, 95, 763–770, 2010.

194. Fu, T., Wang, X.L., Wang, Y.-Z., Flame-responsive aryl ether nitrile structure towards multiple fire hazards suppression of thermoplastic polyester. *J. Hazard. Mater.*, 403, 123714, 2021.

195. Jing, X.K., Wang, X.S., Guo, D.M., Zhang, Y., Zhai, F.Y., Wang, X.L., Chen, L., Wang, Y.-Z., The high-temperature self-crosslinking contribution of azobenzene groups to the flame retardance and anti-dripping of copolyesters. *J. Mater. Chem. A*, 1, 9264–9272, 2013.

196. Zhao, H.B., Chen, L., Yang, J.C., Ge, X.G., Wang, Y.-Z., A novel flame-retardant-free copolyester: cross-linking towards self extinguishing and non-dripping. *J. Mater. Chem.*, 22, 19849–19857, 2012.

197. Zhan, J., Wang, L., Hong, N., Hu, W., Wang, J., Song, L., Hu, Y., Flame-retardant and anti-dripping properties of intumescent flame-retardant polylactide with different synergists. *Polym.-Plast. Technol. Engin.*, 53, 387–394, 2014.

198. Pawlowski, K.H. and Schartel, B., Flame Retardancy Mechanisms of Triphenyl Phosphate, Resorcinol Bis(diphenyl phosphate) and Bisphenol A Bis(diphenyl phosphate) in Polycarbonate/ Acrylonitrile-Butadiene-Styrene Blends. *Polym, Int.*, 56, 1404–1414, 2007.

199. Kempel, F., Schartel, B., Marti, J.M., Butler, K.M., Rossi, R., Idelsohn, S.R., Oñate, E., Hofmann, A., Modelling the Vertical UL 94 test: Competition and Collaboration between Melt Dripping, Gasification and Combustion. *Fire Mater.*, 39, 570–584, 2015.

200. Camino, G. and Delobel, R., Intumescence, in: *Fire Retardancy of Polymeric Materials*, A.F. Grand and C.A. Wilkie (Eds.), pp. 217–243, Marcel Dekker, New York, 2000, chap 7.

201. Harashina, H., Tajima, Y., Itoh, T., Synergistic effect of red phosphorus, novolac and melamine ternary combination on flame retardancy of poly(oxymethylene). *Polym. Degrad. Stab.*, 91, 1996–2002, 2006.

202. Zhong, Y.H., Wu, W., Wu, R., Luo, Q.L., Wang, Z., The flame retarding mechanism of the novolac as char agent with the fire retardant containing phosphorous-nitrogen in thermoplastic poly(ether ester) elastomer system. *Polym. Degrad. Stab.*, 105, 166–177, 2014.

203. Le Bras, M., Bourbigot, S., Felix, E., Pouille, F., Siat, C., Traisnel, M., Characterization of a polyamide-6-based intumescent additive for thermoplastic formulations. *Polymer*, 41, 5283–5296, 2000.

204. Chen, T., Xiao, X., Wang, J.K., Guo, N., Fire thermal and mechanical properties of TPE composites with systems containing piperazine pyrophosphate

(PAPP), melamine phosphate (MPP) and titanium dioxide (TiO2). *Plast. Rubber Compos.*, 48, 149–159, 2019.

205. Hu, Z., Zhong, Z.Q., Gong, X.D., Flame retardancy thermal properties, and combustion behaviors of intumescent flame-retardant polypropylene containing (poly) and melamine polyphosphate. *Polym. Adv. Technol.*, 31, 2701–2710, 2020.

206. Feng, C.M., Zhang, Y., Liu, S.W., Chi, Z.G., Xu, J.R., Synthesis of Novel Triazine Charring Agent and Its Effect in Intumescent Flame-Retardant Polypropylene. *J. Appl. Polym, Sci.*, 123, 3208–3216, 2012.

207. Zhu, C.J., He, M.S., Liu, Y., Cui, J.G., Tai, Q.L., Song, L., Hu., Y., Synthesis and application of a mono-component intumescent flame retardant for polypropylene. *Polym. Degrad. Stab.*, 151, 144–151, 2018.

208. Li, B. and Xu, M.J., Effect of a novel charring-foaming agent on flame retardancy and thermal degradation of intumescent flame retardant polypropylene. *Polym. Degrad. Stab.*, 91, 1380–1386, 2006.

209. Hu, X.P., Li, Y.L., Wang, Y.-Z., Synergistic effect of the charring agent on the thermal and flame retardant properties of polyethylene. *Macromol. Mater. Engin.*, 289, 208–212, 2004.

210. Yang, K., Xu, M.J., Li, B., Synthesis of N-ethyl triazine-piperazine copolymer and flame retardancy and water resistance of intumescent flame retardant polypropylene. *Polym. Degrad. Stab.*, 98, 1397–1406, 2013.

211. Liu, H.C., Li, S., Zhang, Z.Y., Li, B., Xu, M.J., An efficient, and convenient strategy toward fire safety and water resistance of polypropylene composites through design and synthesis of a novel mono-component intumescent flame retardant. *Polym. Adv. Technol.*, 30, 1543–1554, 2019.

212. De Chirico, A., Armanini, M., Chini, P., Cioccolo, G., Provasoli, F., Audisio, G., Flame retardants for polypropylene based on lignin. *Polym. Degrad. Stab.*, 79, 139–145, 2003.

213. Cayla, A., Rault, F., Giraud, S., Salaun, F., Fierro, V., Celzard, A., PLA with Intumescent System Containing Lignin and Ammonium Polyphosphate for Flame Retardant Textile. *Polymers*, 8, 331, 2016.

214. Mandlekar, N., Malucelli, G., Cayla, A., Rault, F., Giraud, S., Salaun, F., Guan, J.P., Fire retardant action of zinc phosphinate and polyamide 11 blend containing lignin as a carbon source. *Polym. Degrad. Stab.*, 153, 63–74, 2018.

215. Luda, M.P. and Zanetti, M., Cyclodextrins and Cyclodextrin Derivatives as Green Char Promoters in Flame Retardants Formulations for Polymeric Materials. A Review. *Polymers*, 11, 664, 2019.

216. Alongi, J., Poskovic, M., Frache, A., Trotta., F., Novel flame retardants containing cyclodextrin nanosponges and phosphorus compounds to enhance EVA combustion properties. *Polym. Degrad. Stab.*, 95, 2093–2100, 2010.

217. Li, L., Liu, X.L., Shao, X.M., Jiang, L.C., Huang, K., Zhao, S., Synergistic effects of a highly effective intumescent flame retardant based on tannic acid functionalized graphene on the flame retardancy and smoke suppression

properties of natural rubber. *Compos. Part A: Appl. Sci. Manufact.*, 129, 105715, 2020.

218. Kong, F.-B., He, Q.-L., Peng, W., Nie, S.-B., Dong, X., Yang, J.-N., Eco-friendly flame retardant poly(lactic acid) composites based on banana peel powders and phytic acid: flame retardancy and thermal property. *J. Polym. Res.*, 27, 204, 2020.

219. Gómez-Fernández, S., Günther, M., Schartel, B., Corcuera, M.A., Eceiza, A., Impact of the combined use of layered double hydroxides, lignin and phosphorous polyol on the fire behavior of flexible polyurethane foams. *Ind. Crops Prod.*, 125, 346–359, 2018.

220. Weil, E.D. and Choudhary, V., Flame-retarding Plastics and Elastomers with Melamine. *J. Fire Sci.*, 13, 104–126, 1995.

221. Liu, L.B., Xu, Y., He, Y.T., Xu, M.J., Shi, Z.X., Hu, H.C., Yang, Z.C., Li, B., An effective mono-component intumescent flame retardant for the enhancement of water resistance and fire safety of thermoplastic polyurethane composites Polym. *Polym. Degrad. Stab.*, 167, 146–156, 2019.

222. Levchik, S.V., Levchik, G.F., Balabanovich, A.I., Camino, G., Costa, L., Mechanistic study of combustion performance and thermal decomposition behaviour of nylon 6 with added halogen-free fire retardants. *Polym. Degrad. Stab.*, 54, 217–222, 1996.

223. Turski Silva Diniz, A., Huth, C., Schartel, B., Dripping and decomposition under fire: Melamine cyanurate vs. glass fibres in polyamide 6. *Polym. Degrad. Stab.*, 171, 109048, 2020.

224. Levchik, S.V. and Weil, E.D., Overview of the recent literature on flame retardancy and smoke suppression in PVC. *Polym. Adv. Technol.*, 16, 707–716, 2005.

225. Green, J., Mechanisms for flame retardancy and smoke suppression - A review. *J. Fire Sci.*, 14, 426–442, 1996.

226. Kim, S., Flame retardancy and smoke suppression of magnesium hydroxide filled polyethylene. *J. Polym. Sci. Part B - Polym. Phys.*, 41, 936 944, 2003.

227. Hull, T.R., Wills, C.L., Artingstall, T., Price, D., Milnes, G.J., Mechanisms of smoke and CO suppression from EVA composites, in: *Fire Retardancy of Polymers: New Applications of Mineral Fillers*, M. Le Bras, C.A. Wilkie, S. Bourbigot, S. Duquense, C. Jama (Eds.), pp. 372–385, RSC, Cambridge, 2005, chap 28.

228. Hirschler, M.M., Reduction of Smoke Formation from and Flammability of Thermoplastic Polymers by Metal-oxides. *Polymer*, 25, 405–411, 1984.

229. Levchik, G.F., Vorobyova, S.A., Gorbarenko, V.V., Levchik, S.V., Weil, E.D., Some Mechanistic Aspects of the Fire Retardant Action of Red Phosphorus in Aliphatic Nylons. *J. Fire Sci.*, 18, 172–182, 2000.

230. Sheng, H.B., Zhang, Y., Ma, C., Yang, L., Qiu, S.L., Wang, B.B., Hu, Y., Influence of zinc stannate and graphene hybrids on reducing the toxic gases and fire hazards during epoxy resin combustion. *Polym. Adv. Technol.*, 30, 666–674, 2019.

231. Zhou, K.Q., Yang, W., Tang, G., Wang, B.B., Jiang, S.H., Hu, Y., Gui, Z., Comparative study on the thermal stability, flame retardancy and smoke suppression properties of polystyrene composites containing molybdenum disulfide and graphene. *RSC Adv.*, 3, 25030–25040, 2013.

232. Price, D., Liu, Y., Milnes, G.J., Hull, T.R., Kandola, B.K., Horrocks, A.R., An investigation into the mechanism of flame retardancy and smoke suppression by melamine in flexible polyurethane foam. *Fire Mater.*, 26, 201–206, .

233. Chen, X.L., Jiang, Y.F., Jiao, C.M., Smoke suppression properties of ferrite yellow on flame retardant thermoplastic polyurethane based on ammonium polyphosphate. *J. Hazard. Mater.*, 266, 114–121, 2014.

234. Xu, Q.W., Zhai, H.M., Wang, G.J., Mechanism of smoke suppression by melamine in rigid polyurethane foam. *Fire Mater.*, 39, 271–282, 2015.

235. Shi, Y.Q., Qian, X.D., Zhou, K.Q., Tang, Q.B., Jiang, S.H., Wang, B.B., Wang, B., Yu, B., Hu, Y., Yuen, R.K.K., CuO/Graphene Nanohybrids: Preparation and Enhancement on Thermal Stability and Smoke Suppression of Polypropylene. *Ind. Engin. Chem. Res.*, 52, 13654–13660, 2013.

236. Rao, W.H., Liao, W., Wang, H., Zhao, H.B., Wang, Y.Z., Flame-retardant and smoke-suppressant flexible polyurethane foams based on reactive phosphorus-containing polyol and expandable graphite. *J. Hazard Mater.*, 360, 651–660, 2018.

237. Costa, L. and Camino, G., Thermal-behaviour of Melamine. *J. Therm. Anal.*, 34, 423–429, 1988.

238. Costa, L., Camino, G., Dicortemiglia, M.P.L., Mechanism of Thermal-degradation of fire-retardant Melamine Salts, in: *Fire and Polymers: Hazards Identification and Prevention. Book Series: ACS Symposium Series*, vol. 425, G.L. Nelson (Ed.), pp. 211–238, ACS, Washington, 1990.

239. Braun, U. and Schartel, B., Flame Retardancy Mechanisms of Aluminium Phosphinate in Combination with Melamine Cyanurate in Glass-Fibre Reinforced Poly (1,4-Butylene Terephthalate). *Macromol. Mater. Engin.*, 293, 206–217, 2008.

240. Gaan, S., Sun, G., Hutches, K., Engelhard, M.H., Effect of nitrogen additives on flame retardant action of tributyl phosphate: Phosphorus-nitrogen synergism. *Polym. Degrad. Stabil.*, 93, 99–108, 2008.

241. Neisius, M., Liang, S.Y., Mispreuve, H., Gaan, S., Phosphoramidate-Containing Flame-Retardant Flexible Polyurethane Foams. *Ind. Engin. Chem. Res.*, 52, 9752–9762, 2013.

242. Liang, S.Y., Neisius, M., Mispreuve, H., Naescher, R., Gaan, S., Flame retardancy and thermal decomposition of flexible polyurethane foams: Structural influence of organophosphorus compounds. *Polym. Degrad. Stab.*, 97, 2428–2440, 2012.

243. Neisius, N.M., Lutz, M., Rentsch, D., Hemberger, P., Gaan, S., Synthesis of DOPO-Based Phosphonamidates and their Thermal Properties. *Ind. Engin. Chem. Res.*, 53, 2889–2896, 2014.

244. Markwart, J.C., Battig, A., Zimmermann, L., Wagner, M., Fischer, J., Schartel, B., Wurm, F.R., Systematically Controlled Decomposition Mechanism in Phosphorus Flame Retardants by Precise Molecular Architecture: P–O vs P–N. *ACS Appl. Polym. Mater.*, 1, 1118–1128, 2019.

245. Nazir, R. and Gaan, S., Recent developments in P(O/S)-N containing flame retardants. *J. Appl. Polym, Sci.*, 137, 47910, 2020.

246. Levchik, S.V. and Weil, E.D., Thermal decomposition, combustion and flame-retardancy of epoxy resins - a review of the recent literature. *Polym. Int.*, 53, 1901–1929, 2004.

247. Chen, X.L. and Jiao, C.M., Synergistic effects of hydroxy silicone oil on intumescent flame retardant polypropylene system. *Fire Saf. J.*, 44, 1010–1014, 2009.

248. Ciesielski, M., Burk, B., Heinzmann, C., Döring, M., Fire-retardant high-performance epoxy-based materials, in: *Novel Fire Retardant Polymers and Composite Materials. Book Series: Woodhead Publishing Series in Composites Science and Engineering*, vol. 73, D.-Y. Wang, (Ed.), pp. 3–51, 2017.

249. Bonnet, J., Bounor-Legare, V., Boisson, F., Melis, F., Camino, G., Cassagnau, P., Phosphorus based organic-inorganic hybrid materials prepared by reactive processing for EVA fire retardancy. *Polym. Degrad. Stab.*, 97, 513–522, 2012.

250. Zhu, Z.-M., Xu, Y.-J., Liao, W., Xu, S., Wang, Y.-Z., Highly Flame Retardant Expanded Polystyrene Foams from Phosphorus-Nitrogen-Silicon Synergistic Adhesives. *Ind. Engin. Chem. Res.*, 56, 4649–4658, 2017,.

251. Häublein, M., Peter, K., Bakis, G., Mäkimieni, R., Altstädt, V., Möller, M., Investigation on the Flame Retardant Properties and Fracture Toughness of DOPO and Nano-SiO$_2$ Modified Epoxy Novolac Resin and Evaluation of Its Combinational Effects. *Materials*, 12, 1528, 2019.

252. Qiu, Y., Wachtendorf, V., Klack, P., Qian, L.J., Liu, Z., Schartel., B., Improved flame retardancy by synergy between cyclotetrasiloxane and phosphaphenanthrene/triazine compounds in epoxy thermoset. *Polym, Int.*, 66, 1883–1890, 2017.

253. Wang, X., Hu, Y.A., Song, L., Xing, W.Y., Lu, H.D., Thermal Degradation Behaviors of Epoxy Resin/POSS Hybrids and Phosphorus-Silicon Synergism of Flame Retardancy. *J. Polym. Sci. Part B-Polym. Phys.*, 48, 693–705, 2010.

254. Wu, C.S., Liu, Y.L., Chiu, Y.S., Epoxy resins possessing flame retardant elements from silicon incorporated epoxy compounds cured with phosphorus or nitrogen containing curing agents. *Polymer*, 43, 4277–4284, 2002.

255. Qian, X.D., Song, L., Yu, B., Wang, B.B., Yuan, B.H., Shi, Y.Q., Hu, Y., Yuen, R.K.K., Novel organic-inorganic flame retardants containing exfoliated graphene: preparation and their performance on the flame retardancy of epoxy resins. *J. Mater. Chem. A*, 1, 6822–6830, 2013.

256. Sut, A., Greiser, S., Jäger, C., Schartel, B., Aluminium Diethylphosphinate versus Ammonium Polyphosphate: A Comprehensive Comparison of the

Chemical Interactions during Pyrolysis in Flame-retarded Polyolefine/poly(phenylene oxide). *Thermochim. Act.*, 640, 74–84, 2016.

257. Product Information Dow Corning® 43-821 Additive, Form No. 26-2357-01, © 2017 Dow Corning Corporation, at the internet, 2017.

258. Sanchez-Olivares, G., Rabe, S., Pérez-Chávez, R., Calderas, F., Schartel, B., Industrial-waste agave fibres in flame-retarded thermoplastic starch biocomposites. *Compos. Part B*, 177, 107370, 2019.

259. Gilman, J.W., Harris, R.H., Shields, J.R., Kashiwagi, T., Morgan, A.B., A study of the flammability reduction mechanism of polystyrene-layered silicate nanocomposite: layered silicate reinforced carbonaceous char. *Polym. Adv. Technol.*, 17, 263–271, 2006.

260. Morgan, A.B., Harris, R.H., Kashiwagi, T., Chyall, L.J., Gilman, J.W., Flammability of polystyrene layered silicate (clay) nanocomposites: Carbonaceous char formation. *Fire Mater.*, 26, 247–253, 2002.

261. Wagner, J., Deglmann, P., Fuchs, S., Ciesielski, M., Fleckenstein, C.A., Döring, M., Flame retardant synergism of organic disulfides and phosphorous compounds. *Polym. Degrad. Stab.*, 129, 63–76, 2016.

262. Howell, B.A. and Daniel, Y.G., The impact of sulfur oxidation level on flame retardancy. *J. Fire Sci.*, 36, 518–534, 2018.

263. Balabanovich, A.I. and Engelmann, J., Fire retardant and charring effect of poly(sulfonyldiphenylene phenylphosphonate) in poly(butylene terephthalate). *Polym. Degrad. Stab.*, 79, 85–92, 2003.

264. Macocinschi, A., Grigoriu, A., Filip., D., Aromatic polysulfones for flame retardancy. *Eur. Polym. J.*, 38, 1025–1031, 2002.

265. Braun, U., Knoll, U., Schartel, B., Hoffmann, T., Pospiech, D., Artner, J., Ciesielski, M., Döring, M., Perez-Gratero, R., Sandler, J.K.W., Altstädt, V., Novel phosphorus-containing poly(ether sulfone)s and their blends with an epoxy resin: Thermal decomposition and fire retardancy. *Macromol. Chem. Phys.*, 207, 1501–1514, 2006.

266. Battig, A., Markwart, J.-C., Wurm, F.R., Schartel, B., Sulfur's role in the flame retardancy of thio-ether–linked hyperbranched polyphosphoesters in epoxy resins. *Eur. Polym. J.*, 122, 109390, 2020.

267. Tang, S., Qian, L.J., Qiu, Y., Dong, Y.P., Synergistic flame-retardant effect and mechanisms of boron/phosphorus compounds on epoxy resins. *Polym. Adv. Technol.*, 29, 641–648, 2018.

268. Dogan, M. and Unlu, S.M., Flame retardant effect of boron compounds on red phosphorus containing epoxy resins. *Polym. Degrad. Stab.*, 99, 12–17, 2014.

269. Unlu, S.M., Dogan, S.D., Dogan, M., Comparative study of boron compounds and aluminum trihydroxide as flame retardant additives in epoxy resin. *Polym. Adv. Technol.*, 25, 769–776, 2014.

270. Shen, K.K. and Griffin, T.S., Zinc Borate as a Flame Retardant, Smoke Suppppressant, and Afterglow Surpressant in Polymers, in: *Fire and Polymers:*

Hazards Identification and Prevention. Book Series: ACS Symposium Series, vol. 425, G.L. Nelson (Eds.), pp. 157–177, ACS, Washington, 1990.

271. Shen, K.K., Kochesfahani, H.S., Jouffret, F., Boron-Based Flame Retardants and Flame Retardancy, in: *Fire Retardancy of Polymeric Materials*, 2nd Edition, C.A. Wilkie and A.B. Morgan (Eds.), pp. 207–237, CRC Press, Boca Raton, 2010, chap, 9.

272. Bourbigot, S., Le Bras, M., Leeuwendal, R., Shen, K.K., Schubert, D., Recent advances in the use of zinc borates in flame retardancy of EVA. *Polym. Degrad. Stab.*, 64, 419–425, 1999.

273. Formicola, C., De Fenzo, A., Zarrelli, M., Frache, A., Giordano, M., Camino, G., Synergistic effects of zinc borate and aluminium trihydroxide on flammability behaviour of aerospace epoxy system. *Express Polym. Lett.*, 3, 376–384, 2009.

274. Pawlowski, K.H., Schartel, B., Fichera, M.A., Jäger, C., Flame Retardancy Mechanisms of Bisphenol A Bis(diphenyl phosphate) in Combination with Zinc Borate in Bisphenol A Polycarbonate/Acrylonitrile-Butadiene-Styrene Blends. *Thermochim. Act.*, 498, 92–99, 2010.

275. Karrasch, A., Wawrzyn, E., Schartel, B., Jäger, C., Solid-state NMR on Thermal and Fire Residues of Bisphenol A Polycarbonate/Silicone Acrylate Rubber/Bisphenol A Bis(diphenyl-phosphate)/ (PC/SiR/BDP) and PC/SiR/BDP/Zinc Borate (PC/SiR/BDP/ZnB) — Part I: PC Charring and the Impact of BDP and ZnB. *Polym. Degrad. Stab.*, 95, 2525–2533, 2010.

276. Sut, A., Greiser, S., Jäger, C., Schartel, B., Interactions in Multicomponent Flame-retardant Polymers: Solid-state NMR Identifying the Chemistry Behind it. *Polym. Degrad. Stab.*, 121, 116–125, 2015.

277. Hoffendahl, C., Fontaine, G., Duquesne, S., Taschner, F., Mezger, M., Bourbigot, S., The combination of aluminum trihydroxide (ATH) and melamine borate (MB) as fire retardant additives for elastomeric ethylene vinyl acetate (EVA). *Polym. Degrad. Stab.*, 115, 77–88, 2015.

278. Carty, P. and White, S., Smoke/Char Relationships in PVC Formulations. *Fire Mater.*, 18, 151–166, 1994.

279. Kuljanin, J., Marinovic-Cincovic, M., Zec, S., Comor, M., Nedeljkovic, C., Influence of Fe_2O_3-filler on the thermal properties of polystyrene. *J. Mater. Sci. Lett.*, 22, 235–237, 2003.

280. Aufmuth, W., Levchik, S.V., Levchik, G.F., Klatt, M., Poly(butylene terephthalate) fire retarded by 1,4-diisobutylene-2,3,5,6-tetraxydroxy-1, 4-diphosphine oxide. I. Combustion and thermal decomposition. *Fire Mater.*, 23, 1–6, 1999.

281. Peng, Y., Niu, M., Qin, R.H., Xue, B.X., Shao, M.Q., Study on flame retardancy and smoke suppression of PET by the synergy between Fe2O3 and new phosphorus-containing silicone flame retardant. *High Perform. Polym.*, 32, 871–882, 2020.

282. Laachachi, A., Cochez, M., Ferriol, M., Leroy, E., Lopez Cuesta., J.-M., Thermal degradation and flammability of poly(methyl methacrylate) containing TiO2 nanoparticles and modified montmorillonite, in: *Fire and Polymers IV: Materials and Concepts for Hazard Prevention. Book Series: ACS Symposium Series*, vol. 922, C.A. Wilkie and G.L. Nelson (Eds.), pp. 36–47, ACS, Washington, 2006.

283. Laachachi, A., Cochez, M., Leroy, E., Ferriol, M., Lopez-Cuesta, J.-M., Fire retardant systems in poly (methyl methacrylate): interactions between metal oxide nanoparticles and phosphinates. *Polym. Degrad. Stab.*, 92, 61–69, 2007.

284. Gallo, E., Braun, U., Schartel, B., Russo, P., Acierno, D., Halogen-free Flame Retarded Poly(butylene terephthalate) (PBT) Using Metal Oxides/ PBT Nanocomposites in Combination with Aluminium Phosphinate. *Polym. Degrad. Stab.*, 94, 1245–1253, 2009.

285. Gallo, E., Schartel, B., Braun, U., Russo, P., Acierno, D., Fire Retardant Synergisms between Nanometric Fe_2O_3 and Aluminium Phosphinate in Poly(butylene terephthalate). *Polym. Adv. Technol.*, 22, 2382–2391, 2011.

286. Gallo, E., Schartel, B., Acierno, D., Russo, P., Flame Retardant Biocomposites: Synergism between Phosphinate and Nanometric Metal Oxides. *Eur. Polym. J.*, 47, 1390–1401, 2011.

287. Xu, L.F., Wu, X.D., Li, L.S., Chen, Y.J., Synthesis of a novel polyphosphazene/triazine bi-group flame retardant in situ doping nano zinc oxide and its application in poly (lactic acid) resin. *Polym. Adv. Technol.*, 30, 1375–1385, 2019.

288. Lewin, M. and Endo, M., Catalysis of intumescent flame retardancy of polypropylene by metallic compounds. *Polym. Adv. Technol.*, 14, 3–11, 2003.

289. Duquesne, S., Bachelet, P., Bellayer, S., Bourbigot, S., Mertens, W., Influence of inorganic fillers on the fire protection of intumescent coatings. *J. Fire Sci.*, 31, 258–275, 2013.

290. Gu, J.W., Zhang, G.C., Dong, S.L., Zhang, Q.Y., Kong, J., Study on preparation and fire-retardant mechanism analysis of intumescent flame-retardant coatings. *Surf. Coat. Technol.*, 201, 7835–7841, 2007.

291. Zhou, Y., Hao, J.W., Liu, G.S., Du, J.X., Influencing Mechanism of Transition Metal Oxide on Thermal Decomposition of Ammonium Polyphosphate. *Chin. J. Inorg. Chem.*, 29, 1115–1122, 2013.

292. Mariappan, T., Agarwal, A., Ray, S., Influence of titanium dioxide on the thermal insulation of waterborne intumescent fire protective paints to structural steel. *Prog. Org. Coat.*, 111, 67–74, 2017.

293. Pan, Y.T., Zhang, Z.D., Yang, R.J., The rise of MOFs and their derivatives for flame retardant polymeric materials: A critical review. *Compos. Part B-Engin.*, 199, 108265, 2020.

294. Xu, Z.S., Jia, H.Y., Yan, L., Chu, Z.Y., Zhou, H., Synergistic effect of bismuth oxide and mono-component intumescent flame retardant on the flammability and smoke suppression properties of epoxy resins. *Polym. Adv. Technol.*, 31, 25–35, 2020.

295. Morgan, A.B., Flame retarded polymer layered silicate nanocomposites: a review of commercial and open literature systems. *Polym. Adv. Technol.*, 17, 206–217, 2006.

296. Modesti, M., Lorenzetti, A., Besco, S., Hrelja, D., Semenzato, S., Bertani, R., Michelin, R.A., Synergism between flame retardant and modified layered silicate on thermal stability and fire behaviour of polyurethane-nanocomposite foams. *Polym. Degrad. Stab.*, 93, 2166–2171, 2008.

297. Gomez-Fernandez, S., Ugarte, L., Pena-Rodriguez, C., Corcuera, M.A., Eceiza, A., The effect of phosphorus containing polyol and layered double hydroxides on the properties of a castor oil based flexible polyurethane foam. *Polym Degrad Stabil*, 132, 41–51, 2016.

298. Yan, W., Yu, J., Zhang, M.Q., Qin, S.H., Wang, T., Huang, W.J., Long, L.J., Flame-retardant effect of a phenethyl-bridged DOPO derivative and layered double hydroxides for epoxy resin. *RSC Adv.*, 7, 46236–46245, 2017.

299. Naderi Kalali, E., Montes, A., Wang, X., Zhang, L., Shabestari, M.E., Li, Z., Wang, D.-Y., Effect of phytic acid–modified layered double hydroxide on flammability and mechanical properties of intumescent flame retardant polypropylene system. *Fire Mater.*, 42, 213–220, 2018.

300. Bourbigot, S., LeBras, M., Delobel, R., Tremillon, J.M., Synergistic effect of zeolite in an intumescence process - Study of the interactions between the polymer and the additives. *J. Chem. Soc.-Faraday Transact.*, 92, 3435–3444, 1996.

301. Demir, H., Arkis, E., Balkose, D., Ulku, S., Synergistic effect of natural zeolites on flame retardant additives. *Polym. Degrad. Stab.*, 89, 478–483, 2005.

302. Liu, Y.C., Xu, W.Z., Chen, R., Cheng, C.M., Hu, Y.Z., Effect of different zeolitic imidazolate frameworks nanoparticle-modified beta-FeOOH rods on flame retardancy and smoke suppression of epoxy resin. *J. Appl. Polym., Sci.*, 138, 49637, 2021.

303. Huang, G.B., Chen, S.Q., Song, P.G., Lu, P.P., Wu, C.L., Liang, H.D., Combination effects of graphene and layered double hydroxides on intumescent flame-retardant poly(methyl methacrylate) nanocomposites. *Appl. Clay Sci.*, 88–89, 78–85, 2014.

304. Duquesne, S., Samyn, F., Bourbigot, S., Amigouet, P., Jouffret, F., Shen, K., Influence of talc on the fire retardant properties of highly filled intumescent polypropylene composites. *Polym. Adv. Technol.*, 19, 620–627, 2008.

305. Almeras, X., Le Bras, M., Hornsby, P., Bourbigot, S., Marosi, G., Keszei, S., Poutch, F., Effect of fillers on the fire retardancy of intumescent polypropylene compounds. *Polym. Degrad. Stab.*, 82, 325–331, 2003.

306. Clerc, L., Ferry, L., Leroy, E., Lopez-Cuesta, J.-M., Influence of talc physical properties on the fire retarding behaviour of (ethylene-vinyl acetate copolymer/magnesium hydroxide/talc) composites. *Polym. Degrad. Stab.*, 88, 504–511, 2005.

307. Bertelli, G., Marchetti, E., Camino, G., Costa, L., Locatelli, R., Intumescent Fire Retardant Systems – Effect of Fillers on Char Structure. *Angew. Makromol. Chem.*, 172, 153–163, 1989.

308. Bourbigot, S., Le Bras, M., Duquesne, S., Rochery, M., Recent advances for intumescent polymers. *Macromol. Mater.Eng.*, 289, 499–511, 2004.

309. Huang, G.B., Wang, S.Q., Song, P.A., Wu, C.L., Chen, S.Q., Wang, X., Combination effect of carbon nanotubes with graphene on intumescent flame-retardant polypropylene nanocomposites. *Compos. Part A: Appl. Sci. Manufact.*, 59, 18–25, 2014.

310. Wu, G.M., Schartel, B., Kleemeier, M., Hartwig, A., Flammability of Layered Silicate Epoxy Nanocomposites Combined with Low-Melting Inorganic Ceepree Glass. *Polym. Engin. Sci.*, 52, 507–517, 2012.

311. Kroenke, W.J., Low-melting sulfate glasses and glass ceramics, and their utility as fire and smoke retarder additives for poly(vinyl-chloride). *J. Mater. Sci.*, 21, 1123–1133, 1986.

312. Myers, R.E. and Licursi, E., Inorganic glass forming systems as intumescent flame retardants for organic polymers. *J. Fire Sci.*, 3, 415–431, 1985.

313. Yu, D., Kleemeier, M., Wu, G.M., Schartel., B., Liu, W.Q., Hartwig, A., A Low Melting Organic-inorganic Glass and Its Effect on Flame Retardancy of Clay/Epoxy Composites. *Polymer*, 52, 2120–2131, 2011.

314. Yu, D., Kleemeier, M., Wu, G.M., Schartel., B., Liu, W.Q., Hartwig, A., Phosphorous and Silicon Containing Low-melting Organic-inorganic Glasses Improve Flame Retardancy of Epoxy/Clay Composite. *Macromol. Mater. Engin.*, 296, 952–964, 2011.

315. Wu, G.M., Schartel, B., Yu, D., Kleemeier, M., Hartwig, A., Synergistic Flame Retardancy in Layered-silicate Epoxy Nanocomposite Combined with Low-melting Phenysiloxane Glass. *J. Fire Sci.*, 30, 69–87, 2012.

316. Kashiwagi, T. and Gilman, J.W., Silicon-Based Flame Retardants, in: *Fire Retardancy of Polymeric Materials*, A.F. Grand and C.A. Wilkie (Eds.), pp. 353–390, Marcel Dekker, New York, 2000, chap 10.

317. Hamdani, S., Longuet, C., Perrin, D., Lopez-Cuesta, J.-M., Ganachaud, F., Flame retardancy of silicone-based materials. *Polym. Degrad. Stab.*, 94, 465–495, 2009.

318. Chen, Y.J., Zhan, J., Zhang, P., Nie, S.B., Lu, H.D., Song, L., Hu, Y., Preparation of Intumescent Flame Retardant Poly(butylene succinate) Using Fumed Silica as Synergistic Agent. *Ind. Engin. Chem. Res.*, 49, 8200–8208, 2010.

319. Wei, P., Han, Z., Xu, X., Li, Z., Synergistic flame retardant effect of SiO2 in LLDPE/EVA/ATH blends. *J. Fire Sci.*, 24, 487–498, 2006.

320. Fu, M.Z. and Qu, B.J., Synergistic flame retardant mechanism of fumed silica in ethylene-vinyl acetate/magnesium hydroxide blends. *Polym. Degrad. Stab.*, 85, 633–639, 2004.

321. Schmaucks, G., Friede, B., Schreiner, H., Roszinski, J.Q., Amorphous Silicon Dioxide as Additive to Improve the Fire Retardancy of Polyamides, in: *Fire Retardancy of Polymers: New Strategies and Mechanisms*, T.R. Hull and B.K. Kandola (Eds.), pp. 35–48, RSC, Cambridge, UK, 2009, chapter 3.

322. Gallo, E., Schartel, B., Schmaucks, G., von der Ehe, K., Böhning, M., The Effect of Well-dispersed Amorphous Silicon Dioxide in Flame-retarded Styrene Butadiene Rubber. *Plas. Rubber Compos.*, 42, 34–42, 2013.

323. Laoutid, F., Duriez, V., Brison, L., Aouadi, S., Vahabi, H., Dubois., P., Synergistic flame-retardant effect between lignin and magnesium hydroxide in poly(ethylene-co-vinyl acetate). Flame Retard. *Therm. Stab. Mater.*, 2, 9–18, 2019

324. Cavodeau, F., Otazaghine, B., Sonnier, R., Lopez-Cuesta, J.-M., Delaite, C., Fire retardancy of ethylene-vinyl acetate composites – Evaluation of synergistic effects between ATH and diatomite fillers. *Polym. Degrad. Stab.*, 129, 246–259, 2016.

325. Hoffendahl, C., Duquesne, S., Fontaine, G., Taschner, F., Mezger, M., Bourbigot, S., Decomposition mechanism of fire retarded ethylene vinyl acetate elastomer (EVA) containing aluminum trihydroxide and melamine. *Polym. Degrad. Stab.*, 113, 168–179, 2015.

326. Ye, L., Wu, Q.H., Qu, B.J., Synergistic effects and mechanism of multiwalled carbon nanotubes with magnesium hydroxide in halogen-free flame retardant EVA/MH/MWNT nanocomposites. *Polym. Degrad. Stab.*, 94, 751–756, 2009.

327. Cárdenas, M.A., García-López, D., Gobernado-Mitre, I., Merino, J.C., Pastor, J.M., de D. Martínez, J., Barbeta, J., Calveras., D., Mechanical and fire retardant properties of EVA/clay/ATH nanocomposites – Effect of particle size and surface treatment of ATH filler. *Polym. Degrad. Stab.*, 93, 2032–2037, 2008.

328. Wang, H., Zhou, X., Abro, M., Gao, M., Deng, M., Qin, Z., Sun, Y., Yue, L., Zhang, X., Mussel-inspired General Interface Modification and Its Application in Polymer Reinforcement and as a Flame Retardant. *ACS Omega*, 3. 4891–4898. 2018.

329. Zhang, R., Xiao, X., Tai, Q., Huang, H., Yang, J., Hu, Y., The effect of different organic modified montmorillonites (OMMTs) on the thermal properties and flammability of PLA/MCAPP/lignin systems. *J. Appl. Polym, Sci.*, 127, 4967–4973, 2013.

330. Braun, U. and Schartel, B., Effect of Red Phosphorus and Melamine Polyphosphate on the Fire Behavior of HIPS. *J. Fire Sci.*, 23, 5–30, 2005.

331. Braun, U., Bahr, H., Schartel, B., Fire Retardancy Effect of Aluminium Phosphinate and Melamine Polyphosphate in Glass Fibre Reinforced Polyamide 6. *e-polymers*, 10, 41, 2010.

332. Samyn, F. and Bourbigot, S., Thermal decomposition of flame retarded for-
     mulations PA6/aluminum phosphinate/melamine polyphosphate/organo-
     modified clay: interactions between the constituents? *Polym. Degrad. Stabil.*,
     97, 2217–2230, 2012.
333. Orhan, T., Isitman, N.A., Hacaloglu, J., Kaynak, C., Thermal degradation
     mechanisms of aluminium phosphinate, melamine polyphosphate and zinc
     borate in poly(methyl methacrylate). *Polym. Degrad. Stabil.*, 96, 1780–1787,
     2011.
334. Xu, Z.Z., Huang, J.Q., Chen, M.J., Tan, Y., Wang, Y.-Z., Flame retardant
     mechanism of an efficient flame-retardant polymeric synergist with ammo-
     nium polyphosphate for polypropylene. *Polym. Degrad. Stab.*, 98, 2011–2020,
     2013.
335. Despinasse, M.-C. and Schartel, B., Aryl Phosphate - Aryl Phosphate Synergy
     in Flame-retarded Bisphenol A Polycarbonate/Acrylonitrile-butadiene-
     styrene. *Thermochim. Act.*, 563, 51–61, 2013.
336. Fichera, M.A., Braun, U., Schartel, B., Sturm, H., Knoll, U., Jäger, C.,
     Solid-State NMR Investigations of the Pyrolysis and Thermo-oxidative
     Decomposition Products of a Polystyrene/Red Phosphorus/Magnesium
     Hydroxide System. *J. Anal. Appl. Pyrolys.*, 78, 378–386, 2007.
337. Camino, G., Riva, A., Vizzini, D., Castrovinci, A., Amigouet, P., Bras Pereira,
     P., Effect of hydroxides on fire retardance mechanism of intumescent EVA
     compositions, in: *Fire Retardancy of Polymers: New Applications of Mineral
     Fillers*, M. Le Bras, C.A. Wilkie, S. Bourbigot, S. Duquense, C. Jama, (Eds.),
     pp. 248–263, RSC, Cambridge, 2005, chap 18.
338. Liu, J.C., Peng, S., Zhang, Y.B., Chang, H.B., Yu, Z.L., Pan, B.L., Lu, C., Ma,
     J.Y., Niu, Q.S., Influence of microencapsulated red phosphorus on the flame
     retardancy of high impact polystyrene/magnesium hydroxide composite and
     its mode of action. *Polym. Degrad. Stab.*, 121, 208–221, 2015.
339. Shen, L.G., Shao, C.R., Li, R.J., Xu, Y.C., Li, J.X., Lin, H.J., Preparation and
     characterization of ethylene-vinyl acetate copolymer (EVA)-magnesium
     hydroxide (MH)-hexaphenoxycyclotriphosphazene (HPCTP) composite
     flame-retardant materials. *Polym. Bull.*, 76, 2399–2410, 2019.
340. Savas, L.A., Deniz, T.K., Tayfun, U., Dogan, M., Effect of microcapsulated red
     phosphorus on flame retardant, thermal and mechanical properties of ther-
     moplastic polyurethane composites filled with huntite & hydromagnesite
     mineral. *Polym. Degrad. Stab.*, 135, 121–129, 2017.
341. Döring, M., Ciesielski, M., Heinzmann, C., Synergistic Flame Retardant
     Mixtures in Epoxy Resins, in: *Fire and Polymers VI: New Advances in Flame
     Retardant Chemistry and Science. Book Series: ACS Symposium Series*, vol.
     1118, A.B. Morgan, C.A. Wilkie, G.L. Nelson (Eds.), pp. 295–309, ACS,
     Washington, 2012.
342. Reuter, J., Greiner, L., Schonberger, F., Döring, M., Synergistic flame
     retardant interplay of phosphorus containing flame retardants with

aluminum trihydrate depending on the specific surface area in unsaturated polyester resin. *J. Appl. Polym, Sci.*, 136, 47270, 2019.

343. Yan, W., Wang, K., Huang, W.J., Wang, M., Wang, T., Tu, C.Y., Tian, Q., Synergistic effects of phenethyl-bridged DOPO derivative with Al(OH)(3) on flame retardancy for epoxy resins. *Polym. Plast. Technol. Mater.*, 59, 797–808, 2020.

344. Li, Z.Z. and Qu, B.J., Flammability characterization and synergistic effects of expandable graphite with magnesium hydroxide in halogen-free flame-retardant EVA blends. *Polym. Degrad. Stab.*, 81, 401–408, 2003.

345. Wang, Z.Z., Qu, B.J., Fan, W.C., Huang, P., Combustion characteristics of halogen-free flame-retarded polyethylene containing magnesium hydroxide and some synergists. *J. Appl. Polym, Sci.*, 81, 206–214, 2001.

346. Hu, S., Tan, Z.-W., Chen, F., Li, J.-G., Shen, Q., Huang, Z.-X., Zhang, L.-M., Flame-retardant properties and synergistic effect of ammonium polyphosphate/aluminum hydroxide/mica/silicone rubber composites. *Fire Mater.*, 44, 673–682, 2020.

347. Li, Y.M., Hu, S.L., Wang, D.Y., Polymer-based ceramifiable composites for flame retardant applications: A review. *Compos. Commun.*, 21, 100405, 2020.

348. Modesti, M. and Lorenzetti, A., Flame retardancy of polyisocyanurate-polyurethane foams: use of different charring agents, Polym. *Degrad. Stabil.*, 78, 341–347, 2002.

349. Sypaseuth, F.D., Gallo, E., Çiftci, S., Schartel, B., Polylactic Acid Biocomposites: Approaches to a Completely Green Flame Retarded Polymer. *e-polymers*, 17, 449–462, 2017.

350. Tang, G., Zhang, R., Wang, X., Wang, B., Song, L., Hu, Y., Gong, X., Enhancement of flame retardant performance of bio-based polylactic acid composites with the incorporation of aluminum hypophosphite and expanded graphite. *J. Macromol. Sci. Part A Pure Appl. Chem.*, 50, 255–269, 2013.

351. Yang, S., Wang, J., Huo, S.Q., Wang, M., Wang, J.P., Zhang, B., Synergistic flame-retardant effect of expandable graphite and phosphorus-containing compounds for epoxy resin: Strong bonding of different carbon residues. *Polym. Degrad. Stabil.*, 128, 89–98, 2016.

352. Guo, C.G., Zhou, L., Lv, J.X., Effects of Expandable Graphite and Modified Ammonium Polyphosphate on the Flame-Retardant and Mechanical Properties of Wood Flour-Polypropylene Composites. *Polym. Polym. Compos.*, 21, 449–456, 2013.

353. Zhu, H.F., Zhu, Q.L., Li, J.A., Tao, K., Xue, L.X., Yan, Q., Synergistic effect between expandable graphite and ammonium polyphosphate on flame retarded polylactide. *Polym. Degrad. Stab.*, 96, 183–189, 2011.

354. Wilke, A., Langfeld, K., Ulmer, B., Andrievici, V., Hörold, A., Limbach, P., Bastian, M., Schartel, B., Halogen-free Multicomponent Flame Retardant

Thermoplastic Styrene-Ethylene-Butylene-Styrene Elastomers Based on Ammonium Polyphosphate – Expandable Graphite Synergy. *Ind. Engin. Chem. Res.*, 56, 8251–8263, 2017.

355. Seefeldt, H., Braun, U., Wagner, M.H., Residue Stabilization in the Fire Retardancy of Wood–Plastic Composites: Combination of Ammonium Polyphosphate, Expandable Graphite, and Red Phosphorus. *Macromol. Chem. Phys.*, 213, 2370–2377, 2012.

356. Li, J., Mo, X.H., Li, Y., Zou, H.W., Liang, M., Chen, Y., Influence of expandable graphite particle size on the synergy flame retardant property between expandable graphite and ammonium polyphosphate in semi-rigid polyurethane foam. *Polym. Bull.*, 75, 5287–5304, 2018.

357. Meng, X.Y., Ye, L., Zhang, X.G., Tang, P.M., Tang, J.H., Ji, X., Li, Z.M., Effects of Expandable Graphite and Ammonium Polyphosphate on the Flame-Retardant and Mechanical Properties of Rigid Polyurethane Foams. *J. Appl. Polym, Sci.*, 114, 853–863, 2009.

358. Shi, L., Li, Z.M., Xie, B.H., Wang, J.H., Tian, C.R., Yang, M.B., Flame retardancy of different-sized expandable graphite particles for high-density rigid polyurethane foams. *Polym, Int.*, 55, 862–871, 2006.

359. Modesti, M. and Lorenzetti, A., Halogen-free flame retardants for polymeric foams. *Polym. Degrad. Stab.*, 78, 167–173, 2002.

360. Zhang, L.Q., Zhang, M., Zhou, Y.H., Hu, L.H., The study of mechanical behavior and flame retardancy of castor oil phosphate-based rigid polyurethane foam composites containing expanded graphite and triethyl phosphate. *Polym. Degrad. Stab.*, 98, 2784–2794, 2013.

361. Modesti, M., Lorenzetti, A., Simioni, F., Camino, G., Expandable graphite as an intumescent flame retardant in polyisocyanurate-polyurethane foams. *Polym. Degrad. Stab.*, 77, 195–202, 2002.

362. Thi, N.H., Nguyen, T.N., Oanh, H.T., Trang, N.T.T., Tham, D.Q., Nguyen, H.T., Nguyen, T.V., Hoang, M.H., Synergistic effects of aluminum hydroxide, red phosphorus, and expandable graphite on the flame retardancy and thermal stability of polyethylene. *J. Appl. Polym, Sci.*, 138, 50317, 2020.

363. Sut, A., Metzsch-Zilligen, E., Großhauser, M., Pfaendner, R., Schartel., B., Synergy between melamine cyanurate melamine polyphosphate and aluminum diethylphosphinate in flame retarded thermoplastic polyurethane. *Polym. Test.*, 74, 196–204, 2019.

364. Hollingbery, L.A. and Hull, T.R., The thermal decomposition of huntite and hydromagnesite - A review. *Polym. Degrad. Stab.*, 95, 2213–2225, 2010.

365. Peeterbroeck, S., Alexandre, M., Nagy, J.B., Pirlot, C., Fonseca, A., Moreau, N., Philippin, G., Delhalle, J., Mekhalif, Z., Sporken, R., Beyer, G., Dubois, P., Polymer-layered silicate-carbon nanotube nanocomposites: unique nanofiller synergistic effect Compos. *Sci. Technol.*, 64, 2317–2323, 2004.

366. Schartel, B. and Schmaucks, G., Flame retardancy synergism in polymers through different inert fillers' geometry. *Polym. Engin. Sci.*, 57, 1099–1109, 2017.

367. Schartel, B., Beck, U., Bahr, H., Hertwig, A., Knoll, U., Weise, M., Short Communication: Sub-micrometer Coatings as an IR Mirror: A New Route to Flame Retardancy. *Fire Mater.*, 36, 671–677, 2012.

368. Geoffroy, L., Davesne, A.L., Parent, F., Sanchette, F., Samyn, F., Jimenez, M., Bourbigot, S., Combining Low-Emissivity Thin Coating and 3D-Printed Original Designs for Superior Fire-Protective Performance. *ACS Omega*, 5, 27857–27863, 2020.

369. Jimenez, M., Gallou, H., Duquesne, S., Jama, C., Bourbigot, S., Couillens, X., Speroni, F., New routes to flame retard polyamide 6,6 for electrical applications. *J. Fire Sci.*, 30, 535–551, 2012.

370. Forsth, M., Zhao, S.X., Roos, A., Spectrally selective and adaptive surfaces for protection against radiative heating: ITO and VO2. *Fire Mater.*, 38, 111–124, 2014.

371. Jimenez, M., Duquesne, S., Bourbigot, S., Enhanced fire retardant properties of glass-fiber reinforced Polyamide 6,6 by combining bulk and surface treatments: Toward a better understanding of the fire-retardant mechanism. *Polym. Degrad. Stab.*, 98, 1378–1388, 2013.

372. Bourbigot, S., Jama, C., Le Bras, M., Delobel, R., Dessaux, O., Goudmand, P., New approach to flame retardancy using plasma assisted surface polymerisation techniques. *Polym. Deg. Stab.*, 66, 153–155, 1999.

373. Gallo, E., Fan, Z., Schartel, B., Greiner, A., Electrospun Nanofibre Mat Coating - New Route to Flame Retardancy. *Polym. Adv. Technol.*, 22, 1205–1210, 2011.

374. Schartel, B., Kühn, G., Mix, R., Friedrich, J., Surface Controlled Fire Retardancy of Polymers Using Plasma Polymerisation Macromol. *Mater. Engin.*, 287, 579–582, 2002.

375. Davesne, A.L., Bensabath, T., Sarazin, J., Bellayer, S., Parent, F., Samyn, F., Jimenez, M., Sanchette, F., Bourbigot, S., Low-Emissivity Metal/Dielectric Coatings as Radiative Barriers for the Fire Protection of Raw and Formulated Polymers. *ACS Appl. Polym. Mater.*, 2, 2880–2889, 2020.

376. Bünker, J. and Elfgen, M., Fire-retardant Composite Panel for Shipbuilding. *Lightweight Des. Worldwide*, 11, 36–41, 2018.

377. Hörold, A., Schartel, B., Trappe, V., Korzen, M., Bünker, J., Fire Stability of Glass-fibre Sandwich Panels: The Influence of Core Materials and Flame Retardants. *Compos. Struct.*, 160, 1310–1318, 2017.

378. Xi, W., Qian, L.J., Qiu, Y., Chen, Y.J., Flame-retardant behavior of bi-group molecule derived from phosphaphenanthrene and triazine groups on poly-lactic acid. *Polym. Adv. Technol.*, 27, 781–788, 2016.

379. Chen, L. and Wang, Y.-Z., A review on flame retardant technology in China. Part 1: development of flame retardants. *Polym. Adv. Technol.*, 21, 1–26, 2010.

380. Chen, Y.J., Wang, W., Qiu, Y., Li, L.S., Qian, L.J., Xin, F., Terminal group effects of phosphazene-triazine bi-group flame retardant additives in flame retardant polylactic acid composites. *Polym. Degrad. Stab.*, 140, 166–175, 2017.

381. Sag, J., Kukla, P., Goedderz, D., Roch, H., Kabasci, S., Döring, M., Schonberger, F., Synthesis of Novel Polymeric Acrylate-Based Flame Retardants Containing Two Phosphorus Groups in Different Chemical Environments and Their Influence on the Flammability of Poly (Lactic Acid). *Polymers*, 12, 778, 2020.

382. Müller, P. and Schartel, B., Melamine Poly(metal phosphates) as Flame Retardant in Epoxy Resin: Performance, Modes of Action and Synergy. *J. Appl. Polym, Sci.*, 133, 43549, 2016.

383. Halpern, Y., Mott, D.M., Niswander, R.H., Fire Retardancy of Thermoplastic Materials by Intumescence. *Ind. Engin. Chem. Prod. Res. Dev.*, 23, 233–238, 1984.

384. Wang, X., Hu, Y., Song, L., Xing, W.Y., Lu, H.D.A., Lv, P., Jie, G.X., Flame retardancy and thermal degradation mechanism of epoxy resin composites based on a DOPO substituted organophosphorus oligomer. *Polymer*, 51, 2435–2445, 2010.

385. Wang, C., Wu, Y.C., Li, Y.C., Shao, Q., Yan, X.R., Han, C., Wang, Z., Liu, Z., Guo, Z.H., Flame-retardant rigid polyurethane foam with a phosphorus-nitrogen single intumescent flame retardant. *Polym. Adv. Technol.*, 29, 668–676, 2018.

386. Ma, H.Y., Tong, L.F., Xu, Z.B., Fang, Z.P., Jin, Y.M., Lu, F.Z., A novel intumescent flame retardant: Synthesis and application in ABS copolymer. *Polym. Degrad. Stab.*, 92, 720–726, 2007.

387. Shao, Z.B., Deng, C., Tan, Y., Chen, M.J., Chen, L., Wang, Y.-Z., An efficient monocomponent polymeric intumescent flame retardant for polypropylene: preparation and application. *ACS Appl. Mat. Interfac.*, 6, 7363–7370, 2014.

388. Lv, Q., Huang, J.Q., Chen, M.J., Zhao, J., Tan, Y., Chen, L., Wang Y.-Z., An Effective Flame Retardant and Smoke Suppression Oligomer for Epoxy Resin. *Ind. Engin. Chem. Res.*, 52, 9397–9404, 2013.

389. Deng, C., Yin, H., Li, R.-M., Huang, S.-C., Schartel, B., Wang, Y.-Z., Modes of action of a mono-component intumescent flame retardant MAPP in polyethylene-octene elastomer. *Polym. Degrad. Stab.*, 138, 142–150, 2017.

390. Yan, J.L. and Xu, M.J., Design, synthesis and application of a highly efficient mono-component intumescent flame retardant for non-charring polyethylene composites. *Polym. Bull.*, 78, 643–662, 2021.

391. Liu, L.B., Xu, Y., He, Y.T., Xu, M.J., Wang, W., Li, B., A facile strategy for enhancing the fire safety of unsaturated polyester resins through introducing an efficient mono-component intumescent flame retardant. *Polym. Adv. Technol.*, 31, 1218–1230, 2020.

392. Xia, S.Y., Zhang, Z.Y., Leng, Y., Li, B., Xu, M.J., Synthesis of a novel mono-component intumescent flame retardant and its high efficiency for flame retardant polyethylene. *J. Anal. Appl. Pyrolys.*, 134, 632–640, 2018.

393. Shao, Z.B., Deng, C., Tan, Y., Chen, M.J., Chen, L., Wang, Y.-Z., Flame retardation of polypropylene via a novel intumescent flame retardant:

ethylenediamine modified ammonium polyphosphate. *Polym. Degrad. Stabil.*, 106, 88–96, 2014.

394. Shao, Z.B., Deng, C., Tan, Y., Yu, L., Chen, M.J., Chen, L., Wang, Y.-Z., Ammonium polyphosphate chemically-modified with ethanolamine as an efficient intumescent flame retardant for polypropylene. *J. Mat. Chem. A*, 2, 13955–13965, 2014.

395. Zhu, H.B., Peng, Z.M., Chen, Y.M., Li, G.Y., Wang, L., Tang, Y., Pang, R., Khan, Z.U., Wan, P.Y., Preparation and characterization of flame retardant polyurethane foams containing phosphorus-nitrogen-functionalized lignin. *RSC Adv.*, 4, 55271–55279.

396. Costes, L., Laoutid, F., Aguedo, M., Richel, A., Brohez, S., Delvosalle, C., Dubois, P., Phosphorus and nitrogen derivatization as efficient route for improvement of lignin flame retardant action in PLA. *Eur. Polym. J.*, 84, 652–667, 2016.

397. Yu, Y., Fu, S., Song, P., Luo, X., Jin, Y., Lu, F., Wu, Q., Ye, J., Functionalized lignin by grafting phosphorus-nitrogen improves the thermal stability and flame retardancy of polypropylene. *Polym. Degrad. Stab.*, 97, 541–546, 2012.

398. Prieur, B., Meub, M., Wittemann, M., Klein, R., Bellayer, S., Fontaine, G., Bourbigot, S., Phosphorylation of lignin to flame retard acrylonitrile butadiene styrene (ABS). *Polym. Degrad. Stab.*, 127, 32–43, 2016.

399. Zhang, R., Xiao, X., Tai, Q., Huang, H., Hu, Y., Modification of lignin and its application as char agent in intumescent flame-retardant poly(lactic acid). *Polym. Engin. Sci.*, 52, 2620–2626, 2012.

400. Xing, W.Y., Yuan, H.X., Zhang, P., Yang, H.Y., Song, L., Hu, Y., Functionalized lignin for halogen-free flame retardant rigid polyurethane foam: preparation, thermal stability, fire performance and mechanical properties. *J. Polym Res.*, 20, 234, 2013.

401. Zhang, T., Yan, H.Q., Shen, L., Fang, Z.P., Zhang, X.M., Wang, J.J., Zhang, B.Y., Chitosan/Phytic Acid Polyelectrolyte Complex: A Green and Renewable Intumescent Flame Retardant System for Ethylene Vinyl Acetate Copolymer. *Ind. Engin. Chem. Res.*, 53, 19199–19207, 2014.

402. Cheng, X.W., Guan, J.P., Tang, R.C., Liu, K.O., Phytic acid as a bio-based phosphorus flame retardant for poly(lactic acid) nonwoven fabric. *J. Clean. Produc.*, 124, 114–119, 2016.

403. Fang, F., Huo, S.Q., Shen, H.F., Ran, S.Y., Wang, H., Song, P.G., Fang, Z.P., A bio-based ionic complex with different oxidation states of phosphorus for reducing flammability and smoke release of epoxy resins. *Compos. Commun.*, 17, 104–108, 2020.

404. Gao, Y.Y., Deng, C., Du, Y.Y., Huang, S.C., Wang., Y.-Z., A novel bio-based flame retardant for polypropylene from phytic acid. *Polym. Degrad. Stab.*, 161, 298–308, 2019.

405. Alongi, J., Andrea Carletto, R., Di Blasio, A., Carosio, F., Bosco, F., Malucelli, G., DNA: a novel, green, natural flame retardant and suppressant for cotton. *J. Mater. Chem. A*, 1, 4779–4785, 2013.

406. Carosio, F., Di Blasio, A., Alongi, J., Malucelli., G., Green DNA-based flame retardant coatings assembled through Layer by Layer. *Polymer*, 54, 5148–5153, 2013.

407. Jin, X., Cui, S., Sun, S., Gu, X., Li, H., Liu, X., Tang, W., Sun, J., Bourbigot, S., Zhang, S., The preparation of a bio-polyelectrolytes based core-shell structure and its application in flame retardant polylactic acid composites. *Compos. Part A Appl. Sci. Manuf.*, 124, 105485, 2019.

408. Chang, B.P., Thakur, S., Mohanty, A.K., Misra, M., Novel sustainable biobased flame retardant from functionalized vegetable oil for enhanced flame retardancy of engineering plastic. *Sci. Rep.*, 9, 15971, 2019.

409. Chen, Z.Y., Xiao, P., Zhang, J.M., Tian, W.G., Jia, R.N., Nawaz, H.F., Jin, K., Zhang, J., A facile strategy to fabricate cellulose-based, flame-retardant, transparent and anti-dripping protective coatings. *Chem. Engin. J.*, 379, 122270, 2020.

410. Zhang, Y., Yu, B., Wang, B., Liew, K.M., Song, L., Wang, C., Hu, Y., Highly effective P–P synergy of a novel DOPO-based flame retardant for epoxy resin. *Ind. Engin. Chem. Res.*, 56, 1245–1255, 2017.

411. Peng, H.Q., Zhou, Q., Wang, D.Y., Chen, L., Wang., Y.-Z., A novel charring agent containing caged bicyclic phosphate and its application in intumescent flame retardant polypropylene systems. *J. Ind. Engin. Chem.*, 14, 589–595, 2008.

412. Liang, W.J., Zhao, B., Zhao, P.H., Zhang, C.Y., Liu, Y.Q., Bisphenol-S bridged pent2a(anilino)cyclotriphosphazene and its application in epoxy resins: Synthesis, thermal degradation, and flame retardancy. *Polym. Degrad. Stab.*, 135, 140–151, 2017.

413. Huang, J.Q., Zhang, Y.Q., Yang, Q., Liao, X., Li, G.X., Synthesis and Characterization of a Novel Charring Agent and Its Application in Intumescent Flame Retardant Polypropylene System. *J. Appl. Polym, Sci.*, 123, 1636–1644, 2012.

414. Jian, R.K., Wang, P., Duan, W.S., Wang, J.S., Zheng, X.L., Weng, J.B., Synthesis of a Novel P/N/S-Containing Flame Retardant and Its Application in Epoxy Resin: Thermal Property, Flame Retardance, and Pyrolysis Behavior. *Ind. Engin. Chem. Res.*, 55, 11520–11527, 2016.

415. Huo, S.Q., Wang, J., Yang, S., Chen, X., Zhang, B., Wu, Q.L., Zhang, B., Flame-retardant performance and mechanism of epoxy thermosets modified with a novel reactive flame retardant containing phosphorus, nitrogen, and sulfur. *Polym. Adv. Technol.*, 29, 497–506, 2018.

416. Chen, M.J., Shao, Z.B., Wang, X.L., Chen, L., Wang, Y.-Z., Halogen-Free Flame-Retardant Flexible Polyurethane Foam with a Novel Nitrogen-Phosphorus Flame Retardant. *Ind. Engin. Chem. Res.*, 51, 9769–9776, 2012.

417. Thirumal, M., Khastgir, D., Nando, G.B., Naik, Y.P., Singha, N.K., Halogen-free flame retardant PUF: Effect of melamine compounds on mechanical, thermal and flame retardant properties. *Polym. Degrad. Stab.*, 95, 1138–1145, 2010.

418. Tang, S., Qian, L.J., Liu, X.X., Dong, Y.P., Gas-phase flame-retardant effects of a bi-group compound based on phosphaphenanthrene and triazine-trione groups in epoxy resin Polym. *Degrad. Stab.*, 133, 350–357, 2016.

419. Qiu, Y., Qian, L.J., Xi, W., W Flame-retardant effect of a novel phosphaphenanthrene/triazine-trione bi-group compound on an epoxy thermoset and its pyrolysis behaviour. *RSC Adv.*, 6, 56018–56027, 2016.

420. Mishra, N. and Vasava, D., Recent developments in s-triazine holding phosphorus and nitrogen flame-retardant materials. *J. Fire Sci.*, 38, 552–573, 2020.

421. Allen, C.W., The Use of Phosphazenes as Fire Resistant Materials. *J. Fire Sci.*, 11, 320–328, 1993.

422. Zhou, X., Qiu, S.L., Mu, X.W., Zhou, M.T., Cai, W., Song, L., Xing, W.Y., Hu, Y., Polyphosphazenes-based flame retardants: A review. *Compos. Part B - Engin.*, 202, 108397, 2020.

423. Klinkowski, C., Wagner, S., Ciesielski, M., Döring, M., Bridged phosphorylated diamines: Synthesis, thermal stability and flame retarding properties in epoxy resins. *Polym. Degrad. Stab.*, 106, 122–128, 2014.

424. Schafer, A., Seibold, S., Walter, O., Döring., M., Novel high T-g flame retardancy approach for epoxy resins. *Polym. Degrad. Stab.*, 93, 557–560, 2008.

425. Garth, K., Klinkowski, C., Fuhr, O., Döring, M., Synthesis of a new phosphorylated ethylamine, thereon based phosphonamidates and their application as flame retardants. *Heteroatom Chem.*, 28, e21407, 2017.

426. Tang, S., Qian, L.J., Qiu, Y., Dong, Y.P., High-performance flame retardant epoxy resin based on a bi-group molecule containing phosphaphenanthrene and borate groups. *Polym. Degrad. Stab.*, 153, 210–219, 2018.

427. Zhang, P.K., Fan, H.J., Tian, S.Q., Chen, Y., Yan, J., Synergistic effect of phosphorus-nitrogen and silicon-containing chain extenders on the mechanical properties, flame retardancy and thermal degradation behavior of waterborne polyurethane. *RSC Adv.*, 6, 72409–72422, 2016.

428. Shao, L.S., Xu, B., Ma, W., Wang, J.Y., Liu, Y.T., Qian, L.J., Flame retardant application of a hypophosphite/cyclotetrasiloxane bigroup compound on polycarbonate. *J. Appl. Polym, Sci.*, 137, 48699, 2020.

429. Peng, Y., Xue, B.X., Song, Y.H., Wang, J., Niu, M., Preparation of a novel phosphorus-containing organosilicon and its effect on the flame retardant and smoke suppression of polyethylene terephthalate. *Polym. Adv. Technol.*, 30, 1279–1289, 2019.

430. Luo, H.Q., Rao, W.H., Liu, Y.L., Zhao, P., Wang, L., Yu, C.B., Novel multi-element DOPO derivative toward low-flammability epoxy resin. *J. Appl. Polym, Sci.*, 137, e49427, 2020.

431. Zhang, W.C., Li, X.M., Yang, R.J., Flame retardant mechanisms of phosphorus-containing polyhedral oligomeric silsesquioxane (DOPO-POSS) in polycarbonate composites. *J. Appl. Polym, Sci.*, 124, 1848–1857, 2012.

432. Qian, X.D., Song, L., Yuan, B.H., Yu, B., Shi, Y.Q., Hu, Y., Yuen, R.K.K., Organic/inorganic flame retardants containing phosphorus, nitrogen and silicon: Preparation and their performance on the flame retardancy of epoxy

resins as a novel intumescent flame retardant system. *Mater. Chem. Phys.*, 143, 1243–1252, 2014.

433. Liu, C., Chen, T., Yuan, C.H., Song, C.F., Chang, Y., Chen, G.R., Xu, Y.T., Dai, L.Z., Modification of epoxy resin through the self-assembly of a surfactant-like multi-element flame retardant. *J. Mater. Chem. A*, 4, 3462–3470, 2016.

434. Ai, L., Yang, L., Hu, J., Chen, S., Zeng, J., Liu, P., Synergistic Flame Retardant Effect of Organic Phosphorus-Nitrogen and Inorganic Boron Flame Retardant on Polyethylene P. *Polym. Engin. Sci.*, 60, 414–422, 2020.

435. Yang, S., Zhang, Q.X., Hu, Y.F., Synthesis of a novel flame retardant containing phosphorus, nitrogen and boron and its application in flame-retardant epoxy resin. *Polym. Degrad. Stab.*, 133, 358–366, 2016.

436. Zhang, T., Liu, W.S., Wang, M.X., Liu, P., Pan, Y.H., Liu, D.F., Synthesis of a boron/nitrogen-containing compound based on triazine and boronic acid and its flame retardant effect on epoxy resin. *High Perform. Polym.*, 29, 513–523, 2017.

437. Jin, W.Q., Yuan, L., Liang, G.Z., Gu, A.J., Multifunctional Cyclotriphosphazene/Hexagonal Boron Nitride Hybrids and Their Flame Retarding Bismaleimide Resins with High Thermal Conductivity and Thermal Stability. *ACS App. Mater. Interf.*, 6, 14931–14944, 2014.

438. Song, P.A., Xu, L.H., Guo, Z.H., Zhang, Y., Fang, Z.P., Flame-retardant-wrapped carbon nanotubes for simultaneously improving the flame retardancy and mechanical properties of polypropylene. *J. Mater. Chem.*, 18, 5083–5091, 2008.

439. Yan, L., Xu, Z.S., Deng, N., Chu, Z.Y., Synergistic effects of mono-component intumescent flame retardant grafted with carbon black on flame retardancy and smoke suppression properties of epoxy resins. *J. Therm. Anal. Calorim.*, 138, 915–927, 2019.

440. Yan, L., Xu, Z.S., Wang, X.H., Deng, N., Chu, Z.Y., Preparation of a novel mono-component intumescent flame retardant for enhancing the flame retardancy and smoke suppression properties of epoxy resin. *J. Therm. Anal. Calorim.*, 134, 1505–1519, 2018.

441. Yue, X.P., Li, C.F., Ni, Y.H., Xu, Y.J., Wang, J., Flame retardant nanocomposites based on 2D layered nanomaterials: a review. *J. Mater. Sci.*, 54, 13070–13105, 2019.

442. Fang, F., Ran, S.Y., Fang, Z.P., Song, P.G., Wang, H., Improved flame resistance and thermo-mechanical properties of epoxy resin nanocomposites from functionalized graphene oxide via self-assembly in water. *Compos. Part B-Engin.*, 165, 406–416, 2019.

443. Bao, C.L., Guo, Y.Q., Yuan, B.H., Hu, Y., Song, L., Functionalized graphene oxide for fire safety applications of polymers: a combination of condensed phase flame retardant strategies. *J. Mater. Chem.*, 22, 23057–23063, 2012.

444. Guo, Y.Q., Bao, C.L., Yuan, B.H., Hu, Y., In Situ Polymerization of Graphene, Graphite Oxide, and Functionalized Graphite Oxide into Epoxy Resin and

Comparison Study of On-the-Flame Behavior. *Ind. Engin. Chem. Res.*, 50, 7772–7783, 2011.

445. Cai, W., Feng, X.M., Wang, B.B., Hu, W.Z., Yuan, B.H., Hong, N.N., Hu, Y., A novel strategy to simultaneously electrochemically prepare and functionalize graphene with a multifunctional flame retardant. *Chem. Engin. J.*, 316, 514–524, 2017.

446. Cai, W., Guo, W.W., Pan, Y., Wang, J.L., Mu, X.W., Feng, X.M., Yuan, B.H., Wang, B.B., Hu, Y., Polydopamine-bridged synthesis of ternary h-BN@PDA@SnO2 as nanoenhancers for flame retardant and smoke suppression of epoxy composites. *Compos. Part A: Appl. Sci. Manuf.*, 111, 94–105, 2018.

10

Other Non-Halogenated Flame Retardants and Future Fire Protection Concepts & Needs

Alexander B. Morgan¹*, Paul A. Cusack²** and Charles A. Wilkie³**

¹Center for Flame Retardant Materials Science, University of Dayton Research Institute, Dayton, OH, USA
²ITRI Innovation Ltd, St. Albans, Herts AL2 2DD, United Kingdom
³Marquette University, Milwaukee, WI, USA

Abstract

In the pursuit of non-halogenated flame retardant solutions, there are large swaths of the periodic table which have not been fully studied for their flame retardant potential. In this chapter, these other elements are discussed including known flame retardant effects of transition metals, sulfur compounds, and tin compounds. Carbon based flame retardant solutions including cross-linking chemical groups, carbonates, and bio-based materials are also discussed. Some discussion on engineering solutions based upon metal and chemical coatings which enable non-halogenated flame retardancy is included as well in this chapter. Finally, thoughts on future directions in non-halogenated flame retardancy, especially experimental design and environmental considerations, conclude the chapter.

Keywords: Transition metals, sulfur, carbon, bio-based, tin, coatings, future solutions

10.1 The Periodic Table of Flame Retardants

Modern flame retardant chemistry, where specific elements and classes of flame retardant additives began to be used in earnest, occurred in the early

---

**Corresponding author*: alexander.morgan@udri.udayton.edu; Original Chapter Author for 1st Edition and only author for 2nd Edition.
**Original Chapter Co-Author for 1st Edition. This chapter has been updated with new content by Alexander Morgan.

Alexander B. Morgan (ed.) Non-Halogenated Flame Retardant Handbook 2nd edition, (475–554)
© 2022 Scrivener Publishing LLC

20th century with the broad utilization of halogenated flame retardants, meaning those compounds with halogen (group 17) of the periodic table present somewhere in their chemical structure. More specifically, chlorine and bromine were found to be very effective at providing flame retardancy against a wide range of fire threats and fuel sources. At this time, group 17 (halogen) is the only part of the periodic table where all non-radioactive elements in that family (which would exclude astatine and tennessine in this case) have shown some potential flame retardant effect, namely vapor phase flame retardancy. There may be other groups that, in general, have an effect, but either the trend in that group has not been fully explored to prove this, or, several exceptions have been found to suggest the group likely does not have a universal flame retardant effect. This observation leads to a question about which elements in the periodic table show flame retardant effects? The answer is that only a select group of elements to date have shown flame retardant effects and even then, specific chemical structures with that element may be needed to impart flame retardancy, just as is sometimes seen with halogen. In Figure 10.1 a periodic table showing elements where flame retardancy has been reported (in the open literature as of the writing of this chapter) is shown. Colors are used to indicate if the predominant mechanism of flame retardancy for that element has been identified, and if so, the predominant mechanism (vapor phase, endothermic cooling, condensed phase) for that element is shown. Some elements do show more than one color because they are effective in more than one way, eg., phosphorus as a function of their chemical structure and how they interact with a particular fuel (combustible solids/polymers for the purpose of this chapter). Halogens can abstract hydrogen from a polymer, thus possibly leading to

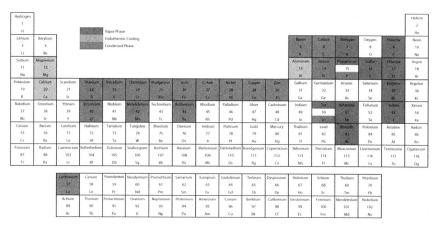

Figure 10.1 Periodic table of flame retardants.

the formation of double bonds and hence charring, meaning some activity in the condensed phase, but by far the largest effect is in the vapor phase and thus they are denoted herein only as vapor phase additives. Only the central elements which impart flame retardancy are shown; other elements may be in the chemical structure of a flame retardant chemical/material, but they are not the main element responsible for the flame retardant effect.

While there are many elements that have shown flame retardant effects, some of those elements have very narrow and specific effects. In general, Mg and Al are mostly only responsible for flame retardancy as hydroxides, but more recently these common ions are used in combination with other salts to impart some improved flame retardancy effects when combined with other elements. To put it another way, some elements by themselves have a very minor role when used as flame retardants, but combined with other elements they may have more potential. So when thinking about non-halogenated flame retardants, one must look quite far and wide across the periodic table beyond group 17, and when this is done, there is a lot of potential but unvalidated chemistry that could be used. Therefore new flame retardant chemistries could be quite varied if one considers the entire periodic table. Some elements can obviously be eliminated from consideration due to toxicity, radioactivity, or rarity/cost, which still leaves quite a large number of elements to combine. Even then, practical chemistry further narrows the choices which leads to the main themes of this chapter; non-halogenated flame retardants that to date show commercial performance (even if limited to select polymers), or show some potential as the likely flame retardants of the future. Indeed, if one looks at the overall book contents that include this chapter, the specific chapters in this book represent the current practical non-halogenated flame retardants that are available. Still, there is value in revisiting the periodic table and conducting basic research in this area to discover new potential flame retardants. With this in mind, this chapter focuses on the other flame retardant chemistries not fully elucidated in the other chapters of this book. This chapter will not cover all of the known examples of these new and novel flame retardants, but it will include examples of relevance and insight into how they work so that they could be used by the readers of this chapter, or, at least the reader will understand their limitations and how they are used today.

Assuming the author has read the contents of this book, especially the previous chapter, it should be obvious that the path forward for non-halogenated flame retardants, and indeed, any flame retardant, is to combine multiple chemistries together to obtain the best balance of properties and enhancements in flame retardant effects. This likely means that the reader should consider combining multiple flame retardant mechanisms and chemistries while avoiding antagonistic interactions, and this may mean that the chemistries of this chapter will find more use in fulfilling that multi-component flame retardant approach.

10.2 Transition Metal Flame Retardants

Transition metals have been studied for flame retardant effects via both vapor phase and condensed phase mechanisms. In theory, transition metal hydroxides may have an endothermic cooling mechanism of flame retardancy, provided the transition metal hydroxide in question decomposes to an oxide with release of water at a temperature within the temperature ranges of decomposing polymers. However, this mechanism of relying upon transition metal hydroxide decomposition is not widely exploited to date and does not appear to be a potent route to flame retardancy except with the metals present in double layered hydroxides [1–3]. Even then, transition metals in double layered hydroxides tend to be of such high molecular weight that they cannot volatilize into the flame, and so if capable of condensed phase char formation/catalysis reactions, double layered hydroxides could correctly be classified as condensed phase flame retardants. Considering this, the endothermic mechanism can mostly be eliminated, such that the current science points to transition metals as being primarily vapor phase or condensed phase flame retardants.

10.2.1 Vapor Phase Transition Metal Flame Retardants

Metals and combustion has been studied since the 1920s when organometallic compounds like tetraethyllead were used as anti-knock compounds for fuels. More recently, work conducted at the US National Institute of Standards and Technology (NIST) focusing on Halon extinguishing agent replacement found that various transition metals inhibited vapor phase combustion [4–6]. Specifically, iron carbonyl ($Fe(CO)_5$) was found to be the most effective flame inhibitor, followed by methylcyclopentadienylmanganese tricarbonyl and tetramethyltin when compared to bromotrifluoromethane ($CBrF_3$). While the mechanism is not fully understood, there appears to be complex interactions between combustion free radicals and metal oxidation states which inhibit vapor phase combustion reactions, thus allowing extinguishment to occur. These compounds however were not investigated in polymers due to their volatility or toxicity, and so short of someone finding a way to successfully encapsulate these additives, they likely will never be flame retardants for polymers. However, this result suggests that if non-toxic non-volatile Fe, Mn, or Sn compounds could be made and compounded into polymers, and these materials can volatilize under fire conditions, then these molecules may impart some vapor phase flame retardancy to those polymers. To this end, there has been some limited work with ferrocene as a flame retardant for polymers [7], and metallocenes for paper [8, 9], but more work needs to be done in this area. One word of caution though with metal based flame retardants needs to be mentioned, and that is the product requirements for environmental performance of plastics. As was discussed in Chapter 1, certain metals are banned from use in plastics due to their toxicity/environmental effects, and so these regulations and known issues with these metals will prevent them from being used, even if they do show effectiveness in polymers. Metals such as chromium, nickel, cadmium, and lead are often a target for removal from use in polymers, and other metals with negative environmental or toxicity profiles will not be allowed for use, nor should they be used. Researchers studying vapor phase flame retardants should look closely at these toxicity profiles of the transition metals before searching for new metal-based vapor phase flame retardants. Failure to do so will likely result in wasted effort. It may perhaps yield some interesting basic chemistry, but that chemistry will not yield any new non-halogenated technologies if the toxicity profile is unacceptable. Certain metals due to their cost/rarity are also likely to be impractical unless they show strong flame retardant effects at catalytic levels.

10.2.2   Condensed Phase Transition Metal Flame Retardants

Far more work has been done to date on condensed phase transition metal flame retardants as char enhancers or smoke reducing agents than has been accomplished with transition metals as vapor phase flame retardants. This is partly a matter of practicality in that metal salts, oxides, and ligand complexes are generally not volatile and can be more easily inserted into polymers for flame retardant studies via traditional polymer formulation techniques such as melt compounding or dispersing of additives in monomer prior to polymerization. Further, results from the metal catalysis field, where transition metals enable carbon-carbon bond formation at elevated temperatures, suggest that transition metals could be flame retardant char formation catalysts. This potential for catalysis has led to the general approach, as will be discussed, of combining metals with other flame retardants to gain some improved fire performance benefits.

Transition metal flame retardants for condensed phase effects fall into two broad classes: metal oxides and metal complexes. There has been one report on the use of nanoscale metal particles providing improved fire performance in the presence of phosphorus [10], but otherwise the metals used to date have been metal oxides or metal ions/metal ion complexes. To some extent, it does not make sense to use metals as an additive for flame retardancy as metals can be oxidized under fire conditions, and this oxidation can be highly exothermic as the metal goes from an unoxidized state towards its final oxide. A review of this overall field was conducted in 2009 [11], so this chapter will focus more on newer advances and insight into the practicality of using such metal compounds for flame retardancy.

10.2.2.1   Metal Oxides

Oxides cover a wide range of chemical formulae, especially when considering the differing oxidation states that exist for many of the transition metals. For flame retardancy, metal oxides are most often combined with other flame retardants, but occasionally they are used by themselves to impart some additional flame retardancy to an engineering plastic or a low-flammability plastic. Examples include the use of mixed oxide fly ash in polycarbonate [12], mixed metal oxide glasses for PVC [13], mixed metal oxides as charring catalysts for polyolefins [14], and one report of various transition metal oxides added directly to poly(butylene terephthalate) [15]. When the oxides are used alone with no other flame retardants, the underlying hypothesis, mostly proven by these papers, is that the metal oxide assists in additional char formation by catalyzing C-C bond formation in the condensed phase during the burning of the polymers. For the PVC example, some C-Cl bond activation and cross-linking may have occurred based upon known metal oxide chemistry, but there is no mechanistic evidence to prove this hypothesis at this time. The above mentioned paper [14] which used mixed-metal oxide catalysts to make polyolefins into charring polymers during burning, with some success, suggests that oxides do have an important role to play as flame retardants, but they have not been successful in imparting regulatory levels of fire safety, and so it seems more likely that metal oxides will have to be combined with other flame retardants.

Building upon the goal that metal oxides help with char formation/char thermal stability to impart flame retardant effects, the majority of studies on transition metal oxides have been when the metal oxide is combined with other flame retardants. Specifically, the goal was to use the transition metal oxide to enhance the flame retardant effect and/or char thermal stability when used with other flame retardants. Examples include mixing the oxides with silicon compounds [15], phosphorus compounds [16–18] and intumescent compounds [19–28]. In all of these cases, the metal oxides help with char formation or provide thermal stability to the char such that it can provide stronger/more robust flame retardancy. Some metal oxides are more effective than others in these various papers, and the reasons for these differences are sometimes clear (obvious chemical reaction between flame retardant and metal oxide at the right time in the fire) or not at all clear (no mechanistic studies yet performed). When metal oxides are used, they are used in small amounts as a synergist or flame retardant enhancer. These results, plus those provided by studies of just metal oxide and polymer flammability with no additional flame retardants, suggest that metal oxides could be a potent source of non-halogen flame retardant synergism in the future. They would be used in small amounts to enhance existing systems, or perhaps enable a non-halogenated flame retardant system to achieve robust performance to replace an existing halogenated flame retardant system. However, because there are no well-defined guidelines for how much oxide to add with a particular flame retardant in a particular polymer, future work in this area will be very empirical and exploratory as scientists study this further. Despite the lack of clear data on how to use metal oxides, the existing data shows great promise, and metal oxides should be explored as parts of future non-halogenated flame retardant systems.

On a final note, toxicity of the metal oxide to be considered as a flame retardant must be studied in any future system. As with potential metal based vapor phase flame retardants, if the metal in the oxide has unacceptable toxicity or falls under existing environmental protocols (see Chapter 1 of this book), such a metal oxide would not be acceptable for use in a polymer no matter how good its flame retardant effects. Some metals with obvious toxicity concerns (Cr, Ni, Hg, Cd, Pb, Os, etc.) should not be considered as flame retardants unless environmental and extraction studies are conducted to show that the particular oxide is safe for use.

10.2.2.2   Metal Complexes

Next to metal oxides, most of the studies on transition metals for flame retardants has been with metals complexed to various ligands or counter ions. The goal behind the use of metal complexes as flame retardants is to utilize metal + organic ligand combinations to impart (A) better dispersion and compatibility with the polymer matrix and (B) carbon-carbon bond formation/cross-linking reactions under pre-ignition and post-ignition scenarios. The range of chemistries used for this goal has been highly varied, and similar to the metal oxide discussion previously, the metal complexes are combined with other flame retardants to achieve good flame retardancy in a wide range of polymers. Examples include smoke suppressants [29–32], polymer + metal chelates for use with mineral fillers [33–35], and metal + ligand complexes [36–46]. Some structures which have been studied to date are shown in Figures 10.2 and 10.3 below.

Choosing metal complexes for flame retardancy is more complicated than that for metal oxides for two main reasons. The first is that like metal oxides, no comprehensive fundamental work has been done, and so the

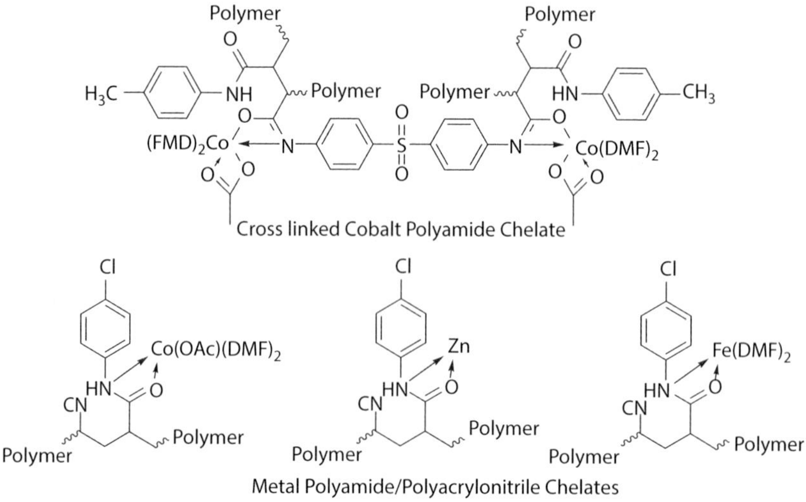

Figure 10.2 Metal + Polymer chelates.

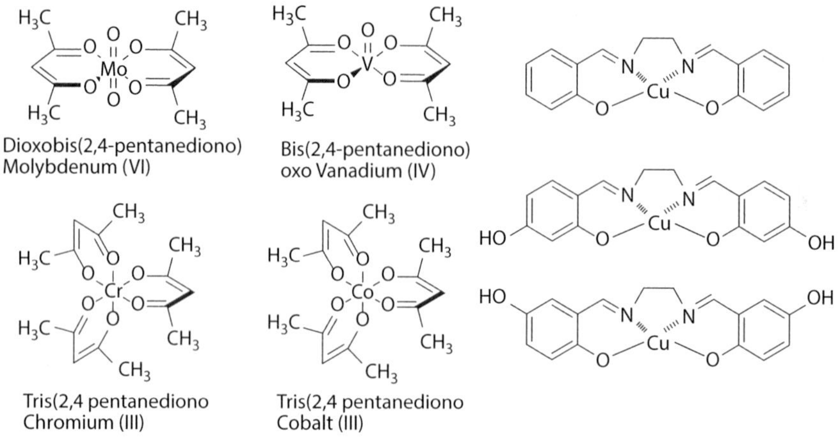

Figure 10.3 Metal complexes reported to have activity as FRs.

choice of which metal to use as a flame retardant is rather unknown. The second is that the organic ligands on the metal must be thermally stable and compatible with the polymer matrix processing conditions and chemistry. For example, a polar ligand for a metal may have difficulty dispersing in a non-polymer polymer, and if that polymer were a thermoplastic, the ligands and complex would need to have a thermal stability above that of the polymer processing temperature, otherwise the metal complex may decompose during processing. Therefore polarity and dispersability of the complex should be considered in experimental design, and thermal stability of the complex (as measured by differential scanning calorimetry or thermogravimetric analysis) should be conducted prior to formulating new non-halogenated flame retardant systems. The only other guidance on using this class of flame retardants is that they should be used with other flame retardants, and used at low loadings as flame retardant enhancers or synergists. Existing data does not seem to indicate these materials can pass regulatory flammability tests by themselves, although it is possible that an example which could pass a specific regulatory test has yet to be discovered. Despite the lack of definite guidance on how to use transition metal complexes as flame retardants, the potential benefits of these flame retardants, especially since only small amounts would need to be used to enhance other non-halogenated flame retardants, suggests that they should be considered in future material design and exploratory studies. One final note, as with the other transition metal flame retardants discussed in this section of the chapter, toxicity of the metal and the ligands should be considered prior to actual use of the compound. Toxic metal ions should be avoided, and any ligands which impart high bioavailability/uptake should also be avoided as well.

There are two final notes to consider in regards to metals. First, there have been some reports of more simple ionic metal species providing some flame retardant benefit. Several metal nitrates (Cu, Zn, Fe, Al) have shown some flame retardant benefit when combined with alumina trihydrate in poly(ethylene-co-vinyl acetate) (EVA) resin [47], and zinc triflate $(CF_3SO_3^{-1})$ combined with poly(acrylonitrile-butadiene-styrene) (ABS) [48]. For the nitrates, the mechanism of flame retardancy in this case appears to be the enhancement of release of non-flammable gases. However, the metal salt solubility in water likely would limit the use of this material, but there may be additional benefit provided by the transition metal if a metal nitrate could be utilized which has low solubility in water. Enhanced char formation was reported when potassium nitrate was used in polyamide-6 [49], but again, water solubility of the potassium salt may limit use of the approach. For the triflate salt in ABS, the mechanism was condensed phase with enhancements in char yield. It's worthwhile to note that nitrates are oxidizers while the triflate is a Lewis-acid, and other Lewis acids as metal halides have shown flame retardant effects [50]. Further, there is the increasing use of different metals in combination with alkyl phosphinate salts with and without melamine as part of the overall metal salt complex [51–53]. Metal salts with alkyl phosphinates are covered in Chapter 2 of this book but the newer combinations with metals and melamine bear attention, as this combination suggests some further experimental space to investigate.

10.3   Sulfur-Based Flame Retardants

To a large extent, sulfur is a forgotten element in flame retardancy. Gay-Lussac proposed in the nineteenth century a mixture of ammonium sulfate, ammonium phosphate and ammonium chloride as a flame retardant for theatre curtains [54].

Currently, sulfur is only used commercially as a flame retardant for polycarbonate, using potassium diphenylsulphonesulphonate, potassium perfluorobutylsulphonate and sodium trichlorobenzenesulfonate. Flame retardancy is thought to occur by promoting decomposition and cross-linking of polycarbonate, inhibiting combustion by generating carbon dioxide and forming a surface char layer [55–61].

Little was heard of sulfur as a putative flame retardant for any other polymers until about the 1990s when Lewin began reporting on the use of ammonium sulfamate as a flame retardant in wool, cellulose and polyamides [62–65]. In wool and cotton, sulfation occurs through an OH group and leads to durable (50 alkaline launderings) flame retardancy. With polyamides, a small amount of pentaerythritol is also required along with the sulfamate and this leads to chain scission of the alkyl-amide bond leading to char formation and its migration to the surface of the PA. A UL-94 V0 rating can be obtained at both 1/16 and 1/32 inch (1.6mm and 0.80 mm respectively).

Based on the previous work and general knowledge of the area, it was felt that sulfur continued to be a neglected element in flame retardancy and that it should continue to be investigated. In work from the Wilkie laboratory, it has been found that three sulfur compounds, ammonium sulfamate, sodium diphenylamine-4-sulfonate and 3-(1-pyridino)-1-propane sulfonate had some flame retardant effect in both PS and PMMA [66]. Perhaps the most interesting is the use of ammonium sulfamate in polystyrene where a reduction in the peak heat release rate of about 50% was obtained. More recently, there has been a report of using sulfonamides and thiocyanurates to provide a condensed phase free-radical reaction that results in polypropylene depolymerizing, which enables the polymer to drip away from a flame [67, 68]. This approach has also been found to work with disulfides [69]. However, this approach of inducing depolymerization to pass a fire test must be used with care, as it often results in an increase in heat release for the material that uses this type of flame retardant mechanism. In another example, sulfonic acid salts were combined with a polymeric conformal coating system to yield significant reductions in heat release for a polyurethane foam [70]. The results from this work suggest that in this case, the conformal coating helps reduce mass loss and prevents the flexible polyurethane foam from collapsing and forming a fuel pool fire, and the sulfur compounds show some vapor phase effect with changes noted in effective heat of combustion when flammability was measured by cone calorimeter. There have been some reports as well of phosphorus-sulfur synergism, but this is a case where sulfur is combined with other chemistries to be effective, rather than being used by itself [71–75].

The key question here is this: Do sulfur compounds have a future in flame retardancy? At this time, the answer is no to this question until someone does a more exhaustive study on sulfur compounds in a variety of polymers to identify the additives that may be used and situations in which they may be useful. One of the hazards that must be borne in mind is the potential release of noxious gases, like SO_2 or H_2S, during the decomposition and fire events. Release of either of these would certainly be considered a negative feature, especially in limited ventilation compartment fires. On the other hand, the wide diversity of sulfur compounds suggests that there may be some gems hidden there, provided the release of irritant sulfur-containing gases from fire events does not become a fundamental issue that cannot be addressed with these materials.

10.4   Carbon-Based Flame Retardants

At first glance, the idea of carbon based flame retardants is rather counterintuitive since carbon burns and combusts in the presence of oxygen. However, flame retardancy is not the absolute prevention of ignition or burning, but rather the slowing/retarding of flame spread and growth. Therefore, carbon-based chemistries which either inhibit vapor phase combustion, allow for endothermic cooling, or help in the formation of thermally stable char can be considered as flame retardants. Of the carbon based flame retardants, there are three broad classes; cross-linking compounds, organic carbonates, and expandable graphite. Cross-linking compounds and expandable graphite are condensed phase flame retardants while organic carbonates are endothermic cooling compounds/fuel diluters. Carbon based flame retardants are not capable of vapor phase flame retardancy because once the carbon radicals are pyrolyzed into the flame front, they will combust as carbon radicals are the primary driver behind carbon oxidation and combustion reactions and any such radicals generated will simply be consumed and help propagate the fire.

10.4.1 Cross-Linking Compounds – Alkynes, Deoxybenzoin, Friedel-Crafts, Nitriles, Anhydrides

If polymeric "fuel" can be converted into thermally stable carbonific char during heating, then flammability should be reduced and flame retardancy will be obtained. This approach has been extensively studied to date via the use of alkynes and deoxybenzoin chemical groups, and through the use of Friedel-Crafts chemistry.

10.4.1.1 Alkynes

Alkynes were originally used in polymeric structures to provide higher use temperatures for aerospace grade composites, and in the late 1990s were studied as flame retardant additives for commodity thermoplastics [76, 77]. More recently, alkyne containing phthalic anhydrides were copolymerized with polyesters for flame retardant effects [78]. In these examples, the alkynes activate at elevated temperatures to form additional char while the polymer is burning such that more of the polymer is converted into glassy carbon/graphitic char which would not easily burn. Alkynes generaly work through the formation of radicals when the alkyne bond breaks at elevated temperatures and forms di-radicals which, in turn, react with each other (or the polymer) to form thermally stable complex polyaromatic hydrocarbons (Figure 10.4). Char formation is a potent method of non-halogenated flame retardancy, but, the approach does not always work with every polymer, especially in cases where the base polymer has very few functional groups available for char formation. Some examples of alkyne-functionalized structures that have been used for flame retardancy are shown in Figure 10.5, and include carbonates, enediynes, and phosphorus-alkyne compounds. Some newer work has utilized an alkyne containing phthalic anhydride (4-phenylethylnylphthalic anhydride) that can react with polyesters or epoxies to enable anti-dripping behavior during burning [79, 80].

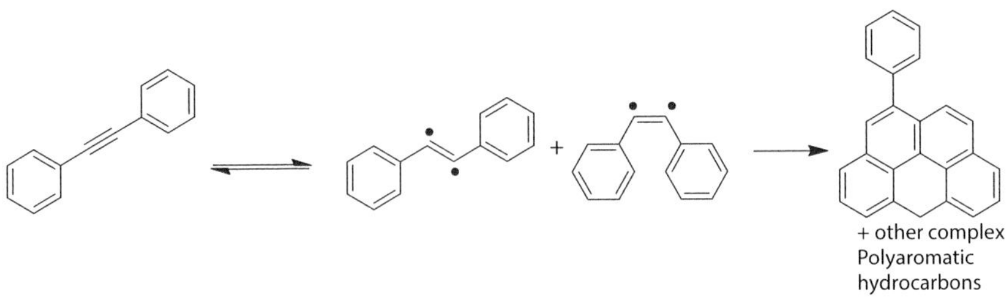

Figure 10.4 General alkyne cross-linking mechanism.

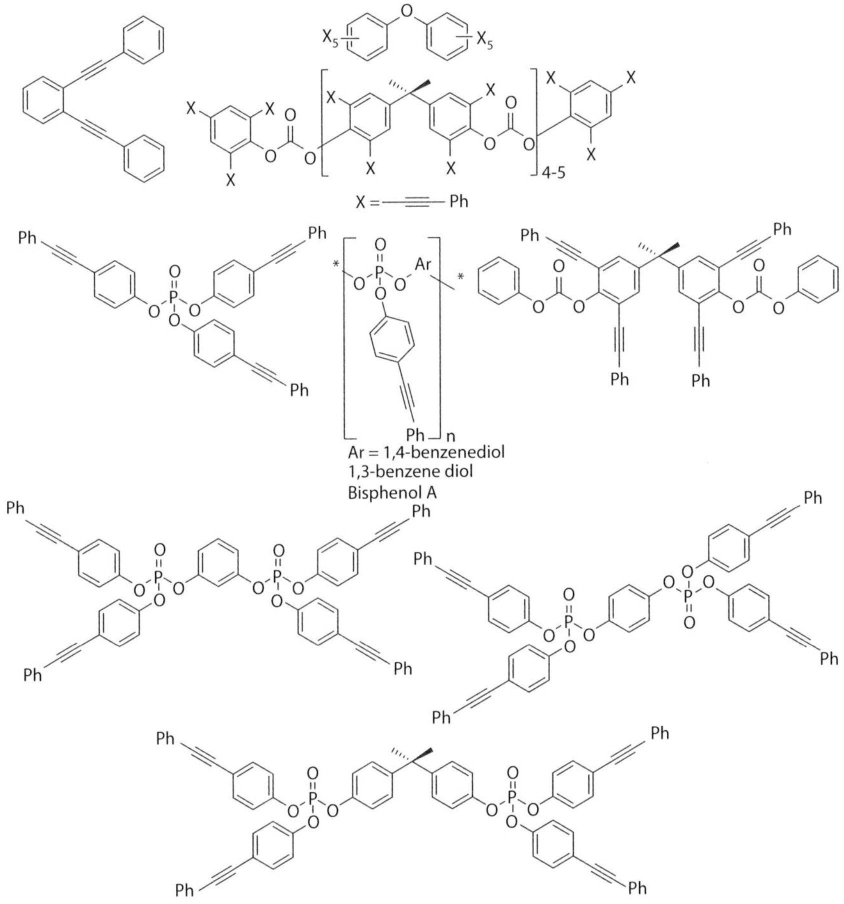

Figure 10.5 Alkyne containing flame retardants.

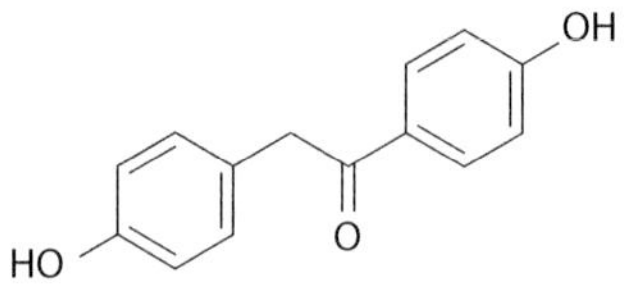

Figure 10.6 Deoxybenzoin monomer.

10.4.1.2   Deoxybenzoin

Deoxybenzoin (Figure 10.6) is a newer molecule that has been shown to greatly reduce heat release in a wide range of polymers when it is used as a co-monomer. Reductions in heat release have been seen in a very wide range of polymers [81–87]. The polymers containing high levels of deoxybenzoin have also been blended into other polymers to impart flame retardancy. The mechanism of action for this molecule appears to be a loss of water and the formation of an alkyne which rapidly rearranges to form a graphitic structure [88]. In the presence of other polymers, these alkynes may induce additional cross-linking and char formation, but more study is needed on this new chemistry to determine its efficacy as a flame retardant additive vs. its success to date as a monomer structure to yield flame retardant polymers.

10.4.1.3   Friedel-Crafts

The Friedel-Crafts reaction is a simple alkylation of an aromatic ring by an alkyl halide (Figure 10.7). As such, one can imagine that this could be used to effectively cross-link aromatic polymers, most likely the styrenic family. As the reaction is most commonly used, anhydrous aluminum chloride is the catalyst, and this presents obvious difficulties in incorporating this material into a polymer. It has been found that a wide variety of Lewis acids as well as proton acids can be used as the catalyst [89].

If this chemistry is to be used, cross-linking of the polymer must not occur in use/processing but only when the polymer is challenged by a fire.

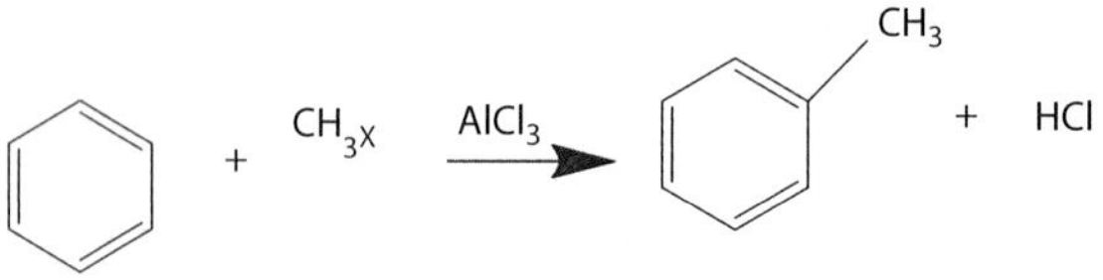

Figure 10.7 The Friedel-Crafts Reaction.

Thus one prefers to have a mixture of the polymer and alkylating agent always present but the catalyst only present at high temperatures. The evolution of HCl is also quite undesirable as this noxious gas will certainly be a problem.

Grassi and Gilks achieved cross-linking of polystyrene using $SnCl_4$ as the catalyst and p-di(chloromethyl)benzene as the alkylating agent [90]. They found that the cross-linked polymers that were produced were less thermally stable than the starting polymer. Brauman used various alkylation and acylation reagents with *in situ* formed $SbCl_3$ as the catalyst. The amount of char formation increased and the amount of volatiles decreased upon alkylation but both char and volatiles increased by acylation [91]. Rabek and Lucki used 1,2-dichloroethane or carbon tetrachloride as the alkylation agent in the presence of anhydrous aluminum chloride [92].

Extensive work on this topic was conducted at Marquette University. In order to eliminate the HCl problem, the alkylating agent that was used was an alcohol, which will eliminate only water. Model studies were conducted using alcohols as the alkylating agents and zeolites as catalysts. With some zeolites the extent of alkylation increased with temperature while with others it decreased. One could utilize 1,4-benzenedimethanol as the alkylating agent with a zeolite in a sealed tube and find that alkylation occurs at 300 °C but not at 200 °C. Unfortunately the alkylating agent will volatilize in an open system so this is not suitable [93].

Pearce showed that one can incorporate chloromethyl functional groups on polystyrene and in the presence of a catalyst, increase the amount of char formation [94–96]. Thus polystyrene was modified by copolymerization of styrene with vinylbenzyl alcohol and 2-ethylhexyl diphenyl phosphate was the Friedel-Crafts catalyst. The phosphate decomposes at about 220 °C by the loss to ethylhexene to give the actual catalyst, diphenylphosphoric acid. Thus alkylation cannot occur until the alkylphosphate is produced. The peak heat release rate is reduced from almost 1000 kW/m^2 in pristine polystyrene to a little less than 400 kW/m^2 when the alkylating agent and catalyst are both present [97–101].

The limitations of this technology are that it can only be used with aromatic polymers and the catalyst must only be active at the temperature at which one wants cross-linking to occur. Thus one must use a catalyst precursor which will form the desired catalyst at the proper temperature. This may be difficult to achieve; on the other hand, it offers the promise of substantial cross-linking which is expected to provide flame retardancy.

Since the work of Wilkie on this type of chemistry, there has only been one other recent study of this type of chemistry, where this chemistry was used with a high density polyethylene/brominated polystyrene/graphene system [102].

10.4.1.4   Nitriles

Nitrile groups, commonly found in ABS and styrene-co-acrylonitrile (SAN) polymers, present a potential cross-linking/char formation route if they can be activated under fire conditions. A report on the use of a zinc chloride catalyst to enable the nitrile groups to cyclotrimerize and form additional char exists [103], but no other catalysts have been found to work besides that zinc salt, which is not practical to use due to its water solubility. If other catalysts could be found to enable these groups to cross-link, it may improve fire performance further. Another approach to getting the nitriles to participate in char formation would be through coordination with boron or other species that complex with nitrogen. There has been one report where regulatory fire performance (UL-94 V) was not obtained, but, the presence of boronic acids in ABS polymer yielded enhanced char formation and an anti-dripping effect in the UL-94 V protocol that was not seen when the nitrile groups were absent in a boronic acid + high impact polystyrene (HIPS) formulation [104]. Other boron salts or organoboron compounds which can complex with nitriles may be a fruitful area for research to further advance the potential char formation of nitrile containing polymers.

10.4.1.5   Anhydrides

Anydrides are often used in polymeric form, with polypropylene-graft-maleic anhydride (PPgMA) and polystyrene-co-maleic anhydride (SMA) being most commonly used as compatiblizers to allow polymers to better mix and interact with a wide range of fillers and other polymers. Other anhydride containing polymers exist as well, and these anhydride groups have potential for reaction with a wide range of groups during melt compounding, and potentially during fire reactions as well. One report has indicated that anhydrides can perform additional cross-linking reactions in the presence of phosphorus and epoxies to result in many different cross-linking reactions during fire conditions [105], but phosphorus may be able to work with just the anhydride itself. More work is needed to determine how to exploit this functional group for flame retardant effects while ensuring cross-linking does not occur during normal polymer processing temperatures. Hydrolytic stability of the anhydride over time could be problematic to ensure that the group remains present in the polymer and activates during fire conditions.

10.4.2   Organic Carbonates

Carbonates as flame retardants are well known when the flame retardant is an inorganic compound (mineral carbonate – see chapter 3 of this book for more details), but less well known when the carbonate is an organic based molecule, with one exception. The one well known organic carbonate with flame retardant potential is polycarbonate, but this example has a caveat. Part of the reason why polycarbonates have lower heat release/flammability when compared to other polymers is that every repeat unit of the molecule has a carbonate structure. Therefore when the polycarbonate decomposes, some CO_2 is always released, thus diluting the amount of fuel available in the vapor phase for combustion. The remaining part of the polycarbonate structure left behind after the carbonate decomposition then forms thermally stable char via Fries rearrangements and other chemical reactions [106]. (Figure 10.8) So one can think of polycarbonate, in a way, as a non-halogenated flame retardant additive in miscible blends with other more flammable polymers. However, polycarbonate by itself does not pass most regulatory tests other than with a minimal rating (UL-94 HB, UL-94 V-2) and so when put into another polymer, it does not enable that other polymer to pass a regulatory test unless that other test has some other defining flammability criteria besides self-extinguishment. This means that polycarbonate can really only be used as part of a flame retardant additive mixture; it cannot be used as the only flame retardant. The most common blend of polycarbonate with another polymer is polycarbonate/poly (acrylonitrile-butadiene-styrene) (PC/ABS), of which there are cases where the ABS is the minor component and PC is the major component. In these cases it can be argued that the ABS is present for cost dilution and mechanical property reinforcement to the PC part of the blend, rather

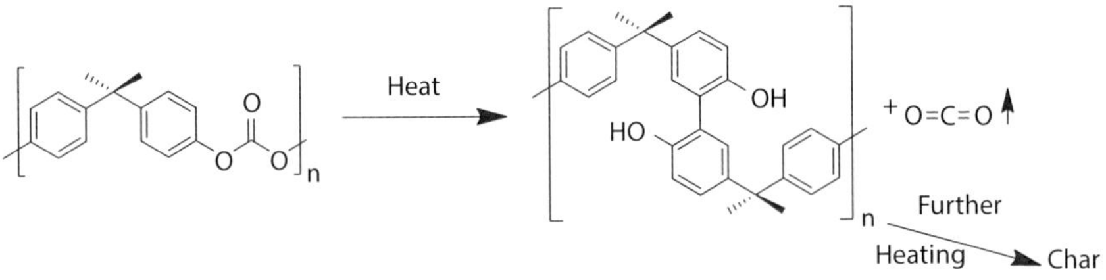

Figure 10.8 Polycarbonate decomposition and char formation.

than the PC helping flame retard the ABS. Still, PC can be defined as a polymeric non-halogenated flame retardant, and this suggests that PC may find additional utility in future non-halogenated flame retardant systems.

The statement that polycarbonate can only be used as part of a flame retardant mixture holds true for other smaller molecule organic carbonates, of which there has been minimal research to date. One can argue that because organic carbonates (and perhaps even inorganic carbonates) have low flame retardant effectiveness at low to moderate loadings, that this is the reason for the lack of reports and studies on small molecule carbonate flame retardants. The only other reports on a small molecule organic carbonate as a flame retardant that the author of this chapter could find was that of a phosphate-carbonate structure used to flame retard polyurethanes [107], and the above mentioned alkyne systems where some of the alkynes had carbonates in their structures (see Figure 10.5) [76, 77]. These two results, along with the findings on polycarbonate, show that small molecule organic carbonates are most effective as non-halogenated flame retardants when combined with other flame retardant chemistries.

10.4.3   Graft Copolymerization

Similar to cross-linking, if one can induce a material on the surface which will insulate the underlying polymer from heat and perhaps also make mass transfer from the bulk to the vapor phase more difficult, one may be able to produce a flame retarded material. Thus, if one can graft copolymerize a material which can char or in some other way deliver a residue onto the surface of a polymer which can offer protection, one may achieve some measure of flame retardancy. The ideal material to attach to the surface of the polymer to be protected is a material which can be effective in a thin coating and offer good protection without sacrificing other properties.

One possibility that has been explored is grafting poly(sodium methacrylate) or poly(sodium acrylate) on the surface of a polymer. These are both reported to produce a mixture of sodium carbonate and elemental carbon upon pyrolysis [108]. Either methacrylic acid can be grafted onto the surface of the polymer and then converted to its sodium salt or one could directly graft copolymerize the salt. It was found to be simpler to graft copolymerize the acid and then convert it. The amount of char that was produced was substantially more than that expected based on the individual components so there is some efficacy to this approach. From cone calorimetry the peak heat release rate fell from 900 kW/m^2 to 260 kW/m^2 while the time to PHRR increased from 530 s to 1130 s. The other cone parameters are also improved [109, 110].

The inherent difficulty in this approach is to identify the properties that are most desirable at the surface and then the material which should be graft copolymerized to achieve this. In some ways, this is similar to putting an intumescent paint onto the surface of a polymer in that that paint offers the protection to the polymer. Thus any coating which is used must be robust both to solvents and abrasion. A substantial difficulty with the above reported work is that the coating is not very adherent. Compatibility is most likely to occur between polymers which are quite similar but this is not likely to produce the best surface to protect the underlying polymer.

10.4.4   Expandable Graphite

Expandable graphite is a carbon-based non-halogenated flame retardant that is used extensively and commercially to flame retard a wide range of materials. While this technology is covered in more detail in Chapter 4, it's important to mention this technology here at a high level. Expandable graphite works by forming a protective barrier of graphitic carbon which in slows heat transfer to the underlying burning material and slows mass loss from the decomposing polymer to the flame front [111, 112]. It has been used in many different polymers, ranging from polyolefins to polyurethanes, and many thermoset materials, especially epoxies. It's also been combined with many other flame retardants to be part of a multi-component flame retardant system, with far too many specific examples to list here, but many of the combinations involve phosphorus or other intumescent flame retardant chemistries. The key practical issue with expandable graphite is that it activates around 200 °C and expands, thus making it impractical for use in high melting thermoplastic materials, or thermoset materials which polymerize at elevated temperatures. For thermoset materials, it is often commercially used as part of a mat/co-cured layer in fiber-reinforced composites [113–116]. Paints containing expandable graphite for protection of steel are also common, with some novel and interesting effects seen when expandable graphite is combined with silicones [117, 118]. Overall, expandable graphite has some interesting versatility to consider, however, it will color materials greatly, thus giving some solutions where the outer color of a material will be black or have a black flecked color. Therefore, it is not appropriate for all applications where colors of the final product other than black or gray is a need, and as mentioned previously, is not appropriate for any material that will be processed above 200 °C before becoming a finished product.

10.5   Bio-Based Materials

Bio-based materials have been studied with greater intensity for flame retardant effects since the 1st edition of this book was put out in 2014. Thanks to their renewable and intrinsic non-toxic properties, bio-based green flame retardants are potentially ideal FR replacements. Recently, bio-resources, such as starch, cyclodextrin, chitosan, cellulose and other polysaccharides, lignin, alginate and protein, etc have been studied, directly or modified, as FRs and/or synergists. Bio-based flame retardants may include other functional elements for flame retardant effect (phosphorus, nitrogen, sulfur), but they stand out as their own broad class due to the above benefits of being sustainable and having relatively benign environmental profiles. Of course, not all things bio-derived are automatically safe, but still, the potential to be bio-compatible has led to a great increase in flame retardant studies on bio-derived chemicals and materials. There are some excellent reviews on the subject to help readers get started on the wide diversity of chemistry available for flame retardant applications [119–122]. Some relevant examples are described below, in no particular order.

Sugars are potential carbonific substitutes, due to their outstanding char-forming ability. Attention has been paid to cyclodextrin (CD) and its derivatives, which have been used in place of the traditional pentaerythritol in an intumescent composition. The presence of reactive hydroxyl groups allows the introduction of a variety of phosphorus and other flame retardant moieties. In addition, CD has a hydrophobic cavity which can contain suitably sized FR molecules [123–137]. Chitosan is another potential candidate as a carbonific; chemical modification can greatly improve the FR properties [138–140]. Chitosan is also often used in layer-by-layer conformal flame retardant barriers (see Chapter 8 for more details). Starch [141–143] and cellulose [144] are also being used in intumescent systems. Alginates have shown as a matrix material to generate low flammability aerogel materials [145, 146], which again reinforces how these complex sugars can be used for flame retardant effect.

Lignin is also used as effective carbonific. After chemical modification and combination with other FRs, lignin can improve thermal stability and flame retardancy [147–156]. Lignin is a very inexpensive and widely available biorenewable feedstock produced from paper processing, and is available worldwide. Polyphenols from lignin have been considered as flame retdrant for nylon fabrics, with some additional chemical functionalization of the polyphenols being carried out to impart additional flame retardant effects [157, 158]. It has great potential as a carbon source for char formation and intumescent reactions, but more work is needed to improve the processing of this material into polymers and to address color issues (dark brown) that this additive brings to many plastics where other colors are desired.

Another approach to bio-derived flame retardants is to use chemicals produced from biological sources as feedstocks/precursors for making flame retardant chemicals. For example, the above mentioned deoxyben-zoin structure can be found partially in a natural compound, daidzein (derived from soybeans) and this molecule also gives high char yields [159]. This one study on daidzein is incomplete in that only vertical burn and Limiting Oxygen Index (LOI) tests have been completed, and not any definitive heat release studies. Some other examples of using biolog-ical feedstocks to make flame retardants include using isosorbide to be functionalized with phosphorus to make a flame retardant additive [160], functionalizing gallic acid (which is obtained from Lignin) to yield a phos-phorus functionalized aromatic molecule [161], and functionalization of castor oil with phosphorus species [162].

Phytic acid (Figure 10.9) is an interesting bio-derived material in that it has phosphate groups already present in its chemical structure, thus set-ting the material up to be a potential flame retardant "as is". Phytic acid has shown effectiveness as a flame retardant when applied to cotton [163, 164], has been combined with chitosan to flame retard cotton in conformal coatings [165], and has shown effectiveness on wool fabrics as well [166]. Due to its chemical structure, it has also been chemically modified in a variety of ways to impart flame retardant effects in combination with other flame retardant chemicals [167–170]. Phytic acid is a low melting solid with limited thermal stability (>200 °C) before it will begin to react and initiate char, so its use can be limited to lower melting thermoplastics and lower temperature polymerization thermoset materials.

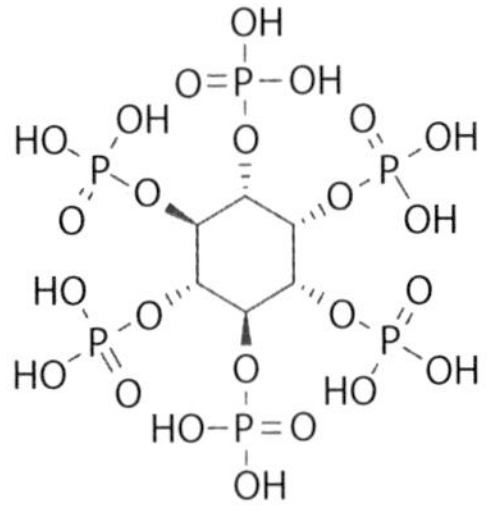

Figure 10.9 Phytic acid.

The final example to discuss in regards to bio-derived flame retardants is deoxyribonucleic acid (DNA). DNA is an interesting molecule in that it contains all the components of a flame retardant in its structure. It has phosphorus, carbon, and nitrogen, and when it decomposes, it will form carbon char, and can act as an intumescent flame retardant due to its structure [171]. The majority of work with DNA has been using the material as a flame retardant treatment for cotton, where is shows good performance in forming char and slowing flame spread [172–175]. It has also shown some effectiveness as a direct additive for polyethylene, with some beneficial effects in horizontal burn rate and heat release reduction when compared against melamine polyphosphate at the same loading levels [176]. However, the data in this study suggests that the DNA was somewhat degraded by the thermal melt compounding process. Related to this study in polyethylene, there was some work using DNA as a coating on a magnesium hydroxide flame retardant to act as a flame retardant compatibilizer for a complex [polyolefin + magnesium hydroxide + keratin fiber] mixed system [177]. When DNA was used as part of the overall system, it did show better overall performance in the cone calorimeter for heat reduction when compared to a system containing magnesium hydroxide with no DNA coating.

Based upon current findings, bio-based flame retardants show a promising future as non-halogenated flame retardants, but their lack of thermal stability will limit their use in some materials. Further investigation of these materials in a wider range of materials, and determining if it is possible to improve their thermal stability, is strongly recommended to fire safety scientists and chemists looking for new materials to develop. It is unclear if any of these materials will become commercial, but the ready availability of many of these chemicals suggest that if there is an appropriate application, these materials could be made commercial.

10.6   Tin-Based Flame Retardants

10.6.1   Introduction

Although tin compounds were reported to exhibit flame-retardant properties as long ago as 1859 [178], and despite the fact that a wide range of inorganic tin and organotin compounds have subsequently been investigated as flame retardants and smoke suppressants [179], only a few tin-based systems have reached commercialization. Tin additives are used as synergists with other flame retardants, and recent work has focused on the development of more cost effective systems, based on proprietary nano-particulate and coating technologies, which still have not yet reached commercialization. A brief review of current knowledge relating to the mode of action of tin flame retardants is included.

10.6.2   Zinc Stannates

Although early research into tin-based flame retardants for plastics focused on tin(IV) oxide (SnO_2) [180, 181], by far the most important tin additives have been the zinc stannates – zinc hydroxystannate (ZHS) and zinc stannate (ZS). Originally developed at International Tin Research Institute (ITRI) during the mid-1980s, these compounds are now being marketed worldwide as flame retardants and smoke suppressants for use in a wide range of polymeric materials [182].

ZHS is manufactured industrially by the aqueous reaction of sodium (or potassium) hydroxystannate with a soluble zinc salt, usually zinc chloride:

$$Na_2Sn(OH)_6 + ZnCl_2 \rightarrow ZnSn(OH)_6 + 2NaCl.$$

The white precipitate is washed free of sodium chloride and dried in air at ca. 105 °C. ZS is manufactured by controlled thermal dehydration of ZHS, usually at a temperature in the range 300 – 400 °C:

$$ZnSn(OH)_6 \rightarrow ZnSnO_3 + 3H_2O$$

Although there is generally little difference in the effectiveness of ZHS and ZS in terms of their flame-retardant properties, ZS is the preferred additive for polymers which are processed at temperatures above ca. 200 °C. Some important properties of ZHS and ZS are given in Table 10.1.

Initially, ZHS and ZS were introduced as safer alternative synergists to antimony trioxide (Sb_2O_3) for use in halogen-containing polymers (such as PVC) and in formulations containing halogenated flame retardants [183, 184]. Although the tin synergists were demonstrated to exhibit a number of technical benefits, including combined flame retardancy and smoke suppression, lower heat release rates and action in both the condensed and vapor phases, their relatively high price (compared to, for example, Sb_2O_3) has somewhat limited their commercial use – current worldwide consumption of ZHS/ZS is estimated to be around 1,000 tons per annum as per a 2006 reference [185]. Factoring in material growth since then, it is likely that ZHS/ZS use has increased since this 2006 reporting. Although industrial applications of ZHS and ZS were originally targeted towards

Table 10.1 Properties of zinc hydroxystannate and zinc stannate.

Property	ZHS	ZS
Chemical formula	$ZnSn(OH)_6$	$ZnSnO_3$
Molecular weight	286.12	232.07
CAS number	12027-96-2	12036-37-2
ELINCS number	404-410-4	405-290-6
TSCA listed	Yes	Yes
Appearance	White powder	White powder
Analysis (typical)	41% Sn 23% Zn < 0.1% Cl < 1% free H_2O	51% Sn 28% Zn < 0.1% Cl < 1% free H_2O
Specific gravity	3.40	4.25
Median particle size (μm)	1 – 2	1 – 2
Decomposition temperature (°C)	> 200°C	> 570°C
Aqueous solubility (at 20°C)	< 0.01%	< 0.01%
Refractive index (at 20°C)	1.9	1.9
Acute oral toxicity	Very low*	Very low*

*LD_{50} (rats) > 5,000 mg/kg.

PVC and other halogenated polymers, recent activity has focused on their considerable potential in halogen-free systems and this has resulted in significant growth in their usage.

10.6.3   Halogen-Free Applications

There has been considerable activity in the electronics industry to develop alternatives to brominated flame retardants, certain of which have been effectively phased out under the EU Restriction of Hazardous Substances (RoHS) Directive. Although some brominated flame retardants are not restricted by RoHS, the related Waste Electrical and Electronic Equipment (WEEE) Directive, requires separation and special handling of plastics containing any brominated flame retardants, and electronics companies are therefore developing alternatives to avoid the extra costs of separation. For additional details on RoHs and WEEE and how these regulations have been driving halogen-free systems, please see Chapter 1 of this book. Furthermore, major Japanese companies have been developing halogen-free polymer systems for well over a decade, primarily for electronic applications, and many of these are known to feature zinc stannates. Many of these electronic/electrical applications also have low smoke requirements as well, where stannates have a role to play, further driving their use. The following sections review applications of tin compounds, including ZHS and ZS, in zero halogen formulations by polymer type.

10.6.3.1   Polyolefins

Traditionally, the major polymer used for cable insulation has been PVC, with polyolefins and elastomers (both halogenated and halogen-free) accounting for considerably smaller volumes. Despite environmental concerns about the production and use of PVC [186], and the resulting demand for low smoke zero halogen cables, PVC remains firmly established in the market and 'low smoke – low acid' PVC compounds are increasingly available. It's important to understand that some of the low smoke/low acid requirements are not necessary environmental, but are based upon fire safety concerns and protection of sensitive electrical equipment. Since many wire and cable systems for retrofitting of buildings are easily put in place through above-ceiling/ductwork areas, smoke from these burning cables can cause issues with visibility limitations for getting to exits, or, smoke spread through the heating/ventilation system of the building. Further, acidic gases can land on sensitive electronics and cause localized corrosion, which often results in circuit boards shorting out and the electronics failing, resulting in computer/electronic shut down and data loss. For these reasons, there is an increased demand in wire and cable systems which not only limit flame spread, but also have much lower smoke and little to no acidic gas release.

Growth in halogen-free cables is clearly evident, especially in the EU and North America, and there is increased demand in Asian markets as well. These compounds are usually based on cross-linked polyethylene, polypropylene, ethylene vinyl acetate (EVA), ethylene ethyl acetate (EEA) or similar materials. In order to achieve appropriate levels of flame retardancy, high loadings of inorganic fillers (mainly ATH or MH, but occasionally others such as magnesium carbonate or huntite/hydromagnesite mixtures), are required and the resulting cables can be difficult to process and often have relatively poor mechanical properties.

Partial substitution of the primary filler by low levels of ZHS or ZS has been found to improve flame-retardant and smoke-suppressant performance, allowing total filler levels to be reduced, with concomitant improvements in processability and mechanical properties. Alternatively, the use of synergists such as ZHS and ZS can allow cable formulations to pass stringent fire tests that would otherwise be difficult to pass. Some key examples of the claimed uses of ZHS and ZS in polyolefin cable compounds are given in Table 10.2.

Very recently, Chinese researchers have demonstrated synergism between ZHS and an intumescent flame-retardant system, comprising ammonium polyphosphate (APP) and pentaerythritol (PER), when used in halogen-free polypropylene [187]. Optimum performance is observed at 1% by weight of ZHS, as evidenced by UL-94, LOI and Cone Calorimeter data.

10.6.3.2   Styrenics

Comparatively little work has been undertaken on zinc stannates in styrenic polymers. Clariant has utilized ZS as a low addition level synergist with their proprietary aluminum diethylphosphinate (DEPAL) flame retardant in ABS co-polymer [199]. Hence, a composition containing 1% ZS + 25% DEPAL achieves V1 classification in the UL-94 test, and an Oxygen Index of 47.0, which is considerably higher than that given by other DEPAL/synergist systems evaluated in the work.

ZS has also been reported as an effective FR synergist when used in conjunction with triphenyl phosphate (TPP) in a halogen-free PC – ABS blend containing huntite – hydromagnesite filler [200]. In addition to increasing Oxygen Index, ZS + TPP systems were found to give good tensile strength and elongation at break properties in the thermoplastic formulation.

10.6.3.3   Engineering Plastics

One of the most encouraging sectors for the use of zinc stannates in halogen-free polymer compositions is that of engineering plastics, primarily comprising polyamides (nylons) and saturated polyesters – both PET and PBT. Since the processing temperatures for these materials are relatively high, ZS is the additive of choice, because ZHS decomposes (by dehydration) at temperatures in excess of ca. 200 °C. Although theoretically ZS could be used as a partial replacement for a thermally-stable hydrated filler such as MH in engineering plastics, in practice, most of the reported use has been in conjunction with phosphorus-based flame retardants, as indicated in Table 10.3.

Table 10.2 ZHS and ZS in polyolefin cable compounds.

FR System*	Polymer(s)	Effect(s)	Company
ZS + MH + nanoclay	PE/PP blends	FR cable with low smoke, corrosivity & heat release rates	Alpha Gary [188, 189]
ZS + MH + AlOOH	Ethylene-octene/EVA blend	FR cable coating with low smoke emission & improved moisture resistance	General Cable [190]
ZHS + ATH + AlOOH + nanoclay	Polyolefins (unspecified)	FR cable jacket material with improved FR & mechanical properties	General Cable [191]
ZHS/ZS + MH + APP	EVA or PE	FR cable	Fujikura [192]
ZS + MH	EVA	FR cable	Hirakawa [193]
ZS + modified MH	EEA or PE	FR cable with low smoke & toxicity	Tateho [194]
ZHS + ATH + nanoclay (optional)	EVA	Partial replacement of ATH gives improved FR & lower smoke	ITRI [195]
ZS + MH + CB	EVA	Wire & cable compound with improved FR & mechanical properties	Hitachi [196]
ZHS + ATH	PE/EVA blends	Cable compound with improved FR (Oxygen Index)	Joseph Storey [197]
ZS + MH + AOM	EVA/PE blends	Wire & cable compound with optimized FR & mechanical properties	LS Cable [198]

*APP = ammonium polyphosphate; CB = calcium borate; AOM = ammonium octamolybdate.

Table 10.3 ZS in engineering plastics.

FR System*	Polymer(s)	Effect(s)	Company/ University
ZS + MPP + aryl phosphate	PA6	Good FR (Oxygen Index, UL-94) with improved tensile property retention	Bolton University [201]
ZS + DEPAL + AlOOH	GF-PA6	UL-94 V0 rating with reduced corrosion on melt processing equipment	Du Pont [202]
ZS + MPP	GF-PA6 or GF-PA66	UL-94 V0 rating with minimal polymer degradation during processing	Clariant [203]
ZS + DEPAL + MC (optional)	GF-PBT	UL-94 V0 rating & Oxygen Index > 35	Clariant [124]
ZS + PHEPAL + MC	GF-PBT	UL-94 V0 rating with minimal polymer degradation during processing	Clariant [204]
ZS + red phosphorus	GF-PBT	High Oxygen Index with excellent mechanical & electrical (tracking resistance) properties	Teijin [205]

*MPP = melamine polyphosphate; DEPAL = aluminum diethylphosphinate; MC = melamine cyanurate; PHEPAL = aluminum monophenylphosphinate.

In addition to the reported work using ZS in engineering plastics, certain tin(II) compounds have been shown to exhibit excellent FR and smoke-suppressant properties when incorporated at levels of 20 – 30% into aromatic polyesters, specifically PBT [206]. Hence, whereas tin(II) oxide, tin(II) oxalate and tin(II) phosphite were all shown to markedly increase flame retardancy (Oxygen Index) and char residue levels in PBT, tin(IV) oxide is almost totally ineffective in the same polymeric substrate, clearly implying a fundamental difference in the mode of action of tin(II) and tin(IV) compounds in halogen-free systems.

Interestingly, hydrous tin(II) oxide has also been claimed as an effective oxidation stabilizer for red phosphorus when applied as a surface coating, enabling the product to be incorporated as a flame retardant in polyamides and other high temperature processing polymers [207].

10.6.3.4 Thermosetting Resins

Zinc stannates (both ZHS and ZS) have been reported to be effective flame retardants and smoke suppressants for halogen-free thermosets, including epoxy resins and unsaturated polyester resins. In most cases, the tin compounds have been used in conjunction with a hydrated filler (usually ATH), although synergy with phosphorus FRs has also been observed. Examples of the use of zinc stannates in epoxy and polyester resins are given in Table 10.4.

Although little work has been carried out to date on tin-based FRs in polyurethanes generally, a recent study using ZHS or ZS in conjunction with a proprietary alkyl aryl phosphate/phosphonate product (Antiblaze 230, Albemarle), reported synergistic FR effects and significantly reduced smoke generation in halogen-free flexible PUR foams [219].

10.6.3.5 Elastomers

Although ZHS and ZS have been investigated extensively as flame retardants and smoke suppressants in halogenated elastomers such as polychloroprene [220] and chlorosulphonated polyethylene [221], less attention has been paid to their activity in halogen-free rubbers. However, introduction of ZHS at a 2.5 weight % level into a natural rubber compound containing carbon black and ATH filler, has been shown to give a small but significant increase in Oxygen Index, and a reduction in smoke density of about one-third [222]. Partial replacement of ATH by ZHS in natural rubber has also been shown to give a small improvement in Oxygen Index, whilst maintaining UL-94 V1 rating [223].

Table 10.4 ZHS and ZS in Thermosetting Resins.

FR System*	Polymer(s)	Effect(s)	Company/ University
ZHS/ZS + ATH	Epoxy	Partial replacement of ATH gives lower heat release rates & smoke emission, but no effect on Oxygen Index	ITRI [208]
ZS + ATH	Epoxy	FR composition with good heat resistance for use in printed circuit boards	Shin Kobe Electric [209]
ZHS/ZS + brucite (MH) + mica	Epoxy	FR composition with low smoke & toxicity, combined with good soldering heat resistance for use in printed circuit boards	Sumitomo Bakelite [210]
ZHS/ZS + ATH + ZM	Epoxy	FR composition suitable for use as a permanent protective coating on a printed circuit board	Mitsubishi Gas Chemical [211]
ZS + DEPAL	Epoxy	FR molding compounds showing UL-94 V0 rating & Oxygen Index > 40	Clariant [212]
ZHS/ZS	Epoxy	Significantly increased Oxygen Index without deterioration of optical properties	Warsaw University [213]
ZHS	Epoxy	Aerospace resin matrix with significantly reduced heat release rates & smoke generation	IMCB – IMAST [214, 215]

(Continued)

Table 10.4 ZHS and ZS in Thermosetting Resins. (*Continued*)

FR System*	Polymer(s)	Effect(s)	Company/ University
ZHS + ATH	Polyester	Marked reductions in smoke density, but no significant effect on Oxygen Index	ITRI [216]
ZHS/ZS	Polyester (GRP)	Marked reductions in smoke density, but no significant effect on Oxygen Index	ITRI [217]
ZHS/ZS + APP + nanoclay	Polyester	Good smoke suppression, but no improvement of flame retardancy	Bolton University [218]

*ZM = zinc molybdate; DEPAL = aluminum diethylphosphinate; APP = ammonium polyphosphate.

Oxygen Index of an ATH-filled ethylene – acrylic rubber formulation is significantly increased (from 27.5 to 33.0) by the incorporation of 2.5 weight % of ZHS, and the improvement shown by the ZHS-containing composition is maintained at elevated temperature (to ca. 200°C) [224].

10.6.3.6   Paints and Coatings

Fire-resistant intumescent coatings are used to protect a range of materials including steel, wood, textiles and polymer composites. ICI Paints have patented a halogen-free FR coating composition comprising a film-forming polymer (such as an acrylic polymer latex) and an inorganic FR additive, the latter comprising a synergistic combination of ZHS + ATH [225].

Work undertaken in 2008 by ITRI and a consortium of industrial and research partners within the EU-funded STEELPROST project [226], has resulted in the development of halogen-free intumescent compositions that exhibit enhanced fire-resistant, chemical and mechanical properties. The coating systems utilize nano-particulate tin FR additives and can be applied as protective paints on steel, composites, wood and reinforced plastics. As expected, the new coatings show significantly reduced levels of smoke and toxic gas emissions when compared to conventional intumescent paints.

10.6.3.7   Textiles

The first reported use of a tin-based flame retardant dates back to 1859, when a process involving the *in situ* precipitation of hydrous tin(IV) oxide was developed to impart FR properties to cotton and other cellulosic materials [97]. Similar processes found commercial use during the first half of the 20th century, but were eventually replaced by more permanent organophosphorus-based FR treatments. However, recent years have seen renewed interest in tin systems, primarily involving impregnation of fibers and textiles with colloidal tin oxide species.

ITRI has developed and patented processes for preparing and utilizing stable aqueous colloidal sols of SnO_2 and related materials [227, 228]. These products, which contain nanometer-scale particulate tin species, are useful precursors for the synthesis of ceramic bodies, powders and coatings, and are finding application in electroconductive materials, catalysts and transparent tin oxide films on glass and other substrates. In the context of flame retardancy, aqueous tin colloids are particularly suitable for treatment of hydrophilic natural fibers. Hence, ITRI has developed processes based on colloidal suspensions of tin(IV) oxide, tin(IV) borate and tin(IV) phosphate for flame-resistant treatments of paper, cotton and wool, the latter substrate being proteinaceous rather than cellulosic in nature. Hence, a number of colloidal tin systems have been shown to match or even outperform a commercial potassium fluorozirconate (PFZ) treatment when applied to a typical aircraft cabin wool – nylon blend fabric [229].

10.6.4   Novel Tin Additives

Despite clear technical benefits, including non-toxicity and excellent smoke suppression, markets for ZHS and ZS have been somewhat limited because of their relatively high price compared to many other flame retardants [110]. Consequently, recent studies at ITRI and elsewhere have focused on the development of more cost-effective tin-based systems. Initial work involved processes for producing ultrafine ZHS and ZS powders with typical particle sizes in the range of 0.1 – 0.3 µm [230]. These novel additives were shown to exhibit a number of performance benefits when compared with commercial ZHS/ZS powders; they do not settle out in thermosetting resins, they can be used in formulations where translucency is required and, most importantly, their FR efficiency is markedly superior, allowing significant reductions in incorporation level to be made without compromising flame retardancy or smoke suppression [147].

10.6.4.1   Coated Fillers

Although ultrafine and colloidal additives have been shown to exhibit significant performance benefits in systems where good dispersion can be achieved, agglomeration of particles and incompatibility with certain polymeric matrices proved to be major drawbacks to their widespread applicability. However, these problems have been largely overcome by coating the active tin species on to the surface of low cost inorganic fillers, which effectively act as carriers and prevent agglomeration when these coated fillers are incorporated into polymeric matrices.

ITRI has developed and patented processes for coating ultrafine ZHS/ZS [231] or nano-particulate tin species [232] on to a range of inorganic fillers, including hydrated fillers (ATH, MH) and 'inert' fillers, such as calcium carbonate and titanium dioxide (Table 10.5). ZHS-coated fillers have been produced with compositions containing up to 50% ZHS by weight, although typical commercial products usually contain 2.5 – 10% ZHS, which appears to be the optimum range, at least as far as cost effectiveness is concerned.

X-ray photoelectron spectroscopy (XPS) and diffuse reflectance infrared Fourier-transform spectroscopy (DRIFTS) have been used to investigate the interaction between the ZHS coating and the hydrated filler surface; although interaction is confirmed, no evidence has been found for condensation reactions occurring between ZHS and ATH (or MH) or the formation of Sn-O-Al (Mg) bonds [233, 234].

Table 10.5 Coated filler types.

Coatings (typically 2.5 – 10% w/w on filler)	Fillers
Zinc hydroxystannate (ZHS)	Alumina trihydrate (ATH)
Zinc stannate (ZS)	Magnesium hydroxide (MH)
Tin(IV) oxide	Calcium carbonate
Tin(IV) phosphate	Huntite/hydromagnesite
Tin(IV) borate	Titanium dioxide
	Silica
	Alumina (anhydrous)
	Zinc borate
	Sodium bentonite
	Nanoclays

Extensive studies in thermoplastic, thermosetting and elastomeric polymers have shown that these coated fillers outperform the fillers themselves or equivalent composition physical mixtures of the tin additive plus filler [98, 235]. Consequently, lower addition levels of the coated grades, compared with uncoated fillers, are required for a given flame-retardant performance and this reduction in filler loading has been shown to lead to better polymer processing and improved physical, mechanical and electrical properties. Fire test data for ZHS-coated fillers in halogen-free ethylene – vinyl acetate (EVA) and ethylene – ethyl acrylate (EEA) formulations are given in Tables 10.6 and 10.7 respectively.

Table 10.6 ZHS-coated fillers in halogen-free EVA [98].

Additive (phr)*	Oxygen Index	Peak Rate of Heat Release (kW/m²)**	Smoke Parameter (MW/kg)**
None	20.2	1404	665
150 CaCO₃	25.7	415	171
150 ATH	34.0	362	114
150 ZHS-coated ATH	37.5	338	99
150 MH	32.9	293	94
150 ZHS-coated MH	33.6	238	74

*Coated filler composition = 10% ZHS : 90% filler w/w.
**Cone Calorimeter operated at 50 kW/m² incident heat flux.

Table 10.7 ZHS-coated ATH in halogen-free EEA [98].

Additive (phr)*	Oxygen Index	Peak Rate of Heat Release (kW/m²)**	Smoke Parameter (MW/kg)**
None	21.0	781	251
175 CaCO₃	26.5	333	136
175 ATH	40.4	177	45
175 ZHS-coated ATH	40.7	167	33

*Coated filler composition = 10% ZHS : 90% ATH w/w.
**Cone Calorimeter operated at 50 kW/m² incident heat flux.

Coated fillers have also been used in halogen-free intumescent paint formulations [150, 151]. In general, superior performance in terms of fire-resistance and smoke emission, is obtained with ZHS-coated ATH than with an equivalent composition blend of ZHS + ATH powders.

10.6.4.2 Tin-Modified Nanoclays

In the 2000-2015 time period, there was a high interest in nano-composite materials, comprising polymer-layered silicate clay intercalated structures, and their flame-retardant properties [236]. It has been shown that the use of a combination of a hydrated filler (preferably ATH) with a nano-clay gives a more coherent char during combustion than using either the hydrated filler or the nanoclay alone [237]. Subsequently, ITRI work on a halogen-free EVA cable formulation has demonstrated that, in addition to improving the performance of the hydrated filler itself, the incorporation of a montmorillonite (MMT) type nanoclay along with ZHS (either as an additive or as a coated filler) leads to further synergistic effects with regard to reducing heat release rates and smoke emission (Table 10.8) [120, 238].

Layered double hydroxides (LDHs) have also attracted interest as non-halogenated flame retardants for polymeric materials, because they combine the features of conventional metal hydroxide fillers and layered silicate type nano-fillers [239]. Recent ITRI studies, carried out within the EU-funded HYBRID project [240], have focused on tin-modified LDH additives as flame retardants in PVC [241] and halogen-free cable materials. Although marginal flame-retardant benefits are observed when a conventional Perkalite -type (Mg – Al) LDH is used as a partial replacement for ATH in a halogen-free EVA cable formulation, generally improved

Table 10.8 ZHS + ATH + nanoclay in halogen-free EVA.

Additive (phr)	Oxygen Index	Peak Rate of Heat Release (kW/m²)*	Smoke Parameter (MW/kg)*
None	20.2	1404	665
100 ATH	23.6	472	177
90 ATH + 10 MMT	27.4	290	161
81 ATH + 9 ZHS + 10 MMT	27.1	228	120

*Cone Calorimeter operated at 50 kW/m² incident heat flux.

Table 10.9 Tin-modified LDH additives in halogen-free EVA [168].

Additive (phr)	Oxygen Index	Peak Rate of Heat Release (kW/m²)*	Smoke Parameter (MW/kg)*
None	21.5	964	356
100 ATH	25.5	391	148
95 ATH + 5 LDH	26.5	387	123
95 ATH + 5 Sn-LDH(1)**	26.0	308	125
95 ATH + 5 Sn-LDH(2)**	26.5	288	127
95 ATH + 5 Sn-LDH(3)**	26.5	268	105

*Cone Calorimeter operated at 50 kW/m² incident heat flux.
**Sn-LDH(1) – (3) contain increasing levels of tin.

properties (especially with regard to reducing heat release rates) are evident when Sn^{4+} ions are incorporated into the LDH lattice structure (Table 10.9). These novel Sn – LDH additives were synthesized using a co-precipitation process [169], and the beneficial effect of increasing tin incorporation level in the LDH is clearly evident.

10.6.4.3   Mechanism of Action

Although the flame-retardant action of tin (and zinc) in halogen-containing polymers has been shown to involve both condensed and vapor phase processes [109, 242, 243], the near quantitative retention of both metals in char residues from burnt halogen-free polymeric formulations is strongly indicative of condensed phase activity only [149]. Thermal analysis of halogen-free polyester resin containing ZHS and ATH indicates that the tin compound exhibits significant char enhancing properties (Table 10.10) [141]. Hence, ZHS is found to markedly increase the weight loss associated with char oxidation at the expense of the initial pyrolysis loss, when compared to the resin containing ATH alone. Further evidence of the char-promoting activity of ZHS is provided by the observed residual yield at 600 °C, which is significantly greater than that expected on the basis of involatile inorganic material (i.e. $ZnSnO_3 + Al_2O_3$) which remains in the char.

In accord with this finding, certain metal oxides are believed to act as dehydrogenation catalysts in halogen-free polymers and proprietary grades of ATH and MH containing small amounts of char-promoting

Table 10.10 Thermal analysis data for halogen-free polyester resin samples.[*]

ZHS (phr)	ATH (phr)	Pyrolysis stage[**]		Char oxidation stage[***]		Residue at 600°C	
		Weight loss (%)	DTG_{max} (°C)	Weight loss (%)	DTG_{max} (°C)	Observed (%)	Calculated (%)
None	None	88.6	363	11.3	544	0.1	0
None	25	77.6	342	9.9	534	12.5	13.1
5	25	67.8	336	14.1	497	18.1	15.7

[*]Heating rate = 10°C per minute in flowing air.
[**]Temperature range = ca. 260 – 450 °C.
[***]Temperature range = ca. 450 – 570 °C.

metal oxides [244] or nanoclays [245] have been developed. In addition to the above processes, the highly endothermic dehydration of ZHS at temperatures above ca. 200 °C may partially account for its flame-retardant activity when used in halogen-free polymer formulations.

In the case of ZHS/ZS-coated fillers, the char-promoting effect of the tin component supplements the endothermic activity of the hydrated filler, which itself involves (a) reduction in heat feedback from the burning gases in the flame to the decomposing polymer beneath, (b) formation of an insulating char layer above the unburnt polymer, and (c) absorption of volatile species and fragments on the very high surface area anhydrous metal oxide residue [246]. Furthermore, the ultrafine or nano-particulate tin species on the coated filler are thought to thermally decompose to form a highly active catalytic surface, thereby maximizing any synergistic effect with the filler.

The flame-retardant mechanism associated with nanoclays has been studied and is likely to involve the formation of a ceramic skin which catalyzes char formation by thermal dehydrogenation of the host polymer to produce a conjugated polyene structure [247]. The nano-composite structure present in the resulting char appears to enhance the performance of the char through reinforcement of the char layer [248]. These effects would explain the apparent flame-retardant synergy observed when nanoclays are incorporated into polymer formulations containing condensed phase fire-retardant systems, including ZHS, ZS, coated fillers and other tin-containing additives.

10.6.4.4   Summary

Certain tin compounds, particularly zinc hydroxystannate (ZHS) and zinc stannate (ZS), are highly effective flame retardants and smoke suppressants, finding use in a wide range of polymeric materials, including plastics, rubbers, coatings, foams, fibers and composites. These additives offer several advantages over many conventional flame retardants, such as combined flame retardancy and smoke suppression, reduced heat release rates, synergism with conventional FR types and low loadings.

Although originally developed as synergists for use in PVC and other halogen-containing polymeric formulations, their use in halogen-free systems is growing steadily. The use of tin additives as components of cost effective, low smoke zero halogen (LSOH) FR systems is expected to experience significant growth in the years ahead.

10.7   Polymer Nanocomposites

This section of the chapter will be short as nancomposites are covered in Chapters 8 and 9 of this book, and, the subject of polymer flame retardants have been extensively studied and reviewed to date [249–251]. However, it is worth mentioning that nanocomposites by themselves do have a flame retardant effect in that they lower heat release and often prevent dripping in thermoplastic materials due to decreases in mass loss rate which is in turn caused by nanoparticle networks setting up increased local melt viscosity, and, creating barriers which slow release of pyrolysis gases. The mechanism by which this flame retardant effect is achieved in polymer nanocomposites has been clearly defined and is now well-understood [252–259]. As discussed in these reviews/books cited above, and covered in Chapter 9, nanocomposites by themselves rarely provide enough flame retardant benefit to meet most regulatory tests, but they do enable other non-halogenated FRs to be used at lower loading levels.

10.8   Engineering Non-Hal FR Solutions

One of the methods to impart flame retardancy to a polymer (or other material requiring fire protection/flame retardant performance) is to use an engineering solution, which is to use design and/or material choices to ensure that a polymer meets fire safety requirements in an end-use application. Some examples of engineering solutions include barrier fabrics for polyurethane containing mattresses, intumescent paint for steel structures, and metal boxes/shields around power supplies to keep an electrical short circuit from arcing onto nearby plastics. Engineering solutions sometimes are a more cost effective way to impart flame retardancy to a finished good than to use flame retardant additives. However, engineering solutions are sometimes equally easy to defeat as they are to utilize. Therefore, users of engineering solutions should understand the potential risks to their product should that solution fail in use, either through accidental failure or deliberate removal. They are not exactly chemical solutions, but they technically are non-halogenated flame retardant solutions. In general, these engineering solutions fall into two main categories, barrier fabrics and coatings. Barrier fabrics are only used with furniture and mattresses, but require some specific discussion to explain how they are different than coatings. Coatings are more universal in how they can be applied to materials to impart non-halogenated flame retardant effects.

10.8.1   Barrier Fabrics

Barrier fabrics to impart fire safety to filled furnishings (upholstered chairs, couches, mattresses) has been in use for some time, and an excellent review on the topic was presented by Nazaré and Davis in 2012 [260]. Barrier fabrics in this case are fabrics/textiles made from a low to non-flammable fiber, and these fabrics are used to wrap/encase the underlying flammable filling material. Example fibers used in barrier fabrics include wool, fiberglass, aramids (Kevlar, Nomex), polbenzimidazole (PBI), melamine-based fibers, and carbon fibers. More exotic fibers could also be used (polyetherimide, basalt) as cost and performance dictate. Depending upon thickness and fiber type, the barrier fabric may provide fire protection against smolder sources, open flames, or even more intense ignition sources. The actual choice of the barrier fabric composition will be driven by fire test need and performance criteria. The underlying concept behind barrier fabrics are to prevent/slow the flame from impinging on the flammable material underneath, which in the case of filled furnishings is typically flexible polyurethane foam, which is a significant fire hazard [261].

Since the 2012 review, the effectiveness of barrier fabrics has been investigated, and the effectiveness of barrier fabrics depends a lot on the specific fabric used, and, the type of ignition source [262–266]. Work by the National Institute of Standards and Technology (NIST) in the United States has carried out the most definitive work on barrier fabrics to date, and the results from their work show that the chemistry and composition of the barrier fabric matters in regards to whether it can or cannot resist smolder or open flame ignition sources [252–255].

The major weakness of barrier fabrics in flame retardancy is when the barrier fabric fails. Such a failure may be due to fabric wear and age, accident, or deliberate damage [267]. Many fabrics do wear out and begin to form holes over time, especially if they should abrade against another fabric during prolonged use. If there is an accidental tear or rip in the fabric, or if the thread holding the seam of the fabric together should fail, there will obviously be an opening in the barrier fabric. Finally, if someone deliberately vandalizes a barrier fabric, or if an animal should scratch/chew through the fabric, there is again an opening through which flames could pass. Once openings are present in the barrier fabric, flames can impinge upon the underlying flammable filled material and flame retardant benefits provided by the fabric over the rest of the furniture/mattress item may be lost. Obviously the size of the opening in the barrier fabric and the intensity of the flame entering through the barrier opening determines rate of flame spread, but once the opening is present, the flame retardant effect brought by the barrier fabric can no longer be fully ensured. This last point is why in some strict fire tests for mass transport applications (subway, rail, ship) there are deliberate "vandalism" tests where the barrier fabric is cut into and an open flame applied to the opening of the fabric. Therefore, the material scientist or fire safety engineer may not want to rely solely upon barrier fabrics (depending upon end-use application), and a "defense-in-depth" may be needed to ensure the finished good still meets fire tests if the barrier fabric fails. One example of this was studied by Southwest Research Institute where a barrier fabric + flame retardant foam was used which resisted ignition from a more intense fire source vs fabric + foam assemblies where the foam was not flame retarded [268]. Or, barrier fabrics with some additional ability to char and seal over the opening when exposed to flame may be needed. Only charring thermoplastic fiber barrier fabrics would fall into this category, although fiberglass and basalt barrier fabrics, since they do not easily burn away, may provide some flame protection provided the cut fibers can still overlap over the cut/opening in the barrier fabric.

10.8.2   Coatings

The use of flame retardant coatings to provide fire protection is well known, and for providing protection to polymers, coatings have been used for some time to provide non-halogenated flame retardant benefits. Excluding intumescent paints, mats, and coatings for polymers, there are four classes of non-halogenated flame retardant coatings for polymers. These four classes are inorganic coatings, metalized coatings, nanoparticle coatings, and layer-by-layer technology. Chapter 8 of this book extensively covers the layer-by-layer technology, but it will still be touched upon briefly in this chapter. With all of the coatings mentioned in this chapter, the concept of "Defense-in-Depth" mentioned above, applies here as well. Coatings that break/fail will not prevent the flames from getting to the non-flame retarded material underneath. Coatings + flame retardant materials underneath provide an overall more robust fire protection system.

10.8.2.1   Inorganic Coatings

Plasma polymerization of silicon based molecules to impart sub-micron to micron levels of an inorganic coating on a substrate has been used to provide flame retardant benefits to thermoplastic and thermoset materials; solid parts and fabrics. Some examples include plasma applied to polyamides [269–272], and fabrics [273–278]. The primary benefit of these plasma coatings is that a conformal inorganic oxide coating (silicon oxide) covers the polymer surface such that when fire is applied to the coated substrate, the fire must burn through the silicon oxide coating before the polymer will ignite. The typical benefit seen with these coatings is a significant delay in time to ignition, along with some additional flame retardant benefits. However, if the underlying polymer foams, melts, softens or causes the silicon oxide layer to break, the benefits of the coating may not hold throughout the entire fire exposure. For this reason these inorganic coatings are often combined with other flame retardants so that the outer inorganic coating only has to provide some of the protection; the underlying flame retardant in the polymer provides fire protection once the coating fails. Of final note, these plasma based silicon oxide coatings likely provide other benefits to the coated fabric or part, including oxidative degradation protection, possible UV protection benefits, and abrasion/surface finish benefits. While there continues to be some research on these silicon based coatings, to date, there have been no known commercializations of the technology. It should be possible to conduct "roll-to-roll" coatings of fabrics with this technology, and injection molded parts should be able to be coated with this technology using commercial batch processes which impart a variety of plasma coatings onto other substrates for aesthetic purposes. Clearly process research and development is needed there to further this technology beyond the laboratory.

Another inorganic coating of note is that brought by geopolymers. Geopolymers are inorganic polymers (typically alkali silicates) of similar chemical composition to concrete. These materials are often water based and when they polymerize, can be made transparent if the geopolymer chemistry is designed properly. Some work on geopolymers has shown that these materials do not support combustion at all, and when applied to the outside of thermoset composites, show some dramatic flame retardant/flame resistance benefits. More specifically, since geopolymers are strong thermal insulators and are in their highest oxidation state, they do not burn when exposed to fire and provide fire protection for prolonged periods when applied to polymer composites [279, 280]. The literature on these materials suggest that geopolymer coatings are much thicker than those seen for plasma based coatings, and this may be the main reason why they have not been more exploited for polymer flame retardant protection. Otherwise, geopolymers remain a laboratory curiosity for flame retarding polymers. Geopolymers used "as is" have inherently no flammability since they are inorganic oxide based materials in their highest oxidation state incapable of suffering further fire damage. So they may be of more value being used in hybrid organic/inorganic composite materials for other fire protection needs [281, 282].

10.8.2.2   IR Reflective Coatings

Infrared reflective coatings are of interest due to some recently published results. Schartel *et al.* published results in 2012 showing that a copper metal mirror applied to plastic provided benefits in ignition resistance [283]. At a heat flux of 50 kW/m^2, the copper metal mirror delayed the time to ignition of the polymer by 350-400 second. The mechanism of protection provided by the metal mirror was one of infrared radiation reflection; the IR radiation was reflected away keeping the underlying plastic cool and from reaching its melting point or thermal decomposition temperature for some time. Eventually enough heat penetrated past the metal coating to ignite the polymer, but this approach suggests that metal coatings added to electronic plastics for electromagnetic interference (EMI) shielding benefit may provide some flame retardancy when radiant heat sources are present. More simply, white colors and optical effects which reflect IR energy due to pigmentation or coatings can provide delays in time to ignition or some other enhancements in flame retardant effects [284, 285], but the coating approach appears to be best suited to delay ignition as an approach to flame retardancy; it cannot prevent ignition if a continual heat source is applied, or if the coating is breached. Beyond this excellent work by Schartel in 2012, there have been no other studies on this concept of metallized reflective coatings for fire protection, nor do there appear to be any commercial examples. Still, the concept is highly promising and worth further consideration.

10.8.2.3   Nanoparticle Coatings

In 2008 work was reported where a nanoparticle rich barrier was applied to the outer layer of a fiber-reinforced thermoset composite for fire protection benefit. The nanoparticle layer was a non-woven made out of agglomerated carbon nanotubes or nanofibers which goes by the nickname "Buckypaper" due to the paper-like films made from "buckytubes" (multi-wall carbon nanotubes). These films were applied to thermoset composites and tested for fire protection in the cone calorimeter [286–290]. The benefit of this particular coating is in providing delays in time to peak heat release, or in reducing initial peak heat release. Once the sample is ignited though, the material will continue to slowly burn, indicating that the buckypaper is only useful for initial fire protection and would need to be combined with other flame retardants for a more robust effect. An additional benefit of these carbon nanoparticle rich outer layers would be in providing surface conductivity that could provide EMI shielding or anti-static benefits to the coated part. Since this work's initial publication in the 2008-2013 time frame, there does not appear to have been any other work on the concept, nor commercialization. Given the limited effectiveness of this technology, this solution likely can only work if combined with other underlying fire protection/flame retardancy using a "defense in depth" type of approach.

10.8.2.4   Conformal/Integrated Coatings

Building upon the concept of protective coatings is the newer work of conformal/integrated coatings on flammable materials. This includes both layer-by-layer technology which provides non-halogenated fire protection to materials which build up single layers of polymer, flame retardant, and/or nanofiller to impart conformal nanometer to micron thick protection on a polymer. This concept has been extensively studied to date [291, 292], and is the main focus of Chapter 8 of this book. Rather than rehashing the contents of Chapter 8 here, it will be simply stated that conformal layer-by-layer technology shows great promise, and while there have been no commercial products to date, the technology is capable of being made commercial, and just requires some process research & development investment to become commercial. Despite the large volume of work conducted to date, there is still a lot of chemistry that could be further studied and developed for flame retardant potential in this area, and research in this area is encouraged.

Similar to the layer-by-layer concept is the use of integrated flame retardant barriers/layers which covalently bond with the outer layers of the polymer. These would not be paints or coatings, but rather, engineering conformal/integrated coatings that become part of the final polymer part structure. This approach currently only lends itself to thermoset composite materials, where the top layers of the composite are made up of a different polymer formulation, rich in flame retardant, but still capable of co-polymerizing with the underlying thermoset polymer + fiber reinforcement layers. Some examples of this technology have already been described in this chapter in the form of intumescent mats that are used in the upper layers of the polymer composite [113, 116]. To date, the intumescent mats appear to be the only commercial system in use for fire protection of composites using this co-polymerization/covalently bonded fire protection layer approach. It's likely that other non-halogenated flame retardants other than intumescent mats could be part of this concept. However, these outer layers would need to be optimized for viscosity and polymerization with the rest of the polymer matrix so that they (a) don't diffuse into the underlying layers of fiber reinforcement, causing mechanical problems and (b) don't interfere with polymerization rates/composite processing.

10.9 Future Directions

The use of flame retardants has always been driven by market trends and regulations that react to new fire threats or concerns. Likewise, flame retardant chemistry has followed suit and these trends continue with the focus of this book and this chapter. Indeed, since the writing of the 1st edition of this book, the trends have been mostly unchanged, with concerns over persistence, bioaccumulation, and toxicity (PBT) of flame retardant chemicals continuing to drive the market. However, the scrutiny on all flame retardant chemicals has increased since the writing of the 1st edition, and Chapter 1 of this book gives some snapshot of the current state of regulations affecting all flame retardants, regardless of halogenated or non-halogenated chemistry. It's important to remember some history of flame retardant chemical use here as that history is driving the regulators, for right or for wrong, on flame retardant chemical use.

Post WWII, as polymers became very common in modern society, fire risk increased and so flame retardants were used to provide product fire safety. Flame retardants developed in the 1950s and 1960s based upon halogen did a superb job meeting the regulatory fire tests; at the time they were developed, sustainability and life cycle assessment of chemicals, let alone final disposition of any chemical, was not under consideration. As chemical environmental incidents began to occur more frequently in the 1960s and 1970s, society became more aware that how chemicals interacted with the environment was an issue, especially in the case of small molecules which were designed to be persistent in a product. Once those persistent chemicals escaped the manufacturing process or leached out of their products over time, they began to have environmental impact, and with new advents in analytical instrumentation that could detect chemicals at the nano to picogram levels, now chemicals of concern began to appear throughout our environment. The detection of halogenated flame retardants in the environment is not in doubt, and it's quite clear these chemicals are present, they are persistent, and the long term effects are still not known. Monitoring of non-halogenated flame retardants in the environment is now beginning, with some chemicals being detected in increasing amounts [293–300].

With the detection of halogenated chemicals, new environmental regulations began to be put in place, first in Europe and later in North America and Japan to slowly phase out the old chemicals and replace them with new ones, or, ban certain classes of chemicals outright. As indicated in Chapter 1, this process remains in place today and new bans are being implemented. One approach (an extreme one) is to ban flame retardants in all applications, which is unreasonable given that if more fires occur, the amount of polyaromatic hydrocarbons, soot, particulates, and dioxins from one large fire will more than overwhelm the environmental impact of using the flame retardant to prevent the fire in the first place [301–305]. Another approach, somewhat as extreme as the first, but more measured, is to ban discrete classes of flame retardants that are highly persistent or can form known toxins upon burning. Currently, both approaches are being used, with most nations going with bans on select chemicals, whereas there have been some attempts in the US (namely California) to attempt to ban/regulate anything with flame retardant potential. The regulatory situation in the US is complicated, and may not sort itself out anytime soon with political talking points drowning out scientific data. In light of the fact that all fires will cause pollution, wholesale bans of flame retardants are unlikely to achieve their intended goal of preventing hazardous chemicals from entering the environment. Halogen (Group VII) of the periodic table is native to our planet. Plastics and products utilized near oceans will automatically detect higher for halogen due to saline deposits from the sea in the air even if they are "halogen free". Further, many land-based plants do contain or pull some halogen into their structures, meaning that halogenated dioxins can and will be found in natural wildfires [306]. Therefore, just about any fire that occurs, if halogen is present, will likely result in toxins being produced along with the normal hazards of fires. Preventing/mitigating the fire emissions from the synthetic built environment will help lower overall toxic emissions into the environment. This is not to say that specific chemicals found to be problematic should continue to be allowed for use. If a specific halogenated or non-halogenated flame retardant is found to be a definite problem, then banning that specific chemistry does make sense. This leads to the saner approach to flame retardant regulation: Should a particular chemical be found to be persistent, bioaccumulative, and toxic, this particular chemical is the one which should be removed from use and banned via regulations, rather than the entire base chemical or class of chemicals to which it belongs. With the knowledge we have about environmental damage caused by man-made processes and chemicals, we as scientists must pay attention to flame retardant chemical design, and must consider environmental impact and the life cycle of that chemical for flame retardant and fire safety purposes.

The remainder of this chapter focuses on solutions to the environmental problem, but also some of the other important fire safety trends and life cycle trends for materials that need to be considered in new material/new flame retardant chemical discovery, development, and commercialization.

10.9.1   Polymeric Flame Retardants and Reactive Flame Retardants

With the awareness that flame retardants (or any industrial chemical/ polymer additive for that matter) can be pollutants, flame retardant & polymer additive manufacturers have instituted new programs and chemistries to address the problem. For new programs, chemical producers are helping with manufacturer awareness on chemical additive use and processing via a new program. This program, the Voluntary Emissions Control Action Programme [307], has been implemented to make manufacturers aware of how the additives they are using can exit the manufacturing process and get into the environment. While the main goal of this program is to address chemical emissions at the manufacturing level, it has a financial benefit as well for the manufacturers since more of what they paid for is ending up in the product, and the waste streams, for which they must pay disposal fees, are limited. For new chemistries, flame retardant manufacturers are focusing on the use of polymeric and reactive flame retardants based upon their environmental profiles.

Polymeric flame retardants are those additives which are very high molecular weight molecules, and are polymeric chains with flame retardant present in every repeat unit in the polymer structure (or at least many of the repeat units). Polymeric flame retardants have a better environmental profile than small molecules in that they are harder for organisms to ingest/metabolize, and therefore can be considered as less toxic. Further, polymeric additives are typically considered to be easier to add and blend into other polymers since polymer/polymer blends are easier to make and the polymeric flame retardant is less likely to leach out of a product once made. Also, polymeric additives tend to have lower impacts on polymer mechanical properties, but this is not always the case depending upon whether the polymer in question is amorphous or crystalline. Almost all new halogenated flame retardant additives in the past 3–4 years have been polymeric/oligomeric, and among non-halogenated chemistries, some high molecular weight phosphorus containing polymers have begun to enter the market as well in the past 2 years. Some non-halogenated chemistries lend themselves to polymeric incorporation (phosphorus, nitrogen, boron, silicon) while others do not (mineral fillers, Al, Mg, etc.). Provided the organic chemistry can be cost effective and the flame retardant monomer can be polymerized (or grafted onto a polymer), it is very likely that this approach will continue to be used for new non-halogenated additives in the near (and far) future. It's important to note here that polymer blends with inherently low-flammability polymers effectively serves as a non-halogenated flame retardant approach (assuming the polymers are non-halogenated that is). For example, polycarbonate (PC) is of lower flammability than poly(acrylonitrile-butadiene-styrene) copolymer, and so a blend of PC and ABS will be lower flammability than just ABS alone. Likewise, addition of polyphenylene oxide (PPO) to high impact polystyrene (HIPS) and other polymers is effectively using a non-halogenated polymer as a flame retardant additive through the use of a polymer/polymer blend. It's possible that other high melting temperature, low flammability polymers such as polyaryletherketones, polyetherimides, ionomers, and polysulfones may be effective "additives" for other plastics if issues of processing temperature differences and phase compatibility can be addressed.

Reactive flame retardants are flame retardants which react with the polymer, forming covalent bonds with the polymer so that they cannot come back out of the plastic either during use or during end-of-use of the product. Reactive flame retardants include those molecules which can react with monomers during polymer synthesis or react with the already made polymer via grafting reactions. Reactive flame retardants have been around for some time, and some well-known examples include tetrabromobisphenol A for epoxies and 9, 10-dihydro-9-oxa-10-phosphaphenanthrene-10-oxide (DOPO), also used with epoxy. Other examples include vinylsilanes for wire and cable polyolefin compounds and reactive phosphates for use with polyurethanes. The driving premise behind reactive flame retardants is that if they react into the polymer, they cannot get into the environment at all since they are part of a high molecular weight polymer rather than a small molecule. There is a caveat to this statement best discussed in regards to the environmental concern with polycarbonate. This polymer contains bisphenol A (BPA) and the finding that BPA has associated endocrine disorders/estrogen mimic issues has meant that the use of PC has come into question. Since the ester linkage can hydrolyze under certain conditions, it can release BPA. Thus, even though BPA is covalently linked, it can be released, and likewise, so could the reactive flame retardant. This speaks to the importance of end-of-life issues for polymers and ensuring that the flame retardant polymer is disposed of properly, so there is almost no chance it can get into the environment. Similarly, the reactive FR can be released if the polymer breaks down/depolymerizes outside fire conditions. Still, even with this concern, reactive flame retardants are likely to be a significant component of future non-halogenated flame retardant research. However, reactive flame retardants are not without their problems. In the case of reactive flame retardants that mix with monomers during polymer synthesis, the reactive flame retardant needs to be compatible with that polymerization chemistry and processing conditions. A reactive flame retardant which is water sensitive will not work well in a polyurethane foam polymerization, and likewise if a reactive flame retardant changes the polymerization kinetics, polymer molecular weight and properties may be negatively affected. Reactive flame retardants which bind to polymers via grafting may also result in losses in mechanical properties and so care in utilization and optimization of reactivity may be required, in depth, before a reactive flame retardant can be used. If the reactive flame retardant inhibits processing or requires expensive processing equipment before it can be used, it may not be used at all, no matter how good its fire performance. This last point is really true of any flame retardant or polymer additive; fire safety is only one requirement among many for a commercial product, and finding an acceptable balance of cost and performance in a final product is not easy.

Some examples of polymeric and reactive non-halogenated flame retardants, commercial and experimental, are shown in Figure 10.10.

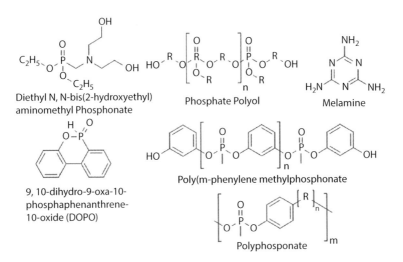

Figure 10.10 Examples of polymeric and reactive non-halogenated flame retardants.

10.9.2   End of Life Considerations For Flame Retardants

The increased emphasis in environmental impact on flame retardant additives also extends to the products and polymers that require fire safety performance. The environmental impacts in the product life cycle, from production to end-of-life and/or reuse, are also affecting flame retardant selection, and will therefore be a future trend to which material scientists must pay attention. What this means in regards to non-halogenated flame retardant use is that the selected non-halogenated flame retardants may need to meet one or more of the following criteria before they can be used as market conditions and regulations mandate:

- Be derived from sustainable or alternative chemical feedstocks. Meaning the starting materials are obtained from biological resources, recycled/chemically reused petrochemical feedstocks, or new petrochemical sources having different chemical compositions than currently obtained oil (shale gas, tar sands, etc.). Potentially, polymers which are pyrolyzed/gasified back into petrochemical feedstocks may be a source of "crude oil" for new flame retardant chemical development [308]. Non-halogenated heteroatoms other than carbon may require special handling and removal steps so that they do not limit the usefulness of waste plastic converted back into crude oil. There is much process chemistry that needs to be worked out with these mixed plastic waste streams.
- Be capable of surviving multiple polymer recycle and regrind thermal cycles while maintaining flame retardant performance. Economic viability in recycling must be attained. There is some concern however that regrind and recycle is no longer viable due to the high costs of "cleaning up" used plastics. This may become less of a concern in the future, but, regulations mandating recycling of plastics are still in place.
- Be compatible with waste-to-energy processes or waste-to-feedstock chemical processes should the flame retardant plastic not be suitable for recycling use. This seems to be more of a viable path forward, as several EU nations already use high energy plastic as a feedstock for waste-to-energy power plants [309, 310]. One can make an argument that with plastics piling up that cannot be recycled or degraded,

waste-to-energy and/or pyrolysis into chemical feedstocks is really the only path forward. However, since many flame retardant systems deliberately lower heat release, there could be an issue where flame retarded plastics must be incinerated/gasified with high heat release non-flame retardant plastics to yield sufficient heat from their destruction to make waste-to-energy viable. This could be worked out with some thought into end-of-life collection of flame retardant plastics, and research needs to be conducted on this important end-of-life consideration for flame retardant materials.

- Be compatible with composting and/or environmental decomposition mechanisms should the flame retardant polymer be disposed of in landfill. Since waste-to-energy is not used widely, and many plastics are now ending up in landfill (due to the above mentioned lack of viable economies for plastic recycling and reuse), unless there is a desire to use them as a future chemical feedstock or energy source in forms of "urban mining" [311, 312], the ability for these plastics to decompose in the environment is required. How flame retardants will affect this decomposition is unknown, especially if the plastic decomposes but the flame retardant does not, which would return us to the very problems of selected flame retardant chemicals today. Research and study on flame retardant chemical environmental decomposition, and design for environmental decomposition, is recommended. While there are standardized methods of determining how plastics decompose in the environment [313, 314], such protocols have not been established for flame retardants or other polymer additives. The use of bio-derived flame retardants, and mentioned earlier in this chapter, may be of increasing utility toward this goal.

The above points benefit from further discussion below.

Derivation from sustainable and alternative chemical feedstocks: Process chemistry used to manufacture flame retardant chemicals, as well as specific chemical structure used in retrosynthetic chemical analysis are the main tools of use to meet this need. By looking at what chemical feedstocks are biologically derived or can obtained from recycle/depolymerization of petrochemical based polymers, flame retardant chemists can determine what can be made to meet this objective. How to handle various non-halogenated chemicals which end up in the plastic "crude" will require process science. Some of them may be easily handled as inorganic "ash" from a pyrolysis process, but other products such as phosphorus and nitrogen containing flame retardants may decompose further and pyrolyze along with the desired hydrocarbon "crude". Further cleanup of this crude product would need to be accomplished, and indeed, even crude oil gets cleaned up before it is fed into chemical plants, so removal of phosphorus, nitrogen, and other elements may not be a difficult task. Recovering these elements for new flame retardant synthesis may also be of great benefit, and may be something achievable with existing process chemistry. Regardless, this approach of deriving chemical feedstocks for non-halogenated flame retardants from biological sources, depolymerized plastics, or other sources does require a very thorough understanding of chemical structure property relationships, especially in the case of a flame retardant which might need to meet other environmental criteria. It is important to note that any new flame retardant made, even if made from bio-derived feedstocks of low to no environmental impact, will have to undergo a full range of chemical tests to prove that it is safe for use under regulatory systems such as REACH [315]. Therefore, scientists may wish to initially focus on taking existing flame retardants known to work and tested under REACH, and focus on new routes for synthesis of those flame retardants with sustainable/alternative feedstocks before deriving whole new chemistries. Admittedly this approach means slower progress on new flame retardant chemistries before they come to market, but using this approach will ensure sale and use of flame retardant materials while new ones are developed, tested, and certified for use.

Flame Retardant Durability for Recycling: Durability of polymers and additives in polymers against multiple recycle runs is not a new concept. Many manufacturers reutilize polymer from "short shot"/mistake runs during injection molding of thermoplastics to save cost in their manufacturing of polymer parts. Further, to improve part production times and cycles, many thermoplastic processes now run hotter to lower polymer viscosity and produce parts faster. Therefore flame retardants are being asked to survive hotter temperatures and multiple thermal cycles. In some cases this has been well studied with some halogenated additives doing well in some cases, and some non-halogenated additives doing well in other cases [316–321]. Those using non-halogenated flame retardants should test the plastics they make for its durability against several regrind/recycle uses to determine how flame retardant performance is maintained or lost as a function of the number of recycle runs. Further, users should check to see how the flame retardant stays/migrates out of the polymer as a function of recycle. The leaching of additives out of a polymer during recycling is a known issue, and it is an issue where polymeric flame retardants or reactive flame retardants are expected to perform better. It should be pointed out that the recycling issues really only hold for thermoplastic flame retardant materials. For thermoset materials, recycling only occurs if the polymer can be chemically recycled (i.e., depolymerized back into monomer). Otherwise, thermoset materials do not lend themselves for recycling at this time and recycling of non-halogenated flame retardant thermosets is not likely to be a future trend. Waste-to-energy is more likely a driver for thermoset materials, especially in the case of circuit boards. This may also apply to structural thermoset composites, such as composites used to produce aircraft, ship, and train parts that are large and complex in shape. To date, these materials at end-of-life are just going to landfill, which is not a long term viable solution. It's possible some of these structures may be reused and repaired with advances in polymer composite repair technology, but even with these advances, eventually the composite structure will wear out and something other than burial in landfill must be considered.

Waste-To-Energy and Flame Retardants: When polymers (or any waste) cannot be recycled, converting that waste into energy is increasingly of interest should the waste have sufficient energetic value. Many plastics do have sufficient energetic value for waste-to-energy (WTE). WTE is not a new technology, and has been in use in Europe and other locations for some time, especially where landfill is not an option. Indeed, WTE can be a significant part of an energy portfolio for a city or municipality, and is quite common in northern Europe and Scandinavia. For polymers to be compatible with WTE, emission testing must be done to determine if the polymer waste generates any emissions of concern, such as polyaromatic hydrocarbons or dioxins. The use of afterburners can address some of these emissions in the case of incinerators, but otherwise gasification and pyrolysis processes which produce syngas and other flammable decomposition products may be needed. Incinerators and acid gas trapping systems are often needed in the case of precious metal recovery from halogen-containing circuit board/electronic waste. For non-halogenated flame retardants, inorganic elements and phosphorus are likely to end up in the "slag" or "fly ash" waste portion of the WTE process, and this particular waste may require some additional disposal considerations. Otherwise, researchers may need to partner with emissions/environmental combustion scientists to determine what potential gaseous emissions would occur from a non-halogenated waste in a WTE process, and how to mitigate these emissions and/or capture them if the emissions hold something of value (example, phosphorus). Finally, as mentioned above, considering the lower energetic content of flame retardant polymers vs. non-flame retardant polymers would need to be considered to ensure the WTE process has a net gain in energy release sufficient to power the plant and make economic sense.

Environmental Decomposition and Flame Retardants: In the event that the flame retardant polymer cannot be recycled, cannot be converted into chemical feedstocks or energy, and cannot be used/disposed of in any other way, it may have to be put into landfill. Or, should the flame retardant product be improperly disposed of (which does happen), it may end up being exposed to soil and water and left to degrade under environmental conditions. Designing for environmental decomposition while maintaining fire safety performance and product criteria is perhaps one of the most difficult future design criteria to address. The reason for this difficulty is the fact that environmental decomposition and degradation of a product is not well understood, as local environmental conditions (soil acidity, humidity, rainfall, salinity, local bacterial/flora/fauna) can all have an effect on polymer decomposition, and mapping all of these effects has not been accomplished. Short of extensive accelerated aging tests in a wide range of potential environmental conditions, it may not be possible for the user of non-halogenated flame retardants to design to this potential future need as flame retardant product durability and safe & quick environmental decomposition of a product are often in conflict with one another in product design. As will be discussed below, the potential need for fire hardening of building materials, especially those in wildland-urban-interface (WUI) areas may require more durability against the elements, not less, making environmental fate even harder to address for some materials. Still, for some disposable electronics and other items, utilizing biodegradable plastics or polymers known to degrade safely over time may be an acceptable starting point, and then any flame retardants added to those plastics should be one with very benign environmental profiles so that when they degrade, the chances of local environmental damage are very low. There are some tools being developed to help with this, with the most well-known one being GreenScreen [322], but some more data on non-halogenated flame retardants may be needed for this tool to be of wider utility. Users of non-halogenated flame retardants should pay attention to this issue in the future and be aware of it, as well as pay attention to new scientific knowledge of chemical environmental decomposition to help guide them on which flame retardants to use.

10.9.3   New and Growing Fire Risk Scenarios

Fire risk scenarios, where a notable fire hazard has been identified, and there is reasonable chance of that hazard occurring with significant loss of life and/or property, drive codes, laws, and regulations. Most of fire protection is reactive to fire threats as they are encountered, as it is hard to predict which potential fire hazards will increase such that they become the new fire risk scenario. As of the writing of this chapter, there are two fire risk scenarios of note that fire safety scientists need to be aware of for flame retardant material development. The first is the increasing threats of wildland-urban interface (WUI) fires, and the second is battery/electrical fires in transportation.

In regards to WUI fires, it should be obvious to the readers that the increasing amount of forest fires worldwide due to climate change, as well as major fire losses caused by said fires entering into towns and cities, is a fire hazard occurring with increasing regularity, making it a fire risk scenario of note. While WUI is heavily studied among fire safety engineers in the past several years, with advances made in understanding how the fire hazard spreads through burning brands [323], there has been no investment or research made on protecting against this fire risk scenario. This is partly because the fire safety engineers are focused on developing "Burning Brand" tests that mimic WUI fires, and so a standard test to test material flammability performance against this threat has yet to be developed. There are, however, numerous existing fire test methods for building materials that look at flame spread or other performance, and it is the opinion of the author that the fire safety engineers need to be looking at exiting tests to determine if the fire protection provided by an existing test provides reasonable protection and can be used now, rather than waiting for the perfect "burning brand" test to be developed. Still, it is being worked on, and hopefully some new test methods will be established and fire safety scientists and flame retardant material scientists can start developing flame retardant materials against WUI threats. Given the intensity of WUI fires, it is likely that flame retardant solutions will have to focus on not only slowing flame spread, but providing some fire hardening against persistent hot winds, radiant heat, and burning brand exposure. Non-halogenated flame retardants, especially those which enable char formation and greatly lowered heat release and protective barriers (such as intumescent materials) should do well in WUI fire scenarios, but this needs to be validated.

The other major fire risk scenario which is just now beginning to be considered is the hazards of all-electric transportation. Specifically, the increased potential for electrical ignition from motors and wires in all-electric cars, trucks, and eventually airplanes, and the increased potential for catastrophic fire events from battery thermal runaway/explosion events. For all-electric ground personal transportation (cars, trucks), these vehicles will have large lithium ion battery packs that present a notable fire hazard in crashes, or when the batteries short circuit/go into thermal runaway [324–331]. The intensity of these events is something that most polymeric materials cannot resist, and they will ignite. Most of the fire protection schemes around these battery packs focus on hardening the battery pack against crashes, or putting in battery/thermal management systems which lower the chances of battery failure. Further, there is research ongoing to lower the flammable components of the batteries themselves such that they will have energy release, but the battery components, being made from inorganic/ceramic materials, will be less likely to be involved in making the fire worse [332–334]. However, while there is ongoing studies on reducing the fire/explosion hazard of the battery pack, there will be increasing use of plastic/light weight polymer composites in the vehicle structure to get to a lighter weight vehicle to increase battery pack range. The fire protection tests for the this material, which can be ignited by electric short circuits from wiring in the car, is still governed by the Federal Motor Vehicle Safety Standard # 302 (FMVSS 302) test, which is a test that is very easy to pass, and can be argued to not really provide any meaningful level of fire safety with modern cars [335]. Whether or not this test method will change is unclear, but, electric ground vehicle fire hazard should be watched to determine what sort of flame retardant materials will be needed. Mostly likely, flame retardants which not only lower flame spread but resist electric arc ignition and thermally induced (direct or indirect heating without spark) ignition will be required. Several non-halogenated flame retardants should do well against these fire risk threats.

Related to the fire threats from batteries in ground vehicles, there is a great interest in all-electric aviation, with an emphasis on small aircraft, either unmanned air vehicles for cargo transportation, or small aircraft that serve as "air taxis" for short range transport. Similar to ground vehicles, the hazards of large battery packs exist, as does electric wiring ignition events located throughout the aircraft. Currently, all manned aviation has strong fire safety requirements under Federal Aviation Regulations (FAR) that are likely to be carried over to electric aviation, but it is unclear if that will carry over to unmanned vehicles. Further, as more lightweight polymers and plastics are used with ground vehicles to extend range, they will definitely be used with all-electric aircraft, although some metal may still be required to address aviation safety needs such as lightning strike/specific bird-strike impact requirements. Still, with an increase in flammable polymers, more fire protection, not less, may be required for all-electric aircraft even though the post-crash fuel pool fire threat has been removed with all-electric aircraft.

10.9.4 Experimental Methodology for Flame Retardant Screening

It is good to close a chapter on new non-halogenated flame retardant solutions for the future with guidance on what experimental methodology will be needed in the future to develop viable and effective non-halogenated flame retardants. Certainly existing experimental techniques and methods can and will be used, but given the new emphasis on environmental impact, as well a need for new chemistries in a very short period of time, it is worthwhile to comment on what new methods are needed and what new methods will be required in the future.

As mentioned in this chapter, environmental impact of flame retardant materials is of increasing importance. Therefore new experimental methods are needed to test the potential impact of new non-halogenated flame retardants before they are used. However, since environmental science is sometimes still not well understood, flame retardant scientists may need to stay abreast of new environmental chemical science, or partner with those scientists who understand this field before they begin their synthesis and testing of flame retardant materials. It can be argued that if the material shows no flame retardant benefit, then it is not worth considering for environmental screening since it will not be used for a flame retardant. While this statement is true, R&D funds spent on a flame retardant which shows fire safety potency but then a negative environmental profile can also be argued to be wasted R&D funds since the flame retardant cannot be used. Material scientists will need to make educated guesses and take balanced approaches in their research, but to assist with this more difficult task, some suggestions are made here for future experimental methods to deliver successful flame retardants with good environmental profiles.

Suggestion #1: Flame retardant chemists should partner with environmental scientists and toxicologists to understand chemical transport mechanisms in the environment and its potential fate/impact to organisms. By gaining understanding of which chemical groups promote rapid transport of chemicals through the food chain or into soil/water streams, the chemist can try to avoid certain chemical groups which may cause environmental damage.

Suggestion #2: Building off the future trend of polymeric and reactive flame retardants mentioned earlier in this chapter, scientists should focus on reactive and polymeric flame retardants in their flame retardant material development.

Suggestion #3: Researchers should consider developing test methods to evaluate flame retardant leachability/migration, and durability at the small scale. In this way, flame retardants with a tendency to come out of the plastic over time (and therefore get into the environment) can be found early in the R&D process and eliminated from future consideration.

Suggestion #4: The scientist should consider product lifecycle in flame retardant design. Working alongside engineers and marketing specialists, the team should think about how the product containing the flame retardant will be dealt with at the end of its lifetime. Will it be recycled, converted into energy/chemicals, or left to degrade in landfill? The team can then take this into effect and make choices early on about which polymer + flame retardant combinations to try and which ones to perhaps avoid for that particular product.

These four suggestions are based upon the current knowledge we have for flame retardant materials and environmental impacts of polymers and chemicals. As that knowledge increases, some of these suggestions may become invalid, or, may need to be expanded upon. But considering these suggestions early on in flame retardant product design is likely to save money and time in the future, even if it may delay new product release due to the complexity of balancing fire safety vs. product performance vs. environmental impact. Admittedly all of these criteria for a flame retardant product, or any product, can be in conflict with one another with seemingly little solution in sight. However, by thinking about these criteria early on, rather than as an afterthought or as a public relations response, can save money in the long run and deliver viable and acceptable fire safety for products.

References

1. Zammarano, M., Franceschi, M., Bellayer, S., Gilman, J.W., Meriani, S., Preparation and flame resistance properties of revolutionary self-extinguishing epoxy nanocomposites based on layered double hydroxides. *Polymer, 46,* 9314–9328, 2005.
2. Manzi-Nshuti, C., Wang, D., Hossenlopp, J.M., Wilkie, C.A., The role of the trivalent metal in an LDH: Synthesis, characterization and fire properties of thermally stable PMMA/LDH systems. *Polym. Degrad. Stab., 94,* 705–711, 2009.
3. Manzi-Nshuti, C., Chen, D., Su, S., Wilkie, C.A., The effects of intralayer metal composition of layered double hydroxides on glass transition, dispersion,

45. Cheng, L., Wu, W., Meng, W., Xu, S., Han, H., Yu, Y., Qu, H., Xu, J., Application of metallic phytates to poly(vinyl chloride) as efficient biobased phosphorus flame retardants. *J. App. Polym. Sci.*, *135*, 2018, https://onlinelibrary.wiley.com/doi/10.1002/app.46601

46. Naik, A.D., Bourbigot, S., Bellayer, S., Touati, N., Tayeb, K.B., Vezin, H., Fontaine, G., Salen Complexes as Fire Protective Agents for Thermoplastic Polyurethane: Deep Electron Paramagnetic Resonance Spectroscopy Investigation. *ACS Appl. Mater. Interfaces*, 10, 24860–24875, 2018.

47. Zhu, W. and Weil, E.D., A paradoxical flame-retardant effect of nitrates in ATH-filled ethylene–vinyl acetate copolymer. *J. Appl. Polym. Sci.*, 56, 925–933, 1995.

48. Feng, J., Carpanese, C., Fina., A., Thermal decomposition investigation of ABS containing Lewis-acid type metal salts. *Polym. Degrad. Stab.*, *129*, 319–327, 2016.

49. Levchik, S.V., Levchik, G.F., Camino, G., Costa, L., Lesnikovich, A.I., Fire-retardant action of potassium nitrate in polyamide 6. *Angew. Makromol. Chem.*, 245, 24–35, 1997.

50. Cullis, C.F. and Hirschler, M.M., Char Formation From Polyolefins: Correlations With Low-Temperature Oxygen Uptake and With Flammability in the Presence of Metal-Halogen Systems. *Eur. Polym. J.*, 20, 53-60, 1984.

51. Muller, P. and Schartel, B., Melamine poly(metal phosphates) as flame retardant in epoxy resin: Performance, modes of action, and synergy. *J. App. Polym. Sci.*, 133, 2016, https://onlinelibrary.wiley.com/doi/10.1002/app.43549

52. Liu, L., Huang, Y., Yang, Y., Ma, J., Yang, J., Yin, Q., Preparation of metal-phosphorus hybridized nanomaterials and the action of metal centers on the flame retardancy of epoxy resin. *J. App. Polym. Sci.*, 134, 2017, https://www.onlinelibrary.wiley.com/doi/full/10.1002/app.45445

53. Naik, A.D., Fontaine, G., Samyn, F. *et al.*, Melamine integrated metal phosphates as non-halogenated flame retardants: Synergism with aluminium phosphinate for flame retardancy in glass fiber reinforced polyamide 66. *Polym. Degrad. Stab.*, 98, 2653–2662, 2013.

54. Lyons, J.W., *The Chemistry and Uses of Fire Retardants*, Wiley-Interscience, Hoboken, New Jersey, USA, p. 166, 1970.

55. W. Nouvertne, US Patent 3 775 367, to Bayer, 1973.

56. V. Mark, US Patent 3 940 366, to General Electric, 1976.

57. Ballistreri, A., Montaudo, G., Scamporrino, E., Puglisi, C., Vitalini, D., Cucinella, S., Intumescent flame retardants for polymers. IV. The polycarbonate–aromatic sulfonates system. *J. Polym Sci. Part A Polym Chem.*, 26, 2113–2127, 1988.

58. Huang, X., Ouyang, X., Ning, F., Wang, J., Mechanistic study on flame retardance of polycarbonate with a small amount of potassium perfluorobutane sulfonate by TGA–FTIR/XPS. *Polym. Degrad. Stab.*, 91, 606–613, 2006.

thermal and fire properties of their PMMA nanocomposites. *Thermochimica Acta*, *495*, 63–71, 2009.

4. Pitts, W.M., Nyden, M.R., Gann, R.G., Mallard, W.G., Tsang, W., *NIST Technical Note 1279*, August 1990, US National Institute of Standards and Technology (NIST), Gaithersburg, MD. USA.

5. Linteris, G.T., Knyazev, V.D., Babushok, V.I., Inhibition of premixed methane flames by manganese and tin compounds. *Combust. Flame*, 129, 221–238, 2002.

6. Linteris, G.T., Katta, V.R., Takahashi, F., Experimental and numerical evaluation of metallic compounds for suppressing cup-burner flames. *Combust. Flame*, 138, 78–96, 2004.

7. Hirasawa, T., Sung, C.-J., Yang, Z., Joshi, A., Wang, H., Effect of ferrocene addition on sooting limits in laminar premixed ethylene–oxygen–argon flames. *Combust. Flame*, 139, 288–299, 2004.

8. Koshiba, Y., Takahasi, Y., Ohtani, H., Flame suppression ability of metallocenes (nickelocene, cobaltcene, ferrocene, manganocene, and chromocene). *Fire Saf. J.*, *51*, 10–17, 2012.

9. Koshiba, Y., Agata, S., Takahashi, T., Ohtani, H., Direct comparison of the flame inhibition efficiency of transition metals using metallocenes. *Fire Saf. J.*, *73*, 48–54, 2015.

10. Antonov, A., Yablokova, M., Costa, L., Balabanovich, A., Levchik, G., Levchik, S., The Effect of Nanometals on the Flammability and Thermooxidative Degradation of Polymer Materials. *Mol. Cryst. Liq. Cryst.*, 353, 203–210, 2000.

11. Morgan, A.B., *ACS Symposium Series 1013 – Fire and Polymers V: Materials and Concepts for Fire Retardancy*, pp. 312–328, Oxford University Press, Oxford, United Kingdom, 2009.

12. Soyama, M., Inoue, K., Iji, M., Flame retardancy of polycarbonate enhanced by adding fly ash. *Polym. Adv. Technol.*, 18, 386–391, 2007.

13. Tian, C., Wang, H., Liu, X., Ma, Z., Guo, H., Xu, J., Flame retardant flexible poly(vinyl chloride) compound for cable application. *J. App. Polym. Sci.*, 89, 3137–3142, 2003.

14. Song, R., Fu, Y., Li, B., Transferring noncharring polyolefins to charring polymers with the presence of Mo/Mg/Ni/O catalysts and the application in flame retardancy. *J. App. Polym. Sci.*, 129, 138–144, 2013.

15. Ishikawa, T., Ueno, T., Watanabe, Y., Mizuno, K., Takeda, K., Flame Retardancy of Polybutylene Terephthalate Blended with Various Oxides. *J. App. Polym. Sci.*, 109, 910–917, 2008.

16. Laoutid, F., Ferry, L., Lopez-Cuesta, J.M., Crespy, A., Red phosphorus/aluminium oxide compositions as flame retardants in recycled poly(ethylene terephthalate). *Polym. Degrad. Stab.*, 82, 357–363, 2003.

17. Gallo, E., Braun, U., Schartel, B., Russo, P., Acierno, D., Halogen-free flame retarded poly(butylene terephthalate) (PBT) using metal oxides/

PBT nanocomposites in combination with aluminium phosphinate. *Polym. Degrad. Stab.*, 94, 1245–1253, 2009.

18. Laachaci, A., Chochez, M., Leroy, E., Ferriol, M., Lopez-Cuesta, J.M., Fire retardant systems in poly(methyl methacrylate): Interactions between metal oxide nanoparticles and phosphinates. *Polym. Degrad. Stab.*, 92, 61–69, 2007.

19. Friederich, B., Laachachi, A., Sonnier, R. *et al.*, Comparison of alumina and boehmite in (APP/MPP/metal oxide) ternary systems on the thermal and fire behavior of PMMA. *Polym. Adv. Technol.*, 23, 1369–1380, 2012.

20. Friederich, B., Laachachi, A., Ferriol, M. *et al.*, Investigation of fire-resistance mechanisms of the ternary system (APP/MPP/TiO2) in PMMA. *Polym. Degrad. Stab.*, 97, 2154–2161, 2012.

21. Wu, N. and Yang, R., Effects of metal oxides on intumescent flame-retardant polypropylene. *Polym. Adv. Technol.*, 22, 495–501, 2011.

22. Xinlong, W., Lianghu, W., Jin, L., A Study on the Performance of Intumescent Flame-retarded Polypropylene with Nano-ZrO2. *J. Fire Sci.*, 29, 227–242, 2011.

23. Lin, M., Li, B., Li, Q. *et al.*, Synergistic effect of metal oxides on the flame retardancy and thermal degradation of novel intumescent flame-retardant thermoplastic polyurethanes. *J. App. Polym. Sci.*, 121, 1951–1960, 2011.

24. Li, Y., Li, B., Dai, J., Jia, H., Gao, S., Synergistic effects of lanthanum oxide on a novel intumescent flame retardant polypropylene system. *Polym. Degrad. Stab.*, 93, 9–16, 2008.

25. Jing, W., Yuan, H., Lei, S., Wenjie, K., Synergistic Effect of Lanthanum Oxide on Intumescent Flame-Retardant Polypropylene-based Formulations. *J. Fire Sci.*, 26, 399–414, 2008.

26. Gao, W., Yu, Y., Chen, T., Zhang, Q., Enhanced flame retardancy of unsaturated polyester resin composites containing ammonium polyphosphate and metal oxides. *J. App. Polym. Sci.*, 137, 2020, https://onlinelibrary.wiley.com/doi/abs/10.1002/app.49148.

27. Sheng, Y., Chen, Y., Bai, Y., Catalytically synergistic effects of novel LaMnO3 composite metal oxide in intumescent flame-retardant polypropylene system. *Polym. Composites*, 35, 2390–2400, 2014.

28. Shen, Y., Gong, W., Zheng, B., Synergistic effect of Ni-based bimetallic catalyst with intumescent flame retardant on flame retardancy and thermal stability of polypropylene. *Polym. Degrad. Stab.*, 129, 114–124, 2016.

29. Sharma, S.K., Flame Retardance and Smoke Suppression of Poly(vinyl chloride) Using Multicomponent Systems. *Fire Technol.*, 39, 247–260, 2003.

30. Sharma, S.K., Saxena, N.K., Surface Coating and Metal-Based Organic Additives—Approaches for FRSS PVC Cables. *J. Fire Sci.*, 25, 447–466, 2007.

31. Starnes, W.H., Jr, Pike, R.D., Cole, J.R., Doyal, A.S., Kimlin, E.J., Lee, J.T., Murray, P.J., Quinlan, R.A., Zhang, J., Cone calorimetric study of copper-promoted smoke suppression and fire retardance of poly(vinyl chloride). *Polym. Degrad. Stab.*, 82, 15–24, 2003.

32. Pike, R.D., Starnes, W.H., Jr, Jeng, J.P., Bryant, W.S., Kourtesis, P., Adam, C.W., Low-Valent Metals as Reductive Cross-Linking Agents: A New Strategy for Smoke Suppression of Poly(vinyl chloride). *Macromolecules*, 30, 6957–65, 1997.

33. Shehata, A.B., Hassan, M.A., Darwish, N.A., Kaolin modified with new resin-iron chelate as flame retardant system for polypropylene. *J. App. Polym. Sci.*, 92, 3119–3125, 2004.

34. Shehata, A.B., A new cobalt chelate as flame retardant for polypropylene filled with magnesium hydroxide. *Polym. Degrad. Stab.*, 85, 577–582, 2004.

35. Hassan, M.A. and Shehata, A.B., The effect of some polymeric metal chelates on the flammability properties of polypropylene. *Polym. Degrad. Stab.*, 85, 733–740, 2004.

36. Fontaine, G., Turf, T., Bourbigot, S., *ACS Symposium Series 1013 – Fire and Polymers V: Materials and Concepts for Fire Retardancy*, pp. 329–340, Oxford University Press, Oxford, United Kingdom, 2009.

37. Cui, J., Zhang, Y., Wang, L., Liu, H. *et al.*, Phosphorus-containing Salen-Ni metal complexes enhancing the flame retardancy and smoke suppression of epoxy resin composites. *J. App. Polym. Sci.*, 137, 2020, https://www.onlinelibrary.wiley.com/doi/full/10.1002/app.48734

38. Costes, L., Laoutid, F., Dumazert, L., Lopez-cuesta, J.-M., Brohez, S., Delvosalle, C., Dubois, P., Metallic phytates as efficient bio-based phosphorus flame retardant additives for poly(lactic acid). *Polym. Degrad. Stab.*, 119, 217–227, 2015.

39. Zhang, Y., Cui, J., Wang, L., Liu, H., Phosphorus-containing Salen-metal complexes investigated for enhancing the fire safety of thermoplastic polyurethane (TPU). *Polym. Adv. Technol.*, 31, 1150–1163, 2020.

40. Wang, D.-Y., Liu, Y., Wang, Y.-Z., Artiles, C.P., Hull, T.R., Price, D., Fire retardancy of a reactively extruded intumescent flame retardant polyethylene system enhanced by metal chelates. *Polym. Degrad. Stab.*, 92, 1592–1598, 2007.

41. Hou, Y., Hu, W., Gui, Z., Hu, Y., A novel Co(II)-based metal-organic framework with phosphorus-containing structure: Build for enhancing fire safety of epoxy. *Compos. Sci. Technol.*, 152, 231–242, 2017.

42. Lavrenyuk, H., Kochubei, V., Mykhalichko, O., Mykhalichko, B., Metal-coordinated epoxy polymers with suppressed combustibility. Preparation technology, thermal degradation, and combustibility test of new epoxy-amine polymers containing the curing agent with chelated copper(II) carbonate. *Fire Mater.*, 42, 266–277, 2018.

43. Holdsworth, A.F., Horrocks, A.R., Kandola, B.K., Price, D., The potential of metal oxalates as novel flame retardants and synergists for engineering polymers. *Polym. Degrad. Stab.*, 110, 290–297, 2014.

44. Holdworth, A.F., Horrocks, A.R., Kandola, B.K., Synthesis and thermal analytical screening of metal complexes as potential novel fire retardants in polyamide 6, 6. *Polym. Degrad. Stab.*, 144, 420–433, 2017.

59. Innes, J.D. and Innes, A., Flame retardants for polycarbonate - new and classical solutions. *Plast. Add. Comp.*, 8, 26–29, 2006.

60. Nodera, A., Kanai, T., Thermal decomposition behavior and flame retardancy of polycarbonate containing organic metal salts: Effect of salt composition. *J. Appl. Polym. Sci.*, 94, 2131–2139, 2004.

61. Dang, X., Bai, X., Zhang, Y., Thermal degradation behavior of low-halogen flame retardant PC/PPFBS/PDMS. *J. Appl. Polym. Sci.*, 119, 2730–2736, 2011.

62. Lewin, M., Flame Retarding of Polymers with Sulfamates. I. Sulfation of Cotton and Wool. *J. Fire Sci.*, 15, 263–276, 1997.

63. Lewin, M., Brozek, J., Martens, M.M., The system polyamide/sulfamate/dipentaerythritol: flame retardancy and chemical reactions. *Polym. Adv. Tech.*, 13, 1091–1102, 2002.

64. Lewin, M., Zhang, J., Pearce, E., Gilman, J., Flammability of polyamide 6 using the sulfamate system and organo-layered silicate. *Polym. Adv. Tech.*, 18, 737–745, 2007.

65. Lewin, M., Flame retarding polymer nanocomposites: Synergism, cooperation, antagonism. *Polym. Degrad. Stab.*, 96, 256–269, 2011.

66. He, J., Cai, G., Wilkie, C.A., The effects of several sulfonates on thermal and fire retardant properties of poly(methyl methacrylate) and polystyrene. *Polym. Adv. Technol.*, 25, 160–167, 2014.

67. Tirri, T., Aubert, M., Pawelec, W., Holappa, A., Wilen, C.-E., Structure-Property Studies on a New Family of Halogen Free Flame Retardants Based on Sulfenamide and Related Structures. *Polymers*, 8, 360, 2016.

68. Honhe, C.-C., Posern, C., Bohme, U., Eichler, F., Kroke, E., Dithiocyanurates and thiocyamelurates: Thermal thiyl radical generators as flame retardants in polypropylene. *Polym. Degrad. Stab.*, 166, 17–30, 2019.

69. Pawelec, W., Holappa, A., Tirri, T., Aubert, M., Hoppe, H., Pfaendner, R., Wilen, C.-E., Disulfides – Effective radical generators for flame retardancy of polypropylene. *Polym. Degrad. Stab.*, 110, 447–456, 2014.

70. Laufer, G., Kirkland, C., Morgan, A.B., Grunlan, J.C., Exceptionally Flame retardant Sulfur-Based Multilayer Nanocoating for Polyurethane Prepared from Aqueous Polyelectrolyte Solutions. *ACS Macro Lett.*, 2, 361–365, 2013.

71. Hou, S., Zhang, Y.J., Jiang, P., Phosphonium sulfonates as flame retardants for polycarbonate. *Polym. Degrad. Stab.*, 130, 165–172, 2016.

72. Wagner, J., Degimann, P., Fuchs, S., Ciesielski, M., Fleckenstein, C.A., Doring, M., A flame retardant synergism of organic disulfides and phosphorus compounds. *Polym. Degrad. Stab.*, 129, 63–76, 2016.

73. Perez, R.M., Sandler, J.K.W., Alstadt, V., Hoffmann, T., Pospiech, D., Ciesielski, M., Doring, M., Braun, U., Balavanovich, A.I., Schartel, B., Novel phosphorus-modified polysulfone as a combined flame retardant and toughness modified for epoxy resins. *Polymer*, 48, 778–790, 2007.

74. Yuan, L., Feng, S., Hu, Y., Fan, Y., Effect of char sulfonic acid and ammonium polyphosphate on flame retardancy and thermal properties of epoxy resin and polyamide composites. *J. Fire Sci.*, 35, 521–534, 2017.

75. Wang, P., Xiao, H., Duan, C., Wen, B., Li, Z., Sulfathiazole derivative with phosphaphenanthrene group: Synthesis, characterization and its high flame-retardant activity on epoxy resin. *Polym. Degrad. Stab.*, *173*, 109078, 2020.

76. Morgan, A.B. and Tour, J.M., Synthesis and Testing of Nonhalogenated Alkyne-Containing Flame-Retarding Polymer Additives. *Macromolecules*, *31*, 2857–2865, 1998.

77. Morgan, A.B. and Tour, J.M., Synthesis and testing of nonhalogenated alkyne/phosphorus-containing polymer additives: Potent condensed-phase flame retardants. *J. App. Polym. Sci.*, *73*, 707–718, 1999.

78. Zhao, H.-B., Chen, L., Yang, J.-C., Ge, X.-G., Wang, Y.-Z., A novel flame-retardant-free copolyester: cross-linking towards self extinguishing and non-dripping. *J. Mater. Chem.*, *22*, 19849–19857, 2012.

79. Kimura, H., Ohtsuka, K., Yonekawa, M., Epoxy resins with high heat resistance and flame retardancy via a new process. *Polym. Adv. Technol.*, 32, 474–483, 2021.

80. Liu, B.-W., Zhao, H.-B., Tan, Y., Chen, L., Wang, Y.-Z., Novel crosslinkable epoxy resins containing phenylacetylene and azobenzene groups: From thermal crosslinking to flame retardance. *Polym. Degrad. Stab.*, *122*, 66–76, 2015.

81. Ranganathan, T., Cossette, P., Emirck, T., Halogen-free, low flammability polyurethanes derived from deoxybenzoin-based monomers. *J. Mater. Chem.*, *20*, 3681–3687, 2010.

82. Ellzey, K.A., Ranganathan, T., Zilberman, J., Coughlin, E.B., Farris, R.J., Emrick, T., Deoxybenzoin-Based Polyarylates as Halogen-Free Fire-Resistant Polymers. *Macromolecules*, *39*, 3553–3558, 2006.

83. Ryu, B.-Y., Moon, S., Kosif, I., Ranganathan, T., Farris, R.J., Emrick, T., Deoxybenzoin-based epoxy resins. *Polymer*, *50*, 767–774, 2009.

84. Hu, X., Wang, Y., Yu, J., Zhu, J., Hu, Z., Nonhalogen flame retarded poly(butylene terephthalate) composite using aluminum phosphonate and phosphorus-containing deoxybenzoin polymer. *J. App. Polym. Sci.*, *134*, 2017, https://onlinelibrary.wiley.com/doi/full/10.1002/app.45537

85. Moon, S., Ku, B.-C., Emrick, T., Coughlin, B.E., Farris, R.J., Flame Resistant Electrospun Polymer Nanofibers from Deoxybenzoin-based Polymers. *J. App. Polym. Sci.*, *111*, 301–307, 2009.

86. Ranganathan, T., Cossette, P., Emirck, T., Halogen-free, low flammability polyurethanes derived from deoxybenzoin-based monomers. *J. Mater. Chem.*, *20*, 3681–3687, 2010.

87. Choudhary, U., Mir, A.A., Emrick, T., Soluble, Allyl-Functionalized Deoxybenzoin Polymers. *Macromolecules*, *50*, 3772–3778, 2017.

88. Ranganathan, T., Beaulieu, M., Zilberman, J., Smith, K.D., Westmoreland, P.R., Farris, R.J., Coughlin, E.B., Emrick, T., Thermal degradation of deoxybenzoin polymers studied by pyrolysis-gas chromatography/mass spectrometry. *Polym. Degrad. Stab.*, *93*, 1059–1066, 2008.

89. Olah, G.A., *Friedel-Crafts and Related Reactions*, vol. 1, G.A. Olah (Ed.), p. 26, Wiley-Interscience, Hoboken, NJ USA, 1963.

90. Grassie, N. and Gilks, J., Thermal analysis of polystyrenes crosslinked by p-di(chloromethyl)benzene. *J. Polym. Sci: Polym. Chem. Ed.*, *11*, 1985–1994, 1973.

91. Brauman, S.K., Friedel–crafts reagents as charring agents in impact polystyrene. *J. Polym. Sci: Polym. Chem. Ed.*, *17*, 1129–1144, 1979.

92. Rabek, J.F. and Lucki, J., Crosslinking of polystyrene under friedel–crafts conditions in dichloroethane and carbon tetrachloride solvents through the formation of strongly colored polymer–AlCl3–solvent complexes. *J. Polym. Sci: Polym. Part A; Chem.*, *26*, 2537–2551, 1988.

93. Li, J. and Wilkie, C.A., Improving the thermal stability of polystyrene by Friedel-Crafts chemistry. *Polym. Degrad. Stab.*, *57*, 293–299, 1997.

94. Khanna, Y.P. and Pearce, E.M., *Flame Retardant Polymeric Materials*, vol. 2, M. Lewin, S. Atlas, E.M. Pearce (Eds.), pp. 43–61, Plenum Press, New York, 1975.

95. Pearce, E.M., *Contemporary Problems in Polymer Science*, vol. 5, E.J. Vandenberg (Ed.), pp. 401–413, Plenum Press, New York, NY USA, 1984.

96. Pearce, E.M., *Proceedings of the Meeting on Recent Advances in Flame Retardancy of Polymeric Materials*, Business Communications Company (BCC), Stamford, CT, USA, pp. 116–120, 1991.

97. Wang, Z., Jiang, D.D., McKinney, M.A., Wilkie, C.A., Cross-linking of polystyrene by Friedel–Crafts chemistry to improve thermal stability. *Polym. Degrad. Stab.*, *64*, 387–395, 1999.

98. Zhu, J., McKinney, M.A., Wilkie, C.A., Stabilization of Polystyrene by Friedel-Crafts Chemistry: Effect of Position of Alcohol and the Catalyst. *Polym. Degrad. Stab.*, *66*, 213–220, 1999.

99. Yao, H., McKinney, M.A., Dick, C., Liggat, J.J., Snape, C.E., Wilkie, C.A., Cross-linking of Polystyrene by Friedel–Crafts Chemistry: Reaction of p-hydroxymethylbenzyl Chloride with Polystyrene. *Polym. Degrad. Stab.*, *72*, 399–405, 2001.

100. Wang, J., Du, J., Yao, H., Wilkie, C.A., XPS characterization of Friedel-Crafts cross-linked polystyrene. *Polym. Degrad. Stab.*, *74*, 321–326, 2001.

101. Yao, H. and Wilkie, C.A., *Fire & Polymers, Materials and Solutions for Hazard Prevention, ACS Symposium Series 797*, G.L. Nelson and C.A. Wilkie (Eds.), pp. 125–135, Oxford University Press, Oxford, United Kingdom, 2001.

102. Ran, S., Guo, Z., Han, L., Fang, Z., Effect of Friedel-Crafts reaction on the thermal stability and flammability of high density polyethylene/brominated polystyrene/graphene nanoplatelet composites. *Polym. Intl.*, *63*, 1835–1841, 2014.

103. Oh, S.Y., Pearce, E.M., Kwei, T.K., *Fire and Polymers II, ACS Symposium Series 599*, pp. 136–158, American Chemical Society, Washington, DC, 1995.

104. Morgan, A.B. and Tour, J.M., Synthesis, flame-retardancy testing, and preliminary mechanism studies of nonhalogenated aromatic boronic acids:

A new class of condensed-phase polymer flame-retardant additives for acrylonitrile–butadiene–styrene and polycarbonate. *J. App. Polym. Sci.*, 76, 1257–1268, 2000.

105. Weil, E.D., *17th BCC Conference on Flame Retardancy*, Stamford, CT, May 23-25, 2004.

106. Pawlowski, K.H. and Schartel, B., Flame retardancy mechanisms of triphenyl phosphate, resorcinol bis(diphenyl phosphate) and bisphenol A bis(diphenyl phosphate) in polycarbonate/acrylonitrile–butadiene–styrene blends. *Polym. Int.*, 56, 1404–1414, 2007.

107. Benin, V., Gardelle, B., Morgan, A.B., Heat release of polyurethanes containing potential flame retardants based on boron and phosphorus chemistries. *Polym. Degrad. Stab.*, 106, 108–121, 2014.

108. McNeil, I.C. and Zulfiqar, M., Preparation and degradation of salts of poly(methacrylic acid). I. Lithium, sodium, potassium, and caesium salts. *J. Polym. Sci. Polym. Chem. Ed.*, 16, 3201–3212, 1978.

109. Suzuki, M. and Wilkie, C.A., The thermal degradation of acrylonitrile-butadiene-styrene terploymer grafted with methacrylic acid. *Polym. Degrad. Stab.*, 47, 223–228, 1995.

110. Wilkie, C.A., Dong, X., Suzuki, M., *Fire and Polymers 2, ACS Symposium Series # 599*, G. Nelson (Ed.), pp. 236–244, American Chemical Society, Washington, DC, 1995.

111. Duquesne, S., Delobel, R., Le Bras, M., Camino, G., A comparative study of the mechanism of action of ammonium polyphosphate and expandable graphite in polyurethane. *Polym. Degrad. Stab.*, 77, 333–344, 2002.

112. Bourbigot, S., Sarazin, J., Bensabath, T., Samyn, F., Jimenez, M., Intumescent polypropylene: Reaction to fire and mechanistic aspects. *Fire Saf. J.*, 105, 261–269, 2019.

113. Chenkai, Z., Jingjing, L., Mandy, C., Xiaosu, Y. *et al.*, The effect of intumescent mat on post-fire performance of carbon fibre reinforced composites. *J. Fire Sci.*, 37, 257–272, 2019.

114. Kandare, E., Chukwudole, C., Kandola, B.K., The use of fire-retardant intumescent mats for fire and heat protection of glass fibre-reinforced polyester composites: Thermal barrier properties. *Fire Mater.*, 34, 21–38, 2010.

115. Li, J., Zhu, C., Zhao, Z., Khalili, P., Clement, M., Tong, J., Liu, X., Yi, X., Fire properties of carbon fiber reinforced polymer improved by coating nonwoven flame retardant mat for aerospace application. *J. App. Polym. Sci.*, 136, 47801, 2019.

116. Zhu, C., Li, S., Li, J., Clement, M., Rudd, C., Yi, X., Liu, X., Fie performance of sandwich composites with intumescent mat protection: Evolving thermal insulation, post-fire performance and rail industry testing. *Fire Saf. J.*, 116, 103205, 2020.

117. Gardelle, B., Duquesne, S., Vandereecken, P., Bourbigot, S., Characterization of the carbonization process of explandable graphite/silicone formulations in a simulated fire. *Polym. Degrad. Stab.*, 98, 1052–1063, 2013.

118. Gardelle, B., Duquesne, S., Vandereecken, P., Bourbigot, S., Resistance to fire of silicone-based coatings: Fire protection of steel against cellulosic fire. *J. Fire Sci.*, *32*, 374–387, 2014.

119. Alongi, J., Carosio, F., Malucelli, G., Current emerging techniques to impart flame retardancy to fabrics: An overview. *Polym. Degrad. Stab.*, *106*, 138–149, 2014.

120. Costes, L., Laoutid, F., Brohez, S., Dubois, P., Bio-based flame retardants: When nature meets fire protection. *Mater. Sci. Eng. R*, *117*, 1–25, 2017.

121. Hobbs, C.E., Recent Advances in Bio-Based Flame Retardrant Additives for Synthetic Polymeric Materials. *Polymers*, *11*, 224, 2019.

122. Watson, D.A.V. and Schiraldi, D.A., Biomolecules as Flame Retardant Additives for Polymers: A Review. *Polymers*, *12*, 849, 2020, doi:10.3390/polym12040849.

123. Feng, J.X., Su, S.P., Zhu, J., An intumescent flame retardant system using β-cyclodextrin as a carbon source in polylactic acid (PLA). *Polym. Adv. Tech.*, *22*, 1115–1122, 2011.

124. Feng, J.X., Zhang, X., Ma, S., Zhu, J., Syntheses of Metallic Cyclodextrins and Their Use as Synergists in a Poly(Vinyl Alcohol)/Intumescent Flame Retardant System. *Ind. Eng. Chem. Res.*, *52*, 2784–2792, 2013.

125. Feng, J.X. and Zhu, J., Ferrocene-ß-Cyclodextrin Inclusion Compound Used in Polystyrene Intumescent Flame Retardant System. *Polym. Mat. Sci. Eng.*, *106*, 85-86, 2012.

126. Feng, J.X., Su, S.P., Zhu, J., *Proceed. 2010 Int. Sym. Flame-Retardant Mat.*, Chengdu, China ISFRMT 2010, September 17-20, 2010, Chengdu, China. Organised by the Chinese Flame Retardant Society and the Sichuan University of Chengdu.

127. Le Bras, M., Bourbigot, S., Le Tallec, Y., Laureyns, J., Synergy in intumescence— application to β-cyclodextrin carbonisation agent in intumescent additives for fire retardant polyethylene formulations. *Polym. Degrad. Stab.*, *56*, 11–21, 1997.

128. Huang, L., Gerber, M., Lu., J., Tonelli, A.E., Formation of a flame retardant-cyclodextrin inclusion compound and its application as a flame retardant for poly(ethylene terephthalate). *Polym. Degrad. Stab.*, *71*, 279–284, 2001.

129. Wang, H. and Li, B., Synergistic effects of β-cyclodextrin containing silicone oligomer on intumescent flame retardant polypropylene system. *Polym. Adv. Tech.*, *21*, 691–697, 2010.

130. Alongi, J., Pošković, M., Frache, A., Trotta, F., Novel flame retardants containing cyclodextrin nanosponges and phosphorus compounds to enhance EVA combustion properties. *Polym. Degrad. Stab.*, *95*, 2093–2100, 2010.

131. Alongi, J., Poskovic, M., Visakh, P.M., Frache, A., Malucelli, G., Cyclodextrin nanosponges as novel green flame retardants for PP, LLDPE and PA6. *Carbohyd. Polym.*, *88*, 1387–1394, 2012.

132. Enescu, D., Alongi, J., Frache, A., Evaluation of Nonconventional Additives as Fire Retardants on Polyamide 6, 6: Phosphorous-Based Master Batch,

a-Zirconium Dihydrogen Phosphate, and b-Cyclodextrin Based Nanosponges. *J. Appl. Polym. Sci.*, *123*, 3545–3555, 2012.

133. Wang, X., Xing, W., Wang, B., Wen, P., Song, L., Hu, Y., Zhang, P., Comparative Study on the Effect of Beta-Cyclodextrin and Polypseudorotaxane As Carbon Sources on the Thermal Stability and Flame Retardance of Polylactic Acid. *Ind. Eng. Chem. Res.*, *52*, 3287–3294, 2013.

134. Luda, M.P. and Zanetti, M., Cyclodextrins and Cyclodextrin Derivatives as Green Char Promoters in Flame Retardants Formulations for Polymeric Materials. A Review, *Polymers*, *11*, 664, 2019.

135. Zhang, N., Shen, J., Pasquinelli, M.A., Formation and characterization of an inclusion complex of triphenyl phosphate and β-cyclodextrin and its use as a flame retardant for polyethylene terephthalate. *Polym. Degrad. Stab.*, *120*, 244–250, 2015.

136. Vehabi, H., Shabanian, M., Aryanasab, F., Laoutid, F., Benali, S., Saeb, M.R., Seidi, F., Kandola, B.K., Three in one: β-cyclodextrin, nanohydroxyapatite, and a nitrogen-rich polymer integrated into a new flame retardant for poly (lactic acid). *Fire Mater.*, *42*, 593–602, 2018.

137. Shan, X., Jiang, K., Li, J., Song, Y., Han, J., Hu, Y., Preparation of β-Cyclodextrin Inclusion Complex and Its Application as an Intumescent Flame Retardant for Epoxy. *Polymers*, *11*, 71, 2019, https://doaj.org/article/5f0854230830444181bb36ba46961cf8.

138. Husseinsyah, S., Amri, F., Husin, K., Ismail, H., Mechanical and thermal properties of chitosan-filled polypropylene composites: The effect of acrylic acid. *J. Vinyl Add. Tech.*, *17*, 125–131, 2011.

139. Hu, S., Song, L., Pan, H., Hu, Y., Gong, X., Thermal properties and combustion behaviors of flame retarded epoxy acrylate with a chitosan based flame retardant containing phosphorus and acrylate structure. *J. Anal. Appl. Pyrol.*, *97*, 109–115, 2012.

140. Hu, S., Song, L., Pan, H.F., Hu, Y., Thermal Properties and Combustion Behaviors of Chitosan Based Flame Retardant Combining Phosphorus and Nickel. *Ind. Eng. Chem. Res.*, *51*, 3663–3669, 2012.

141. Réti, C., Casetta, M., Duquesne, S., Bourbigot, S., Delobel, R., Flammability properties of intumescent PLA including starch and lignin. *Polym. Adv. Tech.*, *19*, 628–635, 2008.

142. Nie, S., Song, L., Guo, Y., Wu, K., Xing, W., Lu, H., Hu, Y., Intumescent Flame Retardation of Starch Containing Polypropylene Semibiocomposites: Flame Retardancy and Thermal Degradation. *Ind. Eng. Chem. Res.*, *48*, 10751–10758, 2009.

143. Wang, X., Hu, Y., Song, L., Xuan, S., Xing, W., Bai, Z., Lu, H., Flame Retardancy and Thermal Degradation of Intumescent Flame Retardant Poly(lactic acid)/Starch Biocomposites. *Ind. Eng. Chem. Res.*, *50*, 713–720, 2010.

144. Fox, D.M., Lee, J., Citro, C.J., Novy, M., Flame retarded poly(lactic acid) using POSS-modified cellulose. 1. Thermal and combustion properties of intumescing composites. *Polym. Degrad. Stab.*, *98*, 590–596, 2013.

145. Chen, H.-B., Shen, P., Chen, M.-J., Zhao, H.-B., Schiraldi, D.A., Highly Efficient Flame Retardant Polyurethane Foam with Alginate/Clay Aerogel Coating. *ACS Appl. Mater. Interfaces*, 8, 32557–32564, 2016.

146. Shang, K., Liao, W., Wang, J., Wang, Y.-T., Wang, Y.-Z., Schiraldi, D.A., Nonflammable Alginate Nanocomposite Aerogels Prepared by a Simple Freeze-Drying and Post-Cross-Linking Method. *ACS Appl. Mater. Interfaces*, 8, 642–650, 2016.

147. Zhang, R., Xiao, X.F., Tai, Q.L., Huang, H., Hu, Y., Modification of lignin and its application as char agent in intumescent flame-retardant poly(lactic acid). *Polym. Eng. Sci.*, 52, 2620–2626, 2012.

148. Zhang, R., Xiao, X.F., Tai, Q.L., Huang, H., Yang, J., Hu, Y., Preparation of lignin–silica hybrids and its application in intumescent flame-retardant poly (lactic acid) system. *High Perf. Polym.*, 24, 738–746, 2012.

149. Fu, R. and Cheng, X.S., Synthesis and Flame Retardant Properties of Melamine Modified EH Lignin. *Adv. Mater. Res.*, 236, 482–485, 2011.

150. Prieur, B., Meub, M., Wittemann, M., Klein, R., Bellayer, S., Fontaine, G., Bourbigot, S., Phosphorylation of lignin to flame retard acrylonitrile butadiene styrene. *Polym. Degrad. Stab.*, 127, 32–43, 2016.

151. Ferry, L., Dorez, G., Taguet, A. *et al.*, Chemical modification of lignin by phosphorus molecules to improve the fire behavior of polybutylene succinate. *Polym. Degrad. Stab.*, 113, 135–143, 2015.

152. Verdolotti, L., Oliviero, M., Lavorgna, M., Iannace, S., Camino, G., Vollaro, P., Frache, A., On revealing the effect of alkaline lignin and ammonium polyphosphate additives on the fire retardant properties of sustainable zein-based composites. *Polym. Degrad. Stab.*, 134, 115–125, 2016.

153. Hu, W., Zhang, Y., Qi, Y., Wang, H., Improved Mechanical Properties and Flame Retardancy of Wood/PLA All-Degradable Biocomposites with Novel Lignin-Based Flame Retardant and TGIC. *Macromol. Mater. Eng.*, 305, 1900840, 2020.

154. Zhou, S., Tao, R., Dai, P., Luo, Z., Two-step fabrication of lignin-based flame retardant for enhancing the thermal and fire retardancy properties of epoxy resin composites. *Polym. Compos.*, 41, 2025–2035, 2020.

155. Chollet, B., Lopez-Cuesta, J.-M., Laoutid, F., Ferry, L., Lignin Nanoparticles as A Promising Way for Enhancing Lignin Flame Retardant Effect in Polylactide. *Materials*, 12, 2132, 2019, https://doi.org/10.3390/ma12132132.

156. Mandlekar, N., Cayla, A., Rault, F., Giraud, S., Salaun, F., Malucelli, G., Guan, J., Thermal Stability and Fire Retardant Properites of Polyamide 11 Microcomposites Containing Different Lignins. *Ind. Eng. Chem. Res.*, 56, 13704–13714, 2017.

157. Xia, Z., Kiratitanavit, W., Facendola, P., Thota, S., Yu, S., Kumar, J., Mosurkal, R., Nagarajan, R., Fire resistant polyphenols based on chemical modification of bio-derived tannic acid. *Polym. Degrad. Stab.*, 153, 227–243, 2018.

158. Yu, S., Xia, Z., Kiratitanavit, W., Thota, S., Kumar, J., Mosurkal, R., Nagarajan, R., Unusual role of labile phenolics in imparting flame resistance to polyamide. *Polym. Degrad. Stab.*, 175, 109103, 2020.

159. Dai, J., Peng, Y., Teng, N., Liu, Y., Liu, C., Shen, X., Mahmud, S., Zhu, J., Liu, X., High-Performing and Fire-Resistant Biobased Epoxy Resin from Renewable Sources. *ACS Sustain. Chem. Eng.*, 6, 7589–7599, 2018.

160. Daniel, Y.G. and Howell, B.A., Phosphorus flame retadrants from isosorbide bis-acrylate. *Polym. Degrad. Stab.*, 156, 14–21, 2018.

161. Howell, B.A., Oberdorfer, K.L., Ostrander, E.A., Phosphorus Flame Retardants for Polymeric Materials from Gallic Acid and Other Naturally occurring Multihydroxybenzoic Acids. *Int. J. Polym. Sci.*, 2018, https://www.hindawi.com/journals/ijps/2018/7237236.

162. Howell, B.A. and Ostrander, E.A., Flame-retardant compounds for polymeric materials from an abundantly available, renewable biosource, castor oil. *Fire Mater.*, 44, 242–249, 2020.

163. Thota, S., Somisetti, V., Kulkarni, S., Kumar, J., Nagarajan, R., Mosurkal, R., Covalent functionalization of cellulose in cotton and a nylon-cotton blend with phytic acid for flame retardant properties. *Cellulose*, 2019, DOI 10.1007/s10570-019-02801-6.

164. Barbalini, M., Bartoli, M., Tagliaferro, A., Malucelli, G., Phytic Acid and Biochar: An Effective All-Biosourced Flame Retardant Formulation for Cotton Fabrics. *Polymers*, 12, 811, 2020, https://pubmed.ncbi.nlm.nih.gov/32260336.

165. Laufer, G., Kirkland, C., Morgan, A.B., Grunlan, J.C., Intumescent Mulitlayer Nanocoating, Made with Renewable Polyelectrolytes, for Flame-Retardant Cotton. *Biomacromolecules*, 13, 2843–2848, 2012.

166. Cheng, X., Guan, J., Chen, G., Yang, X., Tang, R., Adsorption and Flame Retardant Properties of Bio-Based Phytic Acid on Wool Fabric. *Polymers*, 8, 122, 2016, https://core.ac.uk/display/243749454.

167. Kalali, E.N., Zhang, L., Shabestari, M.E., Croyal, J., Wang, D.-Y., Flame-retardant wood polymer composites (WPCs) as potential fire safe bio-based materials for building products: Preparation, flammability and mechanical properties. *Fire Saf. J.*, 107, 210–216, 2019.

168. Zhu, Z.-M., Shang, K., Wang, L.-X., Wang, J.-S., Synthesis of an effective bio-based flame-retardant curing agent and its application in epoxy resin: Curing behavior, thermal stability and flame retardancy. *Polym. Degrad. Stab.*, 167, 179–188, 2019.

169. Costes, L., Laoutid, F., Dumazert, L., Lopez-cuesta, J.-M., Brohez, S., Delvosalle, C., Dubois, P., Metallic phytates as efficient bio-based phosphorus flame retardant additives for poly(lactic acid). *Polym. Degrad. Stab.*, 119, 217–227, 2015.

170. Laoutid, F., Vahabi, H., Shabanian, M., Aryanasab, F., Sarrintaj, P., Saeb, M.R., A new direction in design of bio-based flame retardants for poly(lactic acid). *Fire Mater.*, 42, 914–924, 2018.

171. Alongi, J., Di Blasio, A., Milnes, J., Maiucelli, G., Bourbigot, S., Kandola, B., Camino, G., Thermal degradation of DNA, an all-in-one natural intumescent flame retardant. *Polym. Degrad. Stab.*, 113, 110–118, 2015.

172. Alongi, J., Carletto, R.A., Di Blassio, A., Carosio, F., Bosco, F., Malucelli, G., DNA: a novel, green, natural flame retardant and a suppressant for cotton. *J. Mater. Chem. A.*, *1*, 4779–4785, 2013.

173. Alongi, J., Milnes, J., Malucelli, G., Bourbigot, S., Kandola, B., Thermal degradation of DNA-treated cotton fabrics under different heating conditions. *J. Anal. Appl. Pyrol.*, *108*, 212–221, 2014.

174. Carosio, F., Di Blasio, A., Alongi, J., Malucelli, G., Green DNA-based flame retardant coatings assembled through Layer by Layer. *Polymer*, *54*, 5148–5153, 2013.

175. Carosio, F., Cuttica, F., Di Blasio, A., Alongi, J., Malucelli, G., Layer by layer assembly of flame retardant thin films on closed cell PET foams: Efficiency of ammonium polyphosphate vs. DNA. *Polym. Degrad. Stab.*, *112*, 189–196, 2015.

176. Isarov, S.A., Lee, P.W., Towslee, J.H., Hoffman, K.M., Davis, R.D., Maia, J.M., Pokorski, J.K., DNA as a flame retardant additive for low-density polyethylene. *Polymer*, *97*, 504–514, 2016.

177. Albite-Ortega, J., Sanchez-Valdes, S., Ramirez-Vargas, E., Nunez-Figueredo, Y., de Valle, L.F.R., Martinez-Colunga, J.G., Graciano-Verdugo, A.Z., Sanchez-Martinez, Z.V., Espinoza-Martinez, A.B., Rodriguez-Gonzalez, J.A., Castaneda-Flores, M.E., Influence of keratin and DNA coating on fire retardant magnesium hydroxide dispersion and flammability characteristics of PE/EVA blends. *Polym. Degrad. Stab.*, *165*, 1–11, 2019.

178. F. Versmann and A. Oppenheim, Eng. Patent 2077, 1859.

179. Cusack, P.A., *Proceedings of High Performance Fillers*, Rapra Technology, Cologne, Germany, March 2005, Paper 6.

180. Cusack, P.A., Smith, P.J., Kroenke, W.J., A 119mSn Mössbauer study of tin/molybdenum oxidic systems as flame retardants and smoke suppressants for rigid PVC. *Polym. Degrad. Stab.*, *14*, 307, 1986.

181. Cusack, P.A., An investigation of the flame-retardant and smoke-suppressant properties of tin (IV) oxide in unsaturated polyester thermosets. *Fire Mater.*, *10*, 41, 1986.

182. Cusack, P.A., *Tin Int.*, *72*, 4, 1999.

183. Anon, ITRI Zinc Stannate Technical Bulletins Nos. 1 – 4, , International Tin Association, St. Albans, United Kingdom, 1999.

184. Cusack, P.A., *Tin Chemistry – Fundamentals, Frontiers & Applications*, A.G. Davies, M. Gielen, K. Pannel, E. Tiekink (Eds.), p. 339, John Wiley & Sons, New York, NY, USA, 2008.

185. Anon, *Market Study – Flame Retardants (UC 405E)*, CERESANA Research, Konstanz, Germany, July 2006.

186. Anon, *Environmental Issues of PVC*, EU Commission Green Paper, Brussels, Belgium, 26th July 2000.

187. Su, X., Yi, Y., Tao, J., Qi, H., Synergistic effect of zinc hydroxystannate with intumescent flame-retardants on fire retardancy and thermal behavior of polypropylene. *Polym. Degrad. Stab.*, 97, 2012, 2128-2135.

188. S. Ebrahimian and M.A. Jozokos, US Patent 6492453, 2002.

189. S. Ebrahimian and M.A. Jozokos, US Patent 7978452, 2006.
190. T.J. Clancy, US Patent Appl. 2009/0238957, 2009.
191. C.W. Hills, US Patent Appl. 2011/0290527, 2011.
192. K. Ishida and A. Suzuki, Japan Patent 02/332384, 2002.
193. H. Takahashi and K. Mizuno, Japan Patent 01/72824, 2001.
194. Y. Namiki, Y. Kato, M. Hanai, Y. Kitano, H. Kuriso, US Patent 5726231, 1998.
195. Cross, M.S., Cusack, P.A., Hornsby, P.R., Effects of tin additives on the flammability and smoke emission characteristics of halogen-free ethylene-vinyl acetate copolymer. *Polym. Degrad. Stab.*, *79*, 309, 2003.
196. K. Segawa, K. Fujimoto, K. Shishido, H. Kimura, A. Suzuki, US Patent Appl. 2012/0003473, 2012.
197. Anon, ITRI Zinc Stannates in Halogen-Free Polymer Formulations Technical Bulletin No. 2, International Tin Association, St. Albans, United Kingdom, 1999.
198. Kim, I.H., Nam, G.J., Lee, G.J., *ANTEC 2006 (64th SPE Annual Conf.)*, Society for Plastics Engineers, p. 1819, May 2006.
199. E. Schlosser, B. Nass, W. Wanzke, US Patent 6547992, 2003.
200. Jung, H.C., Kim, W.N., Lee, C.R., Suh, K.S., Kim, S.R., Properties of flame-retarding blends of polycarbonate and poly (acrylonitrile-butadiene-styrene). *J. Polym. Eng.*, *18*, 115, 1998.
201. Horrocks, A.R., Smart, G., Kandola, B., Price, D., Zinc stannate interactions with flame retardants in polyamides; Part 2: Potential synergies with non-halogen-containing flame retardants in polyamide 6 (PA6). *Polym. Degrad. Stab.*, *97*, 645, 2012.
202. Y. Yin, Intern. Patent Appl. PCT WO 2010/002403, 2010.
203. H. Bauer, F. Eisentraeger, S. Hoerold, W. Krause, US Patent Appl. 2006/0226404, 2006.
204. S. Hoerold and H. Bauer, Intern. Patent Appl. PCT WO 2012/113520, 2012.
205. K. Hironaka and M. Suzuki, US Patent 6248814, 2001.
206. S.L. Tondre, A.S. Yeung, V. Jansons, US Patent 5908887, 1999.
207. H. Staendeke and U. Thummler, US Patent 5093199, 1992.
208. Anon, *Development of Halogen-Free Flame-Retardant Systems for Printed Circuit Boards and Encapsulated Electronic Components (HALFREE)*, UK DTI Project No. STI/3/030.
209. S. Osawa, Japan Patent 02/60592, 2002.
210. K. Ikegai and H. Suwabe, Japan Patent 98/146917, 1998.
211. N. Ikeguchi, T. Omori, K. Ishii, T. Harada, US Patent 6528552, 2003.
212. S. Knop, M. Sicken, S. Hoerold, US Patent Appl. 2005/0101708, 2005.
213. Brzozowski, Z.K., Staszczak, S., Hadam, L., Zatorski, W., Rupinski, S., Bogdal, D., Fireproof, solid state epoxy resins. *Polimery*, *52*, 29, 2007.
214. De Fenzo, A., Formicola, C., Antonnuci, V., Zarreli, M., Giordano, M., Effects of zinc-based flame retardants on the degradation behaviour of an aerospace epoxy matrix. *Polym. Degrad. Stab.*, *94*, 1354, 2009.

215. Formicola, C., De Fenzo, A., Zarelli, M., Giordano, M., Antonucci, V., Zinc-based compounds as smoke suppressant agents for an aerospace epoxy matrix. *60*, 304–311, 2011.

216. Cusack, P.A., *Plastics Additives: An A – Z Reference*, G. Pritchard (Ed.), p. 339, Chapman & Hall, London, UK, 1998.

217. Cusack, P.A., Smith, P.J., Arthur, L.T., *J. Fire Retardant Chem.*, 7, 9, 1980.

218. Nazare, S., Kandola, B.K., Horrocks, A.R., Smoke, CO, and CO2 Measurements and Evaluation using Different Fire Testing Techniques for Flame Retardant Unsaturated Polyester Resin Formulations. *J. Fire Sci.*, *26*, 215, 2008.

219. Yang, F. and Nelson, G.L., *Fire and Polymers VI: New Advances in Flame Retardant Chemistry and Science*, A.B. Morgan, C.A. Wilkie, G.L. Nelson (Eds.), p. 139, Am. Chem. Soc., Washington DC, USA, 2012.

220. Hornsby, P.R., Mitchell, P.A., Cusack, P.A., Flame retardance and smoke suppression of polychloroprene containing inorganic tin compounds. *Polym. Degrad. Stab.*, *32*, 299, 1991.

221. Hornsby, P.R., Winter, P., Cusack, P.A., Flame retardancy and smoke suppression of chlorosulphonated polyethylene containing inorganic tin compounds. *Polym. Degrad. Stab.*, *44*, 177, 1994.

222. Cusack, P.A. and Karpel, S., *Tin & Its Uses*, 165, 1, 1991.

223. Kind, D.J. and Hull, T.R., A review of candidate fire retardants for polyisoprene. *Polym. Degrad. Stab.*, *97*, 201, 2012.

224. Cusack, P.A. and Fontaine, P.I., *Speciality Chemicals*, 9, 194, 1989.

225. S.M. Horley and W.M. Worthington, US Patent 7638572, 2009.

226. Anon, *Innovative Fire Protective Coatings for Steel Structures (STEELPROST)*, EU FP7-SME-2008-2 Project No. 243574, https://cordis.europa.eu/project/id/243574 (accessed 06/24/21).

227. S.J. Blunden, P.A. Cusack, A.J. Wallace, UK Patent Appl. 96/13073, 1996.

228. S.J. Blunden, P.A. Cusack, A.J. Wallace, US Patent 6372360, 2002.

229. Cusack, P.A., *Tin Intern.*, 75, 11, 2002.

230. P.A. Cusack and J.A. Pearce, Intern. Patent Appl. PCT WO 90/09962, 1990.

231. P.A. Cusack, B. Patel, M.S. Heer, R.G. Baggaley, US Patent 6150447, 2000.

232. P.A. Cusack and D.R. Clack, Patent Appl. PCT WO 03/97735, 2003.

233. Mohai, M., Toth, A., Hornsby, P.R., Cusack, P.A., Cross, M.S., Marosi, G., XPS analysis of zinc hydroxystannate-coated hydrated fillers. *Surf. Interface Anal.*, *34*, 735, 2002.

234. Hornsby, P.R., Cusack, P.A., Cross, M.S., Toth, A., Zelei, B., Marosi, G., Zinc hydroxystannate-coated metal hydroxide fire retardants: Fire performance and substrate-coating interactions. *J. Mater. Sci.*, *38*, 2893, 2003.

235. Cusack, P.A. and Hornsby, P.R., Zinc stannate–coated fillers: Novel flame retardants and smoke suppressants for polymeric materials. *J. Vinyl Additive Technol.*, *5*, 21, 1999.

236. Kandola, B.K. and Horrocks, A.R., *Fire Retardant Materials*, A.R. Horrocks and D. Price (Eds.), p. 182, Woodhead Publishing, Cambridge, UK, 2000.

237. S.C. Brown, M.L. David, K.A. Evans, J.P. Garcia, Intern. Patent Appl. PCT WO 00/66657, 2000.

238. M.S. Cross and P.A. Cusack, UK Patent Appl. 02/12306, 2002.

239. Jiang, D.D., *Fire Retardancy of Polymeric Materials*, 2nd edition, C.A. Wilkie and A.B. Morgan (Eds.), p. 261, CRC Press, Boca Raton, FL, USA, 2010.

240. Anon, *Novel Tin – Layered Double Hydroxide Hybrid Fire Retardants and their Application in Nano-composite Cable Formulations (HYBRID)*, Eurostars Project No. E4746, https://www.era-learn.eu/network-information/networks/eurostars/eurostars-cut-off-2/novel-tin-layered-double-hydroxide-hybrid-fire-retardants-and-their-application-in-nano-composite-cable-formulations (accessed 06/24/21).

241. Zheng, X. and Cusack, P.A., Tin-containing layered double hydroxides: Synthesis and application in poly(vinyl chloride) cable formulations. *Fire Mater.*, 37, 35, 2012.

242. Cusack, P.A., Monk, A.W., Pearce, J.A., Reynolds, S.J., An investigation of inorganic tin flame retardants which suppress smoke and carbon monoxide emission from burning brominated polyester resins. *Fire Mater.*, 14, 23, 1989.

243. Cusack, P.A., Heer, M.S., Monk, A.W., Zinc hydroxystannate: A combined flame retardant and smoke suppressant for halogenated polyesters. *Polym. Degrad. Stab.*, 32, 177, 1991.

244. S. Miyata, US Patent 5571526, 1996.

245. Lee, D.A. and Herbiet, R., *Proceedings of Polyolefins 2006, SPE Conf.*, Houston, TX, USA, February 2006, Paper 33.

246. Brown, S.C., *Plastics Additives: An A – Z Reference*, G. Pritchard (Ed.), p. 287, Chapman & Hall, London, UK, 1998.

247. Camino, G., Rossetti, F., Manferti, C., *Proceedings of Flame Retardants for Electrical Applications, Eur. Plastics News / Plastics & Rubber Weekly*, Crain Communcations Group, Brussels, Belgium, March 2003.

248. Kashiwagi, T. and Gilman, J.W., *Fire Retardancy of Polymeric Materials*, A.F. Grand and C.A. Wilkie (Eds.), p. 353, Marcel Dekker, New York, NY, USA, 2000.

249. Morgan, A.B., Flame Retarded Polymer Layered Silicate Nanocomposites: A Review Of Commercial and Open Literature Systems. *Polym. Adv. Technol.*, 17, 206–217, 2006.

250. A.B. Morgan and C.A. Wilkie (Eds.), *Flame Retardant Polymer Nanocomposites*, Book published by John Wiley & Sons, Hoboken, NJ, 2007.

251. Kiliaris, P. and Papaspyrides, C.D., Polymer/layered silicate (clay) nanocomposites: An overview of flame retardancy. *Prog. Polym. Sci.*, 35, 902–958, 2010.

252. Zhu, J., Uhl, F.M., Morgan, A.B., Wilkie, C.A., *Chem. Mater.*, 13, 4649–4654, 2001.

253. Gilman, J.W., Jackson, C.L., Morgan, A.B., Harris, R., Manias, E., Giannelis, E.P., Wuthenow, M., Hilton, D., Phillips, S.H., Flammability Properties of

Polymer Layered-Silicate Nanocomposites. Polypropylene and Polystyrene Nanocomposites. *Chem. Mater.*, *12*, 1866–1873, 2000.

254. Morgan, A.B., Harris, R.H., Kashiwagi, T., Chyall, L.J., Gilman, J.W., Flammability of Polystyrene Layered Silicate (Clay) Nanocomposites: Carbonaceous Char Formation. *Fire Mater.*, *26*, 247–253, 2002.

255. Gilman, J.W., Harris, R.H., Shields, J.R., Kashiwagi, T., Morgan, A.B., A study of the flammability reduction mechanism of polystyrene-layered silicate nanocomposite: layered silicate reinforced carbonaceous char. *Polym. Adv. Technol.*, *17*, 263–271, 2006.

256. Rahetekar, S.S., Zammarano, M., Matko, S., Koziol, K.K., Windle, A., Nyden, M., Kashiwagi, T., Gilman, J.W., Effect of carbon nanotubes and montmorillonite on the flammability of epoxy nanocomposites. *Polym. Degrad. Stab.*, *95*, 870–879, 2010.

257. Cipiriano, B.H., Kashiwagi, T., Raghavan, S.R., Yang, Y., Grulke, E.A., Yamamoto, K., Shields, J.R., Douglas, J.F., Effects of aspect ratio of MWNT on the flammability properties of polymer nanocomposites. *Polymer*, *48*, 6086–6096, 2007.

258. Kashiwagi, T., Du, F., Winey, K.I., Groth, K.M., Shields, J.R., Bellayer, S.P., Kim, H., Douglas, J.F., Flammability properties of polymer nanocomposites with single-walled carbon nanotubes: effects of nanotube dispersion and concentration. *Polymer*, *46*, 471–481, 2005.

259. Bartholmai, M. and Schartel, B., Layered silicate polymer nanocomposites: new approach or illusion for fire retardancy? Investigations of the potentials and the tasks using a model system. *Polym. Adv. Technol.*, *15*, 355–364, 2004.

260. Nazare, S. and Davis, R.D., A review of fire blocking technologies for soft furnishings. *Fire Sci. Rev.*, *1*, 1–23, 2012.

261. Morgan, A.B., Revisiting flexible polyurethane foam flammability in furniture and bedding in the United States. *Fire Mater.*, 2021, Vol 45, Pages 68-80, DOI: 10.1002/fam.2848.

262. Storesund, K., Steen-Hansen, A., Bergstrand, A., SP Fire Research Report SPFR Report A15 20124:2, December 12, 2015, DOI:10.13140/ RG.2.2.20464.87041, Published by SP, Sweden.

263. Nazare, S., Pitts, W.M., Shields, J., Knowlton, E., De Leon, B., Zammaranon, M., Davis, R., Assessing fire-blocking effectiveness of barrier fabrics in the cone calorimeter. *J. Fire Sci.*, *37*, 340–375, 2019.

264. Nazare, S., Pitts, W.M., Matko, S., Davis, R.D., Evaluating smoldering behavior of fire-blocking barrier fabrics. *J. Fire Sci.*, *32*, 539–562, 2014.

265. Nazare, S., Pitts, W., Flynn, S., Shields, J.R., Davis, R.D., Evaluating fire blocking performance of barrier fabrics. *Fire Mater.*, *38*, 695–716, 2014.

266. Nazare, S., Pitts, W., Shields, J., Davis, R., Factors for Consideration in an Open-Flame Test for Assesing Fire Blocking Performance of Barrier Fabrics. *Polymers*, *8*, 342, 2016, doi: 10.3990/polym8090342.

267. Lock, A., *Memoranda on Full-Scale Upholstered Furniture Testing, 2014-2015,* US Consumer Product Safety Commission, April 2016, https://www.cpsc.gov/s3fs-public/FY14_Chair_Study_Memos.pdf (accessed 06/22/21).

268. Blais, M. and Carpenter, K., Flexible Polyurethane Foams: A Comparative Measurement of Toxic Vapors and Other Toxic Emissions in Controlled Combustion Environments of Foams With and Without Fire Retardants. *Fire Technol.,* 51, 3–18, 2015.

269. Jimenez, M., Gallou, H., Duquesne, S., Jama, C., Bourbigot, S., Couillens, X., Speroni, F., New routes to flame retard polyamide 6,6 for electrical applications. *J. Fire Sci.,* 30, 535–551, 2012.

270. Schartel, B., Kuhn, G., Mix, R., Friedrich, J., Surface Controlled Fire Retardancy of Polymers Using Plasma Polymerisation. *Macromol. Mater. Eng.,* 287, 579–582, 2002.

271. Bourbigot, S., Jama, C., Le Bras, M., Delobel, R., Dessaux, O., Goudmand, P., New approach to flame retardancy using plasma assisted surface polymerisation techniques. *Polym. Degrad. Stab.,* 66, 153–155, 1999.

272. Quede, A., Mutel, B., Supiot, P., Jama, C., Dessaux, O., Delobel, R., Characterization of organosilicon films synthesized by N2-PACVD. Application to fire retardant properties of coated polymers. *Surf. Coat. Technol.,* 180-181, 265–270, 2004.

273. Totolin, V., Sarmadi, M., Manolache, S.O., Denes, F.S., Atmospheric pressure plasma enhanced synthesis of flame retardant cellulosic materials. *J. App. Polym. Sci.,* 117, 281–289, 2010.

274. Horrocks, A.R., Nazaré, S., Masood, R. *et al.,* Surface modification of fabrics for improved flash-fire resistance using atmospheric pressure plasma in the presence of a functionalized clay and polysiloxane. *Polym. Adv. Technol.,* 22, 22–29, 2011.

275. Totolin, V., Sarmadi, M., Manolache, S.O., Denes, F.S., Environmentally Friendly Flame-Retardant Materials Produced by Atmosphereic Pressure Plasma Modifications. *J. App. Polym. Sci.,* 124, 116–122, 2012.

276. Tata, J., Alongi, J., Frache, A., Optimization of the procedure to burn textile fabrics by cone calorimeter: part II. Results on nanoparticle-finished polyester. *Fire Mater.,* 36, 527–536, 2012.

277. Hilt, F., Boscher, N.D., Duday, D., Esbenoit, N., Levaloss-Grutzmacher, J., Choquet, P., Atmospheric Pressure Plasma-Initiated Chemical Vapor Deposition (AP-PiCVD) of Poly(diethylallyphosphate) Coating: A Char-Forming Protective Coating for Cellulosic Textile. *ACS Appl. Mater. Interfaces,* 6, 18418–18422, 2014.

278. Hilt, F., Gherardi, N., Duday, D., Berne, A., Choquet, P., Efficient Flame Retardant Thin Films Synthesized by Atmospheric Pressure PECVD through the High Co-deposition Rate of Hexamethyldisiloxane and Triethylphosphate on Polycarbonate and Polyamide-6 Substrates. Efficient Flame Retardant Thin Films Synthesized by Atmospheric Pressure PECVD through the High Co-deposition Rate of Hexamethyldisiloxane and Triethylphosphate on

Polycarbonate and Polyamide-6 Substrates, *ACS Appl. Mater. Interfaces*, 8, 12422–12433, 2016.

279. Lyon, R.E., Balaguru, P.N., Foden, A., Sorathia, U., Davidovits, J., Davidovics, M., Fire-resistant Aluminosilicate Composites. *Fire Mater.*, 21, 67–73, 1997.

280. Giancaspro, J., Papakonstantinou, C., Balaguru, P., Fire resistance of inorganic sawdust biocomposite. *Compos. Sci. Technol.*, 68, 1895–1902, 2008.

281. Sakkas, K., Nomikos, P., Sofianos, A., Panias, D., Inorganic polymeric materilas for passive fire protection of underground construtcions. *Fire Mater.*, 37, 140–150, 2013.

282. Rashid, M.K.A., Sulong, N.H.R., Alengaram, U.J., Fire resistance performance of composite coating with geopolymer-based bio-fillers for lightweight panel application. *J. App. Polym. Sci.*, 137, e49558, 2020.

283. Schartel, B., Beck, U., Bahr, H., Hertwig, A., Knoll, U., Weise, M., Submicrometre coatings as an infrared mirror: A new route to flame retardancy. *Fire Mater.*, 36, 671–677, 2012.

284. Weil, E.D., *Flame Retardancy News*, Fall 2005, available from the author on request.

285. Hshieh, F.Y., Shielding effects of silica-ash layer on the combustion of silicones and their possible applications on the fire retardancy of organic polymers. *Fire Mater.*, 22, 69–76, 1998.

286. Wu, Q., Zhang, C., Liang, R., Wang, B., Fire retardancy of a buckypaper membrane. *Carbon*, 46, 1159–1174, 2008.

287. Zhuge, J., Tang, Y., Gou, J., Chen, R.-H., Ibeh, C., Hu, Y., Flammability of carbon nanofiber-clay nanopaper based polymer composites. *Polym. Adv. Technol.*, 22, 2250–2256, 2011.

288. Zhuge, J., Gou, J., Chen, R.-H. *et al.*, Fire performance and post-fire mechanical properties of polymer composites coated with hybrid carbon nanofiber paper. *J. App. Polym. Sci.*, 124, 37–48, 2012.

289. Zhuge, J., Gou, J., Ibeh, C., Fire performance and post-fire mechanical properties of polymer composites coated with hybrid carbon nanofiber paper. *Fire Mater.*, 36, 241–253, 2012.

290. Knight, C.C., Ip, F., Zeng, C., Zhang, C., Wang, B., A highly efficient fire-retardant nanomaterial based on carbon nanotubes and magnesium hydroxide. *Fire Mater.*, 97, 91–99, 2013.

291. Holder, K.M., Smith, R.J., Grunlan, J.C., A review of flame retardant nanocoatings prepared using layer-by-layer assembly of polyelectrolytes. *J. Mater. Sci.*, 52, 12923–12959, 2017.

292. Lazar, S.T., Kolibaba, T.J., Grunlan, J.C., Flame-retardant surface treatments. *Nat. Rev. Mater.*, 5, pages 259–275, 2020.

293. Greaves, A.K. and Letcher, R.J., Comparative Body Compartment Composition and *In Ovo* Transfer of Organophosphate Flame Retardants in North American Great Lakes Herring Gulls. *Environ. Sci. Technol.*, 48, 7942–7950, 2014, https://pubmed.ncbi.nlm.nih.gov/24905208.

294. Ballesteros-Gómez, A., Aragón, A., Van den Eede, N., de Boer, J., Covaci, A., Impurities of Resorcinol Bis(diphenyl phosphate) in Plastics and Dust Collected on Electric/Electronic Material. *Environ. Sci. Technol.*, 50, 1934–1940, 2016.

295. Sühring, R., Diamond, M.L., Scheringer, M., Wong, F., From clothing to Laundry Water: Investigating the Fate of Phthalates, Brominated Flame Retardants, and Organophosphate Esters. *Environ. Sci. Technol.*, 50, 7409–7415, 2016.

296. Castro-Jiménez, J., González-Gaya., B., Pizarro, M., Casal, P., Pizarro-Álvarez, C., Dachs, J., Organophosphate Ester Flame Retardants and Plasticizers in the Global Oceanic Atmosphere. *Environ. Sci. Technol.*, 50, 12831–12839, 2016.

297. Li, J., Xie, Z., Mi, W., Lai, S., Tian, C., Emeis, K.-C., Ebinghaus, R., Organophosphate Esters in Air, Snow, and Seawater in the North Atlantic and the Arctic. *Environ. Sci. Technol.*, 51, 6887–6896, 2017.

298. Tao, F., Sellström, U., de Wit, C.A., Brominated Flame Retardants and Organophosphate Esters in Preschool Dust and Children's Hand Wipes. *Environ. Sci. Technol.*, 53, 2124–2133, 2019.

299. Tan, H., Yang, L., Yu, Y., Guan, Q., Liu, X., Li, L., Chen, D., Co-Existence of Organophosphate Di- and Tri-Esters in House Dust from South China and Midwestern United States: Implications for Human Exposure. *Environ. Sci. Technol.*, 53, 4784–4793, 2019.

300. Sun, Y., De Silva, A.O., St Pierre, K.A., Muir, D.C.G., Spencer, C., Lehnherr, I., MacInnis, J.J., Glacial Melt Inputs of Organophosphate Ester Flame Retardants to the Largest High Arctic Lake. *Environ. Sci. Technol.*, 54, 2734–2743, 2020.

301. Simonson, M., Blomqvist, P., Boldizar, A., Möller, K., Rosell, L., Tullin, C., Stripple, H., Sundqvist, J.O., Fire-LCA Model: TV Case Study. SP Report 2000:13, SP, Sweden, 2010.

302. Blomqvis, P., Persson, B., Simonson, M., Fire Emissions of Organics into the Atmosphere. *Fire Technol.*, 43, 213–231, 2007.

303. Blomqvist, P., Rosell, L., Simonson, M., Emissions from Fires Part I: Fire Retarded and Non-Fire Retarded TV-Sets. *Fire Tech.*, 40, 39–58, 2004.

304. Blomqvist, P., Rosell, L., Simonson, M., Emissions from Fires Part II: Fire Retarded and Non-Fire Retarded TV-Sets. *Fire Tech.*, 40, 59–73, 2004.

305. Simonson-McNamee, M. and Andersson, P., Application of a Cost-benefit Analysis Model to the Use of Flame Retardants. *Fire Tech.*, 51, 67–83, 2015.

306. Aurell, J. and Gullet, B.K., Emission Factors from Aerial and Ground Measurements of Field and Laboratory Forest Burns in the Southeastern U.S.: PM2.5, Black and Brown Carbon, VOC, and PCDD/PCDF, *Environ. Sci. Technol.*, 47, 8443–8452, 2013.

307. http://www.vecap.info/ (accessed 09/30/20)

308. Miandad, R., Rehan, M., Barakat, M., Aburiazaiza, A.S., Khan, H., Ismail, I.M.I., Dhavamani, J., Gardy, J., Haanpour, A., Nizami, A.-S., Catalytic

Pyrolysis of Plastic Waste: Moving Toward Pyrolysis Based Biorefineries. *Front. Energy Res.*, 7, 2019, https://doi.org/10.3389/fenrg.2019.00027.

309. https://www.cewep.eu/ (accessed 10/03/20)

310. Levaggi, L., Levaggi, R., Marchiori, C., Trecroci, C., Waste-to-Energy in the EU: The Effects of Plant Ownership, Waste Mobility, and Decentralization on Environmental Outcomes and Welfare. *Sustainability*, vol. 12, p. 5743, 2020.

311. https://en.wikipedia.org/wiki/Urban_mining (accessed 10/03/20)

312. https://www.dw.com/en/urban-mining-hidden-riches-in-our-cities/a-42913985(accessed 10/03/20)

313. ASTM D6954 – 18, *Standard Guide for Exposing and Testing Plastics that Degrade in the Environment by a Combination of Oxidation and Biodegradation*, American Society for Testing and Materials (ASTM), ASTM International, 100 Barr Harbor Drive, PO Box C700, West Conshohocken, PA, 19428-2959 USA, 2018.

314. ASTM D5511-18, *Standard Test Method for Determining Anaerobic Biodegradation of Plastic Materials Under High-Solids Anaerobic-Digestion Conditions*, ASTM International, 100 Barr Harbor Drive, PO Box C700, West Conshohocken, PA, 19428-2959 USA, 2018.

315. http://ec.europa.eu/environment/chemicals/reach/reach_intro.htm (accessed 10/07/20)

316. Swoboda, B., Buonomo, S., Leroy, E., Lopez Cuesta, J.M., Fire retardant poly(ethylene terephthalate)/polycarbonate/triphenyl phosphate blends. *Polym. Degrad. Stab.*, 93, 910–917, 2008.

317. Laoutid, F., Ferry, L., Lopez-Cuesta, J.M., Crespy, A., Red phosphorus/aluminium oxide compositions as flame retardants in recycled poly(ethylene terephthalate). *Polym. Degrad. Stab.*, 82, 357–363, 2003.

318. Laoutid, F., Ferry, L., Lopez-Cuesta, J.M., Crespy, A., Flame-retardant action of red phosphorus/magnesium oxide and red phosphorus/iron oxide compositions in recycled PET. *Fire Mater.*, 30, 343–358, 2006.

319. Casetta, M., Delaval, D., Traisnel, M., Bourbigot, S., Influence of the Recycling Process on the Fire-Retardant Properties of PP/EPR Blends. *Macromol. Mater. Eng.*, 296, 494–505, 2011.

320. Marton, A., Anna, P., Marosi, Gy., Szep, A., Matko, Sz., Racz, I., Horsnby, P., Ahmadnia, A., Use of Layered Structures in Recycling of Polymers. *Prog. Rubber Plast. Recycl. Technol.*, 20, 97–104, 2004.

321. Tarantili, P.A., Mitsakaki, A.N., Petoussi, M.A., Processing and properties of engineering plastics recycled from waste electrical and electronic equipment (WEEE). *Polym. Degrad. Stab.*, 95, 405–410, 2010.

322. https://www.cleanproduction.org/programs/greenscreen (accessed 10/07/20)

323. Manzello, S.L. and Suzhuki, S., Experimentally Simulating Wind Driven Firebrand Showers in Wildland-urban Interface (WUI) Fires: Overview of the NIST Firebrand Generator (NIST Dragon) Technology. *Proc. Eng.*, 62, 91–102, 2013.

324. Ribiere, P., Grugeon, S., Morcerette, M., Boyanov, S., Laruelle, S., Marlair, G., Investigation on the fire-induced hazards of Li-Ion battery cells by fire calorimetry. *Energy Environ. Sci.*, 5, 5271, 2012.

325. Liu, X., Wu, Z., Stoliarov, S., I, Denlinger, M., Masias, A., Snyder, K., Heat release during thermally-induced failure of a lithium ion battery: Impact of cathode composition. *Fire Saf. J.*, 85, 10–22, 2016.

326. Sun, P., Huang, X., Bisschop, R., Niu, H., A Review of Battery Fires in Electric Vehicles. *Fire Technol.*, 56, 1361–1410, 2020.

327. Chen, M., DeZhou, C., Wang, J., He, Y., Chen, M., Yuen, R., Experimental Study on the Combustion Characteristics of Primary Lithium Batteries Fire. *Fire Technol.*, 52, 365–385, 2016.

328. Ditch, B., The Impact of Thermal Runaway on Sprinkler Protection Recommendations for Warehouse Storage of Cartoned Lithium-Ion Batteries. *Fire Technol.*, 54, 359–377, 2018.

329. Wang, Z., Ning, X., Zhu, K., Hu, J., Evaluating the thermal failure risk of large-format lithium-ion batteries using a cone calorimeter. *J. Fire Sci.*, 37, 81–95, 2019.

330. https://www.youtube.com/watch?v=2O07SIaxB08 (accessed 10/07/20)

331. Mellert, L.D., Welte, U., Tuchschmid, M., Held, M., Hermann, M., Kompatscher, M., Tesson, M., Nachef, L., https://plus.empa.ch/images/2020-08-17_Brandversuch-Elektroauto/AGT_2018_006_EMob_RiskMin_Undergr_Infrastr_Final_Report_V1.0.pdf (accessed 10/07/20).

332. Zhao, Q., Stalin, S., Zhao, C.-Z., Archer, L.A., Designing solid-state electrolytes for safe, energy-dense batteries. *Nat. Rev. – Mater.*, 5, 229–252, 2020.

333. Zhao, W., Yi, J., He, P., Zhou, H., Solid-State Electrolytes for Lithium-Ion Batteries: Fundamentals, Challenges and Perspectives. *Electrochem. Energy Rev.*, 2, 574–605, 2019.

334. Kumar, B., Thomas, D., Kumar, J., Space-Charge-Mediated Superionic Transport in Lithium Ion Conducting Glass–Ceramics. *J. Electrochem. Soc.*, 156, A506–A513, 2009.

335. Digges, K.H., Gann, R.G., Grayson, S.J., Hirschler, M.M., Lyon, R.E., Purser, D.A., Quintiere, J.G., Stephenson, R.R., Tewarson, A., Human survivability in motor vehicle fires. *Fire Mater.*, 32, 249–258, 2008.

Index

ABS. *see* Acrylonitrile-butadiene-styrene (ABS)
Acetic anhydride, 61
Acicular particles, 115
Acicular silicates, 295
Acid gas emission, 140, 141
Acid gas trapping systems, 527
Acrylamide, 46
Acrylic acid, 58
Acrylics, h-BN platelets in, 326
Acrylonitrile-butadiene-styrene (ABS)
ABS/clay and ABS-g-MAH/clay
    nanocomposites, 286
  flammability, 522
  nitrile group, 490
  PC/ABS blends, 48, 51, 53, 55, 64,
    417, 491
  red phosphorus in, 28
  triflate salt in, 483–484
  zinc triflate with, 483
Acyloxysilanes, 273
Additives,
  as adjuvants, 419–420
  anti-drip, 278–279
  antifungal, 30
  bromine-containing, 50
  co-additives, 29, 40, 44
  DOPO-NQ, 61
  epoxy as, 59
  for flammable materials, 3, 4
  halogenated, 3, 4
  homopolymer as, 57
  inorganic, 101
  novel tin. *see* Novel tin additives

organic, 102
in polyamide fibers, 47
silicon-based, 278
tin, 496
TPP as, 48
Adeka (Japan), 179
Adhesives, silicone, 275
Adjuvants, 421–422, 428–431
Air classifiers, 110
Air separation, 110
Akzo Nobel, 216
Alginates, 363, 494
Aliphatic phosphates, 40–48, 66
Aliphatic polyamides, 32
Alkali borate, 246
Alkaline-earth metal borates
  calcium borates, 317
  magnesium borate, 317
Alkaline metal borates, 314–317
  borax pentahydrate/decahydrate,
    314–315
  disodium octaborate tetraborate,
    315–317
    cellulose/particle boards, 315
    phenolic foam, 315
    polyamide 66 fabrics, 316–317
    RPUF, 315
Alkali silicates, 176, 190, 221
Alkoxysilanes, 273
Alkylating agent, 489
Alkyl diphenyl phosphates, 50, 65
Alkylphenyl phosphates, 42, 50
Alkyl phosphinate salts, 37, 484
Alkylsilane, 125

Alkynes, carbon-based flame
retardants, 485–487
Alumina trihydrate (ATH),
with alkyl silane, 146
artificial marble, 163
BET-ranges, 111
construction product, 148
corrosion resistance of HTV silicon,
151
epoxy cast resin in, 162–163
HRR of, 136
LLDPE using, 144
mineral flame retardants, 103,
107–109, 110–111
phosphorus-based flame retardants,
32, 39–42, 56, 60, 61
pHRR and, 193
in pultrusion, 163
for reduced smoke release, 148
smoke density curves, 133–134
suspension, 112
on tensile-elongation properties,
145
viscosities of UP resin paste, 159,
160
water uptake, 125
wet hydrate, 111
Aluminium composite panels (ACP),
149–150
Aluminium hydrates, 109
Aluminium oxide hydroxide (AOH,
boehmite),
with DOPO, 161
engineering plastics, 152, 153
HFFR compounds, 144
HRR of, 136
intumescent flame retardants, 35
LOI values, 130, 131
mineral filler flame retardants, 106,
108, 109, 112–113, 124–126
Aluminium tri-hydroxide (ATH),
HFFR compounds, 144, 146

LOI values, 130, 131
mineral flame retardants, 103–105,
107–111
polyolefins, 319
PVC, 143
SEM pictures, 116
Aluminium-tris-diethylphosphinates
(DEPAL),
with aluminum phosphite, 38
AOH-DEPAL combination,
153–154
DEPAL/synergist systems, 500
development, 37–38
Firebrake®500 and Firebrake 500,
320
formula, 37–38, 39
MC with, 253
melamine cyanurate and, 431
melamine polyphosphate and, 435
melamine salts and, 38
metal borophosphate, 328
MPP with, 255
phosphorus-based flame retardants,
37–39
phosphorus-sulfur (P-S) synergy,
432
in polyamides and polyesters, 39,
415–416
Aluminosilicates, 221–222, 281, 418
Aluminum alkylphosphinates, 38,
282
Aluminum diisobutylphosphinate,
38
Aluminum diphosphate, 35
Aluminum ethylmethylphosphinates,
37
Aluminum hypophosphite (Al-Hypo,
$Al(HPO_2)_3$), 35–37, 39, 255, 435
Aluminum phosphinate, 181
Aluminum phosphite (AlPO$_3$,
PHOPAL), 27, 38
Aluminum pyrophosphate, 36

American Society for Testing Materials (ASTM) standard,
  ASTM E662, smoke density determination, 133
  fire resistance test, 205–206
Ammonia-based flame retardants, 240–247
  ammonium pentaborate, 246
  ammonium sulfamate, 246–247
  ammonium sulfate, 247
  APP, 241, 242–245
Ammonium and amine phosphates, 29–36
  APP systems. see Ammonium polyphosphate (APP) systems
  ATH and UPE, 32
  EDAP, 34–35
  intumescent flame retardants, 29, 31–35
  MAP and DAP, 29, 30
  melamine. see Melamine
  piperazine phosphates, 35
  PUF industry, 32
Ammonium borophosphate (ABP), 327
Ammonium chloride, 261
Ammonium pentaborate (APB), 241, 246, 313, 325
Ammonium phosphate(s),
  phosphorylate cellulose, 30
  treatment of cellulosic materials, 25
  wood, treatment of, 30
Ammonium polyphosphate (APP),
  ammonia based flame retardants, 241, 242–245
  APP/PER-melamine, 216
  aromatic phosphates with, 55
  CH/APP, 368
  for cotton and cotton polyester, 33
  crystalline forms of, 30
  fire-retardant effect of, 32–33
  intumescent compounds, 31, 174, 181, 182, 200–204
  with melamine, 431–432

melamine phosphates, 34
  phenol-formaldehyde coated, 33
  piperazine pyrophosphate, 35
  in polyolefins, 32, 33, 35, 283, 429
  polyphosphoric acid, 31, 33, 35
  PU-APP, 201–204
  on ramie fabrics, 352
  silica/APP coatings, 368
  silicone surface-treated, 31–32
  structure, 242
  synergistic effects, 33, 42, 200–204
  treatment by PDAC/PAA/PDAC/ APP LbL, 364, 365
  water-insoluble, 29
Ammonium sulfamate, 241, 246–247, 484
Ammonium sulfate, 247
Amorphous boehmites, 112–113
Anhydrides, carbon-based flame retardants, 490–491
Anhydrous borax, 312
Anhydrous boric acid, 314
Antacid, 276
Anti-dripping flame retardants, 188, 197, 419, 429, 486, 490
Antifungal additives, 30
Antimony trioxide (ATO),
  brominated flame retardants, 148
  construction product, 150
  mineral flame retardants, 103–104, 107, 148
  PVC, 141–142, 143
  with stabilized Ca-Hypo, 37
  wire and cables, 141–142, 143
  ZHS and ZS, 497
APB (ammonium pentaborate), 241, 246, 313, 325
APP. see Ammonium polyphosphate (APP)
Aromatic bisphosphates, 39, 54, 55, 56–57, 64, 65
Aromatic phosphates and phosphonates, 48–58
Aromatic phosphinates, 58–62

Artificial marble, 163
Arylphosphate–arylphosphate
    synergism, 435
Ascorbic acid, 356
Asia, flame retardant regulations, 11
Aspect ratio, mineral filler flame
    retardants, 115–116
ATH. see Alumina trihydrate (ATH);
    Aluminium tri-hydroxide (ATH)
Atmospheric plasma, 343
ATO. see Antimony trioxide (ATO)
Australia, flame retardant regulations,
    13
Australian Industrial Chemicals
    Introduction Scheme (AICIS), 13
Autocatalytic hydrothermal
    crystallization, 112
Azoalkanes, 257
Azo and triazine compounds, 260

Backcoatings, phosphorus-based, 33
Barrier fabrics, 513–515
BASF SE (Germany), 179
Bauxite, 104, 105, 111
Bayer process, 104, 109, 111
BDP. see Bisphenol A bis(diphenyl
    phosphate) (BDP)
Bentonite, 285
1,4-Benzenedimethanol, 489
Benzenephosphinic acid, 58
Benzoxazine, 61, 64, 218
Bifunctional silanes, 113
Biobased materials, 494–496
Biofillers, 219
4,4'-Biphenyl bis(diphenyl phosphate),
    53–54
Bismaleidtriazine, 155
Bis-n,n-naphthtylimide, 47
1,2-Bis-(pentabromophenyl)ethane
    (Decabromo-diphenylethane,
    DPDE), 148
Bis(PEPA)phosphate-melamine salt, 187

Bisphenol A (BPA), 522
Bisphenol A bis(diphenyl phosphate)
    (BDP),
    aromatic phosphate, 51–57
    ATH with, 146
    commercial oligomeric, 290
    PC/SiR/BDP, 418
Bisphosphates,
    Al-Hypo with, 37
    aromatic, 23, 39, 47, 54–57, 64, 65
    content of, 52
    flame-retardant action of, 56
    HDP, 57
    impurity in, 66
    melamine pyrophosphates with, 35
    melting point, 53–54
    RDP, 51, 55
    RXP, 53–54
Bisphosphinates, 47
Bitumen, 148
Black phosphorus (BP), 326
Blending, 367
"Blowing out" effect, 425
B-MAP (melamine salt of
    3,9-dihydroxy-2,4,8,10-tetraoxa-
    3,9-diphosphaspiro[5,5]-
    undecane-3,9-dioxide) synthesis,
    186
BMC (bulk moulding compounds),
    157–158, 163, 425–426
Boehmites. see Aluminium oxide
    hydroxide (AOH, boehmite)
Borates, 310–311, 432, 433
Borax, 222, 309
Borax decahydrate, 312, 314–315
Borax pentahydrate, 312, 314–315
Borester, 329–330
Boric acid, 11, 30, 311, 312, 314
Boric acid esters, 329–330
Boric oxide, 311, 312, 314, 432, 433
Boron-based flame retardants,
    309–333

alkaline-earth metal borate,
    calcium borates, 317
    magnesium borate, 317
alkaline metal borate, 314–317
    borax pentahydrate/decahydrate,
        314–315
    disodium octaborate tetraborate,
        315–317
applications, 311–331, 312–313
boric acid/boric oxide, 311, 312, 314
carbon-containing boron/borates,
        329–331
    boric acid esters, 329–330
    boron carbide, 331
    boronic acid, 330–331
    graphene, 329
functions, 310–311
mode of actions, 331–332
nitrogen-containing borates,
        324–327
    ammonium borophosphate, 327
    ammonium pentaborate (APB),
        325
    boron nitride (h-BN), 325–327
    melamine diborate, 324
overview, 309–310
phosphorus-containing borates,
    boron phosphate, 327–328
    metal borophosphate, 328
properties, 310, 312–313
silicon-containing borates,
    borosilicate glass and frits,
        328–329
transition metal borates, 317–324
    Firebrake®500 and Firebrake 500,
        313, 318–324
    other metal borates, 324
    zinc borates, 317–324
Boron carbide, 331
Boronic acids, 330–331, 490
Boron nitride (h-BN), 325–327
    acrylics, 326

natural rubber (NR), 327
    polyurethane foam, 326
    red phosphorus, 326
    unsaturated polyester (UP), 327
Boron phosphate (BPO), 313, 327–328
Boron trioxide, 222
Borosilicate glass, 293, 295, 328–329
Borosiloxane, 293
BPEI. see Branched polyethylenimine
    (BPEI)
Branched ammonium polyphosphate,
    242
Branched polyethylenimine (BPEI)
    cotton treatment,
        by BPEI/PhA LbL, 356
        by BPEI/PSP-complexes, 353, 354
    LbL assembly, 348
    natural fabrics, 348, 352–354, 356
    open cell PU foams, 375, 379
    PET treatment, by BPEI/OSA LbL,
        363, 364, 368, 370
Brominated diphenyl ethers (BDPEs),
    9, 10, 12, 13
Brominated flame retardants, 4, 9, 148,
    498–499
Bromotrifluoromethane, 478
Brucite, 106, 109
Brunauer, Emmett and Teller (BET),
    118
    surface area, 110–112, 118, 124,
        145
Buckypaper, 517–518
Bulk density, mineral filler flame
    retardants, 120–121
Bulk moulding compounds (BMC),
    157–158, 163, 425–426
Bunsen burner, 131
"Burning brand" test, 529
Burnthrough test, 213–214, 223, 224
Butadiene rubber, 62
Butanetetracarboxylic acid, 46
Butoxymethylated bisphenol A, 61

Cable jackets, 31, 35, 39
Cables, thermoplastic and elastomeric applications, 140–147
Calcined kaolin, 282
Calcium borate, 312, 317
Calcium carbonate, 102, 193, 507
Calcium cyanurate, 35
Calcium ethylmethylphosphinates, 37
Calcium glyceorate, 35
Calcium hypophosphate, 36–37
Calcium metasilicate, 280
Canada, FR regulations, 10
Canadian Environmental Protection Act (CEPA), 10
Carbonaceous char, 31, 422, 423
Carbon-based flame retardants, 485–493
  cross-linking compounds, 485–491
    alkynes, 485–487
    anhydrides, 490–491
    deoxybenzoin, 488
    Friedel-Crafts reaction, 488–490
    nitrile groups, 490
  expandable graphite, 493
  graft copolymerization, 492–493
  organic carbonates, 491–492
Carbon black, 420
Carbon-containing boron/borates, 329–331
  boric acid esters, 329–330
  boron carbide, 331
  boronic acid, 330–331
  graphene, 329
Carbon fiber reinforced epoxy, Firebrake®500 and Firebrake 500, 321–322, 323
Carbon fiber reinforced polymer (CFRP), 220–221
Carbon monoxide, 4, 135, 290, 291
Carbon nanoparticles, 427, 438, 518
Carbon nanotubes (CNTs),
  in flame retardancy applications, 426
  layered silicate with, 437
  MWCNT, 217, 280, 379

  nanoparticle coatings, 517
  silicone coatings with, 294
Carbon tetrachloride, 3, 489
Cardanol-benzoxazine monomer (CBz), 329
CASICO (Calcium Silicon Carbonate) system, 291–292
Caustic calcined magnesia (CCM), 104–105, 111
CDP (cresyl diphenyl phosphate), 49, 182, 183
Cellulose/particle boards, 315, 494
Cellulosic fire curve, 206
Cellulosic insulation, 314
Cellulosic material, borax pentahydrate/decahydrate, 314–315
CEPPA, 58, 59
Ceramic frits, 328
Cetyltrimethyl ammonium bromide, 222, 223
Chalk (calcium carbonate), 102, 193
Chang Chun Group (Taiwan), 179
Char,
  carbonaceous, 31, 422, 423
  expanded graphite, 199
  formation, 29, 486
  intumescent, 175–176
    APP with, 181, 182
    enhancement, 194
    formation, 185–186, 189–190, 195, 196
    heat conductivity, 175
    mechanical stability, 173
    piperazine pyrophosphate and, 183
    polyol-based, 188
    structure of, 173, 178
    temperature, 177
    viscoelastic, 176
  promoter, 24, 283, 284, 324, 354
Charring,
  cotton in, 33
  intumescence process, 172–173

of polymer, 39, 54, 427
temperature of, 177
Charring agent(s),
    adjuvant, classes of, 429–430
    APP as, 32
    phenolic novolac as, 37
    phosphorus-based flame retardants,
        24, 26, 28, 31
    piperazine as, 32
    triazine based, 34
Chartek, 216–217
Chemical regulation,
    in Canada, 10
    China, Japan, and Korea, 11
    European Union, 10, 11
Chemical resistance, 124–126
Chemical Safety Act, 9
Chemical Substances Control Law
        (CSCL), 12
Chemical vapor deposition (CVD),
        342
China,
    DMMP in, 40
    flame retardant regulations, 11–12
China clay, 282
Chitosan (CS), 196, 197, 316–317, 353,
        494
Chlorinated paraffins, 148
Chlorinated phosphate esters, 9
Chloroaliphatic phosphates, 66
Chloroalkyl phosphates, 28, 41, 43, 45
Chlorobenzene, 261
Chlorophosphazene, cyclic, 261
Chlorosilanes, 273
Chlorosulphonated polyethylene, 503
Citric acid, 46
Clariant AG (Switzerland), 179, 320
Clay nanocomposites,
    ABS/clay and ABS-g-MAH/clay,
        286
    polymer-, 339–340, 348

PP/clay, 288
CNTs. *see* Carbon nanotubes (CNTs)
Coarser fillers, 159–160
Coated fillers, 507–509
Coating(s),
    conformal, 518–519. *see also*
        Conformal flame retardant
        coatings
    halogen-free tin-based flame
        retardant, 505
    infrared reflective, 517
    inorganic, 515–517
    integrated coatings, 518–519
    intumescent-based flame retardant
        (IFR), 215
    LbL, 518–519
    nanoparticle, 517–518
    paints and, 505
    plasma, 515–516
    silicate-based, 176
    silicon-based flame retardant, 294
Co-kneader twin screw extruder, 137
Colemanite, 312
Comparative tracking index (CTI) test,
        25, 38, 53, 320
Complex phase formation, 344
Composite high voltage insulators, 151
Composite material, 102, 280, 509,
        517, 518
Compound formulation principals,
        138–140
Compounding technology, 136–138
Condensation products, melamine,
        256–257
Condensed phase,
    flame retardants, 485–486
    modes of action, 420
    transition metal flame retardants,
        479–484
        metal complexes, 481–484
        metal oxides, 480–481

Cone calorimetry,
  borosilicate glass and frits, 329
  halloysite, 284
  intumescent systems, 172, 175, 184,
    193, 198, 199, 202
  with ISO 5660/ASTM E1354, 135
  layered silicate nanocomposites,
    286, 287
  open cell PU foams, 374, 375,
    379–380, 382–386
  polystyrene sample, 246
  silsesquioxane, 290
  smoke release rate (SRR), 134
Conformal flame retardant coatings,
    non-halogenated, 337–397
  atmospheric plasma, 343
  categories, 341–342
  chemical vapor deposition (CVD),
    342
  complex phase formation, 344
  fabrics, 346–373
    natural, 347–358
    process equipment and related
      patents, 371–373
    synthetic, 358–371
  future trends, 396–397
  layer-by-layer (LbL) deposition, 342,
    344
  multi-functionality, 344–345
  other substrates, 393–396
  overview, 339–346
  physical vapor deposition (PVD),
    342
  porous materials, 373–392
    LbL deposition process of
      CS/P-CNF, 373
    open cell PU foams, 374–386
    other substrates, 386–389
    process equipment and related
      patents, 391–392
  sol-gel wet-chemical approach,
    343
  tunable process, 344

Consumer Product Safety Commission
    (CPSC), 9, 33
Conveyer belts, 150–151
Copper clad laminates (CCL), 162
Co-rotating twin screw extruder, 137
Corrosive gas release, 13–14
Cot. see Cotton (Cot)
Cotton (Cot),
  Cot/Ny, 369, 370
  Cot/PET, 368, 369
  and cotton polyester, ammonium
    polyphosphate for, 33
  natural fabrics, 346, 347, 348
    BPEI and PSP for, 353
    conformal coating solutions for,
      349–350
    treatment by APP/MMT
      nanocoating, 357
    treatment by BPEI/PhA LbL, 356
    treatment by BPEI/PSP-complexes,
      354
    treatment by PAH/SPS LbL
      assembly, 352
  polyester with, 368
Coupling agents, 113, 138–139, 144
Cresyl diphenyl phosphate (CDP), 49,
    182, 183
Cross-linking compounds, carbon-
    based flame retardant, 485–491
  alkynes, 485–487
  anhydrides, 490–491
  deoxybenzoin, 488
  Friedel-Crafts reaction, 488–490
  nitrile groups, 490
  of polystyrene, 489
Crushing process, 109–110
CS (chitosan), 196, 197, 316–317, 353,
    494
CTE (coefficient of thermal
    expansion), 126, 162
Cyanate esters, 64
Cyanuric-acid based flame retardants,
    263–264

Cyanuric chloride, 34, 263
Cyclic chlorophosphazene, 261
Cyclic neopentyl acid phosphate with propylene oxide, 45
Cyclic phenoxyphosphazenes, 64
Cyclic phosphonate, 46, 47, 197–198
Cyclodextrin (CD), 494
1,4-cyclohexanedimethane terephthalate, 56

Daidzein, 495
Dead burned magnesia (DBM), 104–105
DecaBDE, 8, 11
Decabromo diphenylether (DBDE), 9, 10
Decebromo diphenylethane (DBDPE), 10
Denisov Cycle, 258–259
Deoxybenzoin, 488, 495
Deoxyribonucleic acid (DNA), bio-derived flame retardants, 496
DEPAL. *see* Aluminium-tris-diethylphosphinates (DEPAL)
Dialkylphosphinates, 36–40
4,4-diamino diphenyl methane (DDM), 188
Diammonium hydrogen phosphate (DAP), 29, 30
Diammonium imidobisulfonate, 246
Diammonium phosphate, 372
Dibutyl acid pyrophosphate, 44
1,2-dichloroethane, 489
Dichloroethane, with DOPO, 62
Dicyandiamine, 30
Diethylene glycol, 42–43
Diethylene triamine, 34, 35
Diethyl ethylphosphonate (DEEP), 40, 41
Diethyl hydroxymethyl phosphonate (DEHMP), 44

Diethyl N,N bis(2-hydroxyethyl) aminomethylphosphonate, 43, 44
Diethylphosphinate (DEPZN), 37, 39–40, 320
Diethyl phosphinic acid aluminum salt, 261
Differential scanning calorimetry (DSC), mineral filler flame retardants, 122
Diffuse reflectance infrared Fourier-transform spectroscopy (DRIFTS), 507
Diglycidyl methylphosphonate, 219
Diguanidine hydrogen phosphate, 30
9,10-Dihydro-9-oxa-10-phosphaphenanthrene 10-oxide (DOPO), 161, 162, 522
    aromatic phosphinate, 59–62, 63
    dichloroethane with, 62
    DOPO-BPA, 61
    DOPO-HQ, 60–61
    DOPO-NQ, 61
    ethylene bis-DOPO phosphinate, 62
    ethylene glycol with, 62
    hydrolytic and thermal stability, 63
    intumescent coatings, 219, 220
Dilution,
    fuel, 422, 426, 428, 430, 431, 434, 435, 437, 438
    gas, 208, 249
Dimethyl itaconate, 59
Dimethyl methylphosphonate (DMMP), 40, 41, 42, 315
Dimethyloldihydroxyethyleneurea (DMDHEU), 46
Dimethyl phosphite, 46
Dimethyl propylphosphonate (DMPP), 40, 41, 42
Dimethyl siloxane, 280
Dimethyl terephthalate, 58
Dimethylvinylsiloxy, 280
DINP (di-iso-nonyl-phthalate), 119

1,3,2-Dioxaphosphorian-2,2-oxy-bis-
    (5,5-dimethyl-2-sulphide), 48
Dioxin,
    formation, 3, 4, 8
    halogenated, 4
Dipentaerythritol, 246
Diphenylmethane, 190
Diphenyl phosphate, 50
Diphosphonate, 46–47
Disodium octaborate tetraborate,
    315–317
    cellulose/particle boards, 315
    phenolic foam, 315
    polyamide 66 fabrics, 316–317
    RPUF, 315
Disodium octaborate tetrahdrate, 312
DMMP (dimethyl
    methylphosphonate), 40, 41, 42,
    315
DOP (di-octyl-phthalate), 119
DOPO. see 9,10-Dihydro-9-oxa-10-
    phosphaphenanthrene-10-oxide
    (DOPO)
Double layered hydroxides, 478
Dow Corning, 280, 291
Dripping agents, 428–429
Dry grinding technologies, 110
Durability, for recycling, 526–527

E-beam, 146
Ecolabels, 14
EDAP (ethylenediamine salt of
    phosphoric acid), 34–35, 42
E&E (electric and electronic)
    applications, 152–154, 161–163,
    179, 180
EG. see Expandable graphite (EG)
Elastomeric applications,
    thermoplastic and,
    compound formulation principals,
    138–140
    compounding technology, 136–138
    construction products, 147–150
    engineering plastics, 152–154

special applications, 150–152
wire and cables, 140–147
Elastomers,
    APP in, 32
    Firebrake®500 and Firebrake 500,
        322–323
    halogen-free tin-based flame
        retardants, 503, 505
    intumescent systems, 31
    low efficiency in, 26
    polyurethane, 28
Electric and electronic (E&E)
    applications, 152–154, 161–163,
    179, 180
Electro-fused magnesia (EFM),
    104–105
Electromagnetic interference (EMI)
    shielding, 517, 518
Emergency lighting cables, 140
E-mobility, 126
End of life considerations for flame
    retardants, 524–528
    durability for recycling, 526–527
    environmental decomposition, 525,
        528
    recycling/sustainability design,
        524–528
    sustainable and alternative chemical
        feedstocks, 524, 525–526
    waste-to-energy, 524–525, 527–528
Endothermic cooling,
    compounds, 486
    mechanism of flame retardants,
        478
Endothermic decomposition, 240, 241,
    243, 247
Endothermic heat, mineral filler flame
    retardants, 122
Energy dispersive spectroscopy (EDS),
    borosilicate glass and frits,
        328–329
Engineering non-halogenated flame
    retardant solutions, 513–519
    barrier fabrics, 513, 514–515

conformal/integrated coatings, 518–519
  infrared reflective coatings, 517
  inorganic coatings, 515–517
  layer-by-layer coatings, 518–519
  nanoparticle coatings, 517–518
Engineering plastics,
  for E&E applications, 152–154
  halogen-free tin-based flame retardants, 500, 502, 503
Environmental decomposition, 525, 528
Environmental fate, 65–67
Environmental Protection Agency (EPA), US, 9
Environment friendly use period (EFUP), 12
EPDM. *see* Ethylene-propylene-diene-copolymer (EPDM)
Epoxy,
  APB, 325
  boron-doped graphene oxide, 329
  boronic acid, 330
  cast resin, 162–163
  dispersion in, 27
  encapsulation, 28
  Firebrake®500 and *Firebrake* 500, 321–322, 323
  phosphazenes in, 241
  resin, 27, 28, 39, 56, 57, 59, 60, 151, 161–162, 192
Ethanolamine, 34
Ethyl acrylate (EEA), coated filler, 508
Ethylene bis-DOPO phosphinate, 62
Ethylene-butyl acrylate (EBA) formulations, 190–191
Ethylenediamine salt of phosphoric acid (EDAP), 34–35, 42
Ethylene glycol, 58, 62
Ethylene-propylene-diene-copolymer (EPDM),
  cross-linkable compounds, 145
  halloysite, 284

mineral filler flame retardants, 145, 148, 149, 150
Ethyl ethylene glycol phosphate, 42, 44, 45
Ethyl ethylene glycol polyphosphate, 43
2-Ethylhexyl diphenyl phosphate, 50, 489
European Chemicals Agency (ECHA), 11
European Union (EU), flame retardant regulations, 4, 10–11
European Union Ecolabel, 14
EVA. *see* Poly(ethylene-co-vinyl acetate) (EVA)
Exfoliated nanocomposites, melt viscosity for, 286, 287
Exolit 1312, 328
Exolit® AP 760, 243–244
Expandable graphite (EG),
  carbon-based flame retardants, 493
  intumescence-based flame retardants, 175–176, 183, 193, 199, 218, 220
  phosphorus flame retardants and, 436

Fabric(s), 346–373
  dewatering, 275
  natural, 347–358
    Ag nanowires, 356
    cotton treatment by APP/MMT nanocoating, 357
    cotton treatment by BPEI/PhA LbL, 356
    cotton treatment by BPEI/PSP-complexes, 354
    cotton treatment by PAH/SPS LbL assembly, 352
    hydrophilicity and surface roughness, 347
    main FR conformal coating solutions for, 349–351
    phytic acid, 354–355

polyamidoamines (PAAs), 355
tannic acid, 354
process equipment and related
patents, 371–373
synthetic, 358–371
main FR conformal coating
solutions for, 360–362
overview, 357–359
PET, treatment by BPEI/OSA
LbL, 363, 364, 368, 370
PET, treatment by PDAC/PAA/
PDAC/APP LbL, 364, 365
Federal Aviation Administration
(FAA), 160, 209
Federal Aviation Regulations (FAR),
208, 530
Federal Hazardous Substances Act,
9
Federal Motor Vehicle Safety Standard
302 (FMVSS 302) test, 39, 42,
45, 56, 149, 160, 184, 425, 530
Fiber reinforced plastics (FRP), 159
Filler(s),
biofillers, 219
coarser, 159–160
combination with, 424–428
inert, 101–102, 426, 507
inorganic, 426
loading, hydrated mineral filler
flame retardants, 130–132
powdery, 109–110
synthetic FR, 123
Fire alarm, 140
Firebrake®ZB and Firebrake 500,
318–324
coatings (water-based), 324
commercial boron-based FR, 313
elastomers, 322–323
epoxy, 321–322, 323
molecular structure, 318
polyamides, 319–321
polyolefins, 319
polypropylene (PP), 319
Fire curve, cellulosic, 206

Fire extinguishers, liquid halogenated
solvents in, 3
Fire performance index (FPI), 279
Fireproofing products, 206
Fire protection, future, 475–532
biobased materials, 494–496
carbon-based flame retardants,
485–493
alkynes, 485–487
anhydrides, 490–491
cross-linking compounds,
485–491
deoxybenzoin, 488
expandable graphite, 493
Friedel-Crafts reaction, 488–490
graft copolymerization, 492–493
nitrile groups, 490
organic carbonates, 491–492
end of life considerations for flame
retardants, 524–528
durability for recycling, 526–527
environmental decomposition,
525, 528
recycling/sustainability design,
524–525
sustainable and alternative
chemical feedstocks, 524–526
waste-to-energy, 524–525,
527–528
engineering non-halogenated FR
solutions, 513–519
barrier fabrics, 513–515
conformal/integrated coatings,
518–519
infrared reflective coatings, 517
inorganic coatings, 515–517
layer-by-layer coatings, 518–519
nanoparticle coatings, 517–518
experimental methodology, 531–532
fire risk scenarios, new and growing,
528–530
future directions, 519–532
periodic table of flame retardants,
475–478

polymeric flame retardants, 521–523
polymer nanocomposites, 513
reactive flame retardants, 521–523
sulfur-based flame retardants,
    484–485
tin-based flame retardants, 496–512
    halogen-free applications,
        498–506
    novel tin additives, 506–512
    overview, 496–497
    ZHS, 497–498, 499, 500, 501, 503,
        504–505, 506–512
    ZS, 497–498, 499, 500, 501, 502,
        503, 504–505, 506–512
transition metal flame retardants,
    478–484
    condensed phase, 479–484
    vapor phase, 478–479
Fire protective intumescent materials,
    215–224
Fire residue,
    amount, 425
    in composite, 424
    morphology and properties, 424,
        425
    properties, 422, 424, 425
Fire resistance,
    in buildings, 141
    test, 205–209
Fire retardant paints, 209
Fire risk scenarios, 2, 7
    changes in, 8
    nature of, 13
    new and growing, 528–530
    WUI, 528–530
Fire safety,
    issues, 13–16
    regulations, 6–8
    requirements, 7
Flame propagation,
    in buildings, 141
    hydrated mineral filler flame
        retardants, 130–132
Flame retardants (flame retardants),

boron-based. see Boron-based flame
    retardants
chemicals, 2, 477, 495
    fire safety and, 6–8
    mineral, 106, 107
    with recycling process, 14
    regulation, 9–12, 14–16
    toxicity (PBT) of, 519
intumescent. see Intumescent-based
    flame retardants
mineral filler. see Mineral filler
    flame retardants
multicomponent. see
    Multicomponent flame retardants
NFR. see Nitrogen-based flame
    retardants (NFR)
non-halogenated,
    conformal FR coatings. see
        Conformal FR coatings
    other, future fire protection and.
        see Fire protection, future
    regulations. see Regulations, non-
        halogenated flame retardants
phosphorus-based. see Phosphorus-
    based flame retardants
silicon-based. see Silicon-based
    flame retardants
Flame spread, 184
Flamestab® NOR® 116, 257–259,
    429
Flammability,
    of ABS/clay nanocomposites, 286
    of carpets, 149
    hydrated mineral filler flame
        retardants, 130–132
    PDMS, 292
    PET, 197
    of polypropylene, 198, 279
Flexible polyurethane (PUR) foams,
    43, 148
Floridienne Chimie, 255
Flow rate index (FRI), 120
Fluorohectorite, 285
Fluoromica, 285

FMVSS 302 (Federal Motor Vehicle
    Safety Standard # 302) test, 39, 42,
    45, 56, 149, 160, 184, 425, 530
Foaming agent, 32, 243, 387
Formaldehyde, 27, 28, 30, 31, 33, 46,
    170, 434
Friedel-Crafts reaction, 488–490
Frits, 328–329
Functionalized graphene oxide (fGO),
    315

Gallic acid, 495
Gas phase modes of action, 420
Gelcoats, 163
Geopolymers, 222–223, 516
GFRP. see Glass fibre reinforced
    plastics (GFRP)
Glass fiber(s), 424–425
    halloysite and, 284
Glass fibre reinforced plastics (GFRP),
    electrical equipment, 161
    hand lamination/hand-lay-up, 157
    LOI on, 155
    paste production, 156–157
    public transport applications,
        160–161
    pultrusion, 158
    RTM/RIM, 158–159
    SMC and BMC, 157–158
    surface finish of, 160
Glass-forming substances, as
    synergists, 434
Glass transition temperature,
    DOPO, 60
    epoxy matrix, 161
    multicomponent flame retardants,
        413, 417
    PMP, 57
    of PTFE, 51
    reactive flame retardants, 60
    of thermoset matrix, 159
Global GreenTag, 14

Glow wire ignition temperature, 27,
    183
Graft copolymerization, 492–493
Graphene,
    carbon-containing boron/borates,
        329
    nanofillers, 427
Graphite intercalation compound
    (GIC), 193
Green Seal, 14
Grinding process, 109–110
Guanidine, 240
Guanidine sulfamate, 365
Guanidinium phosphate, 33
Guanyl urea phosphate, 30

Halloysite, inorganic silicon-based
    flame retardants, 284–285
Halogen-free flame retardant (HFFR),
    1–6, 143–147, 150
Halogen-free multicomponent
    systems, 417
Halogen-free tin-based flame
    retardants, 498–506
    elastomers, 503, 505
    engineering plastics, 500, 502, 503
    paints and coatings, 505
    polyolefins, 499–500, 501
    styrenics, 500
    textiles, 506
    thermosetting resins, 503–505
Hand lamination/hand-lay-up, 157
HBCD (hexabromocyclododecane),
    8–13
Heat conductivity (k), 126, 127, 173,
    175, 176, 178, 179
Heating, ventilation, and air
    conditioning (HVAC) systems,
    14, 319
Heat release, 134–136
Heat release rate (HRR),
    developing fire, 184

intumescence-based flame
retardants, 175, 184, 191, 199
melamine, 248, 249
peak of (pHRR), 175, 190, 192–193,
198, 199, 202, 279, 422, 425
polypropylene, 279
vs. time for plasticized PVC,
135–136
Heat-transfer rate inducing system
(H-TRIS), 212–213
Hectorite, 285
Hexaammonium
(nitrilotris(methylene))
trisphosphonate, 30
Hexabromocyclo dodecane, 9
Hexabromocyclododecane (HBCD),
8–13
Hexachlorocyclotriphosphazene
(HCCP), 353
Hexagonal boron nitride (h-BN),
313
Hexakis(4-boronic acid-phenoxy)-
cyclophosphazene, 330
Hexaphenoxy-phosphazene, 261
Hexaphenoxytricyclophosphazene,
27, 64
HFFR (halogen-free flame retardant),
143–147, 150
HFST (horizontal flame spread test),
349, 360–362, 367, 382, 383,
388–389
High density polyethylene (HDPE),
35, 285
High impact polystyrene (HIPS), 48,
55, 285–286, 327, 522
High temperature vulcanizing (HTV)
silicone, 151
HIPS (high impact polystyrene), 48,
55, 285–286, 327, 522
HNBR (hydrated nitrile butadiene
rubber), 145
Horizontal burner test, 213–214

Horizontal flame spread test (HFST),
349, 360–362, 367, 382, 383,
388–389
HRR. *see* Heat release rate (HRR)
Huber Engineered Materials (USA),
179
Huntite, 106, 108, 109, 124, 129, 143,
144
Hydrated magnesium carbonate, 109
Hydrated mineral filler flame
retardants,
filler loading, 130–132
flame propagation, 130–132
flammability, 130–132
heat release, 134–136
physical and chemical processes,
128–130
smoke suppression, 132–134
Hydrated sodium silicate, 182
Hydrocarbon fire test curve, 207
Hydrocarbon modified curve (HCM),
207
Hydrochloric (HCl) acid emission, 142
Hydrogen peroxide, 63, 222
Hydromagnesite, 106, 108, 109, 124,
129, 143, 144
Hydrophobic poly(methylhydrogen
siloxane), 29
Hydroquinone bis(diphenyl
phosphate) (HDP), 52–54, 57
Hydrotalcite, 27, 35
Hydroxyl terminated version, 46
Hyperbranched polyphosphoesters
(hbPPEs), 189
Hypophosphites, 36–40
Hypophosphorous acid, 363

Ignition,
loss on, mineral filler flame
retardants, 122
reaction to fire, 183–184
Imidazolium treatments, 296

Incinerators, 527
Industrial chemical process chain, 104–106
Industrial mineral filler flame retardants,
  market share, 103–104
  natural, 106–107
  synthetic mineral flame retardants, 104–106
Industrial Technology Research Institute (Taiwan), 41
Inert fillers, 101–102, 426, 507
Infrared images, intumescent-based flame retardants, 213
Infrared reflective coatings, 517
Injection molding, engineering plastics, 152
Inorganic coatings, 515–517
Inorganic fillers, 426
Inorganic flame retardants. *see* Mineral filler flame retardants
Inorganic silicon-based flame retardants, 278–290
  halloysite, 284–285
  kaolin, 282
  layered silicate nanocomposites, 285–289
  magadiite, 281
  mica, 283
  sepiolite, 281–282
  silicon dioxide (silica), 278–280
  silsesquioxane, 289–290
  sodium silicate, 289
  talc, 283–284
  wollastonite, 280–281
In-situ coupling process, 138–139
Integrated coatings, 518–519
Intercalated nanocomposites, 509
Intercalation, 175, 176, 193, 315, 331, 509
International Marine Organisation (IMO), 160

International Tin Research Institute (ITRI), 497, 505–507, 509
Intumescent-based flame retardants (IFR), 169–225
  chemical composition, 173, 174
  coatings, 215
  commercial, overview, 29, 31–35
  components of, 31, 174
  defined, 169–170
  fundamentals, 172–179
  future trends, 224–225
  hybrids, 285
  ingredients, 187–188
  on market, 179, 181–183
  mechanical properties, 172–173, 186
  morphology, 178
  overview, 169–172
  paints and mastics, 31
  publications and patents, 170–171
  reaction to fire, 180, 183–204
    evaluation, 180, 183–185
    inorganic polymeric materials, 199
    organic polymeric materials, 185–193
    synergistic effects, 199–204
    textile, 194–199
  resistance to fire, 204–224
    evaluation, 205–209
    fire protective intumescent materials, 215–224
    testing and characterization, 210–215
  rheology, 176–178
  silicate-based coating, 176
  thermal conductivity, 173, 178–179
  use, 170
Intumescent polymer metal laminates (IPML), 223–224
Iron carbonyl, 478
Isocyanate,
  excess of, 40
  melamine with, 249
Isodecyl diphenyl phosphate, 50

Isophthalic acid, 61
Isopropyl- and *tert*-butylphenols, 49
Isopropylphenyl phenyl phosphate, 49, 50
Israel Chemicals Limited (Israel), 179
Itaconic acid, 59
Italmatch Chemicals (Italy), 179
ITRI (International Tin Research Institute), 497, 505–507, 509

Japan, flame retardant regulations, 12
Jet fires, 208, 214–215
Jiangsu Liside (China), 179
JM Huber, 255

Kaolin, 282, 372
Korea, flame retardant regulations, 12–13

Lamellar silicates, 295
Lanxess (Germany), 179
Layer-by-layer (LbL) assembly, coating, 352–353
CS/PVPA LbL assembly, 387, 389
in delivering multifunctional coatings, 354
deposition process of CS/P-CNF, 373
intumescent system for cotton fabrics, 348
intumescent textile, 195–197
layered silicate nanocomposites, 288
of SNP nanoparticles, 359
synthetic fabrics, 363–365, 368
Layer-by-layer (LbL) coatings, 518–519
Layer-by-layer (LbL) deposition, 342, 344, 373
Layered double hydroxides (LDHs), 509–510
Layered silicate minerals, 295

Layered silicate nanocomposites, 285–289
LDHs (layered double hydroxides), 509–510
Levoglucosan, 30
Lewis acids, 30, 484
Lignin, 219, 494
Lignin-modified polyol, 33
Lignin sulfonate, 384
Limiting oxygen index (LOI),
daidzein, 495
dependence of BET-surface area, 131
with DIN EN ISO 4589-2 (ASTM D2863), 130
extinction of material, 185
glass fibre reinforcement on, 155
of PET fibers, 58
phenolic resin, 155
in polyethylene, 28
for selected polymers, 132
values, for ATH, AOH and MDH, 130, 131
VFT and, 197
Linear ammonium polyphosphate, 242
Linear low density polyethylene (LLDPE), 143, 144
Liquid halogenated flame retardants, 3
Lithium ion battery (LIB), 127
LOI. *see* Limiting oxygen index (LOI)
Loss on ignition, mineral filler flame retardants, 122
Low density polyethylene (LDPE), 35
Low smoke flame retardant (LSFR), 142
Low-smoke free-of-halogen (LSFOH) polymer, 143

Magadiite, 281
Magnesium borate, 317
Magnesium calcium carbonate, 109
Magnesium chloride, 111, 112
Magnesium hydrate, 109

Magnesium hydroxide (MDH),
  aromatic phosphates with, 55
  ATH and, 61
  BET surface of, 112
  engineering plastics, 152
  HFFR compounds, 144, 146
  HRR of, 136
  LOI values, 130, 131
  mechanical properties, 124
  mineral flame retardants, 104,
    105–106, 108, 109
  natural, 109
  polyolefins, 319
  PVC, 143
  seawater, 112
  synthetic, 111
  for wire and cable applications, 2
Magnesium oxides, 104, 111
Magnesium silicate, 281
Maleic anhydride, 55, 59, 139
Manganese dioxide, synergistic effects,
  201, 203–204
Mannitol, 186
Market share, industrial mineral flame
  retardants, 103–104
Masterbatches, 27, 51, 102, 139
Mastics, 31
MC. see Melamine cyanurate (MC)
MCA Technologies, 263
MDH. see Magnesium hydroxide
  (MDH)
Medium density fiberboard, 315
Melam, 256–257
Melamine,
  APP/PER-melamine, 216
  blowing agent, 31, 430
  condensation product of, 32, 430
  derivatives, 146
  formaldehyde, 28, 46, 434
  gas-forming compounds, 243
  hydrobromide, 37
  intumescent systems, 31
  Nflame retardants based on, 240

orthophosphate, 34
phosphates, 33–34, 38
polyaromatic, 155
polyphosphates, 34, 36, 38, 39
pyrophosphate, 33–35
salts, 27
trimethylene amine phosphonate
  salts, 34
Melamine-based flame retardants,
  247–257
  condensation products, 256–257
  as flame retardants, 247, 248–250
  MC, 250, 251–254
  melamine salts, 250
  MPP, 250, 254–256
Melamine borate, intumescent
  compounds, 181
Melamine cyanurate (MC), 36, 38,
  39
  in flame retardants, 250, 251–254
  mode of action, 251–252
  with organic phosphates, 253
  in polyamides, 241, 252–253
  structure, 251
Melamine diborate, 313, 324
Melamine phosphates, 55
  intumescent compounds, 181
  intumescent flame retardants based
    on, 197–198
  thermal stability, 250
Melamine polyphosphate (MPP), 181,
  182, 496
  aluminum diethylphosphinate and,
    435
  Firebrake®500 and *Firebrake* 500,
    320
  in flame retardants, 250, 254–256
  thermal stability, 254
Melamine pyrophosphate, 254
Melamine salts, of pentaerythritol
  phosphate, 187
Melem, 256–257
Melon, 256–257

Melt compounding process, 113, 120, 139, 245, 296, 339, 496
Melt flow, 428–429
Melt flow index (MFI), 128
Melt volume rate (MVR), 128, 153–154
Metaboric acid, 311, 314
Metakaolin-based alkaline aluminosilicate, 222
Metal borophosphate, 328
Metal complexes, condensed phase transition metal flame retardants, 481–484
Metal hydrates, 107, 128, 136, 148, 416, 434–436
Metal hydroxides, 27, 107, 134, 136, 146, 150
Metal hypophosphites, phosphites and dialkyl phosphinates, 36–40
Metallocenes, for paper, 479
Metal nitrates, 483–484
Metal oxides, 27, 134
    condensed phase transition metal flame retardants, 480–481
    as synergists, 433
Metal salts, 27
Methacrylic acid, 492
Methylcyclopentadienylmanganese tricarbonyl, 478
Methyldichlorophosphine, 37
2-Methyl-2,5-dioxa-1,2-phospholane, 58
Methylethylphosphinates, 37
Methylmethoxysiloxane, 31–32
Methyl phosphate methylphosphonate ethylene glycol oligomer, 45
Methyl tert-butyl ether, 261
Mica, 283, 285
Mineral filler flame retardants, 101–165
    chemical composition, 107–109
    defined, 101
    hydrated, 128–136

filler loading, 130–132
flame propagation, 130–132
flammability, 130–132
heat release, 134–136
physical and chemical processes, 128–130
smoke suppression, 132–134
industrial importance, 102–107
    market share, 103–104
    natural mineral flame retardants, 106–107
    synthetic mineral flame retardants, 104–106
overview, 101–102, 107–128
physical properties, 114–122
    aspect ratio, 115–116
    BET surface area, 118
    bulk density, 120–121
    endothermic heat, 122
    loss on ignition, 122
    morphology, 115–116
    oil absorption, 118, 119
    particle shape, 115–116
    pH-value, 119
    powder flowability, 120–121
    PSD, 116–117
    sieve residue, 117
    specific conductivity, 119
    thermal stability, 122
polymer material properties, 123–128
    chemical resistance, 124–126
    electrical properties, 127
    mechanical properties, 123–124
    optical properties, 123
    rheological properties, 128
    thermal properties, 126–127
    water uptake, 124–126
processing methods, 109–114
    air separation, 110
    crushing and grinding, 109–110
    precipitation, 110–113
    surface treatment, 113–114

reactive resins/thermoset,
applications, 154–163
construction and industrial
applications, 163
electric and electronic
applications, 161–163
formulation principles, 159–160
overview, 154–156
production processes for GFRP,
156–159
public transport applications of
GFRP, 160–161
thermoplastic and elastomeric
applications,
compound formulation
principals, 138–140
compounding technology,
136–138
construction products, 147–150
engineering plastics, 152–154
special applications, 150–152
wire and cables, 140–147
trends and challenges, 164–165
MMT (montmorillonite), 281, 282,
285–286, 357, 363, 381, 509
Molybdenum disilicide (MoSi$_2$),
217–218
Monoammonium dihydrogen
phosphate (MAP), 29, 30
Monoguanidine dihydrogen
phosphates, 30
Monohydric alcohol, 45
Montmorillonite (MMT), 281, 282,
285–286, 357, 363, 381, 509
Morpholine, 32, 263, 430
MPP. see Melamine polyphosphate
(MPP)
Multicomponent flame retardants,
413–438
adjuvants, 428–431
combinations of different flame
retardants, 435–437
groups in one FR, 437–438

concepts, 419–424
fillers, combination with, 424–428
need for, 413–419
synergists, 428, 431–435
Multi-functionality, conformal FR
coatings, 344–345
Multi-walled carbon nanotube
(MWCNT), 217, 280, 379
Muscovite mica, 283
MVR (melt volume rate), 128, 153–154
MWCNT (multi-walled carbon
nanotube), 217, 280, 379

N-alkoxyamines (NOR), 37, 257, 429
N-alkoxy hindered amines, 257–259,
260, 429
Nanoclays, tin-modified, 509–510
Nanocomposites,
clay. see Clay nanocomposites
for flame retardants, 427
intercalated, 509
layered silicate, 285–289
polymer, 513
Nanocyl, 294
Nanofillers,
graphene, 427
as synergists, 201–202
Nanoparticle(s),
carbon, 427, 438, 518
coatings, 517–518
multicomponent systems, 427–428
SNP, 359
Naphthoquinone (NQ), DOPO-NQ,
61
National Institute of Standards and
Technology (NIST), 478, 514
Natural fabrics, 347–357
Ag nanowires, 356
cotton treatment
by APP/MMT nanocoating, 357
by BPEI/PhA LbL, 356
by BPEI/PSP-complexes, 354
by PAH/SPS LbL assembly, 352

hydrophilicity and surface roughness, 347
main FR conformal coating solutions for, 349–351
phytic acid, 354–355
polyamidoamines (PAAs), 355
tannic acid, 354
Natural fibers, 196, 346, 347, 358, 363, 367, 426, 432
Natural mineral colemanite, 317
Natural mineral flame retardants, 106–107
Natural rubber (NR), 150, 327
NBR (nitrile butadiene rubber), 145, 147, 148, 150
NBS (National Bureau of Standards), 133
Neobor®, 314–315
Neopentyl acid phosphate with propylene oxide, 45
NFR. *see* Nitrogen-based flame retardants (NFR)
Nitrile butadiene rubber (NBR), 145, 147, 148, 150
Nitrile groups, 490
Nitrogen-based flame retardants (NFR), 239–265
  ammonia-based flame retardants, 241–247
  ammonium pentaborate, 246
  ammonium polyphosphate (APP), 241, 242–245
  ammonium sulfamate, 246–247
  ammonium sulfate, 247
  cyanuric-acid based flame retardants, 263–264
  melamine-based flame retardants, 247–257
  condensation products, 256–257
  as flame retardants, 247, 248–250
  MC, 250, 251–254
  melamine salts, 250
  MPP, 250, 254–256
  overview, 239–240

phospham, 261–263
phosphazenes, 261–263
phosphoroxynitride (PON), 261–263
radical generators, 257–260
types, 240–241
Nitrogen-containing borates, 324–327
  ammonium borophosphate, 327
  ammonium pentaborate (APB), 325
  boron nitride (h-BN), 325–327
  acrylics, 326
  natural rubber (NR), 327
  polyurethane foam, 326
  red phosphorus, 326
  unsaturated polyester (UP), 327
  melamine diborate, 324
Non-durable finishes, 29, 30
Non-fire safety issues, 13–16
Non-halogenated conformal FR coatings. *see* Conformal FR coatings
Non-halogenated flame retardants chemistries, 239
  future fire protection. *see* Fire protection, future
  market growth for, 240
  regulations. *see* Regulations, non-halogenated flame retardants
Non-intermeshing counter-rotating twin-screw technology, 137
Non-reinforced thermoset applications, 159
Nordic Swan, 14
Novel tin additives, 506–512
  benefits, 506
  coated fillers, 507–509
  mechanism of action, 510–512
  tin-modified nanoclays, 509–510
Novolac type epoxy, 60, 430
Novolak, 263
Nyglos®, 281
Nylons,
  nylon 6 and 6.6, 198
  phospham in, 263

Octa-ammonium Polyhedral
    Oligomeric Silsesquioxane
    (OAmPOSS), 366
Octamethyl polyhedral silsesquioxane
    (OMPOSS), 202–203
Octamethyltetrasiloxane, 290–291
Octaphenoxytetraphosphazene, 64
Octa(tetramethylammonium) POSS
    (octaTMA-POSS), 290
Oil absorption, mineral filler flame
    retardants, 118, 119
Open cell PU foams, 374–386
    burn-through fire test, 382
    cone calorimetry, 374, 375, 379–380,
        383–385
    conformal coating solutions for,
        376–378
    flame penetration test results, 382,
        383
    real scale chair mockup test, 381
Optibor®, 11
Organic carbonates, 491–492
Organic silicone-based flame
    retardants, 290–293
    polyorganosiloxanes, 290–292
    silanes, 292–293
Organoclays, 253, 289, 296
Organohalogens (Oflame retardants),
    defined, 9
Organo-modified layered double
    hydroxide (OLDH), 218
Organophilic montmorillonite
    (o-MMT), 285–286
Organophosphorus compounds,
    25
Organophosphorus flame retardants,
    exposure to, 65–67
Organosilanes, 139, 273
Organosilicones, 292
Organosiloxane materials, 274
Oxide sodium alginate (OSA), 363,
    364, 368, 370

PA. see Polyamides (PA)
PAA (poly(acrylic acid)), 364, 365
PAAs (polyamidoamines), 355, 375,
    379
PAH (poly(allyl amine)), 348, 352, 366,
    378, 384, 394
Paints,
    halogen-free tin-based flame
        retardants, 505
    intumescent, 31
PAN (poly acrylonitrile), synthetic
    fabrics, 358, 362, 366
Paraffinic oil, 319
Paraffins, 8, 11, 148, 285
Particle shape, mineral filler flame
    retardants, 115–116
Particle size distribution (PSD),
    mineral filler flame retardants,
    116–117, 159
Paste production, 156–157
Patents, process equipment and,
    371–373, 391–392
PBDEs (polybrominated diphenyl
    ethers), 5, 8, 10–11
PBT. see Persistence, bioaccumulation,
    and toxicity (PBT); Polybutylene
    terephthalate (PBT)
PC. see Polycarbonates (PC)
PDMS (polydimethylsiloxanes), 274,
    276, 278, 280, 282, 290–291
PE (polyethylene), 28, 35, 37, 56
Peak of HRR (pHRR), 175, 190,
    192–193, 198, 199, 202, 279,
    422, 425
Pentaerythritol (PER),
    APP/PER, 29, 174, 216, 429
    carbonaceous char, formation, 31
    as 'conventional' char formers, 186
    pentaerythritol di(phosphate acid
        monochloride)s, 188
    pentaerythritol phosphate, 187,
        197–198

pentaerythritol spirobis(benzylphosphonate), 47, 48

pentaerythritol spirobis(methylphosphonate), 47

PLA, preparation, 330

with polyamides, 484

silicone surface-treated APP, 31–32

synergistic product, 243

with thermoplastic polyurethane, 34

PER. *see* Pentaerythritol (PER)

Periodic table of flame retardants, 475–478

Persistence, bioaccumulation, and toxicity (PBT),

engineering plastics, 153

of FR chemicals, 519

issues, 5, 8

MVR values of, 153–154

profiles, 3, 4, 10, 11, 15

properties, SVHC, 11

synergism in, 38

TPU and, 188

Persistent organic pollutants (POP), 8, 65

PET. *see* Polyethylene terephthalate (PET)

P-formylphenylboronic acid, 330

PhA. *see* Phytic acid (PhA)

Phenol-formaldehyde resole resin, 30

Phenolic foam, 315

Phenolics, 27, 37, 55, 154, 155, 161, 315

Phenoxyphosphazenes, cyclic, 64

Phlogopite mica, 283

Phospham, 261–263

Phosphate esters, 27, 28, 148

Phosphates,
aliphatic, 40–48
aromatic, 48–58

Phosphazenes, 64, 161–163, 241, 432

Phosphinates, aromatic, 58–62

Phosphine, 26, 27, 28, 36

Phosphine oxides, 62–64

Phosphites, 36–40

Phosphonates, 40–48
aliphatic, 40–48
aromatic, 48–58
liquid cyclic, 46–47

Phosphonic acid, with ethylene oxide, 45

Phosphoramidates, 432

Phosphoric acid, 30, 197, 429

Phosphorous acids, 26

Phosphoroxynitride (PON), 261–263

Phosphorus-based flame retardants, 23–68
advantages, 25
aliphatic phosphates and phosphonates, 40–48
ammonium and amine phosphates, 29–36
aromatic phosphates and phosphonates, 48–58
aromatic phosphinates, 58–62
classification, 25–26
dialkyl phosphinates, 36–40
disadvantages, 26
environmental fate, 65–67
ethyl ethylene glycol phosphate, 42, 44
future, 67–68
metal hypophosphites, 36–40
organophosphorus flame retardants, exposure to, 65–67
overview, 24
phosphazenes, 64
phosphine oxides, 62–64
phosphites, 36–40
PUF, 28, 32, 33, 40, 42–43
red phosphorus, 26–29
tasks for, 414–415
water-soluble/water-insoluble, 25

Phosphorus-containing borates,
boron phosphate, 327–328
metal borophosphate, 328

Phosphorus flame retardants, 432–433
  expandable graphite and, 436
  metal hydrates and, 435–436
Phosphorus-nitrogen (P-N) synergy,
  431–432
Phosphorus-phosphorus (P-P)
  synergism, 435
Phosphorus-silicone (P-Si) synergy,
  432
Phosphorus-sulfur (P-S) synergy, 432
Phosphorus trichloride, 59
Phosphorylated biopolymers, 438
Phosphorylated vegetable oils, 438
Phosphotungstic acid, 187
Photovoltaic cables, 145, 146
PHRR (peak of HRR), 175, 190,
  192–193, 198, 199, 202, 279,
  422, 425
Phthalate plasticizers, 119
Phthalic anhydrides, 485, 486
Phyllosilicates, 285
Physical vapor deposition (PVD),
  342
Phytic acid (PhA),
  bio-resourced, 196–197, 495
  BPEI/PhA LbL assembly on cotton,
  356
  synthetic fabrics, 365–366
  via LbL assembly, 316–317
  wool treatment by, 354–355
Piperazine, 32, 35, 183, 263, 430
Pittchar (PPG), 216–217
PLA (polylactic acid), 37, 47, 52, 172,
  330–331
Plasma coatings, 515–516
Plasma enhanced chemical vapor
  deposition (PECVD), 342
Plasma technology, 342
Plenum spaces, 142, 143
PMMA (poly(methyl methacrylate)),
  29, 37, 47–48, 52, 279, 291, 484
Politics, fire safety and non-fire safety
  issues, 15

Poly(acrylamide) (PAAm), 349
Poly(acrylic acid) (PAA), 364, 365
Poly acrylonitrile (PAN), synthetic
  fabrics, 358, 362, 366
Poly(allylamine) (PAAm), 195
Poly(allyl amine) (PAH), 348, 352, 366,
  378, 384, 394
Polyamides (PA),
  aliphatic, 32
  DEPAL in, 38, 39
  in electrical engineering, 424
  Firebrake®ZB and Firebrake 500,
  319–321
  glass-filled, 26, 27, 35
  melamine cyanurate in, 241,
  252–253
  PA 66 fabrics, 27–28, 37, 316–317,
  327
  PPO/PA, 327
  requirements for flame retardants,
  36
  synthetic fabrics, 361
  thermal decomposition, 28
Polyamidoamines (PAAs), 355, 375,
  379
Polyaromatic hydrocarbons (PAHs),
  4–5
Polyaromatic melamine, 155
Poly(bisphenol A methylphosphonate)
  (PAMP), 57–58
Polybor®, 312, 315–317
Polybrominated diphenyl ethers
  (PBDEs), 5, 8, 10–11
Polybutylene terephthalate (PBT),
  255
  glass-filled, 36
  PC/PBT blends, 52, 56
  phospham, PON and mixtures with,
  261–262
  transition metal oxides, 480
Polycaprolactone, 56
Polycarbonates (PC),
  aromatic phosphates in, 48–54

arylphosphate–arylphosphate synergism, 435
bio-based, 47
calcium hypophosphite in, 36
defined, 491–492
flammability, 522
FR-PC/ABS blends, 417
PC/ABS blends, 48, 51, 53, 55, 64, 491
PC/PBT blends, 52, 56
PC/PET, 52
PC/PLA, 52
PC/PMMA, 52
PC/SiR/BDP blends, 418
structure, 491
Polycarbosilane (PCS), 293
Polycarboxylic acid, 46
Polychloroprene, 503
Poly(diallyldimethylammonium chloride) (PDAC), 364, 365
Polydimethylsiloxanes (PDMS), 274, 276, 278, 280, 282, 290–291
Polyelectrolyte complex (PEC), 196
Polyester(s), 26
    cotton, 33
    DEPAL, 39, 415–416
    in electrical engineering, 424
    fibers, 197
    requirements for flame retardants, 36
    thermoplastic elastomers, 37
    thermosol treatment of, 46
    unsaturated, 32, 39, 40, 41–42, 327
Polyethylene (PE), 28, 35, 37, 56
Poly(ethylene-co-methacrylic acid) (EMAA), 191
Poly(ethylene-co-vinyl acetate) (EVA), 32, 37
    alumina trihydrate in, 483
    coated fillers, 508
    HFFR wire and cables, 143, 144
    HRR of, 136
    LOI values, 130–131

sepiolite containing, 282
tensile-elongation properties for, 145
Polyethylene terephthalate (PET), aluminum hypophosphite in, 255
    anti-dripping properties, 188
    aromatic phosphinates, 58, 59
    dialkylphosphinates in, 37
    flammability, 197
    intumescent flame retardants, 188, 197
    PC/PET, 52
    phosphorus-based flame retardants, 27, 28, 37, 39, 46, 47, 52, 56, 58
    with red phosphorus, 27, 28
    synthetic commodity textiles with, 358
    synthetic fabrics, 360–361
    textiles, 39, 46, 47, 56
    treatment by BPEI/OSA LbL, 363, 364, 368, 370
    treatment by PDAC/PAA/PDAC/ APP LbL, 364, 365
Polyhedral oligomeric silsesquioxanes (POSS), 191–192, 202–203, 289, 290, 366
Polyhydric alcohols, 243
Poly(2-hydroxy propylene spirocyclic pentaerythritol bisphosphonate) (PPPBP), 197
Polyisocyanurate foams (PIR), 40, 49
Polylactic acid (PLA), 37, 47, 52, 172, 330–331
Polymeric flame retardants, 521–523
Polymer-layered silicate (PLS) nanocomposites, 285
Polymer(s),
    charring, 39, 54, 427
    material properties, mineral filler flame retardants, 123–128
        chemical resistance, 124–126
        electrical properties, 127
        mechanical properties, 123–124

optical properties, 123
rheological properties, 128
thermal properties, 126–127
water uptake, 124–126
nanocomposites, 513
silicon based, 278
synthetic, 277
thermal stability effects, 277
Poly(methyl methacrylate) (PMMA),
29, 37, 47–48, 52, 279, 291, 484
Polymethylsiloxane oil, 35
Polyolefins, 28
aluminum hypophosphite, 37
APP in, 32, 33, 35, 283, 429
construction product, 148, 150
Firebrake®ZB and Firebrake 500,
319
halogen-free tin-based flame
retardants, 499–501
heat release material, 260
HIPS and, 55
intumescent systems, 31
limited compatibility with, 57
low efficiency in, 26
metal hydrates in, 434
Polyol(s),
lignin-modified, 33
of melamine, 248–249
use, 186
Polyorganosiloxanes, 290–292
Polyphenols, 494
Polyphenylene ether (PPE), 61
charring polymer, 54–55, 56,
429–430
decomposition temperature, 56
non-epoxy, 62
phosphazene, 64
polyamide blends, 38
PPE/elastomer blends, 50
PPE/HIPS blends, 48, 50, 51, 54, 55,
56, 182, 430
PPE/SEBS blends, 39, 55
TPP in, 56

Poly(1,3-phenylene
methylphosphonate) (PMP), 57
Polyphenylene oxide (PPO), 327,
429–430, 522
Polyphosphazene, 199
Polyphosphoric acid, 31, 33, 35, 197,
243, 257, 353
Polyphthaliamide (PPA), Firebrake®ZB
and Firebrake 500, 319–320
Polypropylene (PP),
APP with synergists, 190, 243, 244,
263
Firebrake®ZB and Firebrake 500, 319
flammability, 198, 279
glass fiber and halloysite in, 284
intumescence and char formation
by, 243
intumescent flame retardants,
174–176, 190, 198–199
LOI, dependency of, 243, 244
phosphorus-based flame retardants,
28, 32, 34, 35
PP/clay nanocomposite, 288
V-2 rating in, 37
Polypropylene-graft-maleic anhydride
(PPgMA), 490
Polysilastyrene (PSS), 293
Polysiloxanes, 55, 291
Polysilsesquioxane preceramic
polymers, 293
Poly(sodium acrylate), 492
Poly(sodium methacrylate), 492
Poly(sodium phosphate) (PSP), 195,
353
Polystyrene (PS),
ammonium sulfamate in, 484
cone calorimetry, 246
cross-linking of, 489
HIPS, 48, 55, 285–286, 327, 522
HTT 1800 as FR for, 291
polystyrene-co-maleic anhydride
(SMA), 490
Polysulfones, 418, 432, 522

Polytetrafluoroethylene (PTFE), 37,
    51–53, 55, 429
Poly(trimethylene terephthalate), 39,
    56
Polyurea, magadiite in, 281
Polyurethane (PU),
    coating for textiles, 194–195
    composites, oxygen index of, 34
    elastomers, 28
    polyphosphazene and, 199
    PU-APP, 201–204
Polyurethane foam (PUF),
    aliphatic phosphates and
        phosphonates, 40
    boron nitride (h-BN), 326
    chemical structure, 15
    chlorinated phosphate esters, 42
    conformal coating solutions for,
        376–378
    DMMP in, 40
    flammability, 42
    flexible, 43, 148
    in furniture, 15
    heat release of, 16
    open cell, 374–386
    phosphorus-based flame retardants,
        28, 32, 33, 40, 42–43
    roofing applications in hot desert,
        43
Polyvinyl chloride (PVC), 104
    antimony trioxide for, 141–142, 143
    CDP, 49
    construction product, 148–150
    2-ethylhexyl diphenyl phosphate
        in, 50
    fillers in, 119
    isodecyl diphenyl phosphate in, 50
    LSFR PVC compounds, 142
    mineral flame retardants in, 142,
        143
    plasticizer for, 48–50
    production and use, 499
    PVC/NBR, 147

roofing membrane, 148
smoke suppressants in, 430
TCP, 48, 49
waste, 3
wires and cables, 50, 141–142
Poly(vinyl sulfonic acid sodium salt)
    (PVS), 384
PON (phosphoroxynitride), 261–263
Porcelain earth, 282
Porous materials, 373–392
    LbL deposition process of
        CS/P-CNF, 373
    open cell PU foams, 374–386
        burn-through fire test, 382
        conformal coating solutions for,
            376–378
        flame penetration test results,
            382, 383
        real scale chair mockup test, 381
    other substrates, 386–389
    process equipment and related
        patents, 391–392
POSS (polyhedral oligomeric
    silsesquioxanes), 191–192,
    202–203, 289, 290, 366
Potassium fluorozirconate (PFZ)
    treatment, 506
Powder flowability, mineral filler flame
    retardants, 120–121
Powder rheometry, 120, 121
Powdery fillers, 109–110
PP. see Polypropylene (PP)
PPA (polyphthaliamide),
    Firebrake®ZB and Firebrake 500,
    319–320
PPE. see Polyphenylene ether (PPE)
Ppm-Triazine HF™, 263
PPO (polyphenylene oxide), 327,
    429–430, 522
Pre-ceramic compounds, defined,
    289
Precipitation, mineral filler flame
    retardants, 110–113

Printed circuit boards (PCB), 162
Printed wiring boards (PWB), 59
Propylene oxide, 44, 45
PS. *see* Polystyrene (PS)
PSD (particle size distribution),
    mineral filler flame retardants,
    116–117, 159
Pseudo-boehmites, 112–113
Pseudo-synergy, 420
PTFE (polytetrafluoroethylene), 37,
    51, 52, 53, 55, 429
Public transport applications, of GFRP,
    160–161
PUF. *see* Polyurethane foam (PUF)
Pultrusion, 158, 163
PVC. *see* Polyvinyl chloride (PVC)
P-xylenebis(diphenyl phosphine
    oxide), 63, 64
Pyrophosphate, 33–34

Quinone, DOPO with, 60, 61

RABT (Richtlinien für die Ausstattung
    und den Betrieb von strassen
    Tunnels) curves, 207
Radical generators, nitrogen-based,
    257–260
Ramie, 347, 348, 350, 352
RDP (resorcinol bis(diphenyl
    phosphate)), 51, 52, 54–57, 431,
    435, 436
REACH (Registration, Evaluation,
    Authorisation and Restriction of
    Chemicals), 11, 526
Reaction to fire, of intumescent
    materials, 180, 183–204
    evaluation, 180, 183–185
    developing fire, 184
    extinction, 185
    flame spread, 184
    ignition, 183–184
    inorganic polymeric materials, 199
    organic polymeric materials,
    185–193

    inorganic-based systems, 190–193
    synergistic effects, 199–204
    textile, 194–199
Reactive flame retardants, 521–523
Reactive resins/thermoset applications,
    154–163
construction and industrial
    applications, 163
electric and electronic applications,
    161–163
formulation principles, 159–160
overview, 154–156
production processes for GFRP,
    156–159
    hand lamination/hand-lay-up,
    157
    paste production, 156–157
    pultrusion, 158
    RTM/RIM, 158–159
    SMC and BMC, 157–158
public transport applications of
    GFRP, 160–161
Reagents, 346
Recycling/sustainability design,
    524–528
durability for recycling, 526–527
environmental decomposition, 525,
    528
sustainable and alternative chemical
    feedstocks, 524–526
waste-to-energy, 524–525, 527–528
Red phosphorus, 26–29, 286, 326, 415
Reduced graphene (rGO), carbon-
    containing boron/borates, 329
Registration, Evaluation, Authorisation
    and Restriction of Chemicals
    (REACH), 11, 526
Regulations, non-halogenated flame
    retardants, 1–16
    Asia, 11
    Australia, 13
    Canada, 10
    China, 11–12
    European Union (EU), 10–11

fire safety and flame retardant
    chemicals, 6–8
fire safety and non-fire safety issues
    requiring, 13–16
future market drivers, 16
halogenated *vs.* non-halogenated,
    history, 1–6
Japan, 12
Korea, 12–13
United Nations, 8
United States (US), 9
Residue design, 421, 422
Resin injection moulding (RIM),
    158–159
Resin paste manufacturing, with
    GFRP, 156–159
Resin transfer moulding (RTM),
    158–159, 321, 322
Resistance to fire,
    defined, 171
    of intumescent materials, 204–224
        evaluation, 205–209
        fire protective intumescent
            materials, 215–224
        testing and characterization,
            210–215
Resorcinol bis(diphenyl phosphate)
    (RDP), 51, 52, 54–57, 431, 435,
    436
Resorcinol bis(di-2,6-xylyl phosphate)
    (RXP), 53, 54, 63
Resorcinol bis(2,6-xylenol phosphate),
    37, 39
Restriction of Hazardous Substances
    (RoHS), 4, 10–12, 14, 498–499
Rheology of intumescence, 176–178
Rice husk ash (RHA), 219, 222
Rigid polyurethane foam (RPUF), 315
RIM (resin injection moulding),
    158–159
Rin Kagaku Kogyo Co (Japan), 179
RoHS (Restriction of Hazardous
    Substances), 4, 10–12, 14,
    498–499

RTM (resin transfer moulding),
    158–159, 321, 322
Rubber(s),
    butadiene, 62
    natural (NR), 150, 327
    nitrile butadiene (NBR), 145, 147,
        148, 150
    silicon (SiR), 418
RWS (Rijks Water Staat) curve,
    207–208
RXP (resorcinol bis(di-2,6-xylyl
    phosphate)), 53, 54, 63

Saponite, 285
Scanning electron microscopy (SEM)
    alumimium hydrates, 116
    borosilicate glass and frits, 328–329
    particle shape, 115
SCCP (short-chain chlorinated
    paraffins), 8, 11
SciFinder ACS, 310
Scorch, 43
Selective laser sintering (SLS), 320
SEM. *see* Scanning electron
    microscopy (SEM)
Sepiolite, 281–282
Shandong Moris Tech Co (China), 179
Shandong Ruixing (China), 179
Sheet moulding compound (SMC),
    157–158, 163, 425–426
Short-chain chlorinated paraffins
    (SCCP), 8, 11
Shouguang Weidong Chemical Co
    (China), 179
SI (stability index), 120
Sieve residue, mineral filler flame
    retardants, 117
Silanes,
    defined, 273
    organic silicone-based flame
        retardants, 292–293
    structure, 273
Silanols, 113, 273–274
Silica (silicon dioxide), 278–280

Silicate-based minerals, 295–296
Silicate nanocomposites, layered, 285–289
Silicon,
  characteristics, 272–274
  chemistry, basics of, 272–274
  defined, 272
Silicon-based flame retardants, 271–299
  FR materials, 277–294
  future trends, 296–298
  inorganic, 278–290
    halloysite, 284–285
    kaolin, 282
    layered silicate nanocomposites, 285–289
    magadiite, 281
    mica, 283
    sepiolite, 281–282
    silicon dioxide (silica), 278–280
    silsesquioxane, 289–290
    sodium silicate, 289
    talc, 283–284
    wollastonite, 280–281
  mode of actions, 294–296
    silicate-based minerals, 295–296
    silicon dioxide (silica), 294–295
    silicones, 296
  organic, 290–293
    polyorganosiloxanes, 290–292
    silanes, 292–293
  other materials, 293–294
  overview, 271–272
  protective coatings, 294
  silicon chemistry, basics of, 272–274
Silicon-containing borates, borosilicate glass and frits, 328–329
Silicon dioxide (silica), 278–280, 294–295
Silicone(s),
  applications, 277–278
  based-coating, intumescent, 220–221
  -based intumescent paint, 210

condensed phase effect, 296
defined, 272
elastomer, Firebrake®500 and Firebrake 500, 322–323
elastomer tube, 276
flame retardant materials, 277–278
industrial applications of, 274–276
in medical applications, 276
with methyl and phenyl groups, 292
mode of actions, 296
properties, 275
sealants and adhesives, 275
Silicon hydride, structure, 273
Silicon oxide coating, 516
Silicon rubber (SiR), 418
Silicophosphates, formation, 432
Silk, 347, 348, 350–351, 356
Silsesquioxane, 289–290
Silyamines (silazenes), 273
Silylation, defined, 274
SiR (silicon rubber), 418
SMC (sheet moulding compound), 157–158, 163, 425–426
Smoke, 291
  generation, 141
  release, 13–14, 32, 134
  suppressants, 420, 430, 497
  suppression, 132–134
Sodium alginate, 363
Sodium aluminate, 104, 112
Sodium borates, 314–317
Sodium hydroxide, 104, 112
Sodium metasilicate, 289
Sodium silicate, 289
Sodium stannate, 29–30
Sodium trichlorobenzenesulfonate, 484
Sol-gel wet-chemical approach, 343
Solid phase charring reactions, 244
Sorbitol, 186
South Korea, flame retardant regulations, 12–13
Specific conductivity, mineral filler flame retardants, 119

Spirocyclic pentaerythritol phosphoryl chloride (SPDPC), 196
Spirophosphonate, 47
Spray roasting process, 111
Stability index (SI), 120
Standards,
    ASTM,
        fire resistance test, 205–206
        smoke density determination, 133
    Federal Aviation Agency (FAA), 160
    fire safety, 15
    FMVSS 302 test, 39, 42, 45, 56, 149, 160, 184, 425, 530
    NBS (National Bureau of Standards), 133
    in regulating and testing materials, 13
Stannic oxide, 30
Starch, 243, 352–353, 494
Stearic acid, 113, 193
S-triazine, 190
Styrene-co-acrylonitrile (SAN) polymers, 490
Styrene-ethylene-butadiene-styrene (SEBS), 39, 55
Styrenics, 37
    halogen-free tin-based flame retardants, 500
    low efficiency in, 26
Succinic acid, 36
Sulfenamides, 432
Sulfonamides, 484–485
Sulfur-based flame retardants, 484–485
Surface area, BET, 110–112, 118, 124, 145
Surface during combustion, role, 339–346
Surface treatment, mineral filler flame retardants, 113–114
Sustainable and alternative chemical feedstocks, 524–526
Switchboards, 161
Synergistic effect,
    intumescent systems, 199–204
    of MPP, 255

Synergists, 428, 431–435
Synergy/synergism,
    arylphosphate–arylphosphate synergism, 435
    defined, 419
    interaction of different flame retardant, 435–437
    intumescent systems, 199–204
    for multicomponent flame retardants, 419–422
    phosphorus-phosphorus (P-P) synergism, 435
Synthetic fabrics, 358–371
    main FR conformal coating solutions for, 360–362
    overview, 357–359
    PET, treatment,
        by BPEI/OSA LbL, 363, 364, 368, 370
        by PDAC/PAA/PDAC/APP LbL, 364, 365
Synthetic fibers, 346, 358
Synthetic FR fillers, 123
Synthetic magnesium borates, 317
Synthetic mineral flame retardants, 104–106

Talcs, 33, 53, 201, 283–284, 420, 426
Tannic acid, 354
Tarpaulins, 148
Tauramine oxide, 29
TBBPA (tetra-bromo-bisphenol-A), 9, 10, 162
TCEP (tris(chloroethyl phosphate)), 9, 11, 41, 42, 66, 67
TCP (tricresyl phosphate), 48, 49
TCPP (tris(1-chloro-2-propyl) phosphate), 66, 67, 248, 325
TCPP (tris(isopropyl-2-chloro) phosphate), 40–42, 67
Temperature,
    at APP, 245
    intumescent flame retardants, 173, 175–179

failure, 210–211
fire protective materials, 216, 218–219
as function of time, 210
HCM curve, 207
ignition, 183–184
RABT curve, 207
of reaction, 187
RWS curve, 207–208
time-temperature curve, 205
Tensile strength (TS), 145
TEP (triethyl phosphate), 33, 40–42
Tert-butylphenyl phenyl phosphate, 49–51
Tetrabromobisphenol A, 60, 522
Tetra-bromo-bisphenol-A (TBBPA), 9, 10, 162
Tetrakis(hydroxymethyl) phosphonium chloride (THPC), 63
Tetrakis(hydroxymethyl) phosphonium sulfate (THPS), 63
Tetramethyltin, 478
Textile(s),
    backcoatings coated APP in, 33
    with hydrogen peroxide, 63
    industry applications,
        halogen-free tin-based flame retardants, 506
        silicones, 275
    intumescent, 194–199
    non-durable treatment for, 25, 28
    phosphorus based FR for, 46
THEIC (tris(hydroxyethyl) isocyanurate), 32, 216
Thermal conductivity, intumescent flame retardants, 173, 178–179
Thermal stability,
    of melamine, 250
    melamine polyphosphate, 254
    mineral filler flame retardants, 122
ThermoCyl®, 294

Thermo gravimetric analysis (TGA), mineral filler flame retardants, 122
Thermoplastic and elastomeric applications,
    compound formulation principals, 138–140
    compounding technology, 136–138
    construction products, 147–150
    engineering plastics, 152–154
    special applications, 150–152
    wire and cables, 140–147
Thermoplastic elastomers (TPE), 39, 50, 55
Thermoplastic polyurethanes (TPU), 32, 57
    aromatic bisphosphates in, 56
    artificial leathers, 50
    ASTM D6413 test, 39
    BPO, 327
    Ca-Hypo in, 37
    DEPAL, 39
    formulations for cable application, 35
    HFFR compound formulations and properties, 146, 147
    PBT and, 188
    POSS in, 191
Thermoplastic vulcanizate (TPV) materials, 276
Thermosetting resins, 503–505
Thermosol treatment of polyesters, 46
Thiocyanurates, 484–485
THPC (tetrakis(hydroxymethyl) phosphonium chloride), 63
THPS (tetrakis(hydroxymethyl) phosphonium sulfate), 63
Time to ignition (TI), 279
Tin additives, 496
Tin-based flame retardants, 496–512
    halogen-free applications, 498–506
    elastomers, 503, 505
    engineering plastics, 500, 502, 503
    paints and coatings, 505

polyolefins, 499–500, 501
styrenics, 500
textiles, 506
thermosetting resins, 503–505
novel tin additives, 506–512
benefits, 506
coated fillers, 507–509
mechanism of action, 510–512
tin-modified nanoclays, 509–510
overview, 496–497
ZHS, 497–501, 503–505
ZS, 497–498, 499, 500, 501, 502, 504, 504–505
Tin-modified nanoclays, 509–510
Titanium dioxide, 46, 507
Tolerably daily intake (TDI), 67
Toluene sulfonic acid, 256
Total heat release rate (THR), 175
Toxic Chemicals Control Act, 12–13
Toxic Substances Control Act (TSCA), 9, 10
TPE (thermoplastic elastomers), 39, 50, 55
TPP. see Triphenyl phosphate (TPP)
TPS (transient plane source) method, 178–179
TPU. see Thermoplastic polyurethanes (TPU)
Transient plane source (TPS) method, 178–179
Transition metal borates, 317–324
Firebrake®ZB and Firebrake 500, 313, 318–324
coatings (water-based), 324
elastomers, 322–323
epoxy, 321–322, 323
molecular structure, 318
polyamides, 319–321
polyolefins, 319
polypropylene (PP), 319
other metal borates, 324
zinc borates, 317–324

Transition metal flame retardants, 478–484
condensed phase, 479–484
metal complexes, 481–484
metal oxides, 480–481
vapor phase, 478–479
Transition metal hydroxides, 478
Trialkyl phosphates, 65
Triaryl phosphates, 49, 50, 65
Triazine HF®, 263
Triazines, 34, 37, 188–190, 260, 430
Tributyl phosphate, 48, 66
Tricresyl phosphate (TCP), 48, 49
Triethoxybutyl phosphate, 48
Tri-2-ethylhexyl phosphate, 48
Triethyl phosphate (TEP), 33, 40, 41, 42
Triflate, 483, 484
Triglycidyl phosphate, 219
Trimethylolmelamine, 46
Trimethyl phosphate, 41
Trioctahedral alkali aluminum silicate, 283
Triphenyl phosphate (TPP), 57, 66
in halogen-free PC – ABS blend, 500
as "high priority substance" candidates, 9
low melting point, 51
phosphorus-based flame retardants, 32, 48–49, 51, 52, 54–56, 65, 66
Triphosphonate, 46, 47
Tris(2-aminoethyl)amine, 330
Tris(2-butoxyethyl) phosphate, 66
Tris(chloroethyl phosphate) (TCEP), 9, 11, 41, 42, 66, 67
Tris(1-chloro-2-propyl) phosphate (TCPP), 66, 67, 248, 325
Tris(2,3-dibromopropyl)phosphate, 13
Tris(2,3-dichloropryl) phosphate (TDCP), 66
Tris(hydroxyethyl) isocyanurate (THEIC), 32, 216
Trisilanolphenyl POSS (TPOSS), 290

Tris(isopropyl-2-chloro)phosphate (TCPP), 40–42, 67
Tris-(phenylene-1-amino- 2-oxy)-tricyclophosphazene, 261
Trixylyl phosphate, 11
TSCA (Toxic Substances Control Act), 9, 10
Tunable process, 344

UL 94. *see* Underwriters Laboratory 94 (UL 94),
UL 790 roof assembly test, 32–33
Underwriters Laboratory 94 (UL 94)
  anti-dripping effect in, 490
  "blowing out" effect, 425
  flame propagation, 131–132
  polypropylene compression molded specimens, 259, 260
  ratings, 59, 187
    electronic appliances, 152, 153, 161–162
    epoxy cast resin, 162–163
    for epoxy novolac formulations, 161
    in glass-filled polyamides, 38
    HBF rating, 57
    in UPE, 32
  reaction to fire, 184
  synthetic fabrics, 365–366
  US standard, 131
United Nations, flame retardant regulations, 8
United States (US), flame retardant regulations, 9
Unsaturated polyesters (UPE), 32, 39–42, 327
Urea, 27, 29
  derivatives, 240
  formaldehyde, 30
  groups, 190
  phosphate, 30
UV-curable latex, 353

Vapor phase transition metal flame retardants, 478–479
Vegetable oils, phosphorylated, 438
Vertical flame testing (VFT), 195–197
VFT (vertical flame testing), 195–197
Vinylbenzyl alcohol, 489
Vinylesters, 163
Vinylsilanes, 125, 152, 522
Volatile organic compounds (VOC), 45
Voluntary Emissions Control Action Program, 521

Waste Electrical and Electronic Equipment (WEEE), 4, 10–11, 14, 498–499
Waste-to-energy, 524–525, 527–528
Water-based coatings, *Firebrake*®ZB and *Firebrake* 500, 324
Water-borne polyurethane (WPU), boron nitride (h-BN), 326
Water-insoluble APP, 29
Water-soluble ammonium salt, 30
Water-soluble MAP, 29
Water uptake, 124–126
WEEE (Waste Electrical and Electronic Equipment), 4, 10–11, 14, 498–499
Wet grinding, 110
Wet hydrates, 104
White graphite, 325–327
Wildland-urban-interface (WUI) fire scenarios, 524–530
Wires, thermoplastic and elastomeric applications, 140–147
Wollastonite, 280–281
Wood material, borax pentahydrate/decahydrate, 314–315
Wood plastic composite (WPC), 150
Wool, 347, 348, 351
WUI (wildland-urban-interface) fire scenarios, 528–530

X-ray photoelectron spectroscopy (XPS), 507

ZB. *see* Zinc borates (ZB)
Zeolites, 33, 189, 281, 489
Zhejiang Wansheng Co (China), 179
ZHS. *see* Zinc hydroxystannates (ZHS)
Zinc borates (ZB), 32, 33, 37, 38
  commercial boron-based FR, 313
  fire-resistance performance, 217
  mineral flame retardants, 107–109, 130, 134, 142
  PVC, 142, 143
  synergists in multicomponent systems, 432–433
  transition metal borates, 317–324
Zinc borophosphate, 328
Zinc chloride catalyst, 490
Zinc cyanurate, 35
Zinc diethylphosphinate (DEPZN), 37, 39–40, 320
Zinc hydroxystannates (ZHS), 106, 108, 109
  coated fillers, 507–509
  construction product, 150
  electronic/electrical applications, 499
  in halogenated elastomers, 503

HFFR compounds, 144
novel tin additives, 506–512
in polyolefin cable compounds, 500, 501
preparation, 495–498
properties, 498
PVC, 142, 143
in thermosetting resins, 503, 504–505
Zinc oxide, 35, 433
Zinc stannates (ZS), 106, 108, 109, 130, 134, 154
  coated fillers, 507–509
  electronic/electrical applications, 499
  in engineering plastics, 500, 502, 503
  in halogenated elastomers, 503
  novel tin additives, 506–512
  in polyolefin cable compounds, 500, 501
  preparation, 497–498
  properties, 498
  in thermosetting resins, 503, 504–505
Zinc triflate, 483
ZS. *see* Zinc stannates (ZS)

Printed in the USA/Agawam, MA
November 16, 2021

784606.007